Numerical Solution of Convection-Diffusion Problems

APPLIED MATHEMATICS AND
MATHEMATICAL COMPUTATION

Editors

R.J. Knops and K.W. Morton

This series presents texts and monographs at graduate and research level covering a wide variety of topics of current research interest in modern and traditional applied mathematics, in numerical analysis and computation.

(Full details concerning this series, and more information on titles in preparation are available from the publisher.)

Numerical Solution of Convection-Diffusion Problems

K. W. MORTON
Oxford University Computing Laboratory
Oxford , UK

CRC Press
Taylor & Francis Group
Boca Raton London New York

CRC Press is an imprint of the
Taylor & Francis Group, an **informa** business

Contents

Preface

Accurate modelling of the interaction between convective and diffusive processes is the most ubiquitous and challenging task in the numerical approximation of partial differential equations. This is partly because of the problems themselves, their great variety and widespread occurrence, as well as their close association with singular perturbation problems and boundary layer theory. It is also due to the fact that numerical algorithms, and the techniques used for their analysis, tend to be very different in the two limiting cases of elliptic and hyperbolic equations. Thus many different ideas and approaches have been proposed in widely differing contexts to resolve the difficulties — exponential fitting, compact differencing, upwinding, artificial viscosity, streamline diffusion, Petrov–Galerkin and evolution-Galerkin being some examples from the main fields of finite difference and finite element methods.

The main aim of this book is to try to draw together all these ideas, to see how they overlap and how they differ, and to provide the reader with a useful and wide-ranging source of algorithmic concepts and techniques of analysis. The material that will be presented has been drawn both from theoretically-oriented literature on finite difference and finite element methods and also from accounts of practical, large-scale computing, particularly in the field of computational fluid dynamics. Like most useful numerical analysis, it represents the fruitful interaction of practical problem-solving with the establishment of a sound theoretical framework for the algorithms; this combination also reflects my own interests and experience. My initial enthusiasm for finite difference methods, both their theory and use, found fulfilment in the revision of Bob Richtmyer's seminal book (Richtmyer 1957, Richtmyer and Morton 1967,1994), though this has long been tinged with the regret that the concept of practical stability that was introduced there was not applied as it should have been to the convection-diffusion problem. This was followed by a lasting liking for finite element methods; in particular, I have long been fascinated by the idea that the optimal approximation property, that is their forte for self-adjoint problems, could be carried over to convection-diffusion problems by means of the Petrov–Galerkin formulation. And finally, contact with the practical world of computational fluid dynamics over the last dozen years has stimulated an interest in finite volume methods; these combine features from each of the other two established methodologies, and still need much theoretical attention.

It is my hope, therefore, that the book will be accessible and helpful both to engineers and scientists and to mathematicians; and similarly I hope it will appeal both to those engaged in solving real practical problems and to those interested in developing further the theoretical basis for the methods used. To this end, detailed proofs have been given of some of the key theorems for such central techniques as exponential fitting and the use of local Green's functions, Petrov–Galerkin and streamline diffusion methods, and cell vertex finite volume methods. On the other hand, methods have been selected for their general utility and as far as possible presented in a form that can be readily extended — from the basic linear problem in a single unknown, to a nonlinear system of equations. Compressible, viscous fluid flow governed by the Navier–Stokes equations will be the application that is always in mind — even when considering $-\epsilon u'' + u' = S$, which we will often do. Steady problems, in one or more spatial variables, are considered to form the core of the subject and are taken as the starting point for the design and analysis of nearly all the methods discussed. Apart from the introductory chapter, only the last chapter is devoted to unsteady problems; on the other hand, the use of time-stepping methods to solve steady problems is given increasing attention as the book progresses.

The plan of the book falls naturally into three parts, which will affect the order in which readers with differing objectives may approach it. The first chapter is a fairly lengthy introduction to the range of applications leading to convection-diffusion problems, the issues raised in applying standard finite difference and finite element methods to simple one-dimensional model problems, and some indication of the topics to be taken up in later chapters. This is followed by a group of three chapters which present basic material that might well be skipped over by readers who are relatively familiar with the topic. Thus Chapter 2 collects together some key results from mathematical analysis — on maximum principles, asymptotic expansions, *a priori* bounds, coercivity conditions and Green's functions. Classical difference methods for simple problems are analysed in some detail in Chapter 3; and in Chapter 4 a detailed presentation of standard finite element methods is given, partly as an introduction for readers who are more familiar with difference methods, but also in order to present and put into a comparative context the useful difference schemes that result on a uniform mesh.

The core of the book consists of the three Chapters 5, 6, and 7, each of which could be developed into a separate book. Oleg Zienkiewicz, in a brief postscript to Zienkiewicz *et al.* (1975), was the first to point out how the Petrov–Galerkin formulation was the natural development within the finite element framework of exponential fitting and upwinding in finite difference schemes. The substantial body of material that has arisen from this starting point is the subject of Chapter 5; this includes an introduction to streamline diffusion methods, the invention and development of which by Tom Hughes and his collaborators has been so influential in the field, and which in the title SUPG (Streamline Upwind Petrov–Galerkin) embodies many of the key ideas. Chapter 6 is devoted to the growing field of finite volume methods, which are particularly well suited to the modelling of conservation laws, and which combine the geometric flexibility of finite element methods with the basic compactness attainable with finite difference methods. In this chapter, greater consideration is given to the demands posed by more complex problems, and the discussion of time-stepping methods leads naturally to Chapter 7 where unsteady problems are treated. Although, in the interests of brevity, only finite element and finite volume methods are considered in that chapter, the emphasis on using recovery techniques to squeeze more accuracy out of basically piecewise constant representations forms a natural link to the large recent literature on upwind finite difference methods.

The reader might well ask whether there is a conclusion to be drawn from this lengthy story. Although it would be somewhat invidious to try to summarise the contributions of so many people in one short statement, I think there is one simple observation that is worth making. Consider the approximation of either the model steady problem, $-\epsilon u'' + u' = S$, or the model unsteady problem, $u_t + u_x = \epsilon u_{xx}$, on a uniform mesh: much of the literature in either case is concerned with choices between *three-point* schemes that focus attention on the nodal value U_i, or U_i^n and U_i^{n+1} in the unsteady case, and the two neighbouring values U_{i-1} and U_{i+1}; however, the emphasis on what happens in the cell $[x_{i-1}, x_i]$, which is at the heart of some upwind finite difference schemes, the box difference scheme, the cell vertex finite volume scheme and the evolution-Galerkin methods, leads naturally to *four-point* schemes centred on the cell and using the nodal values U_{i-2}, U_{i-1}, U_i and U_{i+1}. It will be shown that the latter approach yields remarkably robust methods for the steady problem, with no tuning required as one varies the key dimensionless parameter, the *mesh Péclet number* which here is given by $\Delta x/\epsilon$; and for the linear advection problem $u_t + u_x = 0$ it gives the third order accurate schemes, with dominant dissipative rather than dispersive errors, that have been independently devised and advocated by many authors.

There is an enormous and widely dispersed literature in this field. No attempt has been made here to compile a comprehensive set of references, but the aim has been to give a representative selection from the wide variety of contributions and sources. The story is still unfinished and the literature is still growing. For example, the twin topics of mesh adaptation and *a posteriori* error analysis will undoubtedly achieve greater prominence in the near future.

The notation has been kept as simple as possible, and is introduced and explained as needed. The general index is used to refer to instances of first usage, rather than a compilation of all notation being provided.

This book would never have been embarked on, and certainly never completed, had it not been for the help and support of many people. The numerous colleagues and students with whom I have collaborated over the last dozen years are largely acknowledged through the references. One exception is Professor Phoolan Prasad, whose kind invitations enabled me to make two fascinating visits to the Indian Institute of Science at Bangalore, where I was first able to start the book and then to write the later chapters. Another is Nick Birkett who carried out all of the computations for the IAHR/CEGB model problems.

Above all, however, my thanks go to two people who have been essential to the project from start to finish: firstly my wife who, in addition to providing the support that is crucial to undertakings of this kind, also carried out the computations and plotting for most of the figures (Paul Crumpton did not dub her Postscript Pat for nothing!); and, secondly, my secretary Joan Himpson who has taken my old-fashioned manuscript and, through many drafts, turned it into camera-ready copy with all of the 54 figures incorporated into the LATEXfile. Finally, thanks are due to the staff of Chapman and Hall, and especially to Achi Dosanjh, for their patience and continual help in this lengthy project.

CHAPTER 1

Introduction and overview

Mathematical models that involve a combination of convective and diffusive processes are among the most widespread in all of science, engineering and other fields where mathematical modelling is important. Very often the dimensionless parameter that measures the relative strength of the diffusion is quite small; so one often meets with situations where thin boundary and interior layers are present and singular perturbation problems arise. In all such circumstances difficulties will be experienced with standard numerical approximations. Thus a very large literature has built up over the last few decades on a variety of techniques for analysing and overcoming these difficulties.

In this first chapter we begin by collecting together ten typical problems where the convection-diffusion phenomenon occurs. In some cases it constitutes the whole problem; more commonly it is embedded in a larger problem and not so readily recognised. We then give a number of model problems which attempt to embody the main features that occur in this collection, and which will be used quite frequently in later chapters. Finally we consider some of the difficulties commonly experienced with numerical methods of approximation, and summarise the main approaches that will be taken to overcome them.

1.1 Typical problems where convection-diffusion occurs

(i) *Pollutant dispersal in a river estuary.* Consider the two-dimensional domain sketched in Fig.1.1. If c is the concentration of a pollutant — which may be sewage, chemical or radioactive waste, temperature (e.g. of power station cooling water) etc. — the typical equation governing its dispersal will be of the form

$$\frac{\partial c}{\partial t} + \mathbf{v}\cdot\nabla c = \nabla\cdot(D\nabla c) + S. \tag{1.1}$$

Here c is the depth-averaged concentration and S its source, which may be very concentrated in space and vary with time; $\mathbf{v}$ is a given horizontal velocity field which may also vary with time, for instance because of tides; $\mathbf{v}$ will usually be assumed incompressible so that $\mathbf{v}\cdot\nabla c$ can be replaced by

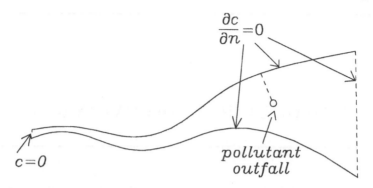

Figure 1.1. *Sketch of estuary with a point source of pollutant*

$\nabla \cdot (\mathbf{v}c)$; and D is a coefficient of diffusivity which will incorporate mixing from vertical motions, which have been averaged out in (1.1), as well as turbulent and molecular diffusion.

Typical dimensions for such a problem would be velocities of from $\frac{1}{2}$ to 3m/sec, a diffusivity of 0.05 to 5m²/sec, and a length scale L of some 100m. Thus the dimensionless *Péclet number*, Pe = $|\mathbf{v}|L/D$, which measures the relative importance of convection compared to diffusion, may vary from 10 to 6×10^3. If D is assumed to be independent of c this is a very representative linear convection-diffusion problem.

(ii)   *Vorticity transport in the incompressible Navier–Stokes equations.* If the velocity field $\mathbf{v}$ in the above problem is also to be computed, it may be assumed to be unaffected by the pollutant concentration and to satisfy the incompressible Navier–Stokes equations,

$$\frac{\partial \mathbf{v}}{\partial t} + \mathbf{v} \cdot \nabla \mathbf{v} + \nabla p = \nu \nabla^2 \mathbf{v} \qquad (1.2a)$$

$$\nabla \cdot \mathbf{v} = 0, \qquad (1.2b)$$

where p is the pressure and ν the kinematic viscosity. Particularly in two Cartesian dimensions, it is common to introduce the vorticity, which is the scalar ζ in two dimensions, and to satisfy (1.2b) by also introducing a stream function ψ. Then if $\mathbf{v} = (u, v) = (\psi_y, -\psi_x)$ and $\zeta = v_x - u_y$, the equations become

$$\frac{\partial \zeta}{\partial t} + \mathbf{v} \cdot \nabla \zeta = \nu \nabla^2 \zeta, \qquad (1.3a)$$

where ψ and hence $\mathbf{v}$ is obtained from solving the compatibility relation

$$\nabla^2 \psi + \zeta = 0. \qquad (1.3b)$$

This is of course a nonlinear pair of equations; but very often (1.3a) and

(1.3b) are solved alternately — not a practice to be widely advocated, however — and then (1.3a) yields another linear convection-diffusion equation for the scalar vorticity.

Here the dimensionless parameter is the *Reynolds number* Re $= |v|L/\nu$. With the typical values given above for $|v|$ and L, and with the viscosity of water equal to about $10^{-6} m^2/\sec$, we obtain Reynolds numbers of around 10^4.

For sufficiently low Reynolds numbers there are steady solutions of the Navier–Stokes equations which are of interest — for example, the flow past an object such as a cylinder. In this last case the wake becomes unsteady when Re, based on the cylinder diameter, is approximately 40, but there is considerable interest in calculating such steady (but unstable) flows for Reynolds numbers up 10^3 — see Fornberg (1993) for a survey. Thus such steady problems can already pose considerable numerical difficulties and have received much attention: for later use it is convenient to write the steady equations as

$$-\nu\nabla^2\zeta + \nabla\cdot(v\zeta) = 0 \qquad (1.4)$$

together with (1.3b).

(iii) *Atmospheric pollution.* The Chernobyl accident focussed worldwide attention on the modelling of atmospheric pollution. In his extensive numerical simulations, Pudykiewicz (1989) showed the importance of using a full three-dimensional model, of the following slightly simplified form for a set of radioactive tracers of specific activity A^i,

$$\frac{\partial A^i}{\partial t} + \mathbf{V}_H\cdot\nabla_H A^i + \frac{\partial}{\partial\sigma}\left(v_\sigma^i A^i\right) = \frac{\partial}{\partial\sigma}\left(K_z\frac{\partial}{\partial\sigma}A^i\right) + S^i. \qquad (1.5)$$

Here a terrain-following vertical coordinate σ is used, so that $\mathbf{V}_H$ is an average horizontal wind vector in the σ system and ∇_H the corresponding gradient operator. The various sources and sinks are gathered in S^i, including a matrix of coefficients describing the radioactive decay and transformation of all the isotopes contained in the model; K_z is a vertical diffusion coefficient which is a function of height; and v_σ^i is composed from the vertical motion in the σ system and the gravitational settling velocity. The wind velocity and other atmospheric parameters were obtained from standard meteorological forecast data.

The horizontal and vertical scales here are quite different. Length scales in the horizontal direction are typically thousands of kilometres and velocities around $10m/\sec$, so that the Reynolds number is in the range $10^{11} - 10^{12}$. In the vertical direction, however, the length scale and velocity are smaller and the turbulent diffusion coefficient much larger: in the atmospheric boundary layer K_z may be $10^3 m^2/\sec$ giving a Péclet number around 10^{-2}; and in the free atmosphere, where K_z is typically $10m^2/\sec$, the Péclet number would be about 10.

(iv) *Fokker–Planck equation.* This equation is derived from the Boltzmann equation of kinetic theory, for the velocity distribution function $f(\mathbf{x}, \mathbf{v}, t)$ of the particles in a gas, by modelling the effect of many weak interactions in the collision integral. Thus the general form of the equation is

$$\frac{\partial f}{\partial t} + \mathbf{v}\cdot\nabla f + \mathbf{a}\cdot\nabla_{\mathbf{v}} f = \left(\frac{\partial f}{\partial t}\right)_c \tag{1.6a}$$

where $\left(\dfrac{\partial f}{\partial t}\right)_c = -\nabla_{\mathbf{v}}\left(f\langle\Delta\mathbf{v}\rangle\right) + \tfrac{1}{2}\nabla_{\mathbf{v}}\nabla_{\mathbf{v}} : \left(f\langle\Delta\mathbf{v}\Delta\mathbf{v}\rangle\right);$ $\qquad$ (1.6b)

here $\mathbf{a}$ is the particle acceleration and $\nabla_{\mathbf{v}}$ is the gradient operator with respect to the velocity variable. Various models are used for the first and second moments of the velocity deflection $\Delta\mathbf{v}$ that appear on the right-hand side of (1.6b); Boyd and Sanderson (1969) give derivations used in plasma kinetic theory which yield expressions of the form

$$\left(\frac{\partial f}{\partial t}\right)_c = \Gamma\left\{-\nabla_{\mathbf{v}}(f\nabla_{\mathbf{v}}H) + \tfrac{1}{2}\nabla_{\mathbf{v}}\nabla_{\mathbf{v}} : (f\nabla_{\mathbf{v}}\nabla_{\mathbf{v}}G)\right\}, \tag{1.6c}$$

where H and G are collision functionals of f. The first term here is clearly a convection term in velocity space, with $\nabla_{\mathbf{v}}H$ called the coefficient of dynamical friction, and in the second the tensor coefficient $\nabla_{\mathbf{v}}\nabla_{\mathbf{v}}G$ is called the coefficient of diffusion. Through their dependence on f the relative magnitudes of these coefficients can vary enormously, but it should be noted that all scales in this example are quite unlike those in the earlier examples; thus length scales will be in velocity space and commonly measured in electron-volts while times will be microseconds. A simple model problem would be where f is independent of $\mathbf{x}$ and $\mathbf{a} = 0$, so the equation reduces to $\partial f/\partial t = (\partial f/\partial t)_c$, and one considers how an initial velocity distribution relaxes to a Maxwellian.

(v) *Semi-conductor equations.* The basic model for electric current through a semi-conductor was developed by van Roosbroeck (1950) in terms of the electric potential ψ, the electron concentration n and the hole concentration p to give a system of equations of the form

$$\epsilon\nabla^2\psi = q(n - p - D) \tag{1.7a}$$

$$-q\frac{\partial n}{\partial t} + \nabla\cdot\mathbf{J}_n = qR \tag{1.7b}$$

$$q\frac{\partial p}{\partial t} + \nabla\cdot\mathbf{J}_p = -qR; \tag{1.7c}$$

here ϵ is the permittivity of the material, q the unit charge, D the background doping concentration and R a recombination term. The currents $\mathbf{J}_n$ and $\mathbf{J}_p$ are commonly given by a drift-diffusion model of the form

$$\mathbf{J}_n = q\mu_n(U_t\nabla n - n\nabla\psi), \tag{1.7d}$$

where μ_n is the electron mobility and U_t represents an internal potential barrier, with a similar expression holding for $\mathbf{J}_p$. It is clear that substitution of (1.7d) into (1.7b) gives a convection-diffusion problem for n, which is nonlinear because of the $\nabla\psi$ coefficient to the convection term. The effective Péclet number is given by V_{app}/U_t, where V_{app} is the applied voltage and this is typically of the order of 10^4, although local values may vary by very large factors.

A great deal of effort has been devoted to the numerical approximation of these equations in recent years — see Mock (1983) for a comprehensive account. In view of the structural similarity to the equations of fluid flow, for example (1.3), it is not surprising that similar techniques such as exponential fitting have been devised in both fields.

(vi) *Groundwater transport.* A typical macroscopic model for the groundwater transport of a substance in solution, of concentration c in an aquifer with moisture content θ, takes the form

$$\frac{\partial(\theta c)}{\partial t} + \nabla\cdot(c\mathbf{q} - \theta D\nabla c) = S(\theta, c). \tag{1.8a}$$

In this abbreviated form, the source term $S(\theta, c)$ represents many processes which include actual addition of the substance and loss through interaction with the water and with the rock or soil. The tensor coefficient D combines molecular diffusion through the aquifer and dispersion due to averaging over microscopic variations of water velocity, and takes account of anisotropy in the aquifer.

The specific discharge of water $\mathbf{q}$, that gives the convection term in (1.8a), can be obtained from a groundwater flow model of the general form

$$\frac{\partial(\rho\theta)}{\partial t} + \nabla\cdot(\rho\mathbf{q}) = 0, \tag{1.8b}$$

where ρ is the density. Then $\mathbf{q}$ and θ are related by a generalisation of Darcy's law,

$$\mathbf{q} = -K\nabla\phi, \tag{1.8c}$$

where K is a matrix of effective hydraulic conductivities and the piezometric head ϕ can be regarded as a given function of θ and the vertical height z. For more details of these models, the reader should consult Bear and Verruijt (1987). Magnitudes of the quantities in these equations vary very considerably, and are often rather uncertain: but a typical set of values would give a velocity of 2m per day, entries in D might be $10\text{m}^2/\text{d}$ and the length scale 100m, giving a Péclet number of 20.

(vii) *Financial modelling.* The widely quoted Black–Scholes model (Black and Scholes 1973) for the value V of a share option, as a function of the underlying asset price S and time t, is given by

$$\frac{\partial V}{\partial t} + \tfrac{1}{2}\sigma^2 S^2 \frac{\partial^2 V}{\partial S^2} + rS\frac{\partial V}{\partial S} - rV = 0, \tag{1.9a}$$

where σ is the volatility of the asset and r the interest rate. This is a backward parabolic equation with a prescribed final condition at the expiry date T: for a simple European call option this would be in the form

$$V(S,T) = \max(S - E, 0), \tag{1.9b}$$

where E is the exercise price; typical boundary conditions would then be

$$V(0,t) = 0, \quad V(S,t) \sim S \quad \text{as} \quad S \to \infty. \tag{1.9c}$$

For typical constant values of $r = 0.1$, $\sigma = 0.2$, the Péclet number relative to the "length" scale ΔS has the value $5\Delta S/S$ so that the convective effect (financial drift) is dominant when the asset price is small.

(viii) *Stefan problem on a variable mesh.* Consider one-dimensional heat flow in liquid and solid regions separated by an interface at $x = x_M(t)$. With the solid on the left, the equations in the two regions take the form

$$c_S \frac{\partial \theta}{\partial t} = \kappa_S \frac{\partial^2 \theta}{\partial x^2} \quad x < x_M; \quad c_L \frac{\partial \theta}{\partial t} = \kappa_L \frac{\partial^2 \theta}{\partial x^2} \quad x > x_M. \tag{1.10}$$

At the fusion point the equations are coupled by a relation

$$\frac{\mathrm{d}x_M}{\mathrm{d}t} = \frac{1}{\rho L}\left\{ \kappa_S \left.\frac{\partial \theta}{\partial x}\right|_{x_M-0} - \kappa_L \left.\frac{\partial \theta}{\partial x}\right|_{x_M+0} \right\} \tag{1.11}$$

where L is the latent heat of fusion and ρ the (common) density. Now suppose this is approximated on a variable mesh over the interval $(0,1)$ with

$$x_j = \frac{j}{M}x_M, \quad 0 \le j \le M;$$

$$x_j = x_M + \frac{j-M}{N-M}(1 - x_M), \quad M \le j \le N. \tag{1.12}$$

Then on the moving mesh we have total derivatives

$$\frac{\mathrm{d}\theta_j}{\mathrm{d}t} = \frac{\partial \theta_j}{\partial x}\frac{\mathrm{d}x_j}{\mathrm{d}t} + \frac{\partial \theta_j}{\partial t} \tag{1.13a}$$

so that in the solid region the heat flow equation becomes

$$\frac{\mathrm{d}\theta_j}{\mathrm{d}t} = \left(\frac{x_j}{x_M}\frac{\mathrm{d}x_M}{\mathrm{d}t}\right)\frac{\partial \theta_j}{\partial x} + \frac{\kappa_S}{c_S}\frac{\partial^2 \theta_j}{\partial x^2}, \tag{1.13b}$$

with a similar equation holding in the liquid region.

Methods based on this formulation are widely used to solve free boundary problems — see for example Murray and Landis (1959). They result in a purely diffusion problem being converted into a convection-diffusion problem; and for some problems, such as the oxygen diffusion problem of Crank and Gupta (1972), the speed at which the interface travels can become very large so that the convection term becomes very significant.

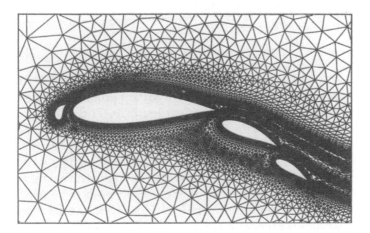

Figure 1.2. *A typical multi-aerofoil, with a triangular Navier–Stokes mesh; general view below, detail at leading edge above, (Courtesy of Professor N.P. Weatherill)*

(ix) *Viscous compressible flow past an aerofoil.* For present day aeronautical design calculations there is an increasing interest in being able to solve the full Navier–Stokes equations around a configuration such as that shown in Fig. 1.2, rather than having to couple boundary-layer calculations with an appropriate inviscid model of the outer flow. In order to be able to deal with transonic flows, we write the equations in conservation law form:

$$\frac{\partial \mathbf{w}}{\partial t} + \text{div} \, (\mathbf{f}, \mathbf{g}, \mathbf{h}) = 0, \quad \mathbf{f} = \mathbf{f}^I + \mathbf{f}^V \text{ etc.} \tag{1.14}$$

where, in terms of the density ρ, pressure p, velocity components (u, v, w) and total energy E, we have

$$\mathbf{w} = \begin{bmatrix} \rho \\ \rho u \\ \rho v \\ \rho w \\ \rho E \end{bmatrix}, \quad \mathbf{f}^I = \begin{bmatrix} \rho u \\ \rho u^2 + p \\ \rho u v \\ \rho u w \\ \rho u H \end{bmatrix}, \quad \mathbf{f}^V = \begin{bmatrix} 0 \\ -\tau_{xx} \\ -\tau_{xy} \\ -\tau_{xz} \\ (-u\tau_{xx} - v\tau_{xy} \\ -w\tau_{xz} + q_x) \end{bmatrix} \tag{1.15a}$$

with similar expressions for $\mathbf{g}^I, \mathbf{h}^I, \mathbf{g}^V, \mathbf{h}^V$; for the inviscid fluxes we also have the relations, for an adiabatic gas with adiabatic constant γ,

$$H = E + p/\rho = \tfrac{1}{2}(u^2 + v^2 + w^2) + \gamma p/(\gamma - 1)\rho; \tag{1.15b}$$

and the deviatoric stresses and heat fluxes in the viscous fluxes are given by

$$\tau_{xx} = \tfrac{2}{3}\mu \left(2\frac{\partial u}{\partial x} - \frac{\partial v}{\partial y} - \frac{\partial w}{\partial z} \right), \quad \tau_{xy} = \mu \left(\frac{\partial u}{\partial y} + \frac{\partial v}{\partial x} \right),$$

$$q_x = -\kappa \frac{\partial T}{\partial x}, \text{ etc.} \tag{1.15c}$$

where $p/\rho = RT$, κ is the conductivity and μ the viscosity. This is a fairly complicated system of equations but the important point is that there are two convection-diffusion problems embedded in it. The first is for the momentum equations where the small parameter is the viscosity μ. For air the kinematic viscosity μ/ρ is 1.5×10^{-5} m^2/sec at 15^0C, so that at a near sonic speed of 500 mph and a wing with a chord length of 5m, we have a Reynolds number of 7.5×10^7. The flow would be highly turbulent under these conditions so that a turbulence model would need to be used (see below) that would add an effective turbulent viscosity of μ_t; thus a realistic value for the Reynolds number for which the solution of these equations might be attempted would be Re $= 10^3$ or 10^4. The second convection-diffusion problem is for the temperature, where the small parameter is the conductivity κ, and from collecting all the coefficients we see that the Péclet number for this process is RePr where Pr is the Prandtl number $(\mu/\kappa)\gamma R/(\gamma - 1)$. For air this has a value of about 0.7 for laminar flows and 0.9 for turbulent flows.

(x) *Turbulence transport.* As indicated above, an engineering turbulence model is very often needed to calculate a turbulent viscosity which may be used in conjunction with the system (1.14)-(1.15). Typical of these is the $k - \epsilon$ model, which comprises two transport equations for the turbulence energy k and dissipation rate ϵ,

$$\frac{\partial(\bar{\rho}k)}{\partial t} + \nabla\cdot(\bar{\rho}\tilde{v}k - \mu_k\nabla k) = S_k \tag{1.16a}$$

$$\frac{\partial(\bar{\rho}\epsilon)}{\partial t} + \nabla\cdot(\bar{\rho}\tilde{v}\epsilon - \mu_\epsilon\nabla\epsilon) = S_\epsilon, \tag{1.16b}$$

where $\bar{\rho}$ is the Reynolds-averaged density, $\tilde{v}$ the mass-averaged velocity and μ_k, μ_ϵ are eddy diffusivities. The values of these quantities vary widely in a typical flow field; and the nonlinearity of the system is greatly enhanced by the source terms, for which many differing models exist. Typical forms, see Launder and Spalding (1972) for details, are as follows

$$S_k = C_1(k^2/\epsilon) - \tfrac{2}{3}\bar{\rho}(\nabla\cdot\tilde{v})k - \bar{\rho}\epsilon - C_2(k/y^2) + C_3 \tag{1.17a}$$

$$S_\epsilon = C_4\bar{\rho}k - C_5(\nabla\cdot\tilde{v})\epsilon - C_6(\epsilon^2/k) - C_7(\epsilon/y^2) + C_8, \tag{1.17b}$$

where the empirical coefficients $C_1 - C_8$ are generally flow and position dependent, and y is the normal distance from the boundary. The turbulent viscosity μ_t, which would then be substituted into the flow equations (1.15), is taken to be proportional to k^2/ϵ. The complete system of equations (1.15)–(1.17) is probably one of the most complicated mathematical models in widespread use and it is clear that its character is dominated by several embedded convection-diffusion problems.

1.2 Model problems

The steady linear problem on a bounded domain in two dimensions will form the centre of attention,

$$Lu := -\epsilon\nabla\cdot(a\nabla u) + \nabla\cdot(bu) + cu = S \text{ in } \Omega \subset \mathbb{R}^2 \tag{1.18a}$$

with boundary conditions

$$u = u_B \text{ on } \partial\Omega_D, \quad \frac{\partial u}{\partial n} = 0 \text{ on } \partial\Omega_N, \tag{1.18b}$$

where $\partial\Omega_D, \partial\Omega_N$ form a partition of the boundary of Ω in which $\partial\Omega_D$ is non-empty. We shall usually assume that the advective velocity field b is incompressible, $\nabla\cdot b = 0$, so that the convective term can also be written $b\cdot\nabla u$, and also that $a(\mathbf{x}) \geq 1, c(\mathbf{x}) \geq 0$ while ϵ is a small positive constant. We shall have little interest in methods which cannot be applied to such problems on a nonrectangular domain, and with ϵ very small.

A particular pair of such problems, which have been widely used to compare different methods, arose from an IAHR/CEGB workshop in 1981 — see Smith and Hutton (1982) for a review of the methods presented there. The domain for both problems is the rectangular region

$$\Omega = \{(x,y) : -1 < x < 1, 0 < y < 1\}; \tag{1.19a}$$

and the incompressible convective velocity field is generated from a stream function $\psi(x, y) = -(1 - x^2)(1 - y^2)$ to give

$$b(x, y) = (2y(1 - x^2), -2x(1 - y^2)), \qquad (1.19b)$$

while $a = 1$ and $c = 0$; $\partial\Omega_N$ consists of just the outflow boundary, $y = 0$ and $0 < x < 1$. In the first problem, a steep inlet temperature profile is specified and the task is to convect this round the 180^0 bend given by the velocity field, for various values of the diffusion coefficient ϵ — see Fig. 1.3. We shall consider only the choice

$$u(x, 0) = 1 + \tanh[\alpha(2x + 1)], \quad -1 \le x \le 0 \qquad (1.19c)$$

with $\alpha = 10$, as shown together with the velocity stream-lines in Fig.1.3. The compatible Dirichlet condition $u = 1 - \tanh 10$ is imposed on the rest of $\partial\Omega_D$.

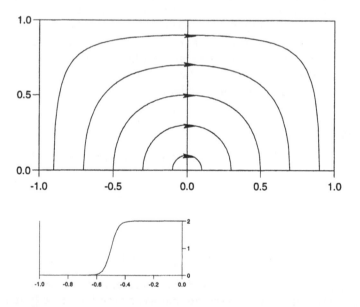

Figure 1.3. *The velocity field and inlet profile for the IAHR/CEGB problems*

In the second problem, the condition $u = 0$ is imposed on the inlet boundary and on the tangential boundary on the left ($x = -1$) and at the top ($y = 1$), while $u = 100$ is imposed on the right ($x = 1$). The task here is to model the developing boundary layer on the right-hand boundary, an asymptotic form for which is given by Scotney (1985).

Some of the difficulties of modelling convection-diffusion can be shown in a simple one-dimensional version of this problem, which will take the form

$$-\epsilon u'' + bu' = S \quad \text{on } (0,1) \tag{1.20a}$$

$$u(0) = 0, \quad u(1) = 1, \tag{1.20b}$$

where b is a positive constant. This is the simplest model problem, especially when $S \equiv 0$, and since we can write down exact solutions of this problem and its numerical approximations on uniform meshes, it will be used very frequently in the following chapters.

However, the distinguishing features of some methods will not be shown up with such a simple problem and a more general one-dimensional model we shall use takes the form

$$Lu := -\epsilon(au')' + bu' + cu = S \quad \text{on } (0,1) \tag{1.21a}$$

$$u(0) = u_L, \quad u(1) = u_R, \tag{1.21b}$$

or with a homogeneous Neumann boundary condition at $x = 1$ if $b(1) \geq 0$,

$$u'(1) = 0; \tag{1.21c}$$

alternatively, it is sometimes more natural to replace (1.21a) by the conservation law form

$$Lu := (-\epsilon au' + bu)' = S \quad \text{on } (0,1). \tag{1.21d}$$

As with (1.18) we shall assume $a(x) \geq 1$ and $c(x) \geq 0$, and also make various assumptions on the function $b(x)$; we shall often taken $a \equiv 1, c \equiv 0$, and turning point problems can be modelled by taking $b = x - \frac{1}{2}$ or $\frac{1}{2} - x$.

A simple set of problems that are useful in emphasising the extra difficulties faced in two dimensions are due to Raithby (1976): they are of the form

$$-\epsilon \nabla^2 u + \mathbf{b} \cdot \nabla u = 0 \quad \text{on } (0,1) \times (0,1), \tag{1.22a}$$

with constant velocity field

$$\mathbf{b} = b(\cos\theta, \sin\theta) \tag{1.22b}$$

for $0 \leq \theta < \frac{1}{2}\pi$, and discontinuous inflow boundary data

$$u(0,y) = \begin{cases} 0, & y \in [0, \frac{1}{2}) \\ 1, & y \in (\frac{1}{2}, 1]; \end{cases} \tag{1.22c}$$

typical conditions on the rest of the boundary consist of setting $\partial u/\partial y = 0$ at the top, $y = 1$, and $u = 0$ on the bottom and on the right, $y = 0$ or $x = 1$. This leads to an internal layer along $y = \frac{1}{2} + x \tan\theta$, and a boundary layer at $x = 1$ for $y > \frac{1}{2} + \tan\theta$ when $\tan\theta < \frac{1}{2}$.

In order to model the effect of nonlinearities, we shall consider scalar equations in the conservation law form

$$L(u) := -\epsilon \nabla^2 u + \nabla \cdot \mathbf{f}(u) = S \quad \text{in } \Omega \subset \mathbb{R}^2, \tag{1.23}$$

together with boundary conditions as in (1.18b); we shall also consider one-dimensional restrictions of this on $(0, 1)$, particularly Burgers' equation in which $f = \frac{1}{2}u^2$. In Chapter 6 we shall consider an example of inviscid one-dimensional gas flow in a nozzle, which reduces to a pair of conservation laws for mass and momentum — see section 6.6.2.

There are a large number of model problems in the literature on methods for hyperbolic equations, which can be readily extended to include diffusion. The simplest is

$$\frac{\partial u}{\partial t} + b\frac{\partial u}{\partial x} = \epsilon\frac{\partial^2 u}{\partial x^2} \quad \text{on } \mathbb{R} \times (0, T] \tag{1.24a}$$

$$u(x, 0) = u^0(x) \quad x \in \mathbb{R}, \tag{1.24b}$$

where b, ϵ are positive constants and $u^0(x)$ is either a Gaussian distribution, or $\sin^{2p} x$ on $[0, \pi]$ and zero elsewhere with $p = 2$ or 4. In two dimensions a 'rotating cone' problem is widely used, namely

$$\frac{\partial u}{\partial t} + \mathbf{b}\cdot\nabla u = \epsilon\nabla^2 u \quad \text{on } \mathbb{R}^2 \times (0, T] \tag{1.25a}$$

$$u(x, y, 0) = u^0(x, y) \quad (x, y) \in \mathbb{R}^2, \tag{1.25b}$$

where $u^0(x, y)$ is a circularly symmetric cone centred at $x = 1, y = 0$ of Gaussian or $\sin^{2p}$ shape and $\mathbf{b}(x, y)$ has circular streamlines given by

$$\mathbf{b}(x, y) = (\omega y, -\omega x) \tag{1.25c}$$

similar to those in (1.19b). The same problem may also be posed on the square $(-1, 1) \times (-1, 1)$, with exact inflow boundary conditions where needed.

1.3 Numerical difficulties with simple difference schemes

Early attempts to approximate the Navier–Stokes equations, in either of the forms (1.2) or (1.3), focussed on low Reynolds number flows. Then the use of central difference approximations was quite successful. Whether solving the steady equation (1.4), or using an implicit time-stepping scheme for (1.2a) or (1.3a), the five-point difference stencil generated for the Laplacian yields a diagonally dominant system of equations which one can solve successfully by standard relaxation techniques; and the results show typical second order accuracy. However, as one becomes more ambitious and attempts to use the same scheme for higher Reynolds numbers, one meets with difficulties which may manifest themselves in three ways:

(i) the relaxation techniques may fail to converge;

(ii) if a solution is obtained for the steady problem it may exhibit oscillations which are physically unrealistic;

(iii) if an explicit time-stepping procedure is used, it may become unstable for unexpectedly small time steps.

As we shall see, these phenomena are all related. They may be exacerbated by effects arising from the nonlinearity of the convective terms, but they are essentially linear in origin.

The core of the difficulties can be seen even with the simple model problems (1.20) and (1.24). Suppose we approximate the former on a uniform mesh of spacing h, giving mesh points $x_j = jh, j = 0, 1, \ldots, J$ with $Jh = 1$. We introduce the difference and divided difference notation that we shall use throughout the book, and use the notation U_j for the approximation to $u(x_j)$:

$$\Delta_+ U_j := U_{j+1} - U_j, \qquad \Delta_- U_j := U_j - U_{j-1} \tag{1.26a}$$

$$D_+ U_j := (U_{j+1} - U_j)/h, \quad D_- U_j := (U_j - U_{j-1})/h \tag{1.26b}$$

$$\Delta_0 := \tfrac{1}{2}(\Delta_+ + \Delta_-), \quad D_0 U_j := \Delta_0 U_j/h = \tfrac{1}{2}(U_{j+1} - U_{j-1})/h \tag{1.26c}$$

$$\delta^2 U_j := U_{j+1} - 2U_j + U_{j-1} \tag{1.26d}$$

$$= (\Delta_+ - \Delta_-)\, U_j = \Delta_+ \Delta_- U_j = \Delta_- \Delta_+ U_j \tag{1.26e}$$

$$D_h^2 U_j := (U_{j+1} - 2U_j + U_{j-1})/h^2 = (\delta^2 U_j)/h^2. \tag{1.26f}$$

All of this is standard notation, except the last. We also note here that we shall use the same notation on a nonuniform mesh in which $h_j = x_j - x_{j-1}$; then (1.26b) becomes

$$D_+ U_j := (U_{j+1} - U_j)/h_{j+1}, \quad D_- U_j := (U_j - U_{j-1})/h_j, \tag{1.27a}$$

(1.26c) becomes

$$D_0 U_j := (U_{j+1} - U_{j-1})/(h_j + h_{j+1}), \tag{1.27b}$$

and (1.26f) becomes

$$D_h^2 U_j := (D_+ U_j - D_- U_j)/\tfrac{1}{2}(h_j + h_{j+1}), \tag{1.27c}$$

while the remainder are unchanged.

1.3.1 Central difference scheme

The central difference approximation to (1.20) is, with $U_0 = 0, U_J = 1$ and writing S_j for $S(x_j)$,

$$L_h^{cd} U_j := -\epsilon \frac{U_{j+1} - 2U_j + U_{j-1}}{h^2} + b \frac{U_{j+1} - U_{j-1}}{2h} = S_j$$

$$\text{or} \qquad -\epsilon D_h^2 U_j + b D_0 U_j = S_j, \; j = 1, 2, \ldots, J - 1. \tag{1.28}$$

The key dimensionless parameter is the *mesh Péclet number*

$$\beta := bh/\epsilon \tag{1.29}$$

which will figure heavily throughout the present book. Collecting terms in (1.28) we obtain

$$(\tfrac{1}{2}\beta - 1)U_{j+1} + 2U_j - (\tfrac{1}{2}\beta + 1)U_{j-1} = (h^2/\epsilon)S_j, \tag{1.30}$$

for which the characteristic equation is

$$(\tfrac{1}{2}\beta - 1)\lambda^2 + 2\lambda - (\tfrac{1}{2}\beta + 1) = 0. \tag{1.31}$$

Its roots are

$$\lambda_0 = 1, \ \lambda_+ = \frac{1 + \tfrac{1}{2}\beta}{1 - \tfrac{1}{2}\beta}. \tag{1.32}$$

Thus when $S(x) \equiv 0$, the solution of (1.30) with $U_0 = 0, U_J = 1$ is

$$U_j = \frac{\lambda_+^j - 1}{\lambda_+^J - 1} \quad \text{for } \beta \neq 2, \tag{1.33}$$

with $U_0 = U_1 = \ldots = U_{J-1} = 0$ for $\beta = 2$, instead of the true solution

$$u(x_j) = \frac{e^{bx_j/\epsilon} - 1}{e^{b/\epsilon} - 1} = \frac{e^{\beta j} - 1}{e^{\beta J} - 1}. \tag{1.34}$$

It is clear that the approximate solution is oscillatory when $\beta > 2$, because $\lambda_+ < 0$ then, while the true solution is a monotone exponential boundary layer. Typical profiles are shown in Fig.1.4 with $J = 20, b = 1$ and $\epsilon = .25, .05, .01$ giving values of $\beta = 0.2, 1.0$ and 5.0. The approximation given in the first two cases is quite good, but in the last case as for any $\beta > 2$ it is not at all representative.

The same situation applies for any source function $S(x)$, since the solution of (1.20) can be written in the form

$$u(x) = \int_0^1 G(x, t)S(t)dt + \frac{e^{bx/\epsilon} - 1}{e^{b/\epsilon} - 1}, \tag{1.35}$$

where $G(x, t)$ is the Green's function for the problem, which is constructed from solutions to the homogeneous equation. A straightforward calculation gives

$$G(x, t) = \begin{cases} \dfrac{(e^{bx/\epsilon} - 1)(e^{b(1-t)/\epsilon} - 1)}{b(e^{b/\epsilon} - 1)} & \text{for } x \leq t \\[4mm] \dfrac{(e^{bt/\epsilon} - 1)(e^{b(1-t)/\epsilon} - e^{b(x-t)/\epsilon})}{b(e^{b/\epsilon} - 1)} & \text{for } x \geq t. \end{cases} \tag{1.36}$$

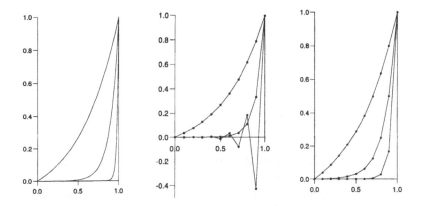

Figure 1.4. *Solutions of the model problem (1.20) with $S \equiv 0$ for $\beta = 0.2, 1.0$ and 5.0; exact (on the left) and approximations with a central (in the centre) and an upwind difference scheme (on the right)*

In the same way, the solution of (1.30) can be written, for $\beta \neq 2$, as

$$U_j = \sum_{k=1}^{J-1} hG_{j,k}S_k + \frac{\lambda_+^j - 1}{\lambda_+^J - 1}, \qquad (1.37)$$

where

$$G_{j,k} = \begin{cases} \dfrac{(\lambda_+^j - 1)(\lambda_+^{J-k} - 1)}{b(\lambda_+^J - 1)} & \text{for } j \leq k \\[4mm] \dfrac{(\lambda_+^k - 1)(\lambda_+^{J-k} - \lambda_+^{j-k})}{b(\lambda_+^J - 1)} & \text{for } j \geq k. \end{cases} \qquad (1.38)$$

Thus the quality of the approximation to $e^{\beta j}$ that is provided by λ_+^j affects any solution of the model problem (1.20) that is approximated by (1.30).

Suppose that in addition we attempted to solve (1.30) by Jacobi iteration,

$$2U_j^{n+1} = (1 - \tfrac{1}{2}\beta)U_{j+1}^n + (1 + \tfrac{1}{2}\beta)U_{j-1}^n + (h^2/\epsilon)S_j. \qquad (1.39a)$$

Writing E_j^n for the error $U_j^n - U_j^\infty$ in the iteration, we have

$$E_j^{n+1} = \tfrac{1}{2}(1 - \tfrac{1}{2}\beta)E_{j+1}^n + \tfrac{1}{2}(1 + \tfrac{1}{2}\beta)E_{j-1}^n, \quad j = 1, 2, \ldots, J - 1 \quad (1.39b)$$

with $E_0^n = E_J^n = 0$. We can use Fourier analysis to determine whether this will converge: the errors can be expanded in a discrete sine series, or equivalently in the Fourier modes $e^{ik x j} = e^{i\xi j}$ for

$$\xi = m\pi/J, \quad m = \pm 1, \pm 2, \ldots, \pm(J - 1). \qquad (1.40)$$

From (1.39b) the amplification or damping factor $\kappa(\xi)$ for each mode is give by

$$\kappa(\xi) = \tfrac{1}{2}[(1 - \tfrac{1}{2}\beta)e^{i\xi} + (1 + \tfrac{1}{2}\beta)e^{-i\xi}]$$
$$= \cos\xi - \tfrac{1}{2}i\beta\sin\xi, \tag{1.41a}$$

so that

$$|\kappa(\xi)|^2 = \cos^2\xi + \tfrac{1}{4}\beta^2\sin^2\xi$$
$$= 1 - (1 - \tfrac{1}{4}\beta^2)\sin^2\xi. \tag{1.41b}$$

Thus all modes are amplified rather than damped if $\beta > 2$, with the $\xi = \tfrac{1}{2}\pi$ mode being amplified most.

Finally, suppose this central difference scheme is used to approximate the unsteady problem (1.24), using the explicit forward time difference, so that

$$U_j^{n+1} = U_j^n + \Delta t[\epsilon D_h^2 U_j^n - b D_0 U_j^n], \tag{1.42}$$

where we use the notation U_j^n for the approximation to $u(x_j, t_n)$ and $t_{n+1} = t_n + \Delta t$. The stability of this scheme can also be analysed with the Fourier analysis used above, except that we need to add the mode $m = 0$ to (1.40) in order to have the full complement of modes which can be distinguished on the discrete mesh; note that $m_1 - m_2 = 2J$ implies that the m_1 and m_2 modes are indistinguishable. Again using the notation $\kappa(\xi)$ for the amplification factor of a mode, but in this case over one time step rather than one step of the Jacobi iteration process, we have after writing $\mu := \epsilon\Delta t/h^2$,

$$\kappa(\xi) = 1 + \mu[(e^{i\xi} - 2 + e^{-i\xi}) - \tfrac{1}{2}\beta(e^{i\xi} - e^{-i\xi})]$$
$$= 1 - 2\mu + 2\mu\cos\xi - i\mu\beta\sin\xi. \tag{1.43}$$

We need to be careful over our use of the term stability here. What we may call *Lax–Richtmyer stability* is that which is necessary and sufficient for a consistent scheme to give a convergent approximation to the solution of a well-posed linear problem, as the mesh is refined in a given way; this is the content of the Lax Equivalence Theorem — see Richtmyer and Morton (1967) or Morton and Mayers (1994). In the present case, we may presume that the mesh is refined with the value of μ fixed. Then stability in this sense depends on

$$|\kappa(\xi)| \leq 1 + O(\Delta t) \quad \text{for } |\xi| \leq \pi, \tag{1.44}$$

with the bound implied in the $O(\Delta t)$ being independent of ξ. From (1.43) we have

$$|\kappa(\xi)|^2 = (1 - 4\mu\sin^2\tfrac{1}{2}\xi)^2 + \mu^2\beta^2\sin^2\xi, \tag{1.45}$$

and $\mu\beta^2 = (b^2/\epsilon)\Delta t$ so that the last term here is $O(\Delta t)$ for any fixed value of μ. Thus this central difference scheme is L–R stable if $\mu \leq \frac{1}{2}$:

$$\text{L–R stability} \iff \mu \equiv \frac{\epsilon \Delta t}{h^2} \leq \tfrac{1}{2}. \tag{1.46}$$

This is exactly the same as the familiar condition obtained when $b = 0$.

However, this is an inadequate stability criterion to apply to any computation carried out for a *fixed value* of h and Δt. Instead, we use the concept of *practical stability* introduced in Richtmyer and Morton (1967). Unfortunately, these authors did not apply the idea to the present problem and left the impression that condition (1.46) was satisfactory. The requirement of practical stability is that the growth rate of any mode in the discrete problem should be limited by the maximum growth rate found for a mode in the differential problem. In the present case, any mode of (1.24) is merely convected and damped and therefore $|\kappa(\xi)|$ should be bounded by unity.

If we continue the computation (1.45) we obtain, with $s = \sin \frac{1}{2}\xi$,

$$\begin{aligned}
|\kappa(\xi)|^2 &= 1 - 8\mu s^2 + 16\mu^2 s^4 + 4\mu^2\beta^2 s^2(1 - s^2) \\
&= 1 - 4\mu(2 - \mu\beta^2)s^2 + 4\mu^2(4 - \beta^2)s^4.
\end{aligned} \tag{1.47}$$

We can check that this quadratic in s^2 always lies in $[0,1]$ by ensuring that the coefficient of s^2 is nonpositive and that $|u^2| \leq 1$ for $s^2 = 1$. The latter leads to the condition $\mu \leq \frac{1}{2}$ and the former to $\mu\beta^2 \leq 2$, or in combination $\mu^2\beta^2 \leq 2\mu \leq 1$, which can also be rewritten in terms of the original parameters,

$$\text{practical stability} \iff \begin{cases} \mu^2\beta^2 \leq 2\mu \leq 1 \\[2mm] \left(\dfrac{b\Delta t}{h}\right)^2 \leq 2\dfrac{\epsilon\Delta t}{h^2} \leq 1. \end{cases} \tag{1.48a}$$

For small values of ϵ, (1.48a) gives a very severe restriction on the *CFL number* $\nu := b\Delta t/h$ — see Fig1.5. In terms of β, we see just how restrictive this is by multiplying (1.48a) throughout by $1/|\nu|$ to get $|\nu| \leq 2/\beta \leq 1/|\nu|$, that is,

$$\text{practical stability} \iff |\nu| \leq \min(\beta/2, 2/\beta). \tag{1.48b}$$

On the other hand, from (1.48a) we see that if we insist on taking a maximal time step in the sense of setting $\mu = \frac{1}{2}$, then we must have $\beta \leq 2$ for practical stability. By comparing (1.41a) with (1.43) we see that this exactly corresponds to the situation found when using Jacobi iteration for solving the steady equations.

The need for this more severe stability condition can be seen by imposing the L–R stability condition $\mu \leq \frac{1}{2}$ and finding the maximum value of (1.47) when the practical stability condition is violated. It is easily checked that taking $\mu = \frac{1}{2}$ is the worst case, which gives just (1.41b). Thus the growth

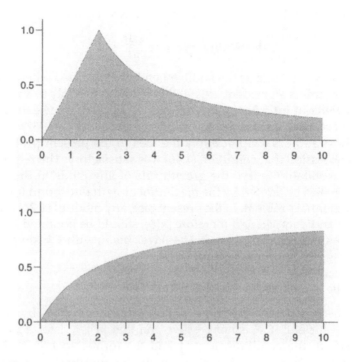

Figure 1.5. *Practical stability regions for the central and upwind difference schemes (top and bottom, respectively) applied to the model problem(1.24); in each case the abscissa is β and the ordinate ν*

of the mode given by $\xi = \frac{1}{2}\pi$ is by a factor of $\frac{1}{2}\beta$ in each time step in this case.

Thus for this central difference scheme all three of the difficulties listed above are closely related and make their appearance for $\beta \geq 2$.

1.3.2 Upwind difference scheme

A commonly advocated remedy for these problems is to use an upwind difference for the convective term in (1.24a) — see Spalding (1972) for an early discussion of the issue, together with the suggestion that a switch to the upwind scheme should be made for $\beta > 2$; and for more controversially expressed views see also Leonard (1979b) and Gresho and Lee (1981). For $b > 0$ an upwind difference corresponds to a backward difference, so that (1.28) is replaced by

$$L_h^{ud}U_j := -\epsilon\frac{U_{j+1} - 2U_j + U_{j-1}}{h^2} + b\frac{U_j - U_{j-1}}{h} = S_j$$

$$\text{or} \quad -\epsilon D_h^2 U_j + bD_- U_j = S_j, \; j = 1, 2, \ldots, J-1. \qquad (1.49)$$

Introducing the mesh Péclet number β, this becomes

$$-U_{j+1} + (2 + \beta)U_j - (1 + \beta)U_{j-1} = (h^2/\epsilon)S_j \qquad (1.50)$$

for which the characteristic equation is

$$\lambda^2 - (2 + \beta)\lambda + (1 + \beta) = 0, \qquad (1.51)$$

with the roots

$$\lambda_0 = 1, \quad \lambda_+ = 1 + \beta. \qquad (1.52)$$

This provides only a $(1,0)$ Padé approximant to e^β so that for small β it is less accurate than the $(1,1)$ Padé approximant (1.32) of the central difference scheme. However, it is positive for all β so that the approximate solution to the homogeneous problem, namely

$$U_j = \frac{\lambda_+^j - 1}{\lambda_+^J - 1}, \qquad (1.53)$$

is never oscillatory but always monotone increasing; on the other hand, it thickens the boundary layer significantly for large β. Both these properties are illustrated by the profiles given in Fig.1.4.

Another very important property of the scheme is that the discrete equation system (1.50) has a diagonally dominant tridiagonal matrix: the coefficients of U_{j+1} and U_{j-1} are both always negative and that of U_j is positive and equal to minus their sum. Thus if Jacobi iteration is used to solve the system in this case, the errors satisfy

$$E_j^{n+1} = \frac{1}{2 + \beta} E_{j+1}^n + \frac{1 + \beta}{2 + \beta} E_{j-1}^n, \quad j = 1, 2, \ldots, J - 1 \qquad (1.54)$$

with $E_0^n = E_j^n = 0$; and a Fourier analysis gives a damping factor for each of the modes (1.40)

$$\kappa(\xi) = [e^{i\xi} + (1 + \beta)e^{-i\xi}]/(2 + \beta)$$

$$= \cos \xi - i\frac{\beta}{2 + \beta} \sin \xi, \qquad (1.55)$$

so that

$$|\kappa(\xi)|^2 = 1 - \left[1 - \frac{\beta^2}{(2 + \beta)^2}\right] \sin^2 \xi. \qquad (1.56)$$

Hence for any β every mode is damped, those for $m = \pm 1$ and $\pm(J - 1)$ having least damping,

$$|\kappa|_{\max}^2 = 1 - \frac{4(1 + \beta)}{(2 + \beta)^2} \sin^2 \pi h. \qquad (1.57)$$

The only effect of the convective term is that the familiar slow convergence rate of $1 - \frac{1}{2}(\pi h)^2$ as $h \to 0$, which one obtains for the purely diffusive case, is slowed even further for large β to

$$|\kappa|_{max} \sim 1 - (2/\beta)(\pi h)^2 \quad \text{as } h \to 0, \ \beta \gg 1. \tag{1.58}$$

At $\beta = 2$ only one third more iterations are needed for a given accuracy; but at $\beta = 20$ five times as many are needed.

Lastly, let us consider the L-R and practical stability criteria required for the upwind scheme when applied to (1.24). From

$$U_j^{n+1} = U_j^n + \Delta t[\epsilon D_h^2 U_j^n - bD_- U_j^n], \tag{1.59}$$

Fourier analysis gives the amplification factor

$$\kappa(\xi) = 1 + \mu[(e^{i\xi} - 2 + e^{-i\xi}) - \beta(1 - e^{-i\xi})]$$
$$= 1 - \mu(2 + \beta)(1 - \cos \xi) - i\mu\beta \sin \xi, \tag{1.60}$$

from which we find, writing ν for $\mu\beta$,

$$|\kappa|(\xi)|^2 = (1 - 2(2\mu + \nu) \sin^2 \tfrac{1}{2}\xi)^2 + \nu^2 \sin^2 \xi. \tag{1.61}$$

As with the central difference scheme, the last term is $O(\Delta t)$ for a fixed μ so we have

$$\text{L–R stability} \iff \begin{cases} 0 \le 2\mu + \nu \le 1 \\[2mm] 0 \le 2\dfrac{\epsilon \Delta t}{h^2} + \dfrac{b\Delta t}{h} \le 1. \end{cases} \tag{1.62}$$

The stronger condition here as compared with (1.46) results from the *enhanced diffusion* that is provided by the upwind scheme. Indeed from the relations (1.26) we see that (1.59) can be rewritten as

$$U_j^{n+1} = U_j^n + \Delta t[\epsilon(1 + \tfrac{1}{2}\beta)D_h^2 U_j^n - bD_0 U_j^n]. \tag{1.63}$$

Comparing this with the central difference scheme (1.42) we see that the analysis for the upwind scheme should be the same as that for the central scheme if μ is replaced by $\mu(1 + \tfrac{1}{2}\beta) = \mu + \tfrac{1}{2}\nu$ while $\nu = \mu\beta$ is kept the same. Thus (1.46) is replaced by (1.62), where the left-hand inequality now has to be given explicitly. Similarly, if we use this to get the practical stability criteria for the upwind scheme, we obtain from (1.48)

$$\text{practical stability} \iff \begin{cases} \nu^2 \le 2\mu + \nu \le 1 \\[2mm] \left(\dfrac{b\Delta t}{h}\right)^2 \le 2\dfrac{\epsilon \Delta t}{h^2} + \dfrac{b\Delta t}{h} \le 1. \end{cases} \tag{1.64}$$

In Fig. 1.5 we have only shown the stability range $\nu \le \beta/(2 + \beta)$ given by these conditions; this is because from our use of the term 'upwind' scheme, we are assuming that $b \ge 0$ when we use it so that the negative values of ν that are allowed in (1.64) should be discounted. With this interpretation, (1.62) and (1.64) are identical, so for this scheme the L–R stability criteria and the practical stability criteria are the same.

It is worth noting that the practical stability conditions of any three-point scheme, e.g. the Lax–Wendroff method, can be deduced by the same analysis as that given above; and in Hindmarsh *et al.* (1984) these conditions are generalised to multi-dimensional schemes.

Finally, note that taking the maximal value $\mu = (2 + \beta)^{-1}$ allowed by (1.64) makes (1.59) identical to the Jacobi iteration (1.54). Since there are no other practical stability conditions if we assume $\mu \geq 0, \nu \geq 0$, this establishes that the iteration is convergent for all values of β.

1.3.3 Truncation error and global error

Suppose that a given difference scheme $L_h U_j^n = S_j$ is scaled so that the difference operator L_h tends formally to the differential operator in the problem $Lu = S$. Denoting by $\{u_j^n\}$ the set of values obtained by restricting the true solution $u(x, t)$ to the nodal points (x_j, t_n), the *truncation error* is defined as the residual obtained when the difference scheme is applied to the true solution values,

$$T_j^n := L_h u_j^n - S_j. \tag{1.65}$$

This applies to problems in one dimension or many, steady or unsteady, linear or nonlinear.

The *global error* is then the difference between the approximate and true solution, which at the nodes is $U_j^n - u_j^n$. The truncation error is relatively easy to estimate in terms of derivatives of the solution; but one of the main tasks in numerical analysis is to estimate the global error in an appropriate norm.

Confining ourselves for the moment to the steady model problem (1.20) and the central and upwind schemes given above, we have in the former case on a uniform mesh

$$\begin{aligned} T_j^{cd} &= -\epsilon D_h^2 u_j + b D_0 u_j - S_j \\ &= -\epsilon (D_h^2 u_j - u_j'') + b(D_0 u_j - u_j'). \end{aligned} \tag{1.66a}$$

Taylor series expansions of u_{j-1} and u_{j+1} about x_j readily give

$$T_j^{cd} = -\tfrac{1}{12}\epsilon h^2 u_j^{iv} + \tfrac{1}{6}bh^2 u_j''' + o(h^2), \tag{1.66b}$$

where the derivatives u_j''' and u_j^{iv} may be estimated through (1.35) and (1.36). If the source function $S(x)$ is sufficiently smooth, it is clear that we shall have a bound for the pth derivative $u^{(p)}$ of the form

$$|u^{(p)}| \leq C_p \epsilon^{-p}. \tag{1.67}$$

Hence we get

$$|T_j^{cd}| \leq C h^2 / \epsilon^3 \tag{1.68}$$

where C is independent of h and ϵ. In principle we could then use the discrete Green's function (1.38) to estimate the global error, by first subtracting (1.65) from (1.28) to get

$$L_h^{cd}(U_j - u_j) = -T_j^{cd}, \qquad (1.69)$$

and then using the formula (1.37), but with zero boundary data, to write

$$U_j - u_j = -\sum_{k=1}^{J-1} hG_{j,k}T_k^{cd}. \qquad (1.70)$$

This approach will work quite well when $\beta < 2$, and will show that $|U_j - u_j| = O(h^2)$ for a fixed ϵ as $h \to 0$. But as we have already seen, for this central difference scheme the Green's function will be oscillatory for $\beta > 2$ and the estimates from (1.70) will be poor and difficult to make. However, there are more basic limitations to this approach. Firstly, the estimates (1.67) and (1.68) are too crude in that they do not show the variation of $u^{(p)}(x)$ with x; in fact, we shall see in Chapter 2 that the ϵ^{-p} behaviour is confined to a boundary layer at the right-hand boundary, and this can be exploited in the summation in (1.70). Secondly, when $\beta < 2$ and (1.70) can be used, then the coefficients in the difference scheme (1.30) are negative for $U^{j\pm 1}$ and positive for U_j; a discrete maximum principle can then be formulated so that $|U_j - u_j|$ can be bounded much more directly from $T_j^{cd\cdot}$. This technique will be exploited in the analysis of three-point schemes in Chapter 3.

Let us now consider the upwind scheme in the same way. Using Taylor series expressions about x_j, we have

$$\begin{aligned}
T_j^{ud} &= -\epsilon(D_h^2 u_j - u_j'') + (bD_- u_j - u_j') \\
&= -\tfrac{1}{12}\epsilon h^2 u_j^{iv} - \tfrac{1}{2}bhu_j'' + o(h).
\end{aligned} \qquad (1.71)$$

However, in view of the second term being $O(h)$ it is better to truncate the expansion of the first term at a lower order to obtain an $O(\epsilon h|u'''|)$ bound. Then from (1.67) we get

$$|T_j^{ud}| \le Ch/\epsilon^2. \qquad (1.72)$$

In this case the Green's function is well-behaved for all β; and in addition the discrete equations as we can see from (1.50) will always satisfy a maximum principle. So from the equation corresponding to (1.69) we shall be able to deduce

$$|U_j - u_j| \le Ch/\epsilon^2, \qquad (1.73)$$

where C is independent of h and ϵ, for all values of β. We shall in fact be able to do better than this in Chapter 3, as indicated above.

However, there are two inherent limitations to the scheme that we cannot remove: one is that we have only $O(h)$ convergence even for a fixed ϵ; and the other is that we cannot show any form of convergence as $h \to 0$ which is uniform in ϵ — for each value of h there is some value of ϵ where the error remains larger than an irreducible minimum. In Chapter 3 we shall consider three-point schemes which, like the upwind scheme, satisfy a maximum principle for all values of β, but which can give fourth order accuracy for a fixed ϵ or second order accuracy uniformly in ϵ.

Before leaving the upwind scheme, we note a small change that would improve its performance. The upwind difference for the convection is what limits the accuracy, but this is second order accurate at the midpoint $x_{j-\frac{1}{2}}$ of x_{j-1} and x_j. So suppose we replace S_j on the right of (1.49) by $\frac{1}{2}(S_{j-1} + S_j)$. Then by expanding in a Taylor series about $x_{j-\frac{1}{2}}$ we obtain

$$
\begin{aligned}
T^{ud}_{j-\frac{1}{2}} &= -\epsilon(D_h^2 u_j - u''_{j-\frac{1}{2}}) + b(D_- u_j - u'_{j-\frac{1}{2}}) \\
&= -\tfrac{1}{2}\epsilon h u'''_{j-\frac{1}{2}} + \tfrac{1}{24}bh^2 u'''_{j-\frac{1}{2}} + o(\epsilon h + h^2).
\end{aligned}
\tag{1.74}
$$

Detailed analysis will show that error terms for the standard upwind scheme which are $O(h)$ will be replaced by terms which are $O(h^2 + \epsilon h)$ with this modified upwind scheme. Moreover we will see that most of the improvements in schemes studied in later chapters result from changing the sampling of the source term from the simple S_j in (1.28) and (1.49). The finite element methods, which we begin to consider in the next section, are a case in point.

These examples, and others that can be given for the unsteady problem, show the care that has to be taken in defining the term *order of accuracy*, which we have so far used without definition. Standard usage assumes that all space and time steps can be given in terms of a single parameter h, and that the global error can be estimated in terms of the truncation error. Then the order of accuracy is p, when p is the largest integer for which we can show that the truncation error is bounded by Ch^p for all sufficiently smooth solutions of the differential equation; any (common) point can be chosen for making the Taylor series expansions, and terms can be cancelled by use of the differential equation.

1.3.4 Graded meshes

So far we have considered only uniform meshes, partly for simplicity and ease of presentation. In practice, however, meshes will often be nonuniform, either because of geometrical constraints which arise in more than one dimension, or to exploit the local behaviour of the truncation error. If we have local *a priori* estimates of the solution derivatives, one possible strategy is to choose the mesh such that the truncation errors, as for example in (1.66b) and (1.71), are of roughly the same size everywhere on the mesh; this is called *equi-distribution of error*. On the other hand, one might make an *a posteriori* estimate of the local truncation error, or of the local contribution to the global error, from a calculation on a uniform mesh and then recalculate on an adjusted mesh; or in an iterative calculation this *mesh adaptation* might occur within the iteration. All of these possibilities will be discussed below and in later chapters when we have a finite element framework and sharper error estimates available.

1.4 Finite elements in one dimension

One of the chief advantages of finite element methods is the natural way in which they deal with nonuniform meshes, for instance triangles and quadrilaterals in two dimensions. So even in this introductory section we will assume that the unit interval is sub-divided into J elements or intervals, with the jth having length h_j from x_{j-1} to x_j, that is

$$0 = x_0 < x_1 < \ldots < x_{J-1} < x_J = 1.$$

We will also introduce some of the terminology and notation that we will need later, and will therefore consider the more general steady problem (1.21) with one of the boundary conditions possibly being a Neumann condition. Note that for notational simplicity we shall generally favour the use of h_j for the interval length, rather than the alternative $h_{j-\frac{1}{2}}$.

1.4.1 *Discrete equations for piecewise linear approximation*

Consider the use of a continuous piecewise linear approximation on $(0, 1)$. The free parameters in such an approximation $W(x)$ are the nodal values W_j at x_j which are not prescribed by Dirichlet boundary conditions. We call this space of approximation functions the *trial space* S_E^h, with the subscript denoting that the *essential boundary conditions*, which in this case are the Dirichlet conditions, have been imposed. In the jth element there are two *element basis functions* associated with the nodes x_{j-1} and x_j,

$$N_1^{(j)}(x) := \frac{x_j - x}{x_j - x_{j-1}}, \quad N_2^{(j)}(x) := \frac{x - x_{j-1}}{x_j - x_{j-1}}, \quad x \in [x_{j-1}, x_j]; \quad (1.75)$$

and for each interior node x_j these combine to give a *global basis function*

$$\phi_j(x) = \begin{cases} N_2^{(j)}(x) & x \in [x_{j-1}, x_j] \\ N_1^{(j+1)}(x) & x \in [x_j, x_{j+1}], \ j = 1, 2, \ldots, J-1, \end{cases} \quad (1.76)$$

as sketched in Fig.1.6. In the end intervals $\phi_0(x) = N_1^{(1)}(x)$ on $[0, x_1]$ and $\phi_J(x) = N_2^{(J)}(x)$ on $[x_{J-1}, 1]$. Then, with the convention that each $\phi_j(x)$ is zero outside the interval where it has been defined, we can represent S_E^h

as

$$S_E^h := \{W(x) = \sum_{j=0}^{J} W_j \phi_j(x) \mid W(x) \text{ satisfies the Dirichlet b.c.'s}\}. \quad (1.77)$$

In the case (1.21b) where both boundaries are Dirichlet, there are $J - 1$

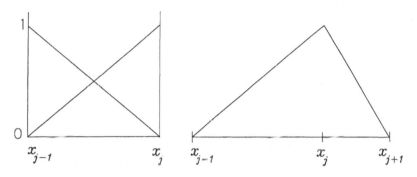

Figure 1.6. *Piecewise linear basis functions, the element basis functions on the left and a global basis function on the right*

free parameters to determine by taking the inner product of (1.21a) with $J - 1$ *test functions* or *weighting functions*. These form the *test space*. In this section we shall consider only *Galerkin methods* which we define to be those in which the test basis functions are drawn from those of the trial space, namely just those which correspond to free parameters in the expansion (1.77). When one of the boundary conditions is of Neumann type, there is an extra free parameter in (1.77) and hence J test functions, so that we always have the correct number of equations. Thus we define S_0^h as

$$S_0^h := \{V(x) = \sum_{j=0}^{J} V_j \phi_j(x) \mid V(x) \text{ satisfies the homogeneous}$$
$$\text{Dirichlet b.c.'s}\}. \quad (1.78)$$

Finally, we need to integrate by parts the first term of (1.21a) in defining a bilinear form over the trial and test spaces. We write

$$B(u, v) := \int_0^1 \left[\epsilon a(x) u' v' + b(x) u' v + c(x) uv \right] \, dx, \quad (1.79)$$

and use the notation

$$(u, v) := \int_0^1 uv \, dx \quad (1.80)$$

for the usual L_2 inner product.

Then we can define the *Galerkin approximation* $U \in S_E^h$ to the problem

(1.21) by the equations

$$B(U, V) = (S, V) \quad \forall V \in S_0^h, \tag{1.81}$$

where by assuming that any Neumann condition is homogeneous we have avoided having to include any Neumann boundary data on the right-hand side. (The more general problem will be treated in Chapter 4, specifically section 4.2.)

This abstract formulation can be made more concrete by substituting the basis functions $\phi_j(x)$ for $V(x)$ in (1.81) and carrying out the integrals. To avoid the complications of numerical quadrature at this stage, we assume the integral on the right can be performed exactly and also assume that we have constant coefficients in $B(\cdot, \cdot)$, namely $a(x) \equiv 1$, $b(x) \equiv b > 0$ and $c(x) \equiv c \geq 0$. Furthermore, an important simplification in formulating finite element methods results from breaking up the integrals into those over each element. In the present case we have only to consider the interactions of the two basis functions (1.75) in order to define a 2×2 *element stiffness matrix*

$$k_{\alpha,\beta}^{(j)} := B(N_\beta^{(i)}, N_\alpha^{(j)}). \tag{1.82}$$

(N.B. In order to conform with the usual matrix notation it would have been better to reverse the order of the arguments in $B(u, v)$. Unfortunately, the present notation is too well entrenched, particularly through the usage in the symmetric case, for us to attempt to change it now.) The derivatives of $N_1^{(j)}$ and $N_2^{(j)}$ are constants $\mp 1/h_j$ and we need the trivially evaluated integrals

$$\int_{x_{j-1}}^{x_j} N_\beta^{(j)'} N_\alpha^{(j)'} \, dx = (-1)^{\alpha+\beta}/h_j, \int_{x_{j-1}}^{x_j} N_\beta^{(j)'} N_\alpha^{(j)} \, dx = \tfrac{1}{2}(-1)^\beta, \tag{1.83a}$$

$$\int_{x_{j-1}}^{x_j} N_\beta^{(j)} N_\alpha^{(j)} \, dx = \tfrac{1}{6}(1 + \delta_{\alpha,\beta})h_j. \tag{1.83b}$$

Then the element stiffness matrix becomes

$$\left(k_{\alpha,\beta}^{(j)} \right) = \frac{\epsilon}{h_j} \begin{bmatrix} 1 & -1 \\ -1 & 1 \end{bmatrix} + \tfrac{1}{2}b \begin{bmatrix} -1 & 1 \\ -1 & 1 \end{bmatrix} + \tfrac{1}{6}ch_j \begin{bmatrix} 2 & 1 \\ 1 & 2 \end{bmatrix}. \tag{1.84}$$

In calculating the bilinear form $B(U, \phi_j)$ this stiffness matrix multiplies the vector of nodal values in the element $(U_{j-1}, U_j)^T$ and, as we see from (1.76), the second row combines with the first row of the stiffness matrix $(k_{\alpha,\beta}^{(j+1)})$ multiplying its vector $(U_j, U_{j+1})^T$. The result is a difference equation which is associated with the jth node. If we introduce the notation $\widehat{S}_j$ to write the right-hand side

$$S_j := (S, \phi_j) =: \tfrac{1}{2}(h_j + h_{j+1})\widehat{S}_j, \quad j = 1, 2, \ldots, J - 1 \tag{1.85}$$

and divide the whole equation by $\frac{1}{2}(h_j + h_{j+1})$, we see by reference to our definitions of (1.27) that this difference equation can be written in the form

$$-\epsilon D_h^2 U_j + b D_0 U_j + \tfrac{1}{6}c \left(\frac{2h_j}{h_j + h_{j+1}} U_{j-1} + 4U_j + \frac{2h_{j+1}}{h_j + h_{j+1}} U_{j+1} \right)$$

$$= \widehat{S}_j \quad j = 1, 2, \ldots, J-1. \tag{1.86}$$

The first two terms are exactly what we would have obtained if we had generalised the central difference scheme (1.28) to a nonuniform mesh. However, the source term is sampled differently here; had we introduced a cu term into (1.20), it would have been approximated in (1.28) by cU_j instead of what we have here.

These differences stemming from the finite element approach are quite significant. But much more significant is the fact that the Galerkin approach will suffer all the disadvantages of the central difference method when the mesh Péclet number is greater than 2. To overcome this we shall need to choose a test function space which is different from the trial function space. We call the resultant class of methods *Petrov–Galerkin methods* and we shall study them in detail in Chapter 5. This will form one of the key topics in this book.

Before we leave these Galerkin equations, however, let us write them in the more usual and more easily generalisable matrix form. The combination, through addition of corresponding entries, of the stiffness matrices of two neighbouring elements is called the *matrix assembly* process and leads to the *global stiffness matrix*. It is at this point that we take account of the boundary conditions. Suppose they are of Dirichlet type at both ends, as in (1.21b). Then the first column of $(k_{\alpha,\beta}^{(1)})$ multiplies the boundary value u_L and the second column of $(k_{\alpha,\beta}^{(J)})$ multiplies the boundary value u_R. In the global matrix system for the vector $\mathbf{U}$ of $J-1$ unknowns $(U_1, U_2, \ldots, U_{J-1})^T$, these quantities should be transferred to the right-hand side so that we obtain

$$K\mathbf{U} = \mathbf{S} - u_L \mathbf{k}_0 - u_R \mathbf{k}_J. \tag{1.87}$$

Here the components S_j of $\mathbf{S}$ are given by (1.85) and those of K, $\mathbf{k}_0$ and $\mathbf{k}_J$ by

$$K_{j,k} = B(\phi_k, \phi_j), \quad (\mathbf{k}_0)_j = B(\phi_0, \phi_j), \quad (\mathbf{k}_J)_j = B(\phi_J, \phi_j)$$

$$j, k = 1, 2, \ldots, J-1. \tag{1.88}$$

Each of these entries is obtained from combinations of the element stiffness matrices, but notice that compared with the difference form (1.86) we have not divided by $\frac{1}{2}(h_j + h_{j+1})$.

1.4.2 Outline error analysis

These systems of equations, (1.81) and (1.87), will have the same form in two and three dimensions. So too will the error analysis which we will outline now. The starting point is to write the problem which is being approximated in the same weak form as (1.81): for (1.21) we need to introduce the Sobolev space $H^1(0,1)$ of functions which together with their first derivatives are square integrable over the interval $(0,1)$; then the bilinear form (1.79) is defined over $H^1(0,1) \times H^1(0,1)$. The solution u will be sought in the space

$$H_E^1 := \left\{ w \in H^1(0,1) \mid w \text{ satisfies the Dirichlet b.c.'s} \right\} \qquad (1.89a)$$

and test functions will be drawn from

$$H_{E_0}^1 := \left\{ v \in H^1(0,1) \mid v \text{ satisfies the homog. Dirichlet b.c.'s} \right\}. \qquad (1.89b)$$

Then the weak form of (1.21) is as follows: find $u \in H_E^1$ such that

$$B(u,v) = (S,v) \quad \forall v \in H_{E_0}^1. \qquad (1.90)$$

We have been careful to ensure that our finite element method is *conforming*; that is, that we have $S_E^h \subset H_E^1$ and $S_0^h \subset H_{E_0}^1$. Then we can clearly put any $V \in S_0^h$ into (1.90) instead of v. Subtracting the result from (1.81) gives the *error projection property*

$$B(U-u,V) = 0 \quad \forall V \in S_0^h. \qquad (1.91)$$

In the absence of convection, $B(\cdot,\cdot)$ is a symmetric positive definite form which defines a norm. Then (1.91) leads directly to an error bound in this norm.

Suppose that when $b \not\equiv 0$ we continue to use this as a norm and write

$$\|u\|_{B_S} := \left\{ \int_0^1 \left[\epsilon a(x)(u')^2 + c(x)u^2 \right] dx \right\}^{\frac{1}{2}}, \qquad (1.92)$$

where it is now clear why we have assumed $c(x) \geq 0$. Then it is readily shown that there are positive constants C_0 and C_1 such that

$$|B(u,v)| \leq C_1 \|u\|_{B_S} \|v\|_{B_S} \quad \forall u,v \in H_{E_0}^1 \qquad (1.93a)$$

$$B(u,u) \geq C_0 \|u\|_{B_S}^2 \quad \forall u \in H_{E_0}^1. \qquad (1.93b)$$

To obtain the upper bound (1.93a) is usually straightforward; it is the lower bound (1.93b), called a *coercivity condition*, which is crucial. A crude error bound can then be derived from (1.91) and these bounds as follows,

$$
\begin{aligned}
C_0 \|U-u\|_{B_S}^2 &\leq B(U-u, U-u) \\
&= B(U-u, W-u) \quad \forall W \in S_E^h \\
&\leq C_1 \|U-u\|_{B_S} \|W-u\|_{B_S}, \qquad (1.94)
\end{aligned}
$$

because $W - U \in S_0^h$. Since W is arbitrary we have

$$\|U - u\|_{Bs} \leq (C_1/C_0) \inf_{W \in S_E^h} \|W - u\|_{Bs}. \tag{1.95}$$

This establishes that the $\|\cdot\|_{Bs}$ norm error of U is within a certain factor of the best error one can obtain from a member of the approximation space S_E^h. It is then purely a problem in approximation theory to estimate the magnitude of this error. All error bounds for finite element methods take this form: but considerable refinements of this basic argument are needed to give practically useful results.

Firstly, even in the self-adjoint case where $b \equiv 0$ and hence $C_0 = C_1 = 1$, (1.95) gives a bound only in one particular norm. Thus for our piecewise linear approximation we obtain an error bound that is dominated by how well u' can be approximated by W' for $W \in S_E^h$, and this can clearly be no better than $O(h)$. The analysis therefore has to be taken further to show that the error in the L_2 norm is actually $O(h^2)$. Secondly, in the above analysis C_1/C_0 depends essentially on the global Péclet number, $(\max b)/\epsilon(\min a)$; however, a more refined analysis will enable a bound of the form (1.95) to be obtained in which the constant depends only on the mesh Péclet number. This bound is sharp and provides an important way of characterising the inadequacy of the Galerkin method. So, thirdly, we can use the objective of a bound of this kind with a constant near to unity as a key criterion in the design of Petrov–Galerkin methods which overcome the mesh Péclet number barrier. All of these points will be taken up in Chapter 5 in the context of two-dimensional as well as one-dimensional problems.

1.4.3 Unsteady problems

Lastly let us consider a piecewise linear finite element approximation to (1.24). More precisely, let us suppose first that the finite element approximation is used in the spatial coordinate only, and the same explicit finite difference method is used to approximate the time derivative as was used in (1.42) and (1.59). We write this, after suitably extending the domain in which each $U^n(x) \in S^h$ is defined and the range of integration in $B(\cdot, \cdot)$, as

$$\left(\frac{U^{n+1} - U^n}{\Delta t}, \phi_j \right) + B\left(U^n, \phi_j\right) = 0 \quad \forall \phi_j \in S^h. \tag{1.96}$$

This involves a *mass matrix* M, given by

$$M_{j,k} = \left(\phi_k, \phi_j\right), \tag{1.97}$$

as well as the stiffness matrix K defined in (1.88); in terms of these, (1.96) can be written as

$$M(\mathbf{U}^{n+1} - \mathbf{U}^n) + \Delta t K \mathbf{U}^n = 0. \tag{1.98}$$

The entries in the mass matrix are just the coefficients of c in (1.86), which arose from the term cuv in the bilinear form (1.79).

In practice, of course, the discretisation will be on a finite interval and (1.98) will be a finite matrix system. But, in order to compare with the difference schemes of section 1.3, we consider an infinite uniform mesh. Then (1.96) will yield the following equations to approximate (1.24),

$$\tfrac{1}{6}\left[\left(U_{j-1}^{n+1} - U_{j-1}^n\right) + 4\left(U_j^{n+1} - U_j^n\right) + \left(U_{j+1}^{n+1} - U_{j+1}^n\right)\right]$$
$$+ \Delta t\left[-\epsilon D_h^2 U_j^n + bD_0 U_j^n\right] = 0. \qquad (1.99a)$$

This can be written more compactly as

$$\left(1 + \tfrac{1}{6}\delta^2\right) U_j^{n+1} = \left(1 + \tfrac{1}{6}\delta^2\right) U_j^n + \Delta t\left[\epsilon D_h^2 U_j^n - bD_0 U_j^n\right], \qquad (1.99b)$$

which can be compared directly with the central difference scheme (1.42). Fourier analysis of stability yields the following amplification factor, since the Fourier transform of δ^2 is $-4\sin^2\tfrac{1}{2}\xi$,

$$\kappa(\xi) = 1 - \left(1 - \tfrac{2}{3}\sin^2\tfrac{1}{2}\xi\right)^{-1}\left[4\mu\sin^2\tfrac{1}{2}\xi + i\nu\sin\xi\right], \qquad (1.100)$$

where as before $\mu = \epsilon\Delta t/h^2$ and $\nu = b\Delta t/h = \mu\beta$. Hence, with $s = \sin\tfrac{1}{2}\xi$,

$$|\kappa(\xi)|^2 = \frac{\left[1 - (\tfrac{2}{3} + 4\mu)s^2\right]^2 + 4\nu^2 s^2(1 - s^2)}{(1 - \tfrac{2}{3}s^2)^2}. \qquad (1.101)$$

As with the difference schemes, we use the fact that $\nu^2 = O(\Delta t)$ to deduce that L–R stability depends only on

$$|1 - \left(\tfrac{2}{3} + 4\mu\right)s^2| \leq 1 - \tfrac{2}{3}s^2 \text{ for } s^2 \in [0,1].$$

The worst case occurs at $s^2 = 1$ and yields $4\mu \leq \tfrac{2}{3}$, so we have

$$\text{L–R stability} \iff \mu \equiv \frac{\epsilon\Delta t}{h^2} \leq \tfrac{1}{6}. \qquad (1.102)$$

The practical stability criterion $|\kappa|^2 \leq 1$, after clearing the denominator becomes

$$-4(2\mu - \nu^2)s^2 + 4(\tfrac{4}{3}\mu + 4\mu^2 - \nu^2)s^4 \leq 0 \quad \forall s^2 \in [0,1]. \qquad (1.103)$$

This is satisfied for $s^2 = 1$ if (1.102) is satisfied and it holds for small s^2 if $2\mu - \nu^2 \geq 0$; these two conditions are sufficient and yield the result

$$\text{practical stability} \iff \begin{cases} \nu^2 \leq 2\mu \leq \tfrac{1}{3} \\ \left(\dfrac{b\Delta t}{h}\right)^2 \leq 2\dfrac{\epsilon\Delta t}{h^2} \leq \tfrac{1}{3} \end{cases}, \qquad (1.104)$$

which conditions are to be compared with (1.48).

This seems to put the finite element approach in a rather poor light: the introduction of the mass matrix has not only required a tridiagonal system of equations to be solved at each time level, but has also considerably reduced the stability range. However, while this simple example has served to introduce the finite element method for unsteady problems, it is too simple an approach to indicate what can be achieved. The form of the mass matrix in (1.99b) clearly suggests that we must consider implicit difference schemes for the time derivative along with the explicit scheme (1.96). Similarly we must explore the use of space-time finite elements. And, as with the steady problems there is the possibility of overcoming the limit on the mesh Péclet number contained in (1.104) by the use of alternative test spaces in a Petrov–Galerkin framework. However, in Chapter 7 where all these approaches will be explored, we shall also present a much more powerful framework for developing approximations to the unsteady convection-diffusion equation. Called the *evolution–Galerkin* approach, it exploits properties of the evolution operator which takes the solution of the differential equation for one time level to the next, and allows alternative projections to be used rather than the L_2 projection implied in (1.96). In this way we obtain a new insight into the phenomenon of upwinding and are able to closely integrate the theory for steady and unsteady problems.

A note on notation Equations for finite difference schemes such as (1.27) or (1.41), and corresponding finite element schemes such as (1.86) and (1.99), have all been given using the subscript j rather than the more usual i. This has been done in part to avoid confusion with the i used in the Fourier analysis. However, to make for simpler notation in mass and stiffness matrices there is an advantage to using i rather than j; so we shall adopt this notation from Chapter 4 onwards, with the set of global trial functions in (1.88) and (1.97) denoted by $\{\phi_j\}$ and the set of test functions by $\{\phi_i\}$ to give typical matrix entries $M_{i,j}$ and $K_{i,j}$.

1.4.4 A posteriori error analysis

The *a priori* error analysis, either for finite difference schemes as in section 1.3.3 or for finite element schemes as in section 1.4.2, suffer from the fact that they depend on estimating high derivatives of the unknown solution either in the truncation error or in the approximation bounds. In *a posteriori* analysis, knowledge of an approximate solution is used to give a *computable error bound*; this may then be used to adapt the method or the mesh so as to eventually yield a guaranteed accuracy. Such techniques are well developed and widely used for linear self-adjoint problems (see Babuška and Rheinboldt (1978a),(1978b) for influential expositions), and are receiving increased attention for convection-diffusion and similar problems. Definitive techniques that might be used for example on the compressible Navier–Stokes equations of (1.14)–(1.15) are not yet available, so that it is premature for a detailed account to be given in this volume: however, the analysis and results for the inviscid Euler equations that are given by Sonar and Süli (1994) show the potential for the approach, and we give a brief outline below which will be referred to in later chapters.

The longest established technique within a finite element framework is relatively direct. Consider a steady problem in the form (1.90) with the error measured in the norm (1.92) and suppose this is dominated by the derivative term. Now with a piecewise linear approximation the gradient of U is only piecewise constant. So suppose a more accurate piecewise linear approximation to the gradient can be constructed and the associated approximate solution is denoted by U^*. Then if we can find constants c_1 and c_2, which are hopefully close to unity, such that

$$c_1 \|U - U^*\|_{B_s} \leq \|U - u\|_{B_s} \leq c_2 \|U - U^*\|_{B_s}, \qquad (1.105)$$

we have a useful *a posteriori error indicator* in $\|U - U^*\|_{B_s}$. The Galerkin approximation (1.81) for a self-adjoint problem in one dimension does give an $O(h^2)$ approximation to the gradient at the midpoint of each interval, from which U^* can be constructed. Similar so-called superconvergence properties hold for higher order approximations and in more dimensions; they form the basis of the Z^2 error estimators developed by Zienkiewicz and Zhu (1992a),(1992b), and the bounding constraints in (1.105) have been estimated for various classes of problems by several authors. The challenge for convection-diffusion problems, as already indicated in regard to the similar bounds occurring in the *a priori* estimates of (1.95), is to devise approximation schemes for which comparable estimates can be shown to hold.

A more general approach, which is probably better suited to the class of problems we are dealing with, is based on the *residual* obtained when the approximate solution is substituted into the original differential equation. Roughly speaking, the approach is the complement of that used in the truncation error analysis of section 1.3.3: with the same notation for the problem $Lu = S$, which might be steady or unsteady, we suppose that the approximation U given by $L_h U = S_h$ can be regarded as in the domain of L so that we can define the residual R as

$$R := LU - S. \qquad (1.106a)$$

Then for a linear problem we have

$$L(U - u) = R, \qquad (1.106b)$$

and to obtain a bound on the global error we have to estimate $L^{-1}R$ from our knowledge of R. Now suppose for some norm $\|\cdot\|_Q$ we define its dual

$\|\cdot\|_{Q^*}$ in the usual way so that we have

$$\|w\|_Q = \sup_v \frac{(w, v)}{\|v\|_{Q^*}},$$

and we denote by L^* the adjoint of the operator L. Then from (1.106b) we have

$$(R, v) = (L(U - u), v) = (U - u, L^*v) \quad \forall v, \tag{1.107a}$$

from which we deduce that

$$\|U - u\|_Q = \sup_v \frac{(R, v)}{\|L^*v\|_{Q^*}}. \tag{1.107b}$$

Various choices of the norm $\|\cdot\|_Q$ and of the way in which the right-hand side of (1.107b) might be estimated have been suggested; note that choosing v from a sufficiently smooth class of functions justifies the definition of the residual in (1.106a) in the sense of distributions.

Suppose now that we apply this technique to a piecewise linear approximation U for the problem (1.20). Then we have

$$\|U - u\|_Q = \sup_v \frac{B(U, v) - (S, v)}{\|\epsilon v'' + bv'\|_{Q^*}}, \tag{1.108}$$

where the norm $\|\cdot\|_Q$ and hence the space from which v is to be taken is as yet unspecified. The most natural choice is perhaps the L_2 norm of the derivative, $\|v\|_Q \equiv \|v'\|$, from which we get

$$\|\epsilon v'' + bv'\|_{Q^*} = \sup_{w \in H_0^1(0,1)} \frac{(\epsilon v' + bv, w')}{\|w'\|} = \|\epsilon v' + bv\|. \tag{1.109a}$$

Now the numerator in (1.108) is zero for every V in the test space, which in the Galerkin case is S_0^h, and also $\|\epsilon v' + bv\|^2 = \epsilon^2 \|v'\|^2 + b^2 \|v\|^2$. Hence (1.108) becomes

$$\|U' - u'\| = \sup_{v \in H^1(0,1)} \inf_{V \in S_0^h} \frac{B(U, v - V) - (S, v - V)}{(\epsilon^2 \|v'\|^2 + b^2 \|v\|^2)^{\frac{1}{2}}}. \tag{1.109b}$$

Various choices for v, for example piecewise quadratic, may then be used to obtain approximations to this bound.

Error estimates in the L_2 norm have been obtained in a wide range of problems by Johnson and his co-workers (see Eriksson *et al.* (1995) for a recent survey). These require an *a priori* bound for the adjoint problem: that is, suppose that for $v \in H_0^1(0, 1)$ with $L^*v = w$ we have the typical bound for an elliptic problem,

$$\|v''\| \le C(\epsilon) \|w\|. \tag{1.110}$$

Then consider $(U - u, L^*v)$, broken up into a sum over the intervals, and integrate by parts twice using the continuity of U, u, u', v, v' and the boundary conditions; we obtain the result, where $x_j^{\pm} \equiv x_j \pm 0$,

$$
\begin{aligned}
(U - u, w) &= \sum_{j=1}^{J} \left\{ \epsilon[U'(x_j^-)v_j - U'(x_{j-1}^+)v_{j-1}] \right. \\
&\quad \left. + \int_{x_{j-1}}^{x_j} L(U - u)v \mathrm{d}x \right\} \\
&= -\epsilon \sum_{j=0}^{J} v_j \Delta U_j' + (R, v),
\end{aligned} \tag{1.111a}
$$

where $\Delta U_j'$ denotes the jump in U' at a node and we suppose that U is extended outside the interval $[0, 1]$ by constants. Now suppose that $V \in S_0^h$ interpolates v at the nodes and we have the approximation property

$$
\|v - V\|_{L_2(I_j)} \le C_2 h_j^2 \|v''\|_{L_2(I_j)} . \tag{1.111b}
$$

Then by (1.91) we can replace v by $(v - V)$ in (1.111a), so the first term drops out, and we have

$$
\begin{aligned}
|(U - u, w)| &\le C_2 \sum_{j=1}^{J} h_j^2 \|R\|_{L_2(I_j)} \|v''\|_{L_2(I_j)} \\
&\le C_2 \|v''\| \|h^2 R\| \le C_2 C(\epsilon) \|w\| \|h^2 R\| ,
\end{aligned} \tag{1.112a}
$$

where the residual has been weighted by h^2 in each interval. The final result is then

$$
\|U - u\| \le C_2 C(\epsilon) \|h^2 R\| , \tag{1.112b}
$$

where in each interval $h^2 R$ is just $h_j^2(bU' - S)$. In a case like that shown in Fig. 1.4 with $\beta \gg 1$, we will have $C(\epsilon) \sim \epsilon^{-1}$ and $R = O(h)$ so that the estimate of error will be $O(h/\epsilon)$; this will certainly give the correct indication of the need for mesh refinement when using the Galerkin method, although a direct calculation of the error from the formulae of (1.32) and (1.33) will give a value which is $O(h/\epsilon^{\frac{1}{2}})$. On the other hand, this type of error analysis is normally only applied to a modified problem, with an enhanced diffusion coefficient having the typical form $\hat{\epsilon} = \max(\epsilon, h^{\frac{3}{2}})$, where the discrepancy is less marked.

At the other problem extreme, let us consider the pure convection problem, so we have $\mathbf{b} \cdot \nabla(U - u) = R$. In the work of Sonar and Süli (1994) already referred to, a graph norm was used, namely

$$
\|u\|_Q^2 := \|u\|^2 + \|\mathbf{b} \cdot \nabla u\|^2 , \tag{1.113a}
$$

together with an *a priori* Gårding inequality

$$\|v\| \le C_G \|\mathbf{b}\cdot\nabla v\| \tag{1.113b}$$

for a problem with zero inflow boundary data. Then, since we have $\|R\| = \|\mathbf{b}\cdot\nabla(U - u)\| \le \|U - u\|_Q$, we obtain the desired two-sided estimate

$$\|R\| \le \|U - u\|_Q \le (1 + C_G^2)^{\frac{1}{2}} \|R\|. \tag{1.113c}$$

If an estimate in the L_2 norm is required, we set $L^*v = w$ for which the *a priori* bound (1.113b) also holds, and obtain

$$\|R\|_{Q^*} = \sup_w \frac{(R, v)}{\|v\|_Q} = \sup_w \frac{(U - u, L^*v)}{\|v\|_Q} = \frac{\|U - u\| \|w\|}{(\|v\|^2 + \|w\|^2)^{\frac{1}{2}}};$$

this immediately gives $\|R\|_{Q^*} \le \|U - u\|$, but it also gives the upper bound on $\|U - u\|$ through the *a priori* inequality on $\|v\|$, so that compared with (1.113c) we have

$$\|R\|_{Q^*} \le \|U - u\| \le (1 + C_G^2)^{\frac{1}{2}} \|R\|_{Q^*}. \tag{1.113d}$$

The penalty for having the convenient norm on the error is that we have to estimate the residual in the awkward dual graph norm.

It is worth pointing out here that this complementary approach, of substituting the approximate solution into the original differential equation, is also used in the analysis of finite difference schemes where it is variously called the *modified equation* approach (Warming and Hyett 1974) or the method of *differential approximation* (Shokin 1983). It has been of great value in distinguishing between alternative treatments of nonlinear terms in hyperbolic equations.

1.5 Iterative solution of discrete equations

For all but the simplest cases, the solution of the resulting discrete equations forms a major part of the computation in the numerical approximation of partial differential equations, and hence has to be a key consideration in the choice of scheme. Elliptic problems lead naturally to large systems of simultaneous equations; and the recent dominance of finite element methods for their approximation has been accompanied by greater emphasis on *direct methods* for solving these systems, rather than the classical *iterative methods* favoured when finite difference schemes on rectangular meshes were used — see Duff *et al.* (1986) for a comprehensive account of the direct methods that have been developed for such sparse systems of equations.

However, the use of iterative methods will be anticipated in much of our consideration of competing discretisations, and there are several reasons for this. Firstly, the dominant convective terms that we shall be mainly concerned with are very often nonlinear so that some iteration is necessary anyway. Moreover, we shall find that the effective handling of these terms even for linear problems often leads to the use of *solution-adaptive* discretisation schemes; that is, the choice between central differencing or some degree of upwinding may be based on a current estimate of the solution, so introducing a degree of nonlinearity. Secondly, steady and unsteady problems are commonly considered together, with considerable overlap in the methods used, particularly since the unsteady problems commonly feature a wide range of timescales which leads to the use of implicit or semi-implicit schemes for their solution. Thirdly, multigrid methods have become a dominant feature in the computing landscape, as their capacity to yield solutions in a time proportional to that required to calculate the algebraic residual has spread from linear elliptic problems on a structured mesh to mixed elliptic/hyperbolic problems on general unstructured meshes — see Mavriplis (1995) for a recent survey. Thus many classical iterative methods have been reassessed as smoothers for an outer multigrid iteration, while multigrid is sometimes used as an inner preconditioner for some other outer iteration. And, finally, for three-dimensional calculations, the feasibility of which developments in computing have made much more widespread, both operation counts and storage requirements favour the use of iterative rather than direct methods.

In standard notation we write the linear system to be solved as

$$A\mathbf{x} = \mathbf{b}, \tag{1.114}$$

where the real $N \times N$ matrix A with entries $A_{i,j}$ may be the result of a linearisation process. Iterative procedures are much simpler to devise if the scheme leading to (1.114) satisfies a maximum principle. This follows from a number of key results which are fully discussed in Varga (1962). Let us denote by D the diagonal matrix composed from the diagonal entries of A; and, if we suppose these are all nonzero, we can write a basic iteration for the sequence $\{\mathbf{x}^n, n = 1, 2, \ldots\}$ as

$$D\mathbf{x}^{n+1} = (D - A)\mathbf{x}^n + \mathbf{b}$$

i.e. $$\mathbf{x}^{n+1} = (I - D^{-1}A)\mathbf{x}^n + D^{-1}\mathbf{b}. \tag{1.115}$$

This is *convergent*, or the matrix $I - D^{-1}A$ is called *convergent*, if $(I - D^{-1}A)^n$ converges to the null matrix as $n \to \infty$, which occurs if and only if the spectral radius of $I - D^{-1}A$ is strictly less than unity. A maximum principle is typically inferred when the diagonal entries of A are positive and dominate the off-diagonal elements which are nonpositive; then both the following lemmas apply.

Lemma 1.5.1 *If $A_{i,j} \leq 0$ for $i \neq j$, the following statements are either both true or both false:*

(i) *A is an M-matrix, i.e. it is nonsingular and $(A^{-1})_{i,j} \geq 0 \ \forall i, j$:*

(ii) *$A_{i,i} > 0 \ \forall i$, and $I - D^{-1}A$ is convergent and its entries non-negative.*

Proof. See Varga (1962) p.84. □

Lemma 1.5.2 *If A is strictly or irreducibly diagonally dominant, then A is non-singular and $I - D^{-1}A$ is convergent. If also $A_{i,i} > 0$ $\forall i$, then all eigenvalues of A are positive.*

Proof. See Varga (1962) p.23 and p.73. □

The simplest example of this situation is provided by the upwind difference scheme (1.49); and in the unsteady case given by (1.58), if the maximum timestep given by (1.61) is used so that $\epsilon(2 + \beta)\Delta t = h^2$, then as already noted the update is equivalent to the basic Jacobi iteration (1.115). Note, too, that since A^{-1} corresponds to the discrete Green's function, the fact that A is an M-matrix means the positivity property of the continuous Green's function is in this case reflected in the discrete approximation.

A wide variety of iterative methods, which generalise (1.115) in several different directions, have been devised over the last fifty years. Reference has already been made to the key rôle played by multigrid methods; but another notable feature has been the growing success of conjugate gradient, minimum residual and other Krylov subspace methods. Nevertheless, we have not deemed it necessary to devote a chapter of this book to these methods, since recent accounts in the context of solving differential equations are already widely available — see, for example, Quarteroni and Valli (1994) and Hirsch (1988). What should be noted here is that application of an iteration method is usually preceded by some manipulation or transformation of the original system of equations. The *splitting* of the matrix into an easily invertible part plus the remainder is already exemplified by (1.115); another widely applied example, which in contrast to (1.115) depends on the ordering of the equations, is the Gauss–Seidel method where the lower triangular section of A is added to D. Of even more importance to convection-diffusion problems, however, is the application of *preconditioning* operators to the system of equations, so that (1.114) is replaced by

$$P^{-1}AQy = P^{-1}b, \quad x = Q^{-1}y. \tag{1.116}$$

The introduction here of an easily invertible matrix P, which is also exemplified by the rôle of D in (1.115), is of most significance: with the wide variation of mesh sizes that are typical in modelling boundary layers, a preconditioning that takes this into account is of great importance in the application of conjugate gradient methods — see Wathen and Silvester (1993); and the lack of symmetry in A that arises from the convection terms can often be partially remedied by a preconditioning which is closely related to the choice of test space in the Petrov–Galerkin methods described in Chapter 5, particularly those in section 4.3.1.

A typical iteration scheme that results for (1.114), which is then easily generalised to a nonlinear system and to a Newton or quasi-Newton iteration, can be written as

$$P(\mathbf{x}^{n+1} - \mathbf{x}^n) = \omega \mathbf{r}^n, \quad \mathbf{r}^n := \mathbf{b} - A\mathbf{x}^n, \tag{1.117a}$$

where ω is an acceleration parameter that may depend on n. This is often called a *preconditioned Richardson (relaxation) method*, and is clearly closely related to the update in an unsteady problem. In choosing P and ω to give rapid convergence, we must be acutely aware of the fact that for unsymmetric operators the spectrum of the operator may be a very poor guide to its convergence properties. If $\mathbf{z}^n := \mathbf{x}^n - \mathbf{x}$, we can write (1.117a) as

$$\mathbf{z}^{n+1} = G\mathbf{z}^n, \quad G \equiv (I - \omega P^{-1} A), \tag{1.117b}$$

where G is the *iteration matrix*. The preconditioner P is generally chosen so that the spectrum of $P^{-1}A$ is as compact as possible, and then ω^{-1} can be chosen to be close to its 'centre'. However, such arguments are implicitly based on the idea that if the spectrum of G is made to lie inside the unit disc, then the iteration (1.117) is not only convergent but is also practically useful in that $\|G^n\|$ is small for reasonable values of n. For convection-diffusion problems this is commonly not the case. Indeed, in their study of the *pseudospectra* of the operator L where $Lu \equiv -\epsilon u'' + u'$, Reddy and Trefethen (1994) have shown that $\|(\lambda I - L)^{-1}\|$ can take on very large values within a whole parabola in the complex plane, while all the eigenvalues of L lie on the real axis. The link between these aspects of the problem, which emphasises the fact that G is not just a single finite matrix but represents a family of matrices of increasing dimension as the mesh size is reduced, is provided by the *Kreiss Matrix Theorem* — see Richtmyer and Morton (1967), pp.73–80: thus $\|G^n\| \leq C_A \; \forall n$ if and only if $\|(\lambda I - G)^{-1}\| \leq C_R(|\lambda| - 1)^{-1} \; \forall \lambda \in \mathbb{C}$ such that $|\lambda| > 1$, and the constants C_A and C_R are related. The precise relation of C_A to C_R, the question of how large $\|G^n\|$ can be when $\rho(G) < 1$, and the difference between the transitory and asymptotic behaviour of the evolution operator e^{-tL} have all received considerable recent attention — apart from the above reference, see also Spijker (1991) and Trefethen (1992).

The upwind scheme (1.49) again provides a simple example of these phenomena; or, as is common practice for large β, let us take a $(\beta-2) : 2$ mix of this scheme with the central difference scheme (1.29). Then the coefficient of U_{j+1} is set to zero and the scheme becomes the pure convection scheme

$$\beta(U_j - U_{j-1}) = (h^2/\epsilon)S_j. \tag{1.118}$$

With P chosen as the diagonal matrix, with entries β, the iteration matrix G has $(1 - \omega)$ on the diagonal, ω on the first subdiagonal and zero elsewhere. Thus its eigenvalues are all equal to $(1 - \omega)$ and lie inside the unit disc for $0 < \omega < 2$, though the best choice is clearly given by $\omega = 1$. However, the iteration corresponds to time-stepping with $\omega = \beta(\epsilon\Delta t/h^2) = b\Delta t/h$, and we know that this is stable only for $0 \leq b\Delta t/h \leq 1$. This conclusion follows from the Fourier mode analysis in sections 1.3.1 and 1.3.2 which was applied to both explicit time-stepping for the unsteady problem and Jacobi iteration for the steady problem. We shall return to this issue in Chapter 6, particularly in section 6.2.2.

1.6 Overview of effective methods

Before embarking on the detailed consideration of various numerical approaches to convection-diffusion in the following chapters, it may be useful to gather together some of the objectives we might set for the resulting schemes, and also to summarise some of the ideas that will be used to address them.

Many of the problems for which the methods are required are quite complicated and set in irregular multi-dimensional domains. Thus the methods should be reasonably compact and applicable to nonuniform meshes, which may have very high aspect ratios: on a topologically rectangular mesh, a 3×3 or a 4×4 difference stencil would seem a reasonable objective; and on a triangular or tetrahedral finite element mesh, piecewise linear or quadratic shape functions provide a similar level of complexity. In view of the target geometries we shall not consider spectral methods.

Few of the problems require predictions to high precision: engineering accuracy of 0.1% to 1% is a typical requirement. Thus methods of second or third order accuracy will be our main concern, since methods of only first order will generally require too fine a mesh and too much computation to attain the 0.1% target, and methods of fourth and higher order will be too complex. However, robustness is a more important criterion, particularly with respect to variations in the Péclet number, the convective field configuration and the location and magnitude of the source function. Very often boundary or interior layers confined to small regions will occur, and predictions of their presence will need to be obtained on a modest mesh so that an adaptive local mesh refinement can be used to capture the details in a subsequent phase of the computation. Hence there will be considerable emphasis on obtaining sharp error bounds for finite meshes, not merely asymptotic estimates as $h \to 0$, and on making these uniform in ϵ, or nearly so.

The most appropriate measure of accuracy is also an issue. For finite difference methods, the accuracy at the mesh points is clearly the main objective aimed at since the approximation is not defined elsewhere. However, many of the problems are formulated in terms of the divergence of a flux and it is sometimes important to predict this flux accurately. Also, although use of the supremum norm and a discrete maximum principle are the natural tools for analysing difference methods, it may not always be appropriate to apply them over the whole domain, since the main objective of a computation may be the accurate prediction of values outside a boundary layer, recognising that a supplementary 'boundary layer' computation is needed to obtain the detailed structure of the layer.

For finite element methods nodal accuracy is not a natural target. The approximations that are sought are global in nature and the variational principles used in their construction are naturally oriented towards minimising errors in a global energy norm, as we have already seen. Thus care must be taken in comparing the relative accuracy of finite difference and finite element methods, as well as of finite volume methods whose error properties are much less well understood. It will be appropriate in some cases to select an important parameter whose value can be recovered from any type of calculation and its accuracy compared in an objective way; the thickness of a boundary layer or the total drag on an aerofoil are typical parameters of this kind.

These issues are closely related to the consideration of various qualitative features that might be required of an approximation. Firstly, the differential problem often satisfies an important maximum principle; and there is then a great advantage to using a method which satisfies a similar principle. We shall devote considerable attention to this idea, especially for finite difference and some finite volume methods, and use it to derive sharp error bounds. Secondly, various coercivity conditions which are considerably stronger than (1.93b) can be derived for many of the problems; and these are crucial to the design and analysis of finite element and some finite volume methods. And, finally, there is the conservation principle that arises from the divergence form of the convection-diffusion operator. Here the natural advantage lies with finite volume methods which are derived by applying conservation principles over compact, nonoverlapping subregions using consistent approximations to the diffusive and convective fluxes.

In the following chapters, a number of distinct design approaches to these various objectives will become apparent, which we can group under four headings.

(i) *Exponential fitting*, which was the first device introduced specifically to deal with the steady convection-diffusion problem, by Allen and Southwell (1955); based on the fact that a boundary layer is usually exponential in form, the idea is to define the coefficients in a difference scheme so as to fit an exponential rather than the usual polynomial function. For finite element methods, the idea generalises to using a local Green's function or some solution to the adjoint equation as a test function.

(ii) *Symmetrization*, where the choice of test function in a Petrov-Galerkin formulation is directed towards the symmetrization of the bilinear form $B(U, V)$. If successful this will yield optimal error bounds of the form

(1.95) with C_1/C_0 approximately equal to unity.

(iii) *Upwinding*, whereby some aspect of a scheme, whether of finite difference, finite element or finite volume type, is modified to take account of the local flow direction. Actual procedures vary widely, but in virtually all cases the idea is to make the approximate solution conform more closely with a property of the exact solution, typically either a maximum principle or a coercivity property. For unsteady problems, this idea leads to the use of the characteristics associated with the hyperbolic part of the problem, or combining the Eulerian time derivative with the convective spatial derivative to form the Lagrangian time derivative which is then approximated.

(iv) *Least squares regularisation*, by which minimising the square of the convective residual, $\|\mathbf{b}\cdot\nabla U - S\|^2$, is used in some way to guide or control an approximation; thus note that, if U has an expansion in global basis functions $\{\phi_j\}$, then this minimisation would give relations of the form $(\mathbf{b}\cdot\nabla U - S, \mathbf{b}\cdot\nabla\phi_j) = 0$ which effectively introduces diffusion in the streamwise direction.

The most effective methods may well combine several of these ideas, and there is also some overlap in the outcome from applying each approach to a given problem. To illustrate this point, and to indicate how different from any of the schemes described in sections 1.3 and 1.4 are those which are most successful, consider the cell vertex finite volume which is introduced in Chapter 6 and the evolution-Galerkin methods described in Chapter 7. A key feature of both is that they start from emphasising the convective part of the problems (1.20) and (1.24), and result in four-point schemes centred on an interval rather than a three-point scheme centred on a mesh point. The cell vertex scheme is a generalisation of the well-known box difference scheme and, when applied to the steady problem (1.20), is based on a piecewise linear approximation which yields a nodal function value U_j and a nodal gradient U'_j obtained from U_{j-1}, U_j and U_{j+1}; together, these approximate the total flux by $-\epsilon U'_j + bU_j$ and imposing flux balance over an interval (or cell) gives a four-point scheme. It will be shown that this highly effective scheme can be regarded as a high Péclet number limit of schemes arising from the two approaches (i) and (ii); moreover, in order to solve the resulting equation system, use will need to be made of the ideas in (iii) and (iv). Similarly, when the evolution-Galerkin method based on piecewise linear approximations is applied to problem (1.24), the drawing back of a characteristic from the point (x_j, t_{n+1}) when the CFL number lies in the interval $(0, 1)$ will give a point in the interval (x_{j-1}, x_j), and hence lead to the use of a four-point stencil at this time level; the dependence on approach (iii) is clear, and it will be shown that there is also a close link to (i) and (ii).

CHAPTER 2

Selected results from mathematical analysis

2.1 Introduction

To cover the possibility of anisotropic diffusion, we may on occasion need to consider a differential equation of the form

$$Lu := -\epsilon \nabla \cdot (A \nabla u) + \mathbf{b} \cdot \nabla u + cu = S \quad \text{in } \Omega, \tag{2.1a}$$

where Ω is an open region of $\mathbb{R}^d$, and $A = A(\mathbf{x})$ is a $d \times d$ symmetric matrix which is uniformly positive definite on its closure. Thus L is a *uniformly elliptic operator* for $\epsilon > 0$. Typical boundary conditions will be

$$u = u_B \text{ on } \partial\Omega_D, \quad \mathbf{n} \cdot A \nabla u + \gamma u = g_B \text{ on } \partial\Omega_N \text{ with } \gamma \geq 0, \tag{2.1b}$$

where $\partial\Omega_D, \partial\Omega_N$ form a partition of the boundary of Ω, in which $\partial\Omega_D$ is not empty and includes $\partial\Omega_-$, the *inflow boundary* where $\mathbf{b} \cdot \mathbf{n} < 0$; we assume that the outward normal direction vector $\mathbf{n}$ is uniquely defined at each point of $\partial\Omega_N$.

The homogeneous problem corresponding to (2.1) will have eigensolutions for specific negative values of c. We shall normally wish to avoid such situations and make sufficient assumptions to ensure that the homogeneous problem has only the zero solution. Then one of the main tools of the analysis, which exploits the ellipticity of the problem, is a *maximum principle*; it is used extensively both in this chapter and in the analysis of discrete approximations in later chapters. Under similar assumptions, the inner product of Lu with a function v will yield a bilinear form with positive definiteness properties which is made use of in the weak formulation of the problem in section 2.5 and in the study of finite element and finite volume methods in subsequent chapters.

The most distinctive feature in the analysis of convection-diffusion problems, however, stems from the fact that (2.1) generally gives rise to a *singular perturbation problem* as $\epsilon \to 0$. The problem satisfied by the limit of the solution of (2.1) when $\epsilon \to 0$ is called the *reduced problem*. We shall find that it is given by

$$\mathbf{b} \cdot \nabla u + cu = S \quad \text{in } \Omega \tag{2.2a}$$

$$u = u_B \quad \text{on } \partial \Omega_-. \tag{2.2b}$$

That is, the boundary conditions on the outflow and tangential boundary are lost; this is because in (2.1) a *boundary layer* may form on $\partial \Omega_D \backslash \partial \Omega_-$. Interior layers may also occur; and approximations to these solutions can be obtained by the techniques of singular perturbation theory, such as the use of matched asymptotic expansions. In several sections of this chapter we shall therefore give brief accounts of this theory and its results.

For general reference purposes we end this section with a theorem on the existence and smoothness of solutions to second order elliptic problems of the form (2.1). We present it in the classical form derived by Schauder in 1934 as a generalisation of potential theory. (See Gilbarg and Trudinger (1983) as a general reference.) Firstly, and henceforth unless specifically stated otherwise, we assume that the region Ω is bounded and satisfies the *exterior sphere condition*; that is, for every point $\mathbf{x}$ on $\partial \Omega$ there exists a finite sphere $S_\Omega(\mathbf{x})$ such that $\bar{S}_\Omega(\mathbf{x}) \cap \bar{\Omega} = \mathbf{x}$. To describe smoothness we use the function spaces $C^{l,\alpha}(\Omega)$ consisting of functions which together with their derivatives up to order l are Hölder continuous in Ω with exponent α, where $0 < \alpha \leq 1$ and l is a non-negative integer; we also use the standard notation $C^l(\Omega)$ to denote the space of functions whose derivatives up to order l are continuous in Ω, with $C(\Omega)$ an abbreviation for $C^0(\Omega)$. It is then convenient to rewrite the operator L of (2.1a) as

$$Lu := -\epsilon A \nabla^2 u + \tilde{\mathbf{b}} \cdot \nabla u + cu, \tag{2.3}$$

where the jth component of $\tilde{\mathbf{b}}$ is $\tilde{b}_j = b_j - \epsilon \nabla \cdot \mathbf{A}_j$ and $\mathbf{A}_j$ is the jth column of A. Here and elsewhere we shall assume that all coefficients occurring in L are uniformly bounded on $\bar{\Omega}$. The central theorem is an example of the Fredholm alternative.

Theorem 2.1.1 *Suppose L is uniformly elliptic, the coefficients in (2.3) belong to $C^{0,\alpha}(\Omega)$ and the region Ω satisfies the exterior sphere condition; and consider the Dirichlet problem*

$$Lu = S \text{ in } \Omega, \quad u = u_B \text{ on } \partial \Omega. \tag{2.4}$$

Then

either *(i) the homogeneous problem, (2.4) with $S = 0$ and $u_B = 0$, has only the trivial solution, in which case for every $S \in C^{0,\alpha}(\Omega)$ and $u_B \in C^0(\partial \Omega)$ the inhomogeneous problem (2.4) has a unique solution $u \in C^0(\bar{\Omega}) \cap C^{2,\alpha}(\Omega)$;*

or *(ii) the homogeneous problem has nontrivial solutions which form a finite dimensional subspace of $C^{2,\alpha}(\Omega)$.*

The condition $c \geq 0$ is sufficient to ensure that alternative (i) holds.

Greater smoothness of the solution follows from greater smoothness of the coefficients, the data and the boundary: if $A, \tilde{\mathbf{b}}, c$ and S belong to

$C^{l,\alpha}(\Omega)$ and u is any solution of (2.4), then $u \in C^{l+2,\alpha}(\Omega)$; if $A, \tilde{b}, c$ and S belong to $C^{l,\alpha}(\bar{\Omega})$, the boundary $\partial\Omega$ is defined by $C^{l+2,\alpha}$ functions, and $u_B \in C^{l+2,\alpha}(\bar{\Omega})$, then any solution of (2.4) satisfies $u \in C^{l+2,\alpha}(\bar{\Omega})$.

Similar theorems hold when a derivative boundary condition is applied everywhere on the boundary,

$$\beta \cdot \nabla u + \gamma u = g_B \quad \text{on } \partial\Omega, \tag{2.5}$$

although greater smoothness of the boundary is now required. For example, if $A, \tilde{b}, c \geq 0$ and S belong to $C^{0,\alpha}(\bar{\Omega})$ with Ω a $C^{2,\alpha}$ region, and if in (2.5) we have $\gamma(\mathbf{n} \cdot \beta) > 0$ on $\partial\Omega$ and γ, β and g_B belong to $C^{1,\alpha}(\partial\Omega)$, then we have a unique solution to the problem with $u \in C^{2,\alpha}(\bar{\Omega})$. Comparing (2.5) with the more physically motivated derivative boundary condition in (2.1b), we see that the latter satisfies the constraint $\mathbf{n} \cdot \beta > 0$ on $\partial\Omega_N$ through the positive definiteness of A; but a switch to Dirichlet boundary conditions on part of the boundary corresponding to $\beta = \mathbf{0}$ in (2.5), or allowing $\gamma = 0$, is not covered by the above statement.

Such switches in boundary conditions give rise to singularities in the solution at the point of switching, the details of which can be found in Grisvard (1985). For our present purposes it is sufficient to note, regarding the boundary conditions of (2.1b), that uniqueness of the solution requires that either the Dirichlet condition or the derivative condition with $\gamma > 0$ holds at some point of the boundary; and, secondly, the smoothness of the solution in the interior of the region Ω depends only on the smoothness of the equation coefficients and the source data S.

2.2 Maximum principles

For a linear problem, the existence of a *maximum principle* allows the introduction of *comparison functions*, which are often used to provide solution bounds and error bounds for approximations. Even for nonlinear problems, the modern theory of *viscosity solutions* (Crandall and Lions 1983) utilises similar techniques. The following theorem and its corollaries give the key results for the convection-diffusion problem (2.1).

Theorem 2.2.1 *Suppose that the entries in A, the components of $\mathbf{b}$, and c in (2.1a) are all uniformly bounded on Ω; suppose also that $c(\mathbf{x}) \geq 0$ in Ω. Then if the function $v(\mathbf{x})$ satisfies*

$$Lv \leq 0 \text{ in } \Omega, \tag{2.6}$$

it can attain a non-negative maximum M at an interior point of Ω only if $v(\mathbf{x}) \equiv M$ in Ω. Moreover, if v attains its maximum M at a boundary point P, which also lies on the boundary of a sphere contained in Ω, and if v is continuous in $\Omega \cup P$ and an outward directional derivative $\partial v / \partial n$ exists at P, then $\partial v / \partial n > 0$ at P unless $v(\mathbf{x}) \equiv M$. When $c(\mathbf{x}) \equiv 0$, the results hold for any maximum.
Proof. See Protter and Weinberger (1967). □

Proofs of such theorems are based on obtaining a contradiction between the hypotheses and the conditions holding at an extremum. Thus a *minimum principle* can be obtained under the same hypotheses on the operator L: if $Lv \geq 0$ in Ω, v can attain a nonpositive interior minimum m only if $v(\mathbf{x}) \equiv m$ in Ω.

When the data determining the solution u of the problem (2.1) allows the direct application of Theorem 2.2.1, one obtains a very simple *a priori* bound as follows.

Corollary 2.2.1a *Suppose the operator L satisfies the hypotheses of the theorem and there is a constant $M \geq 0$ such that the data of problem (2.1) satisfies*

$$S \leq 0 \text{ in } \Omega, \quad u_B \leq M \text{ in } \partial\Omega_D, \quad g_B \leq 0 \text{ in } \partial\Omega_N. \tag{2.7}$$

Then the solution u of (2.1) satisfies $u \leq M$ in $\bar{\Omega}$.

More generally, and in order to obtain more precise bounds on u, one may introduce comparison functions as in the following results.

Corollary 2.2.1b *Suppose $Lu = S$ as in (2.1a) with the hypotheses of the theorem satisfied, and u satisfies the boundary conditions (2.1b). Suppose also that there exist two comparison functions $z_1(\mathbf{x})$ and $z_2(\mathbf{x})$, with $z_1(\mathbf{x}) \geq z_2(\mathbf{x})$ on Ω satisfying the following conditions:*

$$Lz_1 \geq S \quad \text{and} \quad Lz_2 \leq S \text{ on } \Omega; \tag{2.8a}$$

$$z_1 \geq u_B \quad \text{and} \quad z_2 \leq u_B \text{ on } \partial\Omega_D; \tag{2.8b}$$

$$\frac{\partial z_1}{\partial n} + \gamma z_1 \geq g_B \quad \text{and} \quad \frac{\partial z_2}{\partial n} + \gamma z_2 \leq g_B \text{ on } \partial\Omega_N. \tag{2.8c}$$

Then it follows that

$$z_2 \leq u \leq z_1 \text{ in } \bar{\Omega}. \tag{2.9}$$

Proof. Apply Corollary 2.2.1a to $u - z_1$ and $z_2 - u$ with $M = 0$. ☐

Corollary 2.2.1c *Suppose u satisfies the same hypotheses as in Corollary 2.2.1b and there is a comparison function $z(\mathbf{x}) \geq 0$ satisfying the conditions*

$$Lz \geq |S| \text{ on } \Omega \tag{2.10a}$$

$$z \geq |u_B| \text{ on } \partial\Omega_D \tag{2.10b}$$

$$\frac{\partial z}{\partial n} + \gamma z \geq |g_B| \text{ on } \partial\Omega_N. \tag{2.10c}$$

Then it follows that

$$|u| \leq z \text{ on } \bar{\Omega}. \tag{2.11}$$

Proof. Apply Corollary 2.2.1b with $z_1 = z$, $z_2 = -z$. ☐

Note that if $c(\mathbf{x}) < 0$ only rather weaker results can be derived directly: in Theorem 2.2.1, $Lv \leq 0$ in Ω precludes only a nonpositive maximum in the interior; and $Lv \geq 0$ in Ω precludes only a non-negative minimum in the interior. However, it is often possible to transform an equation with $c < 0$ to obtain more general results. An important example occurs in one dimension, for the equation

$$-\epsilon u'' + bu' + cu = S. \tag{2.12a}$$

Suppose that we have constants $b_\phi > 0$ and $c_\phi < 0$ such that

$$b(x) \geq b_\phi > 0, \quad c(x) \geq c_\phi \quad \text{and} \quad b_\phi^2 + 4\epsilon c_\phi > 0. \tag{2.12b}$$

Then the transformation $u = ve^{\sigma x}$, for constant σ, yields the equation

$$-\epsilon v'' + (b - 2\epsilon \sigma)v' + (c + b\sigma - \epsilon \sigma^2)v = e^{-\sigma x}S. \tag{2.12c}$$

Now choose σ as

$$\sigma := [b_\phi - (b_\phi^2 + 4\epsilon c_\phi)^{\frac{1}{2}}]/2\epsilon, \tag{2.12d}$$

for which we have $0 < \sigma < b_\phi/2\epsilon$. Then in the equation for v, both the coefficient of v' and that of v are non-negative, so that the maximum principle theorems as well as the boundary layer estimates given in the next section can be applied to v rather than u.

2.3 Boundary and interior layers in one dimension

Let us consider the model problem of (1.21), namely

$$Lu := -\epsilon(au')' + bu' + cu = S \text{ on } (0,1) \tag{2.13a}$$

$$u(0) = u_L, \quad u(1) = u_R, \tag{2.13b}$$

where we assume a, b, c and S are independent of ϵ and u, but may depend on x; also, so that we can make use of the maximum principles in section 2.2, we shall initially suppose that

$$a(x) \geq 1, \quad c(x) \geq 0. \tag{2.14}$$

It is clear from Theorem 2.1.1 that, with some added regularity conditions on the coefficients, a unique solution to (2.13) exists for each $\epsilon > 0$ and that u, u', u'' are all bounded. However, as $\epsilon \to 0$ there may be regions where u or its derivatives become unbounded; if these regions shrink to zero as $\epsilon \to 0$, they are called *boundary layers* or *interior layers* according to their position. There are two important cases to consider: when $b(x) \neq 0$ in $[0,1]$, and when b has at least one zero in the interval, corresponding to a *turning point* in the velocity field.

We distinguish these two cases in applying the maximum principle to obtain *a priori* bounds on the solution. In the first case, suppose $b(x) \geq b_\phi > 0$ and consider the comparison function

$$z(x) := \max(u_L, u_R, 0) + K \int_0^x \frac{1}{a(t)} dt, \tag{2.15}$$

where K is a positive constant. We clearly have $Lz \geq Kb/a$, so that it is straightforward to apply Corollary 2.2.1b to obtain the bound

$$u(x) \leq \max(u_L, u_R, 0) + \sup_{(0,1)} \left(\frac{a}{b} S \right). \tag{2.16}$$

Note that this bound is independent of ϵ.

However, in the general case that includes turning points this is not so, as we have had to make use of the second operator in L to apply the maximum principle. A crude bound is obtained with the commonly used comparison function

$$z(x) := \max(u_L, u_R, 0) + K \int_x^1 \frac{e^{\alpha t}}{a(t)} dt, \tag{2.17a}$$

where K and α are positive constants. It follows that

$$Lz \geq K(\alpha \epsilon - b/a)e^{\alpha x},$$

which leads to the choice

$$\alpha = \epsilon^{-1} \sup_{(0,1)} (|b|/a) + 1, \quad K = \epsilon^{-1} \sup_{(0,1)} \left(S e^{-\alpha x} \right) \tag{2.17b}$$

in order to ensure that $u(x) \geq z(x)$. Thus the bound has a strong dependence on ϵ, not only through the $1/\epsilon$ factor in K, but also through the exponential.

A more specific comparison function which could be used for the turning point problems discussed below is the following; we suppose $a(x) \equiv 1$ and set

$$z(x) := \max(u_L, u_R, K_1) + K_2 B(x) \tag{2.18a}$$

where $B(x)$ is a non-negative indefinite integral of b. Then a number of different cases emerge from noting that

$$Lz \geq S \quad \text{if} \quad K_1 c + K_2(b^2 - \epsilon b') \geq S. \tag{2.18b}$$

If $c > 0$ whenever $S > 0$ it is clear that an *a priori* bound which is independent of ϵ is always possible. So suppose $c \equiv 0$: then if $b' < 0$, such as for example $b(x) = \frac{1}{2} - x$, a bound which is inversely proportional to ϵ is obtained and even this might be avoided if $S = 0$ at points where $b = 0$; on the other hand, if $b' > 0$, such as for example $b(x) = x - \frac{1}{2}$, no useful bound can be obtained and comparison functions like (2.17) must be used.

A point at which $b = 0$, $b' < 0$ is sometimes called a *converging flow* or *compression* turning point: in general, when $c = 0$ and $S \neq 0$, we may expect an interior layer of a magnitude which becomes unbounded as $\epsilon \to 0$. A point at which $b = 0$, $b' > 0$ is correspondingly called a *diverging flow* or *expansion* turning point: it is less clear that there should be an interior layer but it is more difficult to obtain a sharp *a priori* bound. However, it should be pointed out that in the conservation form of the equation, $-\epsilon u'' + (bu)' = S$, it is in the latter case that we have $c = b' > 0$ so that (2.18a) gives an *a priori* bound independent of ϵ. In inviscid fluid dynamics, the converging flow turning point corresponds to a *shock point*, and the diverging flow turning point to a *sonic point* — see section 6.6 where these are considered.

The reader is referred to Abrahamsson (1977) for a more detailed presentation of *a priori* bounds for turning point problems; in particular, it is shown there that a bound independent of ϵ is possible if and only if $c > 0$ at the turning point. It is also shown there how the value of $c/|b'|$ at the turning point plays a key rôle in determining the behaviour of the solution.

2.3.1 Problems with no turning points

Returning to the first case, we assume $b(x) > 0$ on $[0, 1]$. Then it is easy to solve the reduced equation corresponding to (2.2), which forms the starting point for the method of *matched asymptotic expansions*. In this method for studying singular perturbation problems one identifies outer and inner regions which should overlap: in the outer region, where the solution is well behaved, it is expanded in a series which is variously called the *outer*, *regular* or *global expansion* and typically takes the form

$$u_{outer}(x, \epsilon) = \sum_{j=0}^{m} \epsilon^j u_j(x) + o(\epsilon^m); \tag{2.19a}$$

in the inner region the equation coefficients are replaced by local values and an approximation, called the *inner* or *local approximation*, is sought in a stretched coordinate ξ,

$$u_{inner}(x, \epsilon) \equiv w(\xi, \epsilon) = \sum_{j=0}^{m} \epsilon^j w_j(\xi) + o(\epsilon^m); \tag{2.19b}$$

and the two are matched in a region where they are both valid.

Solution of (2.13a) with constant coefficients yields a positive exponential solution in x/ϵ, which shows that in general there will be a boundary layer at the right of the interval. Thus in the present case we take the inner region on the right and introduce the stretched coordinate

$$\xi = (1 - x)/\epsilon. \tag{2.20a}$$

Substitution of (2.19b) into equation (2.13a), expansion of the coefficients in local Taylor series, and collection of terms of given order in ϵ yields the expansion

$$-\epsilon^{-1}\left[a(1)w_{\xi\xi} + b(1)w_\xi\right]$$
$$+ \left[\xi a'(1)w_{\xi\xi} + (\xi b'(1) + a'(1))w_\xi + c(1)w - S(1)\right] + \cdots = 0. \tag{2.20b}$$

To the leading order in ϵ, and after applying the boundary condition

$w(0) = u_R$, we therefore get

$$w_0(\xi) = u_R + C\left[1 - e^{-b(1)\xi/a(1)}\right], \tag{2.20c}$$

for some constant C.

For the outer expansion, substitution of (2.19a) into (2.13a) and application of the boundary condition $u_{outer}(0, \epsilon) = u_L$ gives

$$bu_0' + cu_0 = S, \qquad u_0(0) = u_L, \tag{2.21a}$$
$$bu_1' + cu_1 = a(u_0')', \qquad u_1(0) = 0, \tag{2.21b}$$

and so on. Matching takes place on the right of the outer region, where $x \to 1$ in (2.21), and on the left of the inner region where $\xi \to \infty$ in (2.20). Hence we have to leading order

$$u_0(1) = u_R + C, \tag{2.22}$$

which determines the unknown C from the solution $u_0(x)$ of the reduced equation. The leading order inner and outer solutions and the exact solution for a typical example are shown in Fig. 2.1, namely for $a = b = c = 1$, $S = 0$ and $\epsilon = 0.1$.

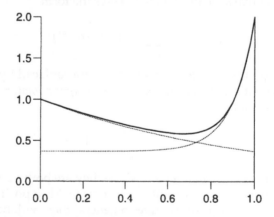

Figure 2.1. *Inner and outer asymptotic solutions (dashed lines) and the exact solution (full line) for* $-\epsilon u'' + u' + u = 0, u(0) = 1, \; u(1) = 2$ *for* $\epsilon = 0.1$

2.3.2 Turning point problems

To consider a turning point we suppose that there is just one, at $x = \frac{1}{2}$, in the neighbourhood of which $b(x) = \alpha(x - \frac{1}{2})$ with $\alpha = 1$ or -1, and for simplicity we set $a(\frac{1}{2}) = 1$. Moreover, we initially make the simplifying assumptions that $S(\frac{1}{2}) = 0$ and $c(\frac{1}{2}) = \gamma > 0$. The behaviour near $x = \frac{1}{2}$ can then be investigated by introducing a stretched coordinate $\xi = (x - \frac{1}{2})/\delta$, for some suitably chosen δ, and replacing $u(x)$ by $w(\xi)$; with the local values of a, c and S, we get for the leading term in the inner expansion

$$-(\epsilon/\delta^2)w_{\xi\xi} + \alpha\xi w_\xi + \gamma w = 0. \tag{2.23}$$

Thus to obtain a solution as ϵ, δ and $\xi \to 0$ we choose $\delta = \epsilon^{\frac{1}{2}}$ and, after the substitution $w(\xi) = v(\xi)\exp(\frac{1}{4}\alpha\xi^2)$, find that v satisfies

$$v'' + (\tfrac{1}{2}\alpha - \gamma - \tfrac{1}{4}\alpha^2\xi^2)v = 0. \tag{2.24}$$

Solutions of this equation are the parabolic cylinder functions $D_\nu(\xi)$; see Bender and Orszag (1978) for their use in this context. We can write the general solution in the form

$$v(\xi) = C_1 D_\nu(\xi) + C_2 D_\nu(-\xi), \tag{2.25}$$

where $\nu = \frac{1}{2}(\alpha - 1) - \gamma$ and C_1, C_2 are constants, since $D_\nu(\xi)$ and $D_\nu(-\xi)$ are linearly independent functions unless $\nu = 0, 1, 2, \ldots$ and we have assumed $\gamma > 0$ so as to avoid this situation. The key properties of the parabolic cylinder functions are that

$$D_\nu(\xi) \sim \frac{(2^\nu \pi)^{1/2}}{\Gamma(\frac{1}{2} - \frac{1}{2}\nu)} \quad \text{as } \xi \to 0 \tag{2.26a}$$

$$D_\nu(\xi) \sim \xi^\nu e^{-\xi^2/4} \quad \text{as } \xi \to +\infty \tag{2.26b}$$

$$D_\nu(-\xi) \sim \frac{(2\pi)^{1/2}}{\Gamma(-\nu)}\xi^{-\nu-1}e^{\xi^2/4} \quad \text{as } \xi \to +\infty. \tag{2.26c}$$

Now we need to distinguish the two cases $\alpha = \pm 1$.

When $\alpha = -1$, the converging flow or shock point case which corresponds to convection towards $x = \frac{1}{2}$ from both sides, we have $\nu = -1 - \gamma$, $w = ve^{-\xi^2/4}$ and hence obtain

$$w(\xi) \sim C_1\xi^{-1-\gamma}e^{-\xi^2/2} + C_2\frac{(2\pi)^{1/2}}{\Gamma(1+\gamma)}\xi^\gamma \quad \text{as } \xi \to +\infty, \tag{2.27a}$$

$$w(\xi) \sim C_1\frac{(2\pi)^{1/2}}{\Gamma(1+\gamma)}(-\xi)^\gamma + C_2(-\xi)^{-1-\gamma}e^{-\xi^2/2} \quad \text{as } \xi \to -\infty. \tag{2.27b}$$

Thus in the general situation there is an internal layer that decays like $\exp[-(x - \frac{1}{2})^2/2\epsilon]$; the coefficient C_1 is obtained by matching with the outer expansion in $[0, \frac{1}{2})$ and C_2 with that in $(\frac{1}{2}, 1]$. We therefore introduce $u_\pm(x)$, the solutions of the two following reduced problems,

$$0 \le x < \tfrac{1}{2}: \quad b(x)u'_- + c(x)u_- = S(x), \quad u_-(0) = u_L \tag{2.28a}$$

$$\tfrac{1}{2} < x \le 1: \quad b(x)u'_+ + c(x)u_+ = S(x), \quad u_+(1) = u_R. \tag{2.28b}$$

The matching for C_1 requires that

$$C_1 \frac{(2\pi)^{1/2}}{\Gamma(1+\gamma)} \left(\frac{\frac{1}{2} - x}{\epsilon^{1/2}} \right)^\gamma \sim u_-(x) \quad \text{as } x \to \tfrac{1}{2}-, \tag{2.29}$$

which is possible since, in that limit, our assumptions on S and c imply that the reduced equation becomes $(\frac{1}{2} - x)u'_- + \gamma u_- = 0$; a similar matching is possible for C_2. Note that, since the solutions to the reduced problems are independent of ϵ, the strength of the interior layer determined from C_1 and C_2 is proportional to $\epsilon^{\gamma/2}$.

In the alternative turning point problem, the expansion case with $\alpha = 1, \nu = -\gamma$ and $w = v e^{\xi^2/4}$, we have

$$w(\xi) \sim C_2 \frac{(2\pi)^{1/2}}{\Gamma(\gamma)} \xi^{\gamma-1} e^{\xi^2/2} \quad \text{as } \xi \to +\infty, \tag{2.30a}$$

$$w(\xi) \sim C_1 \frac{2\pi}{\Gamma(\gamma)} (-\xi)^{\gamma-1} e^{\xi^2/2} \quad \text{as } \xi \to -\infty. \tag{2.30b}$$

However, in general there will be boundary layers at both $x = 0$ where $b = -\frac{1}{2}$ and at $x = 1$ where $b = \frac{1}{2}$. These can be matched to outer expansions in $0 < x < \frac{1}{2}$ and $\frac{1}{2} < x < 1$ respectively, as in the nonturning point problem; but in doing so we will find the matching to (2.25) gives $C_1 = C_2 = 0$ because of the exponential behaviour in (2.30). Thus in this diverging flow or sonic point case, under our assumptions $S(\frac{1}{2}) = 0$ and $c(\frac{1}{2}) > 0$, we obtain two boundary layers and no interior layer.

It is fairly clear from the above analysis and the behaviour of the terms in (2.27) and (2.30) that the matching process is quite delicate, and this lay behind our original simplifications regarding $S(\frac{1}{2})$ and $c(\frac{1}{2})$. If, alternatively, $S(\frac{1}{2}) = 0$ and $c(\frac{1}{2}) \equiv \gamma = 0$, it is still easy to follow through the analysis in the case $\alpha = -1$; or even if $S(\frac{1}{2}) \neq 0$ but $c(\frac{1}{2}) > 0$, we can in this case write

$$S - cu = (S - cK) - c(u - K),$$

and choose the constant $K = S(\frac{1}{2})/c(\frac{1}{2})$, so as to carry out the analysis. However, if both $S(\frac{1}{2}) \neq 0$ and $c(\frac{1}{2}) = 0$, then the process with $\alpha = \nu = -1$ breaks down and a more sophisticated analysis is necessary, which in some cases will lead to values in the interior layer with logarithmic singularities. The situation in the alternative case with $\alpha = 1$ is similar if $S(\frac{1}{2}) \neq 0$ and $c(\frac{1}{2}) > 0$; but whenever $c(\frac{1}{2}) = 0$ we now have $\nu = 0$, and the parabolic cylinder functions $D_0(\xi)$ and $D_0(-\xi)$ are no longer linearly independent so that the analysis has to be modified. This shows that again there is no interior layer if $S(\frac{1}{2}) = 0$, but that $S(\frac{1}{2}) \neq 0$ may give an interior layer in this case too.

The results shown in Fig. 2.2, for $-\epsilon u'' - (x - \frac{1}{2})u' = S$, and in Fig. 2.3, for $-\epsilon u'' + (x - \frac{1}{2})u' = S$, both with $\epsilon = 0.01$ and $S = 0$ or 1, illustrate this analysis; they also bear out the point made earlier that *a priori* bounds for turning point problems depend on having $c > 0$ at the turning point.

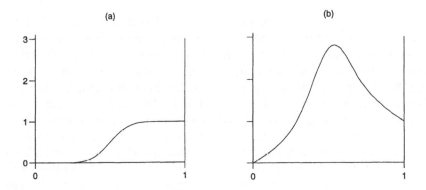

Figure 2.2. *Exact solutions for* $-\epsilon u'' - (x - \frac{1}{2})u' = S$ *with* $\epsilon = 0.01, u(0) = 0, u(1) = 1$ *and (a)* $S = 0$, *(b)* $S = 1$

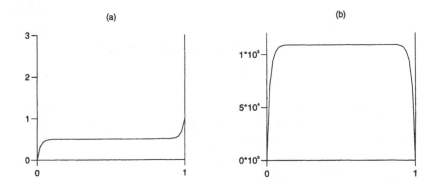

Figure 2.3. *Exact solutions of* $-\epsilon u'' + (x - \frac{1}{2})u' = S$ *with* $\epsilon = 0.01, u(0) = 0, u(1) = 1$ *and (a)* $S = 0$, *(b)* $S = 1$

2.4 Derivative bounds in a one-dimensional boundary layer

In calculating truncation errors and similar quantities, we shall often require bounds on the derivatives of solutions in the neighbourhood of boundary layers — particularly their dependence on the perturbation parameter ϵ. A number of useful bounds were given in Kellogg and Tsan (1978) as part of the authors' analysis of difference schemes, and we derive here similar results. They relate to the problem of (1.21) with $a = 1$, namely

$$Lu := -\epsilon u'' + b(x)u' + c(x)u = S(x, \epsilon) \quad \text{on } (0,1) \qquad (2.31a)$$

$$u(0) = u_L, \quad u(1) = u_R. \qquad (2.31b)$$

Notice that the source function is allowed to depend on ϵ. There is actually a greater need for this when carrying out corresponding estimates for difference equations, where the truncation error gives the source function. Since the difference equation analysis follows that for the differential equation, this is already adequate justification for this extra complication; but there are others. An error analysis based on the 'modified equation' rather than the truncation error would give the differential equation the role of the difference equation; and this is the approach used in much *a posteriori* error analysis — see section 1.4.4. Also, treating such source terms allows the possibility of obtaining bounds for multidimensional boundary layer problems by transferring derivatives or differences in the tangential directions to the right-hand side of the equation and treating them as source terms. In any case, we always choose the hypothesis on S so that it does not weaken the conclusion, as compared with assuming merely $S(\cdot) \in C^p[0,1]$.

The assumptions we shall need on the given functions in (2.31a) are:

$(i) \quad b(x) \geq b_\phi > 0, \quad c(x) \geq 0;$ \hfill (2.32a)

$(ii) \quad b(\cdot), c(\cdot) \in C^p[0,1];$ \hfill (2.32b)

(iii) for some positive constant K, independent of ϵ,

$$|S^{(s)}(x, \epsilon)| \leq K \left[1 + \frac{1}{\epsilon^{s+1}} e^{-(b_\phi/2\epsilon)(1-x)} \right], 0 \leq s \leq p, \qquad (2.32c)$$

with (2.32b) and (2.32c) holding for the same non-negative integer p. These conditions with $p = 0$ are sufficient to ensure that the solution u is bounded, and generally give the following result.

Theorem 2.4.1 *Under the assumptions (2.32), the solution $u(x)$ of (2.31) satisfies the following bounds:*

$$\epsilon |u^{(s+1)}(x)| + |u^{(s)}(x)| \leq C \left[1 + \frac{1}{\epsilon^s} e^{-(b_\phi/2\epsilon)(1-x)} \right], 0 \leq s \leq p+1, \qquad (2.33)$$

where C is a positive constant independent of ϵ.

Note (i) We use C to denote a generic constant, independent of key parameters like ϵ, distinguishing it by a subscript or superscript if we wish to maintain its identity through several expressions. However, we shall generally indicate important dependencies of such constants on the data.

Note (ii) We could replace $\frac{1}{2}b_\phi$ by any $b_0 < b_\phi$, but then C would contain factors $(b_\phi - b_0)^{-1}$.

Before proving this theorem we prove a useful lemma which we shall also need later.

Lemma 2.4.2 *Suppose that* $-\epsilon y'' + by' = g$ *and* $B(x)$ *is an indefinite integral of* $b(x)$. *Then the solution* $y(x)$ *is given by*

$$y(x) = y_g(x) + y(1) - y'(1)\int_x^1 e^{-(1/\epsilon)[B(1)-B(t)]}dt \qquad (2.34a)$$

$$y'(x) = z_g(x) + y'(1)e^{-(1/\epsilon)[B(1)-B(x)]}, \qquad (2.34b)$$

where

$$y_g(x) = -\int_x^1 z_g(t)dt, \quad z_g(x) = \int_x^1 \frac{1}{\epsilon}g(t)e^{-(1/\epsilon)[B(t)-B(x)]}dt. \quad (2.34c)$$

Alternatively, we can write

$$y(x) = y_g(x) + y(1)\frac{\int_0^x e^{-(1/\epsilon)[B(1)-B(t)]}dt}{\int_0^1 e^{-(1/\epsilon)[B(1)-B(t)]}dt}$$

$$+ [y(0) - y_g(0)]\frac{\int_x^1 e^{-(1/\epsilon)[B(1)-B(t)]}dt}{\int_0^1 e^{-(1/\epsilon)[B(1)-B(t)]}dt}. \qquad (2.34d)$$

Suppose further that $b(x) \geq b_\phi > 0$, *and that* $g(x)$ *satisfies*

$$|g(x)| \leq C_g + \frac{C_g'}{\epsilon^{\sigma+1}}e^{-(b_\phi/2\epsilon)(1-x)}, \qquad (2.35)$$

where C_g, C_g' *and* σ *are constants independent of* ϵ. *Then*

$$|z_g(x)| \leq \frac{1}{b_\phi}\left[C_g + \frac{2C_g'}{\epsilon^{\sigma+1}}e^{-(b_\phi/2\epsilon)(1-x)}\right] \qquad (2.36a)$$

$$and \qquad |y_g(x)| \leq \frac{1}{b_\phi}\left[C_g + \frac{4C_g'}{b_\phi\epsilon^\sigma}\right]. \qquad (2.36b)$$

Proof. The first part follows from standard calculus. From the hypothesis on $b(x)$ we obtain the inequality

$$e^{-(1/\epsilon)[B(t)-B(x)]} \leq e^{-(b_\phi/\epsilon)(t-x)} \text{ for } t \geq x, \qquad (2.37)$$

from which we find

$$|z_g(x)| \leq \frac{1}{\epsilon}\int_x^1\left[C_g e^{-(b_\phi/\epsilon)(t-x)} + \frac{C_g'}{\epsilon^{\sigma+1}}e^{-(b_\phi/2\epsilon)(1+t-2x)}\right]dt$$

and hence (2.36a). Integrating this we immediately obtain (2.36b). □

Proof (of Theorem 2.4.1). We first consider the case $s = 0$, by using the maximum principle that is clearly satisfied by (2.31). To construct an appropriate comparison function we use the positive functions

$$\phi_0(x) := 1 + x, \quad \phi_1(x) := e^{-(b_\phi/2\epsilon)(1-x)}, \qquad (2.38a)$$

for which we clearly have

$$L\phi_0 \geq b_\phi, \quad L\phi_1 \geq (b_\phi^2/4\epsilon)\phi_1, \tag{2.38b}$$

and set $z(x) = c_0\phi_0(x) + c_1\phi_1(x)$. Then $L(z \pm u) \geq 0$ if $c_0 \geq K/b_\phi$ and $c_1 \geq 4K/b_\phi^2$; and the boundary data is non-negative if $c_0 \geq |u_L|$ and $2c_0 + c_1 \geq |u_R|$. Hence, by the maximum principle of Corollary 2.2.1c, $|u(x)| \leq z(x)$ and we have proved the main part of (2.33) for $s = 0$, with a constant C whose most significant dependence on the data is the factor K/b_ϕ^2; the bound for $\epsilon u'$ will be deduced as part of the next step.

We prove the complete result by induction on p. To start at $p = 0$, we have already obtained a bound for u; and to continue we write (2.31a) as $-\epsilon u'' + bu' = S - cu$, to which we can apply Lemma 2.4.2 with $y = u, g = S - cu$ and $\sigma = 0$. This establishes that $|y_g(0)| \leq C$ from (2.36b). Also, since $b(x)$ is bounded, we can write $B(1) - B(t) \leq C_b(1 - t)$ for some constant C_b and hence in (2.34a)

$$\int_0^1 e^{-(1/\epsilon)[B(1)-B(t)]}dt \geq \frac{\epsilon}{C_b}\left[1 - e^{-C_b/\epsilon}\right] \geq \frac{\epsilon}{C_b}\left[1 - e^{-b_\phi/\epsilon}\right]; \tag{2.39}$$

the positivity of b_ϕ therefore ensures that (2.34a) yields the bound

$$y'(1) \leq \frac{C}{\epsilon}|y_g(0) + y(1) - y(0)|. \tag{2.40}$$

Applying this to (2.34b), together with the inequality (2.37) and the bound on $z_g(x)$ given by (2.36a), we obtain the bound on $y'(x)$, that is $u'(x)$, which is given by (2.33) with either $s = 0$ or $s = 1$. To complete the case $p = 0$, we deduce that $\epsilon|u''| \leq |S - cu - bu'|$ directly from the differential equation.

In order to continue the induction, we differentiate the differential equation (2.31a) making use of the hypotheses (2.32b) and (2.32c). Setting $y = u'$, we have $-\epsilon y'' + by' = g$ where $g = S' - b'u' - (cu)'$ satisfies (2.35) with $\sigma = 1$. Also $|y(0)|$ and $|y(1)|$ are bounded by C/ϵ from the previous step of the induction. Thus the next step can be completed by application of Lemma 2.4.2 with $\sigma = 1$ to give a sharper bound on $u''(x)$ and a bound on $u'''(x)$; the process is continued as long as the smoothness hypotheses allow. $\square$

A useful decomposition of the solution of (2.31) is given by the following.

Theorem 2.4.3 *Suppose that $b(x)$ and $c(x)$ satisfy (2.32a,2.32b) and that, if $p = 0, b(x)$ satisfies a Lipschitz condition in some neighbourhood of $x = 1$. Suppose also that (2.32c) is replaced by*

$$|S^{(s)}(x,\epsilon)| \leq K\left[1 + \frac{1}{\epsilon^s}e^{-(b_\phi/2\epsilon)(1-x)}\right], \quad 0 \leq s \leq p. \tag{2.41}$$

Then the solution of (2.31) can be written as

$$u(x) = \gamma e^{-b(1)(1-x)/\epsilon} + z(x), \tag{2.42}$$

where $\gamma = \epsilon u'(1)/b(1), |\gamma| \leq C_1$ *and*

$$\epsilon|z^{(s+1)}(x)| + |z^{(s)}(x)|$$

$$\leq C_2 \left[1 + \frac{1}{\epsilon^{s-1}} e^{-(b_\phi/2\epsilon)(1-x)} \right], \quad 0 \leq s \leq p+1, \quad (2.43)$$

where C_1 *and* C_2 *are constants independent of* ϵ.

Proof. The fact that $|\gamma| \leq C_1$ follows from the bound $|u'(1)| \leq C/\epsilon$ given in Theorem 2.4.1 where the hypotheses were weaker than here; and $|z(x)| \leq C$ then follows from $|u(x)| \leq C$, also derived there. To complete the proof of (2.43) for $p = 0$, we apply Lemma 2.4.2 to the equation for z, $-\epsilon z'' + bz' = g$ where

$$g = S - cu + (\gamma/\epsilon)b(1)[b(1) - b(x)]e^{-b(1)(1-x)/\epsilon}. \quad (2.44)$$

By hypothesis (2.41), S is bounded independently of ϵ, and so is u. Thus g satisfies (2.35) with $\sigma = -1$ if the last expression of (2.44) is so bounded. But for each ϵ, suppose this expression attains its maximum value at some x_m. Then if $x_m \equiv x_m(\epsilon) \leq x^* < 1$ for all ϵ, the expression is uniformly bounded; otherwise we can suppose that $x_m \to 1$ as $\epsilon \to 0$. Hence we need consider only ϵ sufficiently small that $|b(1) - b(x)| \leq L(1 - x)$ for some L, by the hypotheses of the theorem; and therefore we need only show that

$$(1/\epsilon)(1 - x)e^{-b(1)(1-x)/\epsilon}$$

is uniformly bounded, which it clearly is.

Having shown g satisfies (2.35) with $\sigma = -1$, we can apply Lemma 2.4.2 to deduce (2.43) with $s = 1$ from (2.36a) and (2.34b) very simply, since $z'(1) = 0$. For $p > 0$, the successive differentiation of the equation for z yields all of (2.43) as in Theorem 2.4.1, since successive derivatives of g satisfy bounds of the form (2.35) with successively larger σ. □

Comparing Theorem 2.4.1 and Theorem 2.4.3, we see that by accounting for the dominant boundary layer explicitly, the remaining part of the solution is smoother, with its derivatives reduced by a factor of ϵ in the boundary layer. The differences in the hypotheses of the two theorems are not significant.

Alternative decompositions of the solution can be deduced from the results of Lemma 2.4.2. For example, (2.34b) gives immediately

$$u'(x) = u'(1)e^{-(1/\epsilon)[B(1)-B(x)]} + z_g(x) \quad (2.45)$$

where $z_g(x)$ is given by (2.34c) with $g = S - cu$. This is the same as (2.42) when b is constant, but otherwise is often less convenient. A much more useful result is based on separating out the solution of the reduced equation; in its simplest and most widely used form we have the following theorem.

Theorem 2.4.4 *Suppose* $c(x) \equiv 0$, b *and* S *lie in* $C^1(0,1)$, *and* $\tilde{u}_L(x)$ *the solution of the reduced problem is defined by*

$$b\tilde{u}'_L = S, \quad \tilde{u}_L(0) = u_L. \tag{2.46}$$

Then the solution of (2.31) satisfies

$$u(x) = \tilde{u}_L(x) + [u_R - \tilde{u}_L(1)]\frac{\int_0^x e^{-(1/\epsilon)[B(1)-B(t)]}\mathrm{d}t}{\int_0^1 e^{-(1/\epsilon)[B(1)-B(t)]}\mathrm{d}t} + O(\epsilon). \tag{2.47}$$

Proof. The difference $y = u - \tilde{u}_L$ satisfies $-\epsilon y'' + by' = g$, where $g = \epsilon\tilde{u}''_L = \epsilon(S/b)'$, with boundary conditions $y(0) = 0$ and $y(1) = u_R - \tilde{u}_L(1)$. Hence it is given by (2.34d) of Lemma 2.4.2 where y_g is defined by (2.34c). By the hypotheses, $|g| \leq C\epsilon$ and therefore by the lemma $|y_g| \leq C\epsilon$ so that (2.47) follows. $\square$

The above result highlights the role of the outflow Dirichlet boundary condition in forming the boundary layer. So let us finally suppose that this is replaced by the homogeneous Neumann condition $u'(1) = 0$. Then it is seen immediately from any of these decompositions that there is no boundary layer; more precisely, u and its derivatives satisfy the bounds for z and its derivatives given by (2.43).

2.5 Weak formulation of the problem

It will be useful here and for later use to introduce the *weak form* for the differential problem (2.1). Multiplying (2.1a) by a suitably smooth test function v and integrating the first term by parts, we obtain

$$-\epsilon \int_{\partial\Omega} v\mathbf{n}\cdot(A\nabla u)\mathrm{d}\Gamma + \epsilon(A\nabla u, \nabla v) + (\mathbf{b}\cdot\nabla u + cu, v) = (S, v), \tag{2.48}$$

where $(\cdot, \cdot)$ is used to denote the L_2 inner product over Ω of either scalar or vector quantities, that is

$$(\mathbf{f}, \mathbf{g}) := \int_\Omega \mathbf{f}\cdot\mathbf{g}\,\mathrm{d}\Omega, \quad \|\mathbf{f}\|^2 := (\mathbf{f}, \mathbf{f}). \tag{2.49}$$

Suppose now that the mixed boundary condition of (2.1b) on $\partial\Omega_N$ is applied to u and v is set to zero on $\partial\Omega_D$. Then the integral of uv over $\partial\Omega_N$ can be combined with the other inner product terms to define a *bilinear form* associated with the problem, namely

$$B_\gamma(u, v) := \epsilon(A\nabla u, \nabla v) + (\mathbf{b}\cdot\nabla u + cu, v) + \epsilon\gamma \int_{\partial\Omega_N} uv\,\mathrm{d}\Gamma; \tag{2.50a}$$

and the integral over the data g_B can be combined with the right-hand side of (2.45) to define a data linear functional

$$l(v) := (S, v) + \epsilon \int_{\partial\Omega_N} g_B v\,\mathrm{d}\Gamma. \tag{2.50b}$$

The bilinear form is well defined for $u, v \in H^1(\Omega)$, the Sobolev space of functions which together with their first derivatives are square integrable over Ω. The solution u must have its trace on $\partial\Omega_D$ equal to u_B and the trace of test function v on $\partial\Omega_D$ must equal zero. So we define

$$H^1_E := \{w \in H^1(\Omega)|\ w = u_B \ \text{on}\ \partial\Omega_D\} \tag{2.51a}$$

$$H^1_{E_0} := \{w \in H^1(\Omega)|\ w = 0 \ \text{on}\ \partial\Omega_D\}. \tag{2.51b}$$

Then the weak formulation of problem (2.1) is as follows: *find* $u \in H^1_E$ *such that*

$$B_\gamma(u, v) = l(v) \quad \forall v \in H^1_{E_0}, \tag{2.52}$$

where $B_\gamma(\cdot, \cdot)$ and $l(\cdot)$ are defined by (2.50a) and (2.50b) respectively.

Most of the time we shall consider simpler forms of the problem (2.1), for which it is convenient to introduce a simplified notation. Thus suppose the matrix A is replaced by the scalar constant a, so that $\mathbf{n} \cdot A\nabla u$ in the boundary condition (2.1b) can be replaced by the outward normal derivative $\partial u/\partial n$, and that both γ and g_B are zero. Then (2.47) reduces to

$$B(u, v) := \epsilon(a\nabla u, \nabla v) + (\mathbf{b}\cdot\nabla u + cu, v) \tag{2.53}$$

and the weak form of the equation for $u \in H^1_E$ can be written

$$B(u, v) = (S, v) \quad \forall v \in H^1_{E_0}. \tag{2.54}$$

We establish the well-posedness of this problem through use of the Lax-Milgram lemma — see Gilbarg and Trudinger (1983).

To apply the lemma it is convenient to make use of the standard Sobolev space notation $H^{-1}(\Omega)$ for the space of source functions S which define the linear functional on $H^1(\Omega)$ forming the right-hand side of (2.54), and $H^{-1/2}(\partial\Omega)$ in which the Dirichlet data u_B must lie — see Gilbarg and Trudinger (1983) also for the definition of these spaces. The space $H^{-1/2}(\partial\Omega)$ is such that any element can be extended to all of Ω to lie in $H^1(\Omega)$. We shall also need the norm on $H^1(\Omega)$ given by

$$\|v\|^2_{H^1(\Omega)} := (\nabla v, \nabla v) + (v, v) \tag{2.55a}$$

and that on $H^{-1}(\Omega)$ given by

$$\|f\|_{H^{-1}(\Omega)} := \sup_{v \in H^1(\Omega)} \frac{(f, v)}{\|v\|_{H^1(\Omega)}}. \tag{2.55b}$$

Theorem 2.5.1 *Suppose that $a, c \in L_\infty(\Omega)$, that $\nabla \cdot \mathbf{b} \in L_\infty(\Omega)$, that $\mathbf{b} \cdot \mathbf{n} \geq 0$ on $\partial\Omega_N$ and that*

$$a \geq a_0 > 0, \quad c - \tfrac{1}{2}\nabla \cdot \mathbf{b} \geq 0 \ \text{on}\ \Omega \tag{2.56}$$

for some constant a_0. Then a unique solution $u \in H^1_E$ to (2.54) exists for $\forall u_B \in H^{-1/2}(\partial\Omega) \supset L_2(\partial\Omega)$, and $\forall S \in H^{-1}(\Omega) \supset L_2(\Omega)$. Moreover, if $\tilde{u}_B \in H^1_E$

is any continuation of the Dirichlet data into a $H^1(\Omega)$ function, then there is a constant $C \equiv C(\epsilon)$ such that

$$\|u\|_{H^1(\Omega)} \leq C \left[\|S\|_{H^{-1}(\Omega)} + \|\tilde{u}_B\|_{H^1(\Omega)} \right]. \tag{2.57}$$

Proof. By writing (2.54) as

$$B(u - \tilde{u}_B, v) = (S, v) - B(\tilde{u}_B, v) \quad \forall v \in H^1_{E_0}$$

we can apply the Lax-Milgram lemma if we show

$$|B(w,v)| \leq C_1 \|w\|_{H^1(\Omega)} \|v\|_{H^1(\Omega)} \quad \forall w, v \in H^1_{E_0} \tag{2.58a}$$

and

$$B(w,w) \geq C_2 \|w\|^2_{H^1(\Omega)} \quad \forall w \in H^1_{E_0} \tag{2.58b}$$

for some positive constants C_1, C_2. The continuity condition (2.58a) follows immediately from the hypotheses on a, **b** and c. For the coercivity condition (2.58b) we use Gauss' theorem to write

$$2(\mathbf{b}\cdot\nabla w, w) = \int_{\partial\Omega_N} w^2 \mathbf{b}\cdot\mathbf{n}d\Gamma - \int_\Omega w^2(\nabla\cdot\mathbf{b})d\Omega, \tag{2.59}$$

so that the result follows from the condition on $\partial\Omega_N$ and (2.56), giving $C_2 \geq \epsilon a_0$. Hence the existence of u follows, and we also have

$$C_2 \|u - \tilde{u}_B\|^2_{H^1(\Omega)} \leq (S, u - \tilde{u}_B) - B(\tilde{u}_B, u - \tilde{u}_B)$$
$$\leq \left[\|S\|_{H^1(\Omega)} + C_1 \|\tilde{u}_B\|_{H^1(\Omega)} \right] \|u - \tilde{u}_B\|_{H^1(\Omega)}$$

from which (2.57) follows. □

Remark There is no loss of generality in assuming $\nabla\cdot\mathbf{b} = 0$; thus in condition (2.56) we need assume only $c \geq 0$. To see this we multiply the original equation by a positive weighting function w, replacing a, c and S by $\tilde{a} = wa, \tilde{c} = wc$ and $\tilde{S} = wS$; we also write

$$-\epsilon w\nabla\cdot(a\nabla u) = -\epsilon\nabla\cdot(wa\nabla u) + \epsilon a(\nabla w)\cdot(\nabla u)$$

and therefore complete the transformation of the problem by setting

$$\tilde{\mathbf{b}} = w\mathbf{b} + \epsilon a\nabla w.$$

It remains to note that imposing the condition $\nabla\cdot\tilde{\mathbf{b}} = 0$ sets up an equation for the weighting function,

$$\epsilon\nabla\cdot(a\nabla w) + \nabla\cdot(\mathbf{b}w) = 0 \quad \text{on } \Omega,$$

which can be solved to give $w > 0$ so long as positive Dirichlet boundary conditions are imposed.

2.6 A priori bounds and boundary layers in two dimensions

Many of the results on bounds and the formation of boundary and interior layers that have been given in sections 2.3 and 2.4 can be extended to two and three dimensions. However, there are significant differences and it is these that we shall highlight here.

It is very common to have an incompressible flow field, that is $\nabla \cdot b = 0$, and we shall make this assumption in what follows here. Indeed, as we have just seen there is no loss of generality in assuming this in establishing the well-posedness of the problem (2.54) in Theorem 2.5.1. This theorem also gives an *a priori* bound (2.57) for the solution, but this is strongly dependent on ϵ. In section 6.5.3 a sharper bound will be derived by introducing a weighting function into the argument, with its rôle being similar to that of the comparison functions used in deriving the bounds in section 2.3 by means of a maximum principle. To take a simple case, assume $u = 0$ on $\partial\Omega$, $a = 1$, $c = 0$, as well as $\nabla \cdot b = 0$, and let $B(\mathbf{x})$ be an indefinite integral of b so that $\mathbf{b} = \nabla B$, and hence B is a harmonic function. Then define $g = e^{-B}$, $q^2 = -\mathbf{b} \cdot \nabla g - \epsilon \nabla^2 g$, so that

$$\nabla g = -g\mathbf{b}, \quad \nabla^2 g = g\,|\mathbf{b}|^2 \quad \text{and} \quad q^2 = (1 - \epsilon)g\,|\mathbf{b}|^2 .$$

It follows from substituting $v - gu$ in (2.54) — see the proof of Theorem 6.5.1 for the details — that

$$\epsilon \left\| g^{\frac{1}{2}} \nabla u \right\|^2 + \tfrac{1}{4} \left\| g^{\frac{1}{2}} \mathbf{b} u \right\|^2 \leq \left\| \frac{g^{\frac{1}{2}} |\mathbf{b}|^{-1}}{(1 - \epsilon)} S \right\|^2 . \tag{2.60}$$

Note that the factor $|\mathbf{b}|^{-1}$ indicates the difficulties that are likely to occur at *stagnation points* where $\mathbf{b} = \mathbf{0}$, as was experienced with turning points in one dimension.

Maximum principle arguments of the type used in section 2.3 can equally well be used in two dimensions, with similar results. We shall make use of them in later chapters, especially for the discrete equations, but will not consider them in more detail here.

The idea of boundary layers, and the use of asymptotic expansions with some sort of matching, has been the key to understanding viscous flows in two and three dimensions . The originator in this field was Prandtl (1905) but the general ideas are older (see van Dyke (1994) for a brief historical survey), and they have been very widely applied and developed in many fields. We will consider first three simple cases in order to point up the features common to, and those that are different from, the one-dimensional cases already considered. A useful general reference is the book by Eckhaus (1979), or the earlier review article, Eckhaus (1972).

In the three simple examples we take the equation to be

$$-\epsilon \nabla^2 u + u_x = S, \tag{2.61}$$

and distinguish the three by the shape of the boundary and the boundary conditions. Suppose first of all that the region Ω is the unit disc and the boundary condition is $u = 0$ on the unit circle boundary $\partial\Omega$. Then the boundary is tangential to the flow field only at the top and bottom of the circle, $x = 0$ and $y = \pm 1$. These points are excluded in the outer or regular expansion which takes the same form as in (2.19a), namely

$$u_{outer}(\mathbf{x}, \epsilon) = \sum_{j=0}^{m} \epsilon^j u_j(\mathbf{x}) \tag{2.62a}$$

in $\Omega_p := \{\mathbf{x} \in \Omega | -\cos p \le y \le \cos p, p > 0\}$, and

$$\frac{\partial}{\partial x} u_0 = S, \quad \frac{\partial}{\partial x} u_j = -\nabla^2 u_{j-1}, \quad j = 1, 2, \ldots, m, \tag{2.62b}$$

with $u_j = 0$ on the inflow boundary $\partial \Omega_-$. An ordinary boundary layer forms at the right-hand boundary of Ω_p, where the behaviour can be studied by transforming to polar coordinates with the stretching

$$\xi = (1 - r)/\epsilon. \tag{2.62c}$$

The inner expansion for $w(\xi, \theta, \epsilon)$ has the same form as (2.19b) and is obtained by substituting $r = 1 - \epsilon \xi$ and matching terms in an expansion in powers of ϵ of the equation

$$-\epsilon \left[\frac{\partial^2 w}{\partial r^2} + \frac{1}{r} \frac{\partial w}{\partial r} + \frac{1}{r^2} \frac{\partial^2 w}{\partial \theta^2} \right] + \left[\cos \theta \frac{\partial w}{\partial r} - \frac{\sin \theta}{r} \frac{\partial w}{\partial \theta} \right]$$

$$\equiv -\epsilon^{-1} \left[\frac{\partial^2 w}{\partial \xi^2} + \cos \theta \frac{\partial w}{\partial \xi} \right] + \frac{1}{r} \left[\frac{\partial w}{\partial \xi} - \sin \theta \frac{\partial w}{\partial \theta} \right] - \frac{\epsilon}{r^2} \frac{\partial^2 w}{\partial \theta^2} = 0. \tag{2.62d}$$

When $|\theta| \le \frac{1}{2}\pi - p$, at each order a set of ordinary differential equations in ξ have to be solved as for the one-dimensional expansion of (2.20b). The leading term is given as in (2.20c), with one boundary condition obtained from $w(0, \theta, \epsilon) = 0$, and the matching with the outer expansion can be carried out in a similar way, with $w(\xi, \theta, \epsilon) \to u_{outer}(\cos \theta, \sin \theta, \epsilon)$ as $\xi \to \infty$ giving the other boundary condition. The only extra difficulty is at the top and bottom of the region in the limit as $p \to 0$. In this particular example one can show that the asymptotic expansions remain valid over the whole region; but in general for such flow-tangency points one has to consider the possibility of *boundary layer birth*.

In the second example we merely change the domain to the unit square, so that flow tangency occurs over the whole of the bottom and top boundaries, $y = 0, 1$, and modify the boundary condition to $u = 1$ on $y = 0$. In the restricted domain $\Omega_p := \{\mathbf{x} \in \Omega | p \le y \le 1 - p, p > 0\}$, the analysis can be carried out exactly as in the first example giving a boundary layer at the right-hand boundary $x = 1$. However, in the lower strip, $0 \le y \le p$, we introduce the stretched variable

$$\xi = y/\delta(\epsilon) \tag{2.63a}$$

to obtain an equation for $w(x, \xi, \epsilon) \equiv u_{\text{inner}}(\mathbf{x}, \epsilon)$ given by

$$-\epsilon \left[\frac{\partial^2 w}{\partial x^2} + \delta^{-2} \frac{\partial^2 w}{\partial \xi^2} \right] + \frac{\partial w}{\partial x} = 0. \qquad (2.63\text{b})$$

To obtain a useful asymptotic expansion here we need to take

$$\delta = \epsilon^{\frac{1}{2}}, \qquad (2.63\text{c})$$

giving what is called a *parabolic boundary layer* in which to leading order we have

$$\frac{\partial w_0}{\partial x} = \frac{\partial^2 w_0}{\partial \xi^2}. \qquad (2.63\text{d})$$

At each order of the expansion a parabolic equation has to be solved, of the form (2.63d) with source terms from the next lower order, for which the boundary conditions are obtained from

$$w_0(0, \xi, \epsilon) = 0, \quad w(x, 0, \epsilon) = 1 \text{ and } w(x, \xi, \epsilon) \rightarrow u_{\text{outer}}(x, p, \epsilon) \quad (2.63\text{e})$$

as $\xi \rightarrow \infty$ and $p \rightarrow 0$. With smooth inflow boundary data there is no problem with the limit $p \rightarrow 0$ in this boundary condition; but there will be a problem as $x \rightarrow 1$ and the outflow ordinary boundary layer interacts with the parabolic layer. The strip, $1 - p \leq y \leq 1$, can be treated in a similar way, but note that since $u(x, 1) = 0$ there will be no boundary layer if there is no source term.

The third example is also on a unit square, with $u = 0$ on the outflow and the tangential boundaries but with discontinuous inflow boundary data, as in the Raithby model problem of section 1.2. Then there will be an internal boundary layer emanating from each discontinuity. Each can be treated in a similar way to the bottom boundary of the second example, where there was in fact a discontinuity in the boundary data at the origin.

To conclude this section, and to indicate some of the difficulties when inner and outer expansions do not always have an overlap region where they are both valid, or where two boundary layers interact as in the second example above, we consider briefly the problem of viscous flow over a flat plate.

Consider the incompressible Navier–Stokes equations given by (1.2), for steady flow in the half-plane $y > 0$, with rigid wall boundary conditions applied on a flat plate at $y = 0$ extending from $x = 0$ to $x = L$; we suppose we have uniform incoming parallel flow, so that the flow vector $\mathbf{v} \equiv (u, v) \rightarrow (U, 0)$ as $|x| + y \rightarrow \infty$, and $\partial v / \partial y = 0$ at $y = 0$ and $x \notin [0, L]$. For the x-component of the momentum equation we have

$$u \frac{\partial u}{\partial x} + v \frac{\partial u}{\partial y} = -\frac{\partial p}{\partial x} + \nu \nabla^2 u.$$

With the pressure gradient treated as a source term, and with $v \ll u$, this is very similar to (2.61). Thus in the viscous boundary layer equations, $\partial^2 u / \partial x^2$ is neglected in comparison with $\partial^2 u / \partial y^2$ and the pressure gradient is regarded as determined by the outer (inviscid) flow; for the flat plate, the pressure gradient is zero so we obtain

$$u\frac{\partial u}{\partial x} + v\frac{\partial u}{\partial y} = \nu\frac{\partial^2 u}{\partial y^2}, \qquad (2.64a)$$

together with the incompressibility equation

$$\frac{\partial u}{\partial x} + \frac{\partial v}{\partial y} = 0. \qquad (2.64b)$$

It is clear from the earlier examples that on the plate we shall need a stretched coordinate of the form $y/\nu^{\frac{1}{2}}$; but with the absence of a source term and the constant boundary condition we can in fact seek a similarity solution. We also satisfy the incompressibility condition (2.64b) by introducing a stream function ψ such that $u = \psi_y$ and $v = -\psi_x$. Thus we write

$$\eta = y/g(x), \quad \psi = Ug(x)f(\eta), \qquad (2.65a)$$

for some functions f and g. It is easy to see then that $u = Uf'(\eta)$ and $v = U(\eta f' - f)g'$; substitution in (2.64a) and some simplification hence leads to the equation

$$f''' + \frac{Ugg'}{\nu}ff'' = 0, \qquad (2.65b)$$

and we choose the stretching by setting $Ugg' = \nu$. The singularity that occurs at the leading edge of the plate is expressed by choosing $g^2 = 2\nu x/U \equiv 2x/\text{Re}$, where Re is the Reynolds number; the boundary conditions on the plate are given by $u = v = 0$, giving $f = f' = 0$ at $\eta = 0$; and matching to the uniform flow is achieved by setting $f'(\eta) \to 1$ as $\eta \to \infty$. Thus we finally have a boundary layer determined by the function $f(\eta)$ for which we have

$$\eta = y(\text{Re}/2x)^{\frac{1}{2}}, \quad f''' + ff'' = 0 \qquad (2.65c)$$
$$f(0) = f'(0) = 0, \quad f'(\infty) = 1. \qquad (2.65d)$$

A plot of $f'(\eta) \equiv u/U$ is shown in Fig. 2.4. To resolve the singularity at the leading edge $x = 0$ needs a different scaling, and at the trailing edge $x = L$ yet another. In addition, the steady boundary layer flow becomes unstable to incoming perturbations at some Reynolds number above about 10^5. *Separation*, in which u changes sign, then occurs and a much more sophisticated analysis is required, typically in solving a "triple-deck" structure. These complications, involving interacting boundary layers, singular points and multiple layer structures, are typical of what can occur in more than one dimension and with nonlinear problems; they are beyond the scope of the present text.

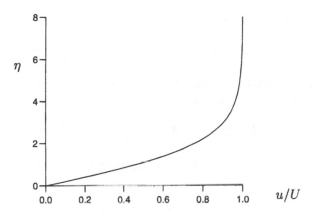

Figure 2.4. *Velocity profile for flow over a flat plate*

2.7 Green's functions

A Green's function $G(\mathbf{x}', \mathbf{x})$ is used to express the dependence of the solution of a problem on its data; for a problem with homogeneous boundary conditions and a source function $S(\cdot)$, this will take the simple form

$$u(\mathbf{x}') = \int_{\Omega} G(\mathbf{x}', \mathbf{x}) S(\mathbf{x}) d\mathbf{x}. \tag{2.66}$$

We can deduce from this form that, as a function of its first argument, $G(\cdot, \cdot)$ should satisfy the same differential equation as u except that its source function will consist of a delta function at $\mathbf{x}$; and it should satisfy the same homogeneous boundary conditions. However, the dependence of $G(\cdot, \cdot)$ on its second argument is more useful as it shows how the source function influences the solution at a given point $\mathbf{x}'$; this is our reason for accenting the first rather than the second argument. Thus suppose we can assume $G(\mathbf{x}', \cdot) \in H^1_{E_0}$. Then combining (2.66) with (2.54) gives

$$u(\mathbf{x}') = (S, G(\mathbf{x}', \cdot)) = B(u, G(\mathbf{x}', \cdot)), \tag{2.67}$$

which holds for any choice of S and hence of u. This expresses the fact that $G(\mathbf{x}', \cdot)$ is the weak solution of an adjoint problem, associated with the bilinear form $B(\cdot, \cdot)$ with the arguments interchanged, and with a delta function at $\mathbf{x}'$ as its source function.

To make these arguments more explicit, we write the problem corresponding to (2.53) and (2.54) as

$$Lu := -\epsilon \nabla \cdot (a \nabla u) + \mathbf{b} \cdot \nabla u + cu = S \quad \text{in } \Omega \tag{2.68a}$$

$$u = u_B \quad \text{on } \partial \Omega_D, \quad \frac{\partial u}{\partial n} = 0 \quad \text{on } \partial \Omega_N. \tag{2.68b}$$

The corresponding adjoint operator is then

$$L^*v := -\epsilon\nabla\cdot(a\nabla v) - \nabla\cdot(\mathbf{b}v) + cv. \tag{2.69}$$

From the definitions of L and L^* we deduce the Green's formula

$$\int_\Omega [vLu - uL^*v]\,\mathrm{d}\Omega = \int_{\partial\Omega}\left[\epsilon a[u\frac{\partial v}{\partial n} - v\frac{\partial u}{\partial n}] + uv\mathbf{b}\cdot\mathbf{n}\right]\mathrm{d}\Gamma. \tag{2.70}$$

Now suppose that u satisfies (2.68) while $v = 0$ on $\partial\Omega_D$ and $\epsilon a\partial v/\partial n + (\mathbf{b}\cdot\mathbf{n})v = 0$ on $\partial\Omega_N$. Then we have

$$\int_\Omega [vS - uL^*v]\,\mathrm{d}\Omega = \int_{\partial\Omega_D}\epsilon au_B\frac{\partial v}{\partial n}\mathrm{d}\Gamma. \tag{2.71}$$

For sufficiently smooth u, the delta function is defined through the property

$$u(\mathbf{x}') = \int_\Omega u(\mathbf{x})\delta(\mathbf{x} - \mathbf{x}')\mathrm{d}\mathbf{x} \quad \forall \mathbf{x}' \in \Omega.$$

So, comparing (2.71) with (2.66) where we assumed $u_B = 0$, we see that the Green's function of the problem (2.68) is completely characterised as follows:

$$L^*G(\mathbf{x}', \cdot) = \delta(\cdot - \mathbf{x}') \quad \text{on } \Omega \tag{2.72a}$$

$$G(\mathbf{x}', \cdot) = 0 \quad \text{on } \partial\Omega_D \tag{2.72b}$$

$$\epsilon a\mathbf{n}\cdot\nabla G(\mathbf{x}', \cdot) + (\mathbf{b}\cdot\mathbf{n})G(\mathbf{x}', \cdot) = 0 \quad \text{on } \partial\Omega_N, \tag{2.72c}$$

where ∇ denotes the gradient with respect to the second argument. Thus the solution of (2.68) can be written

$$u(\mathbf{x}') = \int_\Omega G(\mathbf{x}', \mathbf{x})S(\mathbf{x})\mathrm{d}\mathbf{x}$$

$$- \int_{\partial\Omega_D}\epsilon a(\mathbf{x})u_B(\mathbf{x})(\mathbf{n}\cdot\nabla G(\mathbf{x}', \mathbf{x}))\mathrm{d}\Gamma. \tag{2.73}$$

The behaviour of the Green's function at the singularity $\mathbf{x}' = \mathbf{x}$ can be deduced by applying Green's formula (2.70), with $v(\mathbf{x})$ replaced by $G(\mathbf{x}', \mathbf{x})$, firstly to the whole domain Ω and then to the same domain punctured at $\mathbf{x} = \mathbf{x}'$. Noting that the integral of $G\partial u/\partial n$ over a sphere Γ_δ of radius δ about $\mathbf{x} = \mathbf{x}'$ vanishes as $\delta \to 0$, we obtain an identity corresponding to (2.73)

$$u(\mathbf{x}') = -\lim_{\delta\to 0}\int_{\Gamma_\delta}\epsilon au(\mathbf{n}\cdot\nabla G(\mathbf{x}', \cdot))\Gamma. \tag{2.74}$$

In one dimension, this reduces to specifying a jump in the gradient,

$$d = 1: \quad \left.\frac{\mathrm{d}G(x', x)}{\mathrm{d}x}\right|_{x'+} - \left.\frac{\mathrm{d}G(x', x)}{\mathrm{d}x}\right|_{x'-} = -\frac{1}{\epsilon a(x')}. \tag{2.75a}$$

In two and three dimensions one can deduce from our assumption of an isotropic diffusion coefficient that a fundamental solution of the equation $Lu = 0$ is spherically symmetric at the singularity — see, e.g., Garabedian (1964). This therefore gives the result

$$d = 2, 3: \quad \lim_{\delta \to 0} \left(\delta^{d-1} \frac{\partial G(\mathbf{x'}, \mathbf{x})}{\partial r} \bigg|_{r=\delta} \right) = -\frac{1}{\sigma_d \epsilon a(\mathbf{x'})}, \qquad (2.75b)$$

where r is the radial coordinate from origin $\mathbf{x'}$ to $\mathbf{x}$ and σ_d is the surface measure of the unit sphere in d dimensions, that is $\sigma_2 = 2\pi, \sigma_3 = 4\pi$.

2.7.1 One-dimensional examples

In later chapters it will be useful to refer to specific examples of Green's functions. These can be readily constructed in one dimension from solutions of the equation and its adjoint which satisfy one of the boundary conditions. So let us consider the following pair of problems, (i) and (ii), which differ only in the boundary condition on the right and can be treated together,

$$Lu : = -\epsilon(au')' + bu' + cu = S \quad \text{on } (0, 1) \qquad (2.76a)$$
$$u(0) = u_L \qquad (2.76b)$$
$$(i) \ u(1) = u_R, \ \text{or } (ii) \ u'(1) = 0. \qquad (2.76c)$$

We assume $b \geq 0$ and the conditions of Lemma 2.4.2 are satisfied so that both problems have a unique solution.

Theorem 2.7.1 *Suppose that, for the problems (2.76), the functions u^L and u^R satisfy*

$$Lu^L = 0 \quad \text{and } u^L(0) = 0, \qquad (2.77a)$$
$$Lu^R = 0 \quad \text{and } (i) \ u^R(1) = 0$$
$$\text{or } (ii) \ \frac{du^R(1)}{dx} = 0; \qquad (2.77b)$$

and similarly suppose that v^L and v^R satisfy

$$L^* v^L = 0 \quad \text{and } v^L(0) = 0, \qquad (2.77c)$$
$$L^* v^R = 0 \quad \text{and } (i) \ v^R(1) = 0$$
$$\text{or } (ii) \ \epsilon a(1) \frac{dv^R(1)}{dx} + b(1) v^R(1) = 0. \qquad (2.77d)$$

Then these functions can be so scaled that the Green's function for (2.76) can be written

$$G(x', x) = \begin{cases} v^L(x) u^R(x'), & x < x' \\ u^L(x') v^R(x), & x' < x. \end{cases} \qquad (2.78)$$

Proof. From the properties of v^L and v^R it is clear that $G(x', x)$ satisfies all the conditions of (2.72) away from the point $x = x'$. So first we show that continuity can be imposed at that point. All four functions are determined by (2.77) to within an arbitrary factor; but we fix those in u^L and u^R, and retain those in v^L and v^R as adjustable parameters. Now from the Green's formula applied to u^L and v^L over the interval $(0, x)$, using the boundary conditions in (2.77a) and (2.77c), we have

$$\left[\epsilon a \left(u^L \frac{\mathrm{d}v^L}{\mathrm{d}x} - v^L \frac{\mathrm{d}u^L}{\mathrm{d}x} \right) + b u^L v^L \right]_{x'} = 0. \tag{2.79a}$$

Similarly, by applying the formula to u^R and v^R over the interval $(x', 1)$ we have

$$\left[\epsilon a \left(u^R \frac{\mathrm{d}v^R}{\mathrm{d}x} - v^R \frac{\mathrm{d}u^R}{\mathrm{d}x} \right) + b u^R v^R \right]_{x'} = 0. \tag{2.79b}$$

What we wish to ensure is that

$$v^L(x') u^R(x') = u^L(x') v^R(x'); \tag{2.80}$$

so we consider

$$\frac{\mathrm{d}}{\mathrm{d}x} \left(\frac{v^L}{u^L} \cdot \frac{u^R}{v^R} \right) = \frac{u^L \mathrm{d}v^L/\mathrm{d}x - v^L \mathrm{d}u^L/\mathrm{d}x}{(u^L)^2} \cdot \frac{u^R}{v^R}$$
$$+ \frac{v^R \mathrm{d}u^R/\mathrm{d}x - u^R \mathrm{d}v^R/\mathrm{d}x}{(v^R)^2} \cdot \frac{v^L}{u^L},$$

which it is easily seen from (2.79) is zero at x'. Thus as x' is arbitrary the ratio on the left is constant, and we make use of one of the parameters in v^L and v^R to ensure it is unity so that (2.80) is satisfied.

Now to deal with the jump condition (2.75a) at $x = x'$ we introduce the Wronskian $W^u = u^L \mathrm{d}u^R/\mathrm{d}x - u^R \mathrm{d}u^L/\mathrm{d}x$ and consider the equation $u^L L u^R - u^R L u^L = 0$. We have

$$-\epsilon a \left(u^L \frac{\mathrm{d}^2 u^R}{\mathrm{d}x^2} - u^R \frac{\mathrm{d}^2 u^L}{\mathrm{d}x^2} \right) + (b - \epsilon a') \left(u^L \frac{\mathrm{d}u^R}{\mathrm{d}x} - u^R \frac{\mathrm{d}u^L}{\mathrm{d}x} \right) = 0$$

i.e.,

$$-\epsilon a \frac{\mathrm{d}W^u}{\mathrm{d}x} + (b - \epsilon a') W^u = 0,$$

or

$$-\epsilon (a W^u)' + b W^u = 0. \tag{2.81}$$

We can similarly introduce the Wronskian $W^v = v^L \mathrm{d}v^R/\mathrm{d}x - v^R \mathrm{d}v^L/\mathrm{d}x$ and from $v^L L^* v^R - v^R L^* v^L = 0$ obtain

$$-\epsilon a \left(v^L \frac{\mathrm{d}^2 v^R}{\mathrm{d}x^2} - v^R \frac{\mathrm{d}^2 v^L}{\mathrm{d}x^2} \right) - (b + \epsilon a') \left(v^L \frac{\mathrm{d}v^R}{\mathrm{d}x} - v^R \frac{\mathrm{d}v^L}{\mathrm{d}x} \right) = 0$$

i.e.,

$$\epsilon (a W^v)' + b W^v = 0. \tag{2.82}$$

Together, (2.81) and (2.82) imply that

$$a^2 W^u W^v = \text{const.}. \tag{2.83}$$

Moreover, we can show that $L^* v = 0$ implies $L(aW^u v) = 0$; for from (2.81)

$$
\begin{aligned}
\epsilon(a(aW^u v)')' &= \epsilon(a^2 W^u v')' + \epsilon(av(aW^u)')' \\
&= \epsilon a W^u (av')' + \epsilon(aW^u)' av' + (avbW^u)' \\
&= \epsilon a W^u (av')' + bW^u av' + aW^u (bv)' + bv(aW^u)'
\end{aligned}
$$

and hence

$$
\begin{aligned}
L(aW^u v) &= aW^u[-\epsilon(av')' - (bv)' - bv'] - bv(aW^u)' \\
&\quad + b(aW^u v)' + caW^u v \\
&= aW^u[-\epsilon(av')' - (bv)' + cv] = 0.
\end{aligned}
$$

Thus, since both $u^L(0) = 0$ and $v^L(0) = 0$, there is a constant C_L such that

$$v^L(\cdot) = C_L \frac{u^L(\cdot)}{a(\cdot)W^u(\cdot)}. \tag{2.84a}$$

A similar relationship holds on the right because even for the second problem the condition $du^R(1)/dx = 0$ corresponds to the adjoint condition (2.77d) through $\epsilon(aW^u v)' = W^u(\epsilon av' + bv)$; so we have

$$v^R(\cdot) = C_R \frac{u^R(\cdot)}{a(\cdot)W^u(\cdot)}, \tag{2.84b}$$

and C_L, C_R correspond to the two adjustable parameters introduced in the definition of the Green's function. To satisfy (2.80) we need $C_L = C_R$, so one has been used as we have already seen; and the other we use to satisfy the jump condition (2.75a) which takes the form

$$a(x') \left[u^L \frac{dv^R}{dx} - \frac{dv^L}{dx} u^R \right]_{x'} = -1/\epsilon. \tag{2.85}$$

Substituting for u^R and u^L from (2.84) with $C_L = C_R = C$, we get

$$a^2(x')W^u(x')W^v(x')/C = -1/\epsilon. \tag{2.86}$$

However, we have already shown in (2.83) that the left-hand side is independent of x', so that the constant C can be chosen to satisfy the jump condition (2.85). This completes the construction of the Green's function and shows that it depends only on the two elementary solutions $u^L(\cdot)$ and $u^R(\cdot)$ satisfying (2.77a) and (2.77b). $\square$

Remark 1 We have assumed throughout the proof that $u^L \neq 0$ and $u^R \neq 0$ in $(0,1)$ as well as $W^u \neq 0$. These all follow from the assumptions ensuring that the two problems have unique solutions and therefore that u^L and u^R are independent solutions of $Lu = 0$.

Remark 2 Some of the calculations in the above proof are not strictly necessary but the intermediate results that are derived are of interest in their own right.

In the special case that the coefficients of L are constant, the Green's functions can be written out explicitly in terms of exponential functions. There is no loss in assuming $a = 1$, so that the characteristic equation for L is

$$\epsilon\lambda^2 - b\lambda - c = 0 \tag{2.87}$$

with roots which we write as

$$\lambda = \tilde{b}/\epsilon, \quad -c/\tilde{b} \quad \text{where} \quad \tilde{b} = \tfrac{1}{2}[b + (b^2 + 4\epsilon c)^{\frac{1}{2}}]. \tag{2.88}$$

Then for problem (i), with Dirichlet boundary conditions at both ends of the interval, we can set

$$u^L(x) = e^{\tilde{b}x/\epsilon} - e^{-cx/\tilde{b}} \tag{2.89a}$$
$$u^R(x) = e^{c(1-x)/\tilde{b}} - e^{-\tilde{b}(1-x)/\epsilon} \tag{2.89b}$$

and after a little manipulation find

$$W^u(x) = -(\tilde{b}/\epsilon + c/\tilde{b})(e^{c/\tilde{b}} - e^{-\tilde{b}/\epsilon})e^{\tilde{b}x/\epsilon}e^{-cx/\tilde{b}}. \tag{2.89c}$$

We can confirm immediately from these expressions how (2.84a) leads to the same form for v^L as direct construction from solutions to the adjoint equation, whose characteristic equation has roots $-\tilde{b}/\epsilon, c/\tilde{b}$. It is therefore readily checked that the Green's function for this case, which we distinguish by writing as G^D, has the form

$$G^D(x', x) = \begin{cases} \dfrac{(e^{cx/\tilde{b}} - e^{-\tilde{b}x/\epsilon})(e^{c(1-x')/\tilde{b}} - e^{-\tilde{b}(1-x')/\epsilon})}{\epsilon(\tilde{b}/\epsilon + c/\tilde{b})(e^{c/\tilde{b}} - e^{-\tilde{b}/\epsilon})}, & x \leq x' \\[4mm] \dfrac{(e^{\tilde{b}x'/\epsilon} - e^{-cx'/\tilde{b}})(e^{\tilde{b}(1-x)/\epsilon} - e^{-c(1-x)/\tilde{b}})}{\epsilon(\tilde{b}/\epsilon + c/\tilde{b})(e^{\tilde{b}/\epsilon} - e^{-c/\tilde{b}})}, & x' \leq x. \end{cases} \tag{2.90}$$

This is sketched in Fig. 2.5 for some typical values of the parameters; note the considerable simplification which occurs for $c = 0$, a case which is much studied in later chapters — in particular, see section 3.3.2.

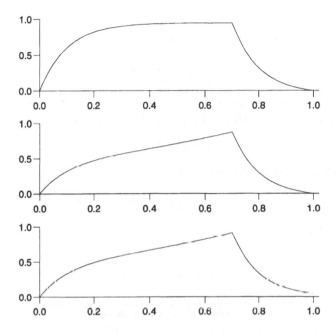

Figure 2.5. *Typical Green's functions $G(x', x)$ on the interval $[0, 1]$: the top two figures are for $G^D(0.7, x)$ with $\bar{b}/\epsilon, c/\bar{b}) = (10, 0)$ and $(10, 1)$ respectively; the bottom figure is for $G^N(0.7, x)$ in the second case*

In the second case, with a Neumann boundary condition on the right, the corresponding Green's function G^N is readily calculated as

$$
G^N(x, x') = \begin{cases}
\dfrac{(e^{cx/\bar{b}} - e^{-\bar{b}x/\epsilon})[(\bar{b}/\epsilon)e^{c(1-x')/\bar{b}} + (c/\bar{b})e^{-\bar{b}(1-x')/\epsilon}]}{\epsilon(\bar{b}/\epsilon + c/\bar{b})[(\bar{b}/\epsilon)e^{c/\bar{b}} + (c/\bar{b})e^{-\bar{b}/\epsilon}]}, \\
\hspace{8cm} x \le x' \\[2mm]
\dfrac{(e^{\bar{b}x'/\epsilon} - e^{-cx'/\bar{b}})[(\bar{b}/\epsilon)e^{\bar{b}(1-x)/\epsilon} + (c/\bar{b})e^{-c(1-x)/\bar{b}}]}{\epsilon(\bar{b}/\epsilon + c/\bar{b})[(\bar{b}/\epsilon)e^{\bar{b}/\epsilon} + (c/\bar{b})e^{-c/\bar{b}}]}, \\
\hspace{8cm} x' \le x.
\end{cases}
\tag{2.91}
$$

This is also sketched in Fig. 2.5 for typical values of the parameters.

2.7.2 *Free-space Green's functions*

In two and three dimensions, the interaction between the singularity in the fundamental solution for the second order diffusion operator, and the exponential form of the Green's function for the one-dimensional convection-diffusion operator, is most simply seen by ignoring boundary effects. If we also assume constant coefficients we can use Fourier analysis to construct the Green's function explicitly.

We write the d-dimensional Fourier transform of $u(\mathbf{x})$ as

$$\hat{u}(\mathbf{k}) = \frac{1}{(2\pi)^{d/2}} \int u(\mathbf{x})e^{-i\mathbf{k}\cdot\mathbf{x}}d\mathbf{x}, \qquad (2.92)$$

where the integral is over $\mathbb{R}^d$. Then the constant coefficient convection-diffusion equation with $a = 1$ and $c = 0$ becomes, with $k = |\mathbf{k}|$,

$$(\epsilon k^2 + i\mathbf{k}\cdot\mathbf{b})\hat{u} = \hat{S}. \qquad (2.93)$$

From the inverse Fourier transform

$$u(\mathbf{x}') = \frac{1}{(2\pi)^{d/2}} \int \hat{u}(\mathbf{k})e^{i\mathbf{k}\cdot\mathbf{x}'}d\mathbf{k}, \qquad (2.94)$$

we then have

$$u(\mathbf{x}') = \frac{1}{(2\pi)^{d/2}} \int \frac{e^{i\mathbf{k}\cdot\mathbf{x}'}}{\epsilon k^2 + i\mathbf{k}\cdot\mathbf{b}} \hat{S}(\mathbf{k})d\mathbf{k}$$

$$= \frac{1}{(2\pi)^d} \int \int \frac{e^{i\mathbf{k}\cdot(\mathbf{x}'-\mathbf{x})}}{\epsilon k^2 + i\mathbf{k}\cdot\mathbf{b}} d\mathbf{k} S(\mathbf{x})d\mathbf{x}, \qquad (2.95)$$

which by comparison with (2.66) gives the Green's function.

It is clear from this form that $\mathbf{b}$ gives a distinguished direction and that the Green's function, as a function of $\mathbf{x} - \mathbf{x}'$, is cylindrically symmetric about $\mathbf{b}$. To evaluate the integrals by the techniques of contour integration the following change of variable is indicated

$$\mathbf{k} = \tfrac{1}{2}(b/\epsilon)(\mathbf{m} - i\mathbf{b}/b).$$

This leads to

$$G(\mathbf{x}',\mathbf{x}) = e^{\mathbf{b}\cdot(\mathbf{x}'-\mathbf{x})/2\epsilon} \cdot \frac{(b/2\epsilon)^{d-2}}{\epsilon(2\pi)^d} \int \frac{e^{i(b/2\epsilon)\mathbf{m}\cdot(\mathbf{x}'-\mathbf{x})}}{1+m^2} d\mathbf{m} \qquad (2.96a)$$

from which we see that

$$G(\mathbf{x}',\mathbf{x}) = (1/\epsilon)(b/2\epsilon)^{d-2}e^{\mathbf{b}\cdot(\mathbf{x}'-\mathbf{x})/2\epsilon} g(b|\mathbf{x} - \mathbf{x}'|/2\epsilon) \qquad (2.96b)$$

for some function $g(\cdot)$. In two dimensions the integral yields the modified Bessel function, so that

$$g(\rho) = (1/2\pi)K_0(\rho) \qquad (2.96c)$$

which has the logarithmic behaviour at $\rho \to 0$ deduced in (2.75b); note too that the coefficient $(2\pi\epsilon)^{-1}$ in $G(\mathbf{x}',\mathbf{x})$ as $\mathbf{x} \to \mathbf{x}'$ is also consistent with that in (2.75b).

Alternatively, the defining equation for the Green's function, written in terms of the second argument with $\mathbf{r} = \mathbf{x} - \mathbf{x}'$ as

$$-\epsilon\nabla^2 G - \mathbf{b}\cdot\nabla G = \delta(\mathbf{r}), \qquad (2.97)$$

can be transformed by the substitution $G = v \exp(-\mathbf{b} \cdot \mathbf{r}/2\epsilon)$ to

$$\nabla^2 v - (b/2\epsilon)^2 v = 0, \ \mathbf{r} \neq \mathbf{0}. \tag{2.98}$$

In cylindrical coordinates this then leads to the modified Bessel function already given. In three dimensions the form is much simpler, giving

$$v(\mathbf{r}) = C(1/r)e^{-br/2\epsilon}$$

and hence

$$G(\mathbf{x}', \mathbf{x}) = \frac{1}{4\pi\epsilon} \frac{1}{|\mathbf{x} - \mathbf{x}'|} e^{-b(1+\cos\theta)|\mathbf{x}-\mathbf{x}'|/2\epsilon}, \tag{2.99}$$

where θ is the angle between $\mathbf{b}$ and $\mathbf{x} - \mathbf{x}'$.

Fig. 2.6 shows a contour plot of a typical two-dimensional Green's function, indicating the influence of upwind data on the solution, but with the logarithmic singularity truncated.

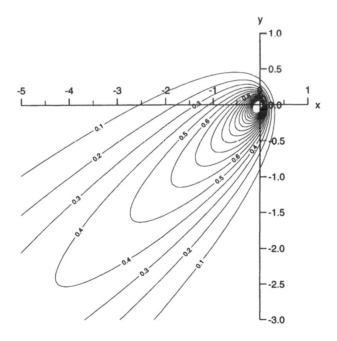

Figure 2.6. *Contour plots for the free-space two-dimensional Green's function for* $b/\epsilon = (5\sqrt{3}, 5)$; *the central contour coincides with the truncation of the logarithmic singularity*

CHAPTER 3

Difference schemes for steady problems

3.1 One-dimensional problems

To highlight the specific schemes developed for convection-diffusion problems, and to study their behaviour through boundary layers, we begin with simple one-dimensional problems. In Chapter 2 we gave bounds for solution derivatives which will be useful in estimating truncation errors for an equation of the following form

$$Lu : = -\epsilon u'' + b(x)u' + c(x)u = S(x) \quad \text{on } (0,1); \tag{3.1a}$$
$$u(0) = u_L, \quad u(1) = u_R. \tag{3.1b}$$

So we start with this problem; the application of the methods to be described to the more general problem (2.13) will be straightforward. The usual conditions on the coefficients, $b(x)$ and $c(x)$, and on the data $S(x)$, are assumed to hold and will be spelt out for the error analysis of section 3.4. In particular, in these first four sections we shall assume that $b(x)$ is strictly positive. Turning-point problems, and more general boundary conditions, will be dealt with in section 3.5. We shall, however, consider only linear problems in this chapter.

In the first instance, we will assume that we have a uniform mesh

$$x_j = jh, \quad j = 0, 1, \ldots, J$$

with $x_0 = 0, x_J = 1$. Then we can use the standard difference notation introduced in section 1.3, which for convenience we repeat here:

$$\Delta_+ U_j := U_{j+1} - U_j, \quad \Delta_- U_j := U_j - U_{j-1}, \quad D_\pm := h^{-1}\Delta_\pm \tag{3.2a}$$
$$\delta^2 := \Delta_+ - \Delta_- = \Delta_+\Delta_-, \quad \Delta_0 := \tfrac{1}{2}(\Delta_+ + \Delta_-), \quad D_0 := h^{-1}\Delta_0. \tag{3.2b}$$

Part of the reason for concentrating on a uniform mesh here is that the construction and analysis of schemes for nonuniform meshes is done more readily in the framework of finite element and finite volume schemes. The presentation of these methods in Chapters 4 and 5 will be in the context of quite general meshes, where issues such as maintaining accuracy and other properties in the face of mesh irregularity will be taken up. However, another justification is that in one dimension it is relatively easy to introduce a graded mesh by a smooth change of variable, after which a uniform mesh can be used in the new variable. Thus suppose we change variable to $y = y(x)$, where $dy/dx > 0$: then we have, if $v(y(x)) \equiv u(x)$,

$$\frac{du}{dx} = \frac{dv}{dy}\frac{dy}{dx}, \quad \frac{d^2u}{dx^2} = \frac{d^2v}{dy^2}\left(\frac{dy}{dx}\right)^2 + \frac{dv}{dy}\frac{d^2y}{dx^2} \qquad (3.3a)$$

and (3.1a) becomes

$$-\epsilon\left(\frac{dy}{dx}\right)^2 v'' + \left(b\frac{dy}{dx} - \epsilon\frac{d^2y}{dx^2}\right)v' + cv = S. \qquad (3.3b)$$

This is of the same form as before and a uniform mesh error analysis can now be applied if the interval in y corresponding to $0 \le x \le 1$ is subdivided uniformly. Note that if y were to satisfy $-\epsilon y'' + by' = 0$ the convective term would be completely eliminated from the equation for v; a more practical choice when $b(x) > 0$ could be based on setting

$$y' = K[1 + (b_\phi/\epsilon)e^{-(b_\phi/\epsilon)(1-x)}]$$

with $y(0) = 0, K$ chosen so that $y(1) = 1$, and b_ϕ a lower bound to $b(x)$. This would place roughly half the mesh points in the boundary layer on the right and spread the other half evenly throughout the interval.

Before leaving this topic it is worth noting that the standard difference operators on a nonuniform mesh are equivalent to differencing the transformed equation in a natural way. This is obvious for the first derivatives, as we have

$$\frac{du}{dx} = \frac{dy}{dx}\frac{dv}{dy} \approx \frac{\Delta y}{\Delta_\pm x}\frac{\Delta_\pm V}{\Delta y} = \frac{\Delta_\pm U}{\Delta_\pm x}; \qquad (3.4a)$$

and we rewrite the second derivative to get

$$\frac{d^2u}{dx^2} = \frac{dy}{dx}\frac{d}{dy}\left(\frac{dy}{dx}\frac{dv}{dy}\right) \approx \frac{\Delta y}{\Delta_0 x}\frac{1}{\Delta y}\left(\frac{\Delta_+ U}{\Delta_+ x} - \frac{\Delta_- U}{\Delta_- x}\right). \qquad (3.4b)$$

Substituting these approximations into (3.1a) clearly gives corresponding approximations to (3.3b); on the other hand, not all difference approximations correspond in this way.

In this chapter we shall primarily consider three-point schemes (and their extensions to higher dimensions), which can therefore be applied at all interior points of the mesh, namely for $j = 1, 2 \ldots, J - 1$, with the conditions $U_0 = u_L, U_J = u_R$ imposed at the boundaries. It is not our main concern to generate highly accurate multi-point methods for one-dimensional problems like (3.1), but to thoroughly explore the effectiveness on these problems of methods which have wider applicability and good basic properties. One of the key tools in analysing the methods will be a discrete maximum principle, as pioneered by Dorr (1970); and there are general theoretical results that guarantee the existence of such principles for sufficiently fine meshes, if they hold for the continuous problem and one is using a so-called *compact* difference scheme, i.e. one that uses the minimum number of points needed to approximate all the derivatives involved — see, for example, Gartland (1991). So there are good theoretical as well as practical reasons for using compact schemes.

Moreover, we shall see in the next section that schemes of fourth order accuracy can be constructed with just the three points. With some ingenuity they can also be so formed that they satisfy a maximum principle uniformly in the mesh Péclet number $\beta = bh/\epsilon$. However, these OCI schemes do not possess error bounds that are uniform in β. To achieve this it is necessary to use the exponentially fitted schemes considered in section 3.3, and analysed together with the OCI schemes in section 3.4.

3.2 Hermitian and Operator Compact Implicit (OCI) schemes

Motivated by Hermite's generalisation of the Taylor series, Collatz (1960), (1973) proposed schemes of the following form for a variety of problems written as $Lu = S$:

$$\sum_{(\nu)} p_{j,\nu} U_{j+\nu} = h^2 \sum_{(\nu)} q_{j,\nu} S_{j+\nu}. \tag{3.5}$$

The coefficients $\{p_{j,\nu}\}$ and $\{q_{j,\nu}\}$ are to be determined by substituting polynomials $w(x)$ into

$$\sum_{(\nu)} p_{j,\nu} w_{j+\nu} - h^2 \sum_{(\nu)} q_{j,\nu} (Lw)_{j+\nu} \tag{3.6}$$

and equating to zero as many coefficients of h^m as possible. For a three-point scheme ν runs over the values $-1, 0, +1$ in each sum and, after normalisation, this leaves five free parameters. One might hope that these can be chosen to leave (3.6) to be $O(h^5)$; indeed, this can be done and actually makes the residual $O(h^6)$. Thus the truncation error after normalisation is $O(h^4)$.

Subsequent to Collatz's work, and independently of it, Swartz (1974) produced the identical fourth order scheme by constructing finite difference analogues of finite element methods. These methods have now been developed and generalised by several authors under the name *Operator Compact Implicit (OCI) schemes* — see, in particular, Berger *et al.* (1980). In the presentation below we shall follow quite closely the treatment in this comprehensive paper.

For the three-point schemes we can simplify the notation so that (3.5) becomes

$$p_j^- U_{j-1} + p_j U_j + p_j^+ U_{j+1} = h^2 [q_j^- S_{j-1} + q_j S_j + q_j^+ S_{j+1}]. \tag{3.7}$$

Furthermore, it is clear from the above outline of the way in which the coefficients are calculated, that the effect of the term $c(x)u$ in the equation (3.1a) can be readily eliminated by pointwise matching of the p and q coefficients. We can state this, without proof, in the following way; in effect one writes $Lu = S$ as $(L - c)u = S - cu$ before designing the difference scheme.

Lemma 3.2.1 *For any choice of the coefficients $\{q_j^-, q_j, q_j^+\}$, the effect of adding the term $c(x)u$ to the operator $-\epsilon u'' + b(x)u'$ is accounted for by making the changes*

$$p_j^- \to p_j^- + h^2 q_j^- c_{j-1}, \quad p_j \to p_j + h^2 q_j c_j, \quad p_j^+ \to p_j^+ + h^2 q_j^+ c_{j+1} \quad (3.8)$$

to the coefficients $\{p_j^-, p_j, p_j^+\}$.

Thus in the derivation of the coefficients we shall assume $c(x) \equiv 0$.

3.2.1 Derivation of the standard scheme

In defining the truncation error we need to normalise the coefficients so that

$$q_j^- + q_j + q_j^+ = 1, \qquad (3.9)$$

although occasionally it will be convenient to relax this. Then we write

$$\begin{aligned} T_j &:= h^{-2}[p_j^- u_{j-1} + p_j u_j + p_j^+ u_{j+1}] \\ &\quad - [q_j^-(Lu)_{j-1} + q_j(Lu)_j + q_j^+(Lu)_{j+1}] \qquad (3.10a) \\ &\equiv T_j^0 u_j + T_j^1 u_j' + \ldots T_j^6 u_j^{vi} + O(h^5), \qquad (3.10b) \end{aligned}$$

after Taylor expansion about the point x_j. The coefficients T_j^m are given by

$$T_j^0 = h^{-2}[p_j^- + p_j + p_j^+] \qquad (3.11a)$$

$$T_j^1 = h^{-1}[(-p_j^- + p_j^+) - h(q_j^- b_{j-1} + q_j b_j + q_j^+ b_{j+1})] \qquad (3.11b)$$

$$T_j^2 = \tfrac{1}{2}(p_j^- + p_j^+) + \epsilon - h(-q_j^- b_{j-1} + q_j^+ b_{j+1}) \qquad (3.11c)$$

$$T_j^3 = \tfrac{1}{6}h[(-p_j^- + p_j^+) + 6\epsilon(-q_j^- + q_j^+) - 3h(q_j^- b_{j-1} + q_j^+ b_{j+1})] \qquad (3.11d)$$

$$T_j^4 = \tfrac{1}{24}h^2[(p_j^- + p_j^+) + 12\epsilon(q_j^- + q_j^+) - 4h(-q_j^- b_{j-1} + q_j^+ b_{j+1})] \qquad (3.11e)$$

with the general form for $m \geq 3$ being

$$T_j^m = \frac{h^{m-2}}{m!}[((-1)^m p_j^- + p_j^+) + m(m-1)\epsilon((-1)^m q_j^- + q_j^+) \\ -mh((-1)^{m-1} q_j^- b_{j-1} + q_j^+ b_{j+1})]. \qquad (3.11f)$$

It is clear that setting $T_j^0 = T_j^1 = T_j^2 = 0$ determines the p coefficients in terms of the q's. Specifically, we have

$$p_j = -(p_j^- + p_j^+) \qquad (3.12a)$$

$$p_j^- = -\epsilon - \tfrac{1}{2}h(3q_j^- b_{j-1} + q_j b_j - q_j^+ b_{j+1}) \qquad (3.12b)$$

$$p_j^+ = -\epsilon - \tfrac{1}{2}h(q_j^- b_{j-1} - q_j b_j - 3q_j^+ b_{j+1}). \qquad (3.12c)$$

Then the standard OCI scheme, which is the same as the Hermitian scheme, is obtained by also setting $T_j^3 = T_j^4 = 0$. Substituting for $p_j^\pm$ from (3.12) gives a pair of equations which are satisfied by taking the q's in the following proportions,

$$q_j^- : q_j : q_j^+ = 6\epsilon^2 + (5b_j - 2b_{j+1})\epsilon h - b_j b_{j+1} h^2$$
$$: 60\epsilon^2 + 16(b_{j-1} - b_{j+1})\epsilon h - 4b_{j-1} b_{j+1} h^2$$
$$: 6\epsilon^2 + (2b_{j-1} - 5b_j)\epsilon h - b_{j-1} b_j h^2. \qquad (3.13)$$

For fixed ϵ, it is clear that $q_j^+ - q_j^- = O(h)$ and $p_j^+ - p_j^- = O(h)$. Hence from (3.11f) we find $T_j^5 = O(h^4)$, and the scheme is at least fourth order accurate.

3.2.2 Properties of OCI and related schemes

To study the properties of the scheme, let us suppose b is constant, so that we can drop the subscript j in the coefficients, and express (3.13) in terms of the mesh Péclet number $\beta = bh/\epsilon$. We have

$$q^- : q : q^+ = 6 + 3\beta - \beta^2 : 60 - 4\beta^2 : 6 - 3\beta - \beta^2, \qquad (3.14)$$

and the sum of these quadratic terms is $6(12 - \beta^2)$. The characteristic equation of any three-point difference scheme is a quadratic with coefficients p^-, p, p^+, namely

$$p^+ \lambda^2 + p\lambda + p^- = 0; \qquad (3.15)$$

and if the consistency condition $T_j^0 = O(h)$ is satisfied by setting $p = -(p^- + p^+)$ as in (3.12a), one root is unity. We denote the other by μ, which is given by p^-/p^+ ; so whenever p^- and p^+ are of opposite signs μ is negative and oscillatory solutions may occur. In the present case, when (3.13) is combined with (3.12b) and (3.12c) we find that

$$\mu = \frac{(1 + \tfrac{1}{2}\beta)(q^- + q + q^+) + \beta(q^- - q^+)}{(1 - \tfrac{1}{2}\beta)(q^- + q + q^+) - \beta(q^- - q^+)}$$
$$= \frac{(2 + \beta)(12 - \beta^2) + 2\beta^2}{(2 - \beta)(12 - \beta^2) - 2\beta^2}. \qquad (3.16)$$

The denominator here changes sign between $\beta = 0$ and $\beta = 2$, so that one obtains an oscillatory solution even earlier than with the second order central difference scheme (1.28) discussed in Chapter 1. At $\beta^2 = 12$ the sum of the q's is zero so that the scheme is not consistent with the differential equation; moreover, for $\beta^2 > 12$ with our normalisation (3.9), we get $q^+ > q^-$ which is totally at variance with the notion of upwinding. Thus this straightforward approach to obtaining fourth order accuracy gives a scheme of very limited value.

The central difference scheme also fits into the above development quite readily; one merely sets $T_j^0 = T_j^1 = T_j^2 = 0$, and obtains the p coefficients by setting $q_j^+ = q_j^- = 0$ in (3.12). From (3.11d) we get $T_j^3 = \frac{1}{6} h^2 b$ so the scheme is second order accurate, but as we saw in Chapter 1, the second root of the characteristic equation becomes negative for $\beta > 2$ and the scheme is widely regarded as valueless for convection-dominated flows — except possibly as a diagnostic aid for determining when the mesh needs to be refined (see Gresho and Lee (1981)).

On the other hand, the upwind scheme abandons even second order accuracy so that only the first of the equations (3.12) is satisfied. Referring back to the relations (3.11), one sets $T_j^0 = T_j^1 = 0$ and ensures first order accuracy with $T_j^2 = O(h)$ by setting $p_j^+ = -\epsilon, p_j^- = -\epsilon - hb_j$ with $q_j^+ = q_j^- = 0$. The payoff is that the second root of the characteristic equation is $1 + \beta$ which is always positive.

Suppose we attempt to do better by taking a weighted average of these two schemes; that is, we choose the p coefficients to ensure no oscillatory solutions and then choose the q coefficients to get as high an accuracy as possible — a similar strategy to that often adopted for stiff initial-value problems. The scheme will have the form

$$-\frac{\epsilon}{h^2} \delta^2 U_j + \frac{b_j}{h}[(1-\alpha)\Delta_0 + \alpha\Delta_-]U_j$$

$$= q_j^- S_{j-1} + (1 - q_j^- - q_j^+)S_j + q_j^+ S_{j+1}, \tag{3.17}$$

which gives for constant b the coefficients

$$p^- = -\epsilon - \tfrac{1}{2}(1+\alpha)bh \qquad p^+ = -\epsilon + \tfrac{1}{2}(1-\alpha)bh \tag{3.18}$$

and hence avoids any oscillatory solution so long as $p^+ \leq 0$, or

$$\alpha \geq 1 - 2/\beta. \tag{3.19}$$

As regards accuracy, from (3.11) we have

$$T_j^2 = -bh(\tfrac{1}{2}\alpha - q^- + q^+)$$

$$T_j^3 = \tfrac{1}{2}bh^2(\tfrac{1}{3} - q^- - q^+) - \epsilon h(q^- - q^+), \tag{3.20}$$

so that second order accuracy requires $q^- - q^+$ to be both $O(h)$ and equal to $\tfrac{1}{2}\alpha + O(h)$. It would seem that we cannot satisfy (3.19) and have second order accuracy when $\beta > 2 + O(h)$. However, a sensible choice would appear to be $q^+ = 0, q^- = \tfrac{1}{2}\alpha, q = 1 - \tfrac{1}{2}\alpha$ giving $T_j^2 = 0$, and then to take α as small as possible consistent with (3.19). This means that $q^- - q^+ = \tfrac{1}{2}\alpha$ is nonzero only for $\beta > 2$, when we can write $\epsilon = bh/\beta$ and get $T_j^3 = O(h^2)$.

So we arrive at the following *adaptive upwind scheme,*

$$-\frac{\epsilon}{h^2}\delta^2 U_j + \frac{b_j}{h}[(1-\alpha)\Delta_0 + \alpha\Delta_-]U_j$$
$$= \tfrac{1}{2}\alpha S_{j-1} + (1-\tfrac{1}{2}\alpha)S_j, \tag{3.21}$$

which has $T_j^2 = 0$, and where α is chosen such that

$$\text{for } \beta \leq 2, \quad \alpha = 0 \text{ and } T_j^3 = \tfrac{1}{6}bh^2 \tag{3.22a}$$

$$\text{for } \beta > 2, \quad \alpha = 1 - 2/\beta \text{ and } T_j^3 = (\tfrac{1}{24} - \tfrac{1}{2}/\beta^2)bh^2. \tag{3.22b}$$

This is a quite effective and simple scheme; examples of its effectiveness will be given in section 4.6.

3.2.3 Generalised OCI schemes

However, we return to our discussion of OCI schemes. Despite the inappropriateness of the standard scheme for other than very small values of the mesh Péclet number, it is a remarkable fact, established in Berger *et al.* (1980), Berger *et al.* (1981), that small changes in T_j^3 and T_j^4 can transform the scheme into one which can be used with confidence over the whole mesh Péclet number range. Their *generalised OCI schemes* still set $T_j^0 = T_j^1 = T_j^2 = 0$, so that the relations (3.12) still hold, but allow $T_j^3 = O(h^4)$ and $T_j^4 = O(h^4)$. Before presenting these schemes we have one or two general observations to make.

Lemma 3.2.2 *For a fixed ϵ, the necessary and sufficient conditions that an OCI scheme is fourth order accurate (in the sense of truncation error) are that $T_j^m = O(h^4)$ for $m = 0,1,2,3,4$. This is the highest order possible.*

Proof. The necessity of these conditions is obvious; and their sufficiency follows by the same argument used for the standard OCI scheme. To prove the last point, suppose $T_j^2 = O(h^4)$ and $T_j^4 = O(h^4)$. Then from (3.9), (3.11c) and (3.11e) we can solve to $O(h)$ for $p_j^- + p_j^+$ and $q_j^- + q_j^+$ in terms of q_j, giving $p_j^- + p_j^+ = -2\epsilon + O(h)$, $q_j^- + q_j^+ = \tfrac{1}{5}q_j + O(h)$ and hence $q_j = 5/6 + O(h)$. Substituting into (3.11f) shows that $T_j^6 = (\epsilon h^4/240) + O(h^5)$. $\square$

For any three-point scheme, it is highly desirable that the resulting tridiagonal system is diagonally dominant, so that it can be solved by simple Gaussian decomposition without pivoting, and that the scheme satisfies a discrete maximum principle. Both of these properties are assured, and oscillatory solutions arising from negative roots to the characteristic equation prevented, if the following conditions are met for $j = 1, 2, \ldots, J - 1$:

$$p_j^- < 0, \quad p_j^+ \leq 0, \quad p_j \geq -(p_j^- + p_j^+) \tag{3.23a}$$

$$q_j^- \geq 0, \quad q_j \geq 0, \quad q_j^+ \geq 0. \tag{3.23b}$$

The first condition in (3.23a) is chosen to ensure strict diagonal dominance, even in the limit $\epsilon \to 0$, because a Dirichlet boundary condition must always be imposed at $x = 0$. The maximum principle takes the following form, which also includes the case of a more general boundary condition on the right — compare with the corollaries to Theorem 2.2.1.

Lemma 3.2.3 *For any difference scheme of the form (3.7) with coefficients satisfying the conditions (3.9) and (3.23), if $S_j \leq 0$ for $j = 0, 1, \ldots J$, (or the right-hand side of (3.7) is nonpositive) and if $U_0 \leq M$, $U_J \leq M$ for some $M \geq 0$, then $U_j \leq M$ for $j = 1, 2, \ldots, J - 1$.*

Proof. Suppose the result is untrue. Then, for some k, the set of values $\{U_j, j = 1, 2, \ldots, J - 1\}$ attains a maximum $U_k > M$. But it follows from (3.7), (3.23) and the hypothesis on $\{S_j\}$ that

$$p_k^- U_{k-1} + p_k U_k \leq -p_k^+ U_{k+1}$$
$$\leq -p_k^+ U_k$$
$$\leq (p_k + p_k^-) U_k$$

and hence that $U_k \leq U_{k-1}$. Hence $U_k = U_{k-1} = \cdots = U_0$ which gives a contradiction. $\square$

Suppose we denote by L^h the properly scaled difference operator approximating the differential operator L,

$$L^h U_j := \frac{1}{h^2}[p_j^- U_{j-1} + p_j U_j + p_j^+ U_{j+1}], \tag{3.24}$$

represented by the tridiagonal matrix considered above. Then the conditions (3.23) ensure that an OCI operator has a bounded inverse, independent of h and ϵ. For consider L^h operating on the comparison function $z(x) = 1 + x$: from $T_j^0 = T_j^1 = 0$ and $b(x) \geq b_\phi > 0$ we get

$$L^h z = (1/h)(p_j^+ - p_j^-) = q_j^- b_{j-1} + q_j b_j + q_j^+ b_{j+1}$$
$$\geq (q_j^- + q_j + q_j^+) b_\phi = b_\phi.$$

Now let us choose a constant C so that Lemma 3.2.3 can be applied to $\pm U - Cz$. For $C \geq |L^h U|/b_\phi$ we have $L^h(\pm U - Cz) \leq 0$; and for $C \geq |U_0|$ and $C \geq \frac{1}{2}|U_J|$ we have $(\pm U - Cz)_0 \leq 0$, $(\pm U - Cz)_J \leq 0$. Hence from the lemma we can deduce that $|U_j| \leq Cz_j$; specifically, in terms of the supremum norm we have

$$\|U\|_\infty \leq C\|z\|_\infty = \max\{(2/b_\phi)\|L^h U\|_\infty, 2|U_0|, U_J\}, \tag{3.25}$$

so proving the assertion made above. In addition, this result will ensure that the $O(h^4)$ truncation error leads to an $O(h^4)$ error in U as $h \to 0$, justifying our description of these schemes as fourth order accurate.

In Berger *et al.* (1980) a family of OCI schemes are considered in which each of the q coefficients is expanded as a polynomial in the local mesh

Péclet number $\beta_j = b_j h / \epsilon$. Conditions on these polynomials are then derived to ensure that (3.23) together with other desirable properties are satisfied. The simplest such scheme, that nevertheless seems to perform as well as other choices, has the following form (assuming $c = 0$)

$$q_j^- : q_j : q_j^+ = 6 + 6\beta_j + 3\beta_j^2 + r_4 \beta_j^3$$

$$: 60 + 30\beta_j + r_3 \beta_j^2 + r_4 \beta_{j-1}\beta_j^2 : 6. \qquad (3.26a)$$

The coefficients r_3 and r_4 are chosen to cope with the variation in b. If we make the reasonable assumption, essentially on the size of h, that

$$2b_j \geq \max(b_{j-1}, b_{j+1}), \qquad (3.26b)$$

then these coefficients reduce to

$$r_3 = \frac{3}{b_j}(4b_j - b_{j+1}) + \frac{1}{32 b_j^2}(3b_{j+1} - 2b_j - b_{j-1})_+^2,$$

$$r_4 = \tfrac{1}{2} \frac{r_3 b_j - 3 b_{j-1}}{b_j + b_{j-1}} \qquad (3.26c)$$

where $(s)_+$ denotes $\max(0, s)$. These are rather complicated but they have been derived in order to achieve the following desirable result.

Theorem 3.2.4 *The difference scheme given by the coefficients in (3.26) and (3.12) is fourth order accurate, as $h \to 0$ for fixed ϵ, and satisfies the maximum principle conditions (3.23) for all ϵ and h, so long as (3.26b) is satisfied and $b(x)$ is sufficiently smooth.*

Proof. It is easily checked that (3.26b) implies that $r_3 \geq 6$ and hence $r_4 \geq 0$, and therefore all of the q coefficients are non-negative. Computation of p_j^+ from (3.12c) gives after considerable simplification

$$p_j^+ = -\epsilon[72 - 3(3\beta_{j+1} - 2\beta_j - \beta_{j-1})$$

$$+ \tfrac{1}{32}(3\beta_{j+1} - 2\beta_j - \beta_{j-1})_+^2] / \sum_j \leq 0, \qquad (3.27)$$

where $\sum_j$ is the sum of the quantities on the right of (3.26a), and the sign of this quadratic form is ensured by the addition of the $(\cdot)_+^2$ term in the expression (3.26c) for r_3 and r_4. The fact that $p_j^- < 0$ follows simply from (3.12b) and the assumption that $b_{j+1} \leq 2b_j$. Thus all the conditions of (3.23) are satisfied.

To check fourth order accuracy, we write T_j^3 and T_j^4 in terms of the q coefficients,

$$T_j^3 = \epsilon h[q_j^+ - q_j^- - \tfrac{1}{3}(q_j^- \beta_{j-1} + q_j^+ \beta_{j+1} - \tfrac{1}{2}q_j\beta_j)] \qquad (3.28a)$$

$$T_j^4 = \tfrac{1}{12}\epsilon h^2[5(q_j^- + q_j^+) - q_j + q_j^- \beta_{j-1} - q_j^+ \beta_{j+1}]. \qquad (3.28b)$$

Substituting from (3.26a) and expanding to $O(h^4)$ we obtain

$$T_j^3 = \tfrac{1}{36}\epsilon h[2\beta_j - \beta_{j-1} - \beta_{j+1} + 2\beta_j(\beta_j - \beta_{j-1}) + O(h^3)]$$

$$T_j^4 = \tfrac{1}{144}\epsilon h^2[\beta_{j-1} - \beta_{j+1} + O(h^2)].$$

Hence both T_j^3 and T_j^4 are $O(h^4)$ if $b(x) \in C^2[0,1]$. $\quad\square$

Thus the generalised OCI scheme shares with the simple upwind scheme (1.49) the property of satisfying a maximum principle for all values of ϵ and all reasonable values of h; but with no more points it manages to achieve fourth order accuracy for a fixed ϵ. As we shall see in section 3.4, this cannot be maintained uniformly over all values of ϵ, and no error bound holds of the form Ch^δ with C independent of ϵ and $\delta > 0$, even though with $\epsilon \leq Ch^{\frac{3}{2}}$ one does get second order accuracy. Moreover, these properties have been achieved at the cost of considerable complexity in the coefficients of the scheme, especially when b is not constant. This bodes ill for nonlinear problems.

There is some simplification where b is a constant, for then r_3 and r_4 reduce to 9 and $\frac{3}{2}$ respectively. As a result we find that

$$p^-/p^+ = 1 + \beta + \tfrac{1}{2}\beta^2 + \tfrac{1}{6}\beta^3 + \tfrac{1}{24}\beta^4, \tag{3.29}$$

the $(4,0)$ Padé approximant to the exponential e^β, which is consistent with the fourth order accuracy. In section 4.6 some numerical results will be given both for the constant b case and for a variable b problem, where results for comparable finite element methods will also be given.

Finally, we should recall from Lemma 3.2.1 that it is easy to extend the above scheme to the full equation (3.1a) with $c \neq 0$. On the other hand, note that $p_j^\pm$ may become positive so that the maximum principle may be lost. This is not particularly serious as we shall see when we consider finite element methods; and in any case it could be restored by increasing r_3, so long as $hc_{j-1} < 2b_{j-1}$ — see Berger et al. (1980).

3.3 Exponential fitting and locally exact schemes

3.3.1 Exponentially fitted schemes

A very different approach to approximating the convection-diffusion operator was adopted by Allen and Southwell (1955), and later independently by Il'in (1969). The former authors were concerned with developing methods to approximate incompressible viscous flow about a cylinder. They used the stream function and potential from the potential flow solution to provide a coordinate system (s, t), as shown in Fig.3.1. Then, in terms of the true stream function ψ and vorticity ζ, the incompressible Navier–Stokes equations become (c.f. section 1.1)

$$\frac{\partial^2\psi}{\partial t^2} + \text{Re }\frac{\partial^2\psi}{\partial s^2} = \frac{\zeta}{\Delta^2} \tag{3.30a}$$

$$\frac{\partial^2 \zeta}{\partial t^2} + \frac{\partial \psi}{\partial s}\frac{\partial \zeta}{\partial t} + \left[\frac{1}{Re}\frac{\partial^2 \zeta}{\partial s^2} - \frac{\partial \psi}{\partial t}\frac{\partial \zeta}{\partial s}\right] = 0, \qquad (3.30b)$$

where Re is the Reynolds number and Δ a scale length. The crucial operator here is that in square brackets, with the other terms in (3.30b) treated as source terms. Thus with their choice of coordinate system Allen and Southwell have encapsulated the essentials of the problem in a one-dimensional convection-diffusion problem.

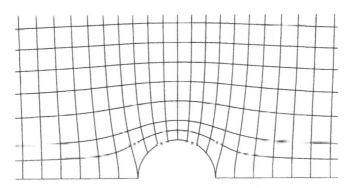

Figure 3.1. *The (s, t) coordinate system used by Allen and Southwell (1955)*

They then argued that as the solution to the homogeneous problem was exponential in form, the difference scheme should be chosen to be exact for that exponential, rather than for a polynomial. Thus suppose one uses a combination of central and upwind differencing for the convection term, and chooses the mixing parameter so that the characteristic equation for the scheme has roots 1 and e^β, where β is the usual mesh Reynolds number. In terms of our standard problem (3.1), with $c = 0$, and with ξ substituted for α in the notation introduced for (3.17), we get

$$-\frac{\epsilon}{h^2}\delta^2 U_j + \frac{b_j}{h}[(1 - \xi)\Delta_0 + \xi\Delta_-]U_j = S_j. \qquad (3.31a)$$

Alternatively, since $\Delta_- = \Delta_0 - \frac{1}{2}\delta^2$, we can write this as

$$-\left(\frac{\epsilon}{h^2} + \frac{\xi b_j}{2h}\right)\delta^2 U_j + \frac{b_j}{h}\Delta_0 U_j = S_j. \qquad (3.31b)$$

Thus the upwinding introduced with the parameter ξ is equivalent to enhancing the diffusion coefficients ϵ by the factor $1 + \frac{1}{2}\xi\beta_j$, where $\beta_j = b_j h/\epsilon$. Writing out the scheme explicitly we have

$$- [1 + \tfrac{1}{2}(\xi + 1)\beta_j]U_{j-1} + [2 + \xi\beta_j]U_j$$
$$- [1 + \tfrac{1}{2}(\xi - 1)\beta_j]U_{j+1} = \frac{h^2}{\epsilon}S_j, \qquad (3.31c)$$

which gives the characteristic equation

$$[1 + \tfrac{1}{2}(\xi - 1)\beta_j]\lambda^2 - [2 + \xi\beta_j]\lambda + [1 + \tfrac{1}{2}(\xi + 1)\beta_j] = 0. \tag{3.32}$$

One root is unity and the other we equate to e^{β_j},

$$[1 + \tfrac{1}{2}(\xi + 1)\beta_j] = e^{\beta_j}[1 + \tfrac{1}{2}(\xi - 1)\beta_j]$$

$$\text{i.e.,} \quad 1 + \tfrac{1}{2}\xi\beta_j = \tfrac{1}{2}\beta_j(e^{\beta_j} + 1)/(e^{\beta_j} - 1) = \tfrac{1}{2}\beta_j \coth \tfrac{1}{2}\beta_j \tag{3.33}$$

$$\text{i.e.,} \quad \xi = \coth \tfrac{1}{2}\beta_j - 2/\beta_j. \tag{3.34}$$

Hence the final scheme becomes

$$-\left(\frac{b_j}{2h} \coth \tfrac{1}{2}\beta_j\right) \delta^2 U_j + \frac{b_j}{h}\Delta_0 U_j = S_j, \tag{3.35a}$$

in which all apparent dependence on ϵ has disappeared: explicitly this is

$$\frac{b_j}{2h}[-(\coth \tfrac{1}{2}\beta_j + 1)U_{j-1} + 2(\coth \tfrac{1}{2}\beta_j)U_j$$

$$- (\coth \tfrac{1}{2}\beta_j - 1)U_{j+1}] = S_j \tag{3.35b}$$

or

$$\frac{b_j}{1 - e^{-\beta_j}} \left[-U_{j-1} + (1 + e^{-\beta_j})U_j - e^{-\beta_j}U_{j+1}\right] = hS_j, \tag{3.35c}$$

where it is clear that the coefficients satisfy the desirable properties (3.23).

The same scheme was proposed by Il'in (1969) who proved that it satisfied an error bound of the form Ch with C independent of h and ϵ. Indeed, we shall show in the next section that the use of exponential coefficients is necessary to achieve such a bound. Otherwise there is typically some combination of ϵ and h for which the error does not decrease as both are reduced.

Subsequently, El-Mistikawy and Werle (1978) proposed an exponentially fitted scheme which uses some of the flexibility in choice of coefficient that was exploited with the OCI schemes. Introducing $b_j^{\pm} := \tfrac{1}{2}(b_j + b_{j\pm1})$ and corresponding Péclet numbers $\beta_j^{\pm}$, we extend the notation used for the OCI schemes by setting $\tilde{p}_j^{\pm} = p_j^{\pm}/h^2$ to allow later generalisation to nonuniform meshes; then without the normalisation to give (3.9), this scheme is given by

$$\tilde{p}_j^- = -\frac{b_j^-}{(1 - e^{-\beta_j^-})h}, \quad \tilde{p}_j^+ = -\frac{b_j^+ e^{-\beta_j^+}}{(1 - e^{-\beta_j^+})h}, \quad p_j = -(p_j^- + p_j^+) \tag{3.36a}$$

$$q_j^- = \tfrac{1}{2}\left[\frac{1}{1 - e^{-\beta_j^-}} - \frac{1}{\beta_j^-}\right], q_j^+ = \tfrac{1}{2}\left[\frac{1}{\beta_j^+} - \frac{e^{-\beta_j^+}}{1 - e^{-\beta_j^+}}\right], q_j = q_j^- + q_j^+. \tag{3.36b}$$

Again the conditions (3.23) are satisfied, with the p coefficients only differing from those of (3.35) when b is not constant, and the q coefficients tending to give equal weight to S_j and S_{j-1} as $\beta_j \to \infty$. This scheme satisfies an error bound of the form Ch^2, with C independent of h and ϵ, which is the highest order achievable for a uniform error bound by a three-point scheme.

3.3.2 Locally exact schemes

These two schemes based on exponential fitting are examples of a wider class of *locally exact schemes* studied by Samarskii (1977), Stoyan (1979), (1984) , Stynes and O'Riordan (1986) and others. Suppose we introduce the Green's function for the operator L of (3.1a), with Dirichlet boundary conditions, on the double interval (x_{j-1}, x_{j+1}). We denote it by $G_j(\cdot, \cdot)$ and write

$$(G_j f)(x) := \int_{x_{j-1}}^{x_{j+1}} G_j(x, \eta) f(\eta) d\eta. \tag{3.37}$$

The problem $Lu = S$ with $u(x_{j-1}) = u_{j-1}$ and $u(x_{j+1}) = u_{j+1}$ could be solved by substituting the Green's function into a general formula of the form given in (2.73); but it is rather simpler to rewrite the problem with homogeneous boundary conditions by setting

$$y(x) := u(x) - u_{j-1}\Phi_j^-(x) - u_{j+1}\Phi_j^+(x), \tag{3.38a}$$

$$\tag{3.38b}$$

where

$$\Phi_j^-(x) = \frac{x_{j+1} - x}{x_{j+1} - x_{j-1}}, \quad \Phi_j^+(x) = \frac{x - x_{j-1}}{x_{j+1} - x_{j-1}}. \tag{3.38c}$$

It then takes the form

$$Ly = S - u_{j-1}L\Phi_j^- - u_{j+1}L\Phi_j^+, \quad y(x_{j-1}) = y(x_{j+1}) = 0 \tag{3.39}$$

and therefore has the solution

$$y = G_j S - u_{j-1}G_j L\Phi_j^- - u_{j+1}G_j L\Phi_j^+. \tag{3.40}$$

Substituting for y from (3.38) then gives u in terms of its data S, u_{j-1} and u_{j+1}; and putting $x = x_j$ gives the locally exact difference scheme satisfied by the three nodal values u_{j-1}, u_j and u_{j+1}.

This can be the basis of a variety of approximate difference schemes which are in some sense locally exact; and note that there is no need to assume a uniform mesh — it can be quite arbitrary. Generally speaking there are three kinds of approximation involved in deriving a practical difference scheme:

(i) taking the local values $b(x_j), c(x_j)$ of the coefficients in L, or some averages of them, when computing the local Green's function;

(ii) splitting the Green's function into separate parts, or in particular, treating the term $c(x)u$ specially by writing $(L - c)u = S - cu$;

(iii) approximating the operator G_j, especially when applied to the source function S.

The Allen and Southwell scheme makes use of all three approximations, with the simplest choice in each case, while the El Mistikawy and Werle scheme takes greater account of the variation of $b(x)$ in approximating G_j and of $S(x)$ in $G_j S$.

As was seen in section 2.7.1, the Green's function can be composed from two solutions to the equation $Lu = 0$, $u^L(x)$ which satisfies $u^L(x_{j-1}) = 0$ and $u^R(x)$ satisfying $u^R(x_{j+1}) = 0$, together with their Wronskian $W^u(x)$. Then from (2.78), (2.80) and (2.84) we obtain

$$G_j(x, \eta) = \begin{cases} Cu^L(x)u^R(\eta)/W^u(\eta) \text{ for } x \le \eta & \text{(3.41a)} \\[2mm] Cu^L(\eta)u^R(x)/W^u(\eta) \text{ for } \eta \le x, & \text{(3.41b)} \end{cases}$$

where the constant C is determined by the condition that the jump in $\partial G_j / \partial \eta$ at $\eta = x$ is $-1/\epsilon$. Suppose we separate out the term $c(x)u$ and denote the Green's function when $c(x) \equiv 0$ by $G_j^0(\cdot, \cdot)$. Then a formula which is entirely equivalent to (3.40) is the following

$$u - u_{j-1}\Phi_j^- - u_{j+1}\Phi_j^+ + G_j^0(L - c)(u_{j-1}\Phi_j^- + u_{j+1}\Phi_j^+)$$
$$= G_j^0(S - cu). \qquad (3.42)$$

Since $\Phi_j^- + \Phi_j^+ \equiv 1$ and $(L - c)\Phi_j^{\pm} = \pm b(x)/(x_{j+1} - x_{j-1})$, we can simplify this relation and then set $x = x_j$ to obtain a useful basis for deriving difference schemes. Denoting by G_{jj}^0 the operator obtained from $G_j^0(x_j, \cdot)$, we get

$$\frac{u_j - u_{j-1}}{x_{j+1} - x_{j-1}} \left(x_{j+1} - x_j + G_{jj}^0 b \right)$$
$$+ \frac{u_j - u_{j+1}}{x_{j+1} - x_{j-1}} \left(x_j - x_{j-1} - G_{jj}^0 b \right) = G_{jj}^0(S - cu). \qquad (3.43)$$

This form ensures that, in our usual notation for the coefficients of a difference scheme, we have $p_j = -(p_j^- + p_j^+)$ apart from the effect of the cu term.

It is also easier to construct approximations to the Green's function $G_j^0(x_j, \cdot)$ than to the more general $G_j(x_j, \cdot)$. Introducing $B(x)$ as an indefinite integral of $b(x)$, the two solutions of $-\epsilon u'' + bu' = 0$ can be written as

$$u^L(x) = \int_{x_{j-1}}^{x} e^{B/\epsilon} ds, \quad u^R(x) = \int_{x}^{x_{j+1}} e^{B/\epsilon} ds. \qquad (3.44a)$$

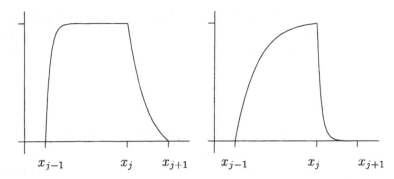

Figure 3.2. *The Green's function $G_j^0(x_j, \eta)$ for constant b corresponding to $\beta = 10$ in (x_{j-1}, x_j) and $\beta = 5$ in (x_j, x_{j+1}), on the left; the values of β are interchanged, on the right*

Their sum is constant, $u^L(x) + u^R(x) = E$ say, and their Wronskian is therefore given by

$$W^u(x) = Ee^{B(x)/\epsilon}. \tag{3.44b}$$

As a function of η, G_j^0 satisfies the adjoint equation $(\epsilon v' + bv)' = 0$; more specifically, from (3.41) and (3.44) we obtain

$$\epsilon G_j^{0'} + bG_j^0 = \begin{cases} -(C\epsilon/E)u^L(x) & \text{for } x \leq \eta \tag{3.45a} \\[2em] (C\epsilon/E)u^R(x) & \text{for } \eta \leq x, \tag{3.45b} \end{cases}$$

with the jump condition at $\eta = x$ determining $C = 1/\epsilon$; see Fig. 3.2 for examples. From the relations (3.45) we can also calculate $G_{jj}^0 b$ exactly. A little algebra leads to

$$G_{jj}^0 b = [(x_j - x_{j-1})u^R(x_j) - (x_{j+1} - x_j)u^L(x_j)]/E, \tag{3.46}$$

from which we can rewrite (3.43) as

$$(u_j - u_{j-1})\frac{u^R(x_j)}{u^L(x_j) + u^R(x_j)}$$

$$+ (u_j - u_{j+1})\frac{u^L(x_j)}{u^L(x_j) + u^R(x_j)} = G_{jj}^0(S - cu). \tag{3.47}$$

It is worth noting here that this could have been obtained directly from the Green's function formula (2.73) by deducing from (3.45) that

$$\epsilon G_j^{0'}(x_j, 0) = u^R(x_j)/E, \quad \epsilon G_j^{0'}(x_j, 1) = -u^L(x_j)/E. \tag{3.48}$$

The crucial simplification that enables one to dispense with the rewriting of the original problem as a homogeneous problem in (3.39) and (3.40) is the use of G_j^0 instead of G_j.

In order to turn (3.47) into a difference scheme in normalised form we need to divide throughout by $G_{jj}^0(1)$ and approximate all of the integrals involved. Splitting the integral for $G_{jj}^0(1)$ into that over the left and right intervals, we can write

$$
\begin{aligned}
\epsilon E G_{jj}^0(1) &= u^R(x_j) \int_{x_{j-1}}^{x_j} \int_{x_{j-1}}^{\eta} e^{(1/\epsilon)[B(s)-B(\eta)]} ds d\eta \\
&\quad + u^L(x_j) \int_{x_j}^{x_{j+1}} \int_{\eta}^{x_{j+1}} e^{(1/\epsilon)[B(s)-B(\eta)]} ds d\eta \\
&=: u^R(x_j) E^L + u^L(x_j) E^R,
\end{aligned}
\tag{3.49}
$$

so defining E^L and E^R. Then the correctly normalised form for (3.47) is, after dropping the argument x_j in $u^L(x_j)$ and $u^R(x_j)$,

$$
\begin{aligned}
(u_j - u_{j-1})\frac{\epsilon u^R}{u^R E^L + u^L E^R} &+ (u_j - u_{j+1})\frac{\epsilon u^L}{u^R E^L + u^L E^R} \\
&= \frac{G_{jj}^0(S - cu)}{G_{jj}^0(1)}.
\end{aligned}
\tag{3.50}
$$

To approximate the integrals in (3.49), suppose b is constant over each interval, taking the values $b_j^{\pm}$, and denote by $h_j^{\pm}$ the interval lengths and by $\beta_j^{\pm}$ the corresponding mesh Péclet numbers. This gives

$$
u^L = \frac{\epsilon}{b_j^-}(1 - e^{-\beta_j^-}), \quad u^R = \frac{\epsilon}{b_j^+}(e^{\beta_j^+} - 1)
\tag{3.51a}
$$

$$
E^L = \frac{\epsilon h_j^-}{b_j^-}\left[1 - \frac{1}{\beta_j^-}(1 - e^{-\beta_j^-})\right], E^R = \frac{\epsilon h_j^+}{b_j^+}\left[\frac{1}{\beta_j^+}(e^{\beta_j^+} - 1) - 1\right].
\tag{3.51b}
$$

Finally, use of the G_j^0-weighted trapezoidal rule over each interval to approximate $G_{jj}^0(S - cu)$ gives

$$
q_j^- = \tfrac{1}{2}\frac{u^R E^L}{u^R E^L + u^L E^R}, \quad q_j^+ = \tfrac{1}{2}\frac{u^L E^R}{u^R E^L + u^L E^R}
\tag{3.52}
$$

and $q_j = q_j^- + q_j^+$. To compare with the El-Mistikawy and Werle scheme of (3.36), we use the same notation and drop the normalisation to obtain from (3.50) and (3.52)

$$
-2q_j^-/\bar{p}_j^- = E_L/\epsilon \quad \text{and} \quad -2q_j^+/\bar{p}_j^+ = E_R/\epsilon.
$$

Substitution from (3.51), and setting $h_j^{\pm} = h$, gives exactly the same expressions as in (3.36). Thus we can regard this scheme as being derived from (3.50) by taking the constant values $b_j^{\pm} = \tfrac{1}{2}(b_j + b_{j\pm1})$ in approximating the integrals of (3.49), and by using the compound trapezoidal rule on the right-hand side. Similarly, the Allen and Southwell scheme could be derived by taking $b_j^{\pm} = b_j$ and using midpoint quadrature for the right-hand side. Other schemes are clearly easy to derive from this general form.

3.4 Error analysis by a maximum principle

We consider schemes in the OCI notation (3.7), where the coefficients satisfy the conditions (3.23) so that the maximum principle set out in Lemma 3.2.3 holds. The analysis technique is based on the use of comparison functions, as developed by Kellogg and Tsan (1978). We summarise a key part in the following two lemmas.

Lemma 3.4.1 *For a difference operator* (3.24) *satisfying* (3.23a), *if* $L^h U_j \leq L^h V_j$ *for* $j = 1, 2, \ldots, J - 1$ *and* $U_0 \leq V_0$ *and* $U_J \leq V_J$, *then* $U_j \leq V_j$ *for* $j = 1, 2, \ldots, J - 1$.

Proof. As in Lemma 3.2.3, suppose the result is untrue. Then for some $k \in \{1, 2, \ldots, J-1\}$ the difference $U_k - V_k$ attains a maximum value which is positive. But from (3.23a) and the hypotheses, $U_k - V_k \leq U_{k-1} - V_{k-1}$; and hence $U_k - V_k = U_{k-1} - V_{k-1} = \ldots = U_0 - V_0 \leq 0$, which is a contradiction. □

Denoting the nodal error by $e_j := U_j - u_j$, with $e_0 = e_J = 0$, assuming henceforth that (3.23) holds, and using a corresponding notation to the $L^h U$ of (3.24) on the right of the difference scheme, we write

$$L^h U_j = R^h S_j, \quad j = 1, 2, \ldots, J - 1. \tag{3.53}$$

Then from the definition (3.10a) of the truncation error, $T_j := L^h u_j - R^h S_j$, we have

$$L^h e_j + T_j = 0, \quad j = 1, 2, \ldots, J - 1. \tag{3.54}$$

Lemma 3.4.2 *For a given difference scheme* (3.53), *suppose there is a set of non-negative functions* $y^{(i)}, z^{(i)}$ *and positive constants* $\rho^{(i)}$ *such that*

$$L^h z^{(i)} \geq \rho^{(i)} y^{(i)} \quad \forall i. \tag{3.55}$$

Suppose also that the truncation error satisfies the bound

$$|T_j| \leq \sum_{(i)} C^{(i)} y_j^{(i)} \quad \forall j \tag{3.56}$$

for certain positive constants $C^{(i)}$. *Then we have*

$$|e_j| \leq \sum_{(i)} (C^{(i)}/\rho^{(i)}) z_j^{(i)} \quad \forall j. \tag{3.57}$$

Proof. From (3.55), (3.54) and (3.56) we have

$$L^h[\sum_{(i)} (C^{(i)}/\rho^{(i)}) z_j^{(i)} \pm e_j] \geq 0. \tag{3.58}$$

The quantity in square brackets is also non-negative at $j = 0$ and $j = J$. Hence the result follows from Lemma 3.4.1. □

3.4.1 The simple upwind scheme

We start the analysis with the simple upwind scheme on a uniform mesh of size h

$$-\frac{\epsilon}{h^2}\delta^2 U_j + \frac{b_j}{h}\Delta_- U_j + c_j U_j = S_j, \quad j = 1, 2, \ldots, J-1 \qquad (3.59)$$

with $U_0 = u_L$ and $U_J = u_R$. A major objective is to estimate the truncation error sufficiently carefully that the dependence of the global error on both ϵ and h is given by the bounds in Lemma 3.4.2.

First we state more carefully the assumptions needed for the problem (3.1) which the schemes approximate, namely

$$b(x) \geq b_\phi > 0 \quad c(x) \geq 0 \quad \forall x \in [0, 1]; \qquad (3.60a)$$

and for some integer $p \geq 1$, and for $0 \leq s \leq p$ and some $K > 0$,

$$b(x), c(x) \in C^p[0, 1]$$

and

$$|S^{(s)}(x, \epsilon)| \leq K \left[1 + \frac{1}{\epsilon^s}e^{-(b_\phi/2\epsilon)(1-x)}\right]. \qquad (3.60b)$$

The assumptions, particularly that on S, in which K is independent of ϵ, have been given before in section 2.2 where they were used to establish *a priori* bounds on the solution u and its derivatives; (3.60b) is sufficient for us to make use of both Theorem 2.4.1 and Theorem 2.4.3. In the following, C will denote a generic constant independent of h and ϵ.

Theorem 3.4.3 *For the upwind scheme (3.59), approximating a problem (3.1) which satisfies (3.60a) and (3.60b) with $p = 1$, we have*

$$|U_j - u_j| \leq Ch\left[1 + \frac{1}{\epsilon}\mu^{-(J-j)}\right] \quad \text{if } h \leq \epsilon, \qquad (3.61a)$$

$$|U_j - u_j| \leq C\left[h + \mu^{-(J-j)}\right] \quad \text{if } \epsilon \leq h, \qquad (3.61b)$$

where $\mu = 1 + \frac{1}{2}b_\phi h/\epsilon$ and C is independent of h and ϵ.

Proof. To estimate the truncation error we shall need the following standard and easily derived results; if $y(x) \in C^3[0, 1]$ then

$$D_- y(x) - y'(x) = -\frac{1}{h}\int_{x-h}^{x}(s + h - x)y''(s)ds \qquad (3.62a)$$

$$= -\tfrac{1}{2}hy''(x) + \tfrac{1}{2}\frac{1}{h}\int_{x-h}^{x}(s + h - x)^2 y'''(s)ds \qquad (3.62b)$$

$$D_+ y(x) - y'(x) = -\frac{1}{h}\int_{x}^{x+h}(s - h - x)y''(s)ds \qquad (3.63a)$$

$$= \tfrac{1}{2}hy''(x) + \tfrac{1}{2}\frac{1}{h}\int_x^{x+h}(s-h-x)^2y'''(s)\mathrm{d}s.$$

(3.63b)

By subtracting the second expressions here, (3.62b) and (3.63b), we get

$$D_h^2 y(x) - y''(x) = \tfrac{1}{2}\frac{1}{h^2}\left[\int_x^{x+h}(s-h-x)^2y'''(s)\mathrm{d}s\right.$$

$$\left. - \int_{x-h}^x (s+h-x)^2y'''(s)\mathrm{d}s\right] \quad (3.64a)$$

$$= \tfrac{1}{6}\frac{1}{h^2}\left[\int_{x-h}^x (s+h-x)^3y^{iv}(s)\mathrm{d}s - \int_x^{x+h}(s-h-x)^2y^{iv}(s)\mathrm{d}s\right],$$

(3.64b)

with the last result requiring $y(\cdot) \in C^4[0,1]$. Hence we have

$$T_j - L^h u_j - S_j = L^h u_j - (Lu)_j$$
$$= -\epsilon[D_h^2 u_j - u_j''] + b_j[D_- u_j - u_j'], \quad (3.65)$$

and thus from (3.62a) and (3.64a)

$$|T_j| \le C \int_{x_{j-1}}^{x_{j+1}}[\epsilon|u'''(s)| + |u''(s)|]\mathrm{d}s. \quad (3.66)$$

This can be used immediately for the case $h \le \epsilon$ by using the *a priori* bounds on u'' and u''' given by Theorem 2.4.1. We also use the fact that in this case, if $\beta_\phi = b_\phi h/\epsilon$, then $\sinh \tfrac{1}{2}\beta_\phi \le C\beta_\phi$. So we get, since $\mu \le e^{\beta_\phi/2}$,

$$|T_j| \le C\int_{x_{j-1}}^{x_{j+1}}\left[1 + \frac{1}{\epsilon^2}e^{-(b_\phi/2\epsilon)(1-x)}\right]\mathrm{d}x$$

$$= C\left[h + \frac{1}{\epsilon}\sinh\tfrac{1}{2}\beta_\phi e^{-(b_\phi/2\epsilon)(1-x_j)}\right]$$

$$\le Ch\left[1 + \frac{1}{\epsilon^2}\mu^{-(J-j)}\right]. \quad (3.67)$$

Thus to apply Lemma 3.4.2 we use comparison functions $(1 + x_j)$ and $\mu^{-(J-j)}$. It is clear that $L^h(1 + x_j) \ge b_\phi \ge \tfrac{1}{2}b_\phi(1 + x_j)$; and also because μ is a root of the characteristic equation for the scheme applied to a problem with $b = \tfrac{1}{2}b_\phi$ and $c = 0$,

$$L^h \mu^{-(J-j)} = \mu^{-J}[(b_j - \tfrac{1}{2}b_\phi)(\mu^j - \mu^{j-1})/h + c_j\mu^j]$$
$$\ge \tfrac{1}{2}b_\phi\mu^{-(J-j)}(\mu - 1)/\mu h$$
$$= \tfrac{1}{2}b_\phi^2\mu^{-(J-j)}/(2\epsilon + b_\phi h). \quad (3.68)$$

To cover the two cases, $h \le \epsilon$ and $\epsilon \le h$, we deduce from (3.68) the important result

$$L^h \mu^{-(J-j)} \geq \frac{\frac{1}{2}b_\phi^2}{2 + b_\phi} \frac{1}{\max(\epsilon, h)} \mu^{-(J-j)}. \tag{3.69}$$

Continuing consideration of the case $h \leq \epsilon$, we apply Lemma 3.4.2 to (3.67), using the fact that $1 \leq 1 + x_j$, and get (3.61a).

For the case $\epsilon \leq h$ we use the decomposition $u = \gamma v + z$ into boundary layer and smooth terms given in Theorem 2.4.3, with the corresponding decomposition $U = \gamma V + Z$ determined from setting $L^h V_j = (Lv)_j$ etc. We shall also need to use the inequality

$$\beta^m \leq C_m e^\beta \tag{3.70}$$

which holds for finite m. Using it for $m = 2$ to deduce that $(h/\epsilon)\mu \leq Ce^{\beta_J}$ and noting that $v_j = e^{-\beta_J(J-j)}$, we get for $j < J$

$$|L^h V_j| = |(Lv)_j| = \epsilon^{-1}| - b_J(b_J - b_j) + \epsilon c_j|e^{-\beta_J(J-j)}$$
$$\leq C\epsilon^{-1} e^{-\beta_J(J-j)} \leq Ch^{-1}\mu^{-(J-j)}. \tag{3.71}$$

We can apply Lemma 3.4.1 to this now, using (3.69) with $\epsilon \leq h$ to get

$$|V_j| \leq C\mu^{-(J-j)}. \tag{3.72}$$

It is clear that $|v_j| \leq \mu^{-(J-j)}$ and hence $|V_j - v_j| \leq C\mu^{-(J-j)}$, which leads to the second term in the bound (3.61b). We apply the truncation error argument used above to z and Z in order to get the first part of the bound in (3.61b): using Theorem 2.4.3 instead of Theorem 2.4.1,

$$|L^h(Z - z)_j| \leq C \int_{x_{j-1}}^{x_{j+1}} \left[\epsilon|z'''(s)| + |z''(s)|\right] ds$$
$$\leq C \int_{x_{j-1}}^{x_{j+1}} \left[1 + \frac{1}{\epsilon}e^{-(b_\phi/2\epsilon)(1-s)}\right] ds$$
$$= 2C \left[h + \sinh\frac{1}{2}\beta_\phi e^{-(b_\phi/2\epsilon)(1-x_j)}\right]$$
$$\leq C \left[h + \mu^{-(J-j-1)}\right]. \tag{3.73}$$

Applying Lemma 3.4.2 with the bound (3.49), we get

$$|Z_j - z_j| \leq Ch \left[1 + \mu^{-(J-j-1)}\right] \leq Ch \tag{3.74}$$

as required. □

The bounds in the theorem correctly reflect the behaviour observed in numerical experiments — see, for instance, Berger *et al.* (1980). For fixed ϵ the errors are $O(h)$ as $h \to 0$; but if we take $\epsilon = h$ and let both go to zero, the maximum error, although it changes its position, does not reduce in magnitude. On the other hand if $\epsilon = O(h^2)$ the error is again $O(h)$ as

$h \to 0$, which is reflected in (3.61b) since $J - j \geq 1$ and $\mu^{-1} = O(\epsilon/h)$ when $\epsilon \leq h$.

As was mentioned in Chapter 1, a modification of this scheme which is more appropriate for high Péclet numbers is the following,

$$-\frac{\epsilon}{h^2}\delta^2 U_j + \frac{b_j + b_{j-1}}{2h}\Delta_- U_j + \tfrac{1}{2}(c_j U_j + c_{j-1}U_{j-1}) = \tfrac{1}{2}(S_j + S_{j-1}).$$

$$(3.75)$$

This was proposed and studied in Abrahamsson *et al.* (1974) where error bounds of the same type as those in Theorem 3.4.3 were established; the same behaviour occurs in the boundary layer but the error is $O(h^2)$ outside it, so that the first terms on the right-hand side of both (3.61a) and (3.61b) become Ch^2.

3.4.2 The Allen and Southwell scheme

Next we consider the Allen and Southwell scheme (3.35), generalised to include a $c_j U_j$ term.

Theorem 3.4.4 *For the Allen and Southwell scheme (3.35) approximating (3.1), and with $p = 1$ in (3.60b), we have the uniform bound*

$$\|U - u\|_\infty \leq Ch$$

$$(3.76a)$$

with C independent of h and ϵ. Furthermore we also have, if (3.60b) is satisfied with $p = 2$, the sharper bound

$$|U_j - u_j| \leq C\left[\frac{h^2}{h+\epsilon} + \frac{h^2}{\epsilon}e^{-(b_\phi/2\epsilon)(1-x_j)}\right].$$

$$(3.76b)$$

Proof. With this scheme we can use the decomposition $u = \gamma v + z$ and $U = \gamma V + Z$ of Theorem 2.4.3 in all cases. We now take $\mu = e^{\beta_\phi/2}$ and as with Theorem 3.4.3 we need a bound of the form (3.69). Since again μ is a root of the characteristic equation for the scheme applied to a problem with $b = \tfrac{1}{2}b_\phi$ and $c = 0$, we obtain

$$L^h\mu^j = \left[-\frac{1}{2h}(b_j \coth \tfrac{1}{2}\beta_j - \tfrac{1}{2}b_\phi \coth \tfrac{1}{4}\beta_\phi)\frac{(\mu - 1)^2}{\mu}\right.$$

$$\left. + \frac{b_j - \tfrac{1}{2}b_\phi}{h}\frac{\mu^2 - 1}{2\mu} + c_j\right]\mu^j.$$

But

$$\tfrac{1}{2}\beta_\phi < \beta_j \Rightarrow \coth \tfrac{1}{4}\beta_\phi - \coth \tfrac{1}{2}\beta_j = \sinh \tfrac{1}{2}(\beta_j - \tfrac{1}{2}\beta_\phi)/ \sinh \tfrac{1}{2}\beta_j \sinh \tfrac{1}{4}\beta_\phi,$$

which is positive, so we have

$$L^h \mu^j \geq \frac{b_j - \frac{1}{2}b_\phi}{2h} \frac{(\mu - 1)^2}{\mu} \left[-\coth \frac{1}{2}\beta_j + \frac{\mu + 1}{\mu - 1} \right] \mu^j$$

$$= 2h^{-1}(b_j - \frac{1}{2}b_\phi) \sinh^2 \frac{1}{2}\beta_\phi [\coth \frac{1}{4}\beta_\phi - \coth \frac{1}{2}\beta_j]\mu^j$$

$$\geq h^{-1}b_\phi \sinh^2 \frac{1}{2}\beta_\phi [\sinh \frac{1}{2}(\beta_j - \frac{1}{2}\beta_\phi)/ \sinh \frac{1}{2}\beta_j \sinh \frac{1}{4}\beta_\phi]\mu^j. \quad (3.77)$$

When $h \leq \epsilon$, we use the bounds $C_1 t \leq \sinh t \leq C_2 t$ for $0 \leq t \leq C_3$ to bound the terms in square brackets by $C(\epsilon/h)$; and when $\epsilon \leq h$ we use the bounds $C_1 e^t \leq \sinh t \leq C_2 e^t$ for $t \geq C_3$ to bound the same terms by $Ce^{-\beta_\phi/2}$. In the first case the lower bound is of the form $Ch^{-1}(h/\epsilon)^2(\epsilon/h) = C\epsilon^{-1}$; and in the second case it is of the form $Ch^{-1}(h/\epsilon) = C\epsilon^{-1}$ also, so

$$L^h \mu^j \geq C\epsilon^{-1}\mu^j \tag{3.78}$$

in all cases.

Then, as with the derivation of (3.73), we need to calculate the truncation error in the approximation of z by Z in order to bound this part of the error. By taking the mean of (3.62a) and (3.63a), we obtain the crude bound

$$|D_0 y(x) - y'(x)| \leq C \int_{x-h}^{x+h} |y''(s)|ds, \tag{3.79}$$

which we shall need together with (3.44a). Now we write

$$L^h(Z - z)_j = (Lz)_j - L^h z_j$$

$$= \epsilon(D_h^2 z_j - z_j'') + \epsilon(\frac{1}{2}\beta_j \coth \frac{1}{2}\beta_j - 1)D_h^2 z_j - b_j(D_0 z_j - z_j')$$

and use the bound $(t \coth t - 1) \leq Ct^2/(1 + t)$, as well as the difference of (3.62a) and (3.63a) to bound $D_h^2 z_j$, in order to bound the middle term by

$$C\left(\frac{b_j h}{2\epsilon}\right)^2 \frac{\epsilon^2}{2\epsilon + b_j h} \frac{1}{h} \int_{x_{j-1}}^{x_{j+1}} |z''(s)|ds \leq C\frac{h}{h+\epsilon} \int_{x_{j-1}}^{x_{j+1}} |z''(s)|ds. \quad (3.80)$$

So we get in total

$$|L^h(Z - z)_j| \leq C \int_{x-h}^{x+h} [\epsilon|z'''(s)| + |z''(s)|]ds; \tag{3.81}$$

and, using the *a priori* bounds of Theorem 2.4.3,

$$|L^h(Z - z)_j| \leq C[h + \sinh \frac{1}{2}\beta_\phi \mu^{-(J-j)}].$$

From (3.78) and Lemma 3.4.2 we finally obtain

$$|Z_j - z_j| \leq C[h + \epsilon \sinh \frac{1}{2}\beta_\phi \mu^{-(J-j)}]$$

$$\leq C\left[h + \epsilon(1 - \mu^{-2})\right] \leq Ch. \tag{3.82}$$

Now we have to bound $|V_j - v_j|$. If b is constant, then $L^h v_j = (Lv)_j$ and $V_j = v_j$ so that (3.76a) follows from (3.82). Otherwise we have, after a little manipulation,

$$\begin{aligned}
L^h(V - v)_j &= (Lv)_j - L^h v_j \\
&= -\epsilon^{-1} b_J (b_J - b_j) v_j \\
&\quad + 2h^{-1} b_j [\coth \tfrac{1}{2}\beta_j \sinh^2 \tfrac{1}{2}\beta_J - \sinh \tfrac{1}{2}\beta_J \cosh \tfrac{1}{2}\beta_J] v_j \\
&= h^{-1}[2b_j \sinh \tfrac{1}{2}\beta_J \sinh \tfrac{1}{2}(\beta_J - \beta_j)(\sinh \tfrac{1}{2}\beta_j)^{-1} \\
&\quad - \beta_J(b_J - b_j)] v_j.
\end{aligned} \tag{3.83}$$

To bound this requires rather more tedious calculation. The two terms in the square brackets would cancel if in each case $\sinh t$ were replaced by t; so we write $\sinh t = t + \sigma(t)$ and use bounds

$$\sigma(t) \leq C \frac{t^3}{1 + t^2} e^t \quad \text{and} \quad C \frac{t}{1 + t} e^t \leq \sinh t \leq C t e^t.$$

Then the key term has the form

$$\left| \frac{(t_1 + \sigma_1)(t_2 + \sigma_2)}{t_3 + \sigma_3} - \frac{t_1 t_2}{t_3} \right| = \left| \frac{t_2 t_3 \sigma_1 + t_3 \sigma_2 \sinh t_1 - t_1 t_2 \sigma_3}{t_3(t_3 + \sigma_3)} \right|$$

which is bounded by

$$C \frac{h^3}{\epsilon^2} \frac{|b_J - b_j|}{h + \epsilon} [1 + e^{(\beta_J - \beta_j)}].$$

We assume that h is sufficiently small that

$$\frac{|b_J - b_j|}{1 - x_j} h \leq \tfrac{1}{2}(b_J - \tfrac{1}{2}b_\phi), \quad \text{i.e.} \quad |\beta_J - \beta_j| \leq \tfrac{1}{2}(\beta_J - \tfrac{1}{2}\beta_\phi)(J - j). \tag{3.84}$$

Thus from (3.83) we have

$$|L^h(V_j - v_j)| \leq C \frac{h^2(1 - x_j)}{\epsilon^2(h + \epsilon)} e^{\frac{1}{2}(\beta_J - \frac{1}{2}\beta_\phi)(J - j)} e^{-\beta_J(J - j)}$$

$$= C \frac{h^2}{\epsilon(h + \epsilon)} \mu^{-(J-j)} \left[\frac{1 - x_j}{\epsilon} e^{-\frac{1}{2}(b_J - \frac{1}{2}b_\phi)(1 - x_j)/\epsilon} \right] \tag{3.85}$$

and the term in square brackets is clearly bounded. We therefore have the bound to which we can apply (3.78) and Lemma 3.4.2 to get finally

$$|V_j - v_j| \leq C \frac{h^2}{h + \epsilon}. \tag{3.86}$$

Combining this with (3.82) gives the result (3.76a).

This result also gives the first term in the bound (3.76b). To obtain the remaining term we recalculate the truncation error for Z using the increased smoothness assumed by taking $p = 2$ in (3.60b). By taking the mean of (3.62b) and (3.63b) we get, instead of (3.79),

$$|D_0y(x) - y'(x)| \leq Ch \int_{x-h}^{x+h} |y'''(s)| ds. \qquad (3.87)$$

Combining this with (3.44b) and (3.80) we get, by use of Theorem 2.4.3,

$$|L^h(Z - z)_j| \leq Ch \int_{x_{j-1}}^{x_{j+1}} \left[\epsilon|z^{iv}(s)| + |z'''(s)| + (h + \epsilon)^{-1}|z''(s)|\right] ds$$

$$\leq Ch \int_{x_{j-1}}^{x_{j+1}} \left[\frac{1}{h+\epsilon} + \frac{1}{\epsilon^2}e^{-(b_\phi/2\epsilon)(1-s)}\right] ds$$

$$\leq C \left[\frac{h^2}{h+\epsilon} + \frac{h}{\epsilon}(\sinh \tfrac{1}{2}\beta_\phi)\mu^{-(J-j)}\right]. \qquad (3.88)$$

Then from (3.78) and Lemma 3.4.2, we have

$$|Z_j - z_j| \leq C \left[\frac{h^2}{h+\epsilon} + h(\sinh \tfrac{1}{2}\beta_\phi)\mu^{-(J-j)}\right]. \qquad (3.89)$$

When $h \geq \epsilon$, the second term can be absorbed in the first because then $h \leq 2h^2/(h+\epsilon)$. When $h \leq \epsilon$, we have $h \sinh \tfrac{1}{2}\beta_\phi \leq Ch^2/\epsilon$. So in both cases we obtain the result needed for (3.76b). $\square$

3.4.3 Exponential fitting and uniform error bounds

At this point we should justify our earlier remark that exponential fitting is necessary for the establishment of a uniform error estimate like (3.76a). It will be sufficient to consider the case where b is constant, $c = 0$ and $u_L = u_R = 0$. We also rescale the p coefficients by bh to write the difference scheme as

$$\hat{p}^- U_{j-1} + \hat{p}U_j + \hat{p}^+ U_{j+1} = (h/b)[q^- S_{j-1} + qS_j + q^+ S_{j+1}]. \qquad (3.90)$$

Lemma 3.4.5 *Consider the difference scheme (3.90) in which the coefficients $\hat{p}^-$, $\hat{p}$ and $\hat{p}^+$ depend only on $\beta = bh/\epsilon$ and the q's sum to unity. Suppose that it provides a uniformly convergent approximation to (3.1) with b and S positive constants, $c = 0$ and $u_L = u_R = 0$; that is,*

$$\|U - u\|_\infty \leq Ch^\delta, \qquad (3.91)$$

where C is independent of h and ϵ, for a fixed $\delta > 0$. Then

$$\hat{p}^- e^{-\beta} + \hat{p} + \hat{p}^+ e^\beta = 0 \qquad (3.92)$$

so that e^β is a root of the characteristic equation for the difference scheme.

Proof. The exact solution, by a special case of Theorem 2.4.3 or by direct computation, is

$$u(x) = u_0(x) - u_0(1)\left(e^{-(b/\epsilon)(1-x)} - e^{-b/\epsilon}\right) / \left(1 - e^{-b/\epsilon}\right), \qquad (3.93)$$

where $u_0(x) = (S/b)x$ is the solution of the reduced problem. We fix β and let h and ϵ both tend to zero. Then for any fixed positive integer k, we obtain from (3.91)

$$\lim_{h \to 0} U_{J-k} = \lim_{h \to 0} u(x_{J-k})$$
$$= \lim_{h,\epsilon \to 0} \left[u_0(x_{J-k}) - u_0(1)(e^{-k\beta} - e^{-b/\epsilon})/(1 - e^{-b/\epsilon}) \right]$$
$$= u_0(1)(1 - e^{-k\beta}). \qquad (3.94)$$

Applying this to the difference equation (3.90), and using the relations $\hat{p}^- + \hat{p} + \hat{p}^+ = 0$, $q^- + q + q^+ = 1$, we get

$$-u_0(1)e^{-k\beta}[\hat{p}^- e^{-\beta} + \hat{p} + \hat{p}^+ e^{\beta}] = \lim_{h \to 0}(h/b)S = 0, \qquad (3.95)$$

which gives (3.92) since $u_0(1) \neq 0$. $\quad\square$

The analysis presented in the proofs of Theorems 3.4.3 and 3.4.4, together with this result, highlight the rôle of a scheme's characteristic equation in determining its accuracy in a thin boundary layer. The emphasis on this aspect and on the establishment of error bounds which are uniform in ϵ stems from the work of Il'in (1969) and was strongly followed up in research on singular perturbation problems in the 1970's and 1980's — see Hemker and Miller (1979), Doolan *et al.* (1980), and the large literature on numerical methods for semi-conductor devices typified by Mock (1983).

In a standard *a priori* error analysis a bound for the truncation error is obtained in the typical form

$$\|T\|_\infty \leq Ch^p \|u^{(p+2)}\|_\infty \qquad (3.96a)$$

where C is independent of ϵ and h, and p is a positive integer. Then when the scheme satisfies the conditions (3.23), so that by Lemma 3.2.3 a maximum principle holds, the difference operator has a bounded inverse; the bound (3.25) is valid for any such scheme which is at least first order accurate, resulting in the global error bound

$$\|U - u\|_\infty \leq (2/b_\phi)\|T\|_\infty \leq Ch^p \|u^{(p+2)}\|_\infty. \qquad (3.96b)$$

Such bounds are easy to establish but are of little value when ϵ and h are of comparable magnitude, because inverse powers of ϵ are generated by taking derivatives of the solution in the boundary layer, as shown in section 2.4 and the theorems that we have made use of in the above analysis.

It is the incorporation of these *a priori* bounds on the solution derivatives into the global error analysis that complicates proofs of the two theorems given above; the sharpness of the bounds depends on the construction of suitable comparison functions and these make use of roots to the characteristic equation, usually for a simplified constant coefficient problem. The results given in (3.61) are typical for any scheme that is not exponentially fitted. They show a bound of the form Ch holding outside the boundary layer for both $h \leq \epsilon$ and $\epsilon \leq h$. Generally, this may be replaced by Ch^p, as in the case of the modified upwind scheme (3.75) where $p = 2$; or inverse powers of ϵ may be involved, as in the case of the generalised OCI scheme of Theorem 3.2.4, where Berger *et al.* (1980) have shown a bound of the form Ch^4/ϵ^2 holding for $h \leq \epsilon$ and Ch^2 for $\epsilon \leq h$ outside the boundary layer. Inside the layer, the general form of error bound is

$$|U_j - u_j| \leq C \min(1, (h/\epsilon)^q)\mu^{-(J-j)} \tag{3.97}$$

where $q \leq p$ and μ is the root of the characteristic equation which would approximate $e^{\beta \phi}$.

Thus from the sufficiency of the conditions given by theorems such as those in this section and the necessity of that given by Lemma 3.4.5, it is clear that the error for any scheme in a boundary layer is determined by how well μ^{-k} approximates $e^{-\beta k}$ for $k = 1, 2, \ldots$; and from the asymptotic form of the layer given in Theorem 2.4.3 it is $b(1)$ that is to be used in calculating μ and β here. If ϵ is sufficiently small there may in effect be no points in the boundary layer; then, as already noted, the simple upwind scheme has an $O(h)$ error if $\epsilon = O(h^2)$, while the modified form (3.75) has an $O(h^2)$ error if $\epsilon = O(h^3)$, and the generalised OCI scheme we have given has an $O(h^2)$ error for $\epsilon = O(h^{3/2})$. But without exponential fitting there is always some combination of values of ϵ and h that prevents uniform convergence over all mesh points as $h \to 0$ — see Fig. 6.8 where the error behaviour of several schemes is shown. The fitting provided by the Allen and Southwell scheme gives a uniform $O(h)$ bound; while a uniform $O(h^2)$ bound was established for the El-Mistikawy and Werle scheme when $c = 0$ in Berger *et al.* (1981) and Hegarty *et al.* (1980), and extended to $c \neq 0$ by Stynes and O'Riordan (1986) — see section 5.8.1 where this proof will be given in a finite element context. A result given in Berger *et al.* (1980) (their Theorem 4.11) strongly suggests that a uniform $O(h^2)$ bound is the best that can be attained with a three-point scheme.

Of course, uniform bounds in weaker norms are possible but this clearly begs the question. A more practical way out of the dilemma is to use a nonuniform mesh which is adapted to the boundary layer. In the problems treated in this section, where the location and asymptotic form of the boundary layer are well defined, a smoothly graded mesh can be introduced and the error analysis extended straightforwardly, as pointed out at the beginning of the chapter. For more general problems, local mesh refinement is attractive and techniques of this kind, together with their analysis, have been developed by Shishkin (1992*a*),(1992*b*) — see Miller *et* *al.* (1995) for a recent account. To adapt the mesh in problems where the location and nature of the boundary layers is not known *a priori* requires an *a posteriori* error analysis of the form outlined in section 1.4.4.

3.5 Turning point problems

The assumptions on the coefficients b and c made in the previous four sections, as well as the restriction to linear problems, were to a large extent directed towards making the error analysis straightforward and sharp. The methods themselves can be applied more generally and we consider here their application to the turning point problems whose asymptotic behaviour was studied in section 2.3.2.

The two basic methods whose error analysis was given in detail in the last section need very little modification. The simple upwind scheme (3.59) is conveniently generalised to

$$-\frac{\epsilon}{h^2}\delta^2 U_j + \left(\frac{b_j + |b_j|}{2h}\Delta_- + \frac{b_j - |b_j|}{2h}\Delta_+\right) U_j + c_j U_j = S_j. \quad (3.98)$$

And for the Allen and Southwell scheme in the form (3.35c), multiplication of numerator and denominator by e^{β_j} shows that no change at all is needed, except that for the leading factor $b_j/(1 - e^{-\beta_j})$ the limiting form has to be used as $b_j \to 0$. In both cases the coefficients always satisfy the conditions (3.23), with p_j^+ and p_j^- always strictly negative.

On the other hand, some of the more complicated schemes may need more modification. From the formulae (3.12) and from (3.13) it is clear that the standard OCI scheme needs no change, in that if all the b_j change sign then q_j^- and p_j^- exchange rôles with q_j^+ and p_j^+. But the construction of the generalised OCI scheme to ensure that the conditions (3.23) are satisfied becomes much more complicated. For example, there is no symmetry between q_j^- and q_j^+ in (3.26a); and conditions like (3.26b) cannot be satisfied near a turning point. If a more accurate scheme without exponential fitting is sought, the adaptive upwind scheme of (3.21) might seem to be a good candidate, with the choice of (3.22) for α generalised to

$$\alpha = (1/\beta_j)\max(|\beta_j| - 2, 0) \quad (3.99a)$$

and

$$q_j^- = \tfrac{1}{4}(|\alpha| + \alpha), \quad q_j^+ = \tfrac{1}{4}(|\alpha| - \alpha). \quad (3.99b)$$

Indeed, this scheme can be regarded as an approximation to the El Mistikawy and Werle scheme of (3.36) where, as for the Allen and Southwell scheme, no change in the definition of the coefficients is needed. In both cases conditions (3.23) are satisfied, except that for the adaptive upwind scheme either one of p_j^- or p_j^+ may be zero rather than strictly negative.

In all of the schemes where p_j^- and p_j^+ are always strictly negative and

$p_j + p_j^- + p_j^+ \geq 0$, a maximum principle as given by Lemma 3.2.3 will hold; and the proof can be carried through with the sequence from the hypothesised maximum running to either the left or the right boundary. However, there is a problem with the adaptive upwind scheme in the case of a diverging flow turning point, for example when $b(x) = x - \frac{1}{2}$. For then if the switch of sign is between b_k and b_{k+1}, it is possible to obtain

$$p_1^- = p_2^- = \cdots = p_k^- = p_{k+1}^+ = \cdots = p_{J-1}^+ = 0.$$

This means that the scheme never picks up the boundary data, the equations for U_k and U_{k+1} may be inconsistent, and there is no maximum principle. To a large extent this difficulty reflects a difficulty with the original problem, since the reduced problem has no boundary data and, indeed, is not well-posed in general — see section 6.6 where an example is discussed in some detail. The remedy in the present case is to approximate the El-Mistikawy and Werle scheme more closely and thus ensure that $\beta_j \alpha > |\beta_j| - 2$ in all cases.

A much more serious problem in establishing error bounds for any of these methods in the presence of turning points is obtaining comparison functions with the key properties (3.56) and (3.57) used in Lemma 3.4.2. This corresponds to the problems met in section 2.3 when seeking *a priori* bounds and asymptotic forms for the exact solutions. When we had $b(x) \geq b_\phi > 0$ it was sufficient to use a linear function and an exponential boundary layer based on β_ϕ: neither are relevant for turning point problems. In Abrahamsson (1977) the discussion of bounds for the continuous problem is closely linked to consideration of the discrete problems obtained when upwind schemes of the type used in Abrahamsson *et al.* (1974) are employed.

3.6 Simple extensions to two dimensions

The accurate treatment of general two-dimensional problems will require the use of flexible meshes, constructed from quadrilaterals or triangles with mesh points lying on the boundary. Then the discretisation is best accomplished by means of a finite element or finite volume formulation; and the latter is sufficiently all-embracing that it covers most recently proposed finite difference methods. Thus in this chapter we will consider only the straight-forward extension of one-dimensional difference schemes to a rectangular mesh, where one has to interpolate for the points where a general boundary crosses the mesh.

In many practical problems, such as pollutant dispersal in a very irregular waterway, a rectangular mesh — with perhaps some local mesh refinement — is quite adequate for the accuracy that can reasonably be aimed at. And it has the advantages that the mesh points are easily stored and ordered, the program takes a simple form, and the discrete equations are readily solved by well-established methods.

We begin with a uniform Cartesian mesh with spacing Δx in the x-direction and Δy in the y-direction. Mesh points are given by

$$(x_i, y_j) = (i\Delta x, j\Delta y), \quad i = 0, 1, \ldots, I, \quad j = 0, 1, \ldots, J \quad (3.100)$$

and approximations at these points denoted by $U_{i,j}$, or U_{ij} where this notation is clear. The domain Ω of the problem is assumed to be contained within this rectangle, but the set of mesh points Ω^h at which difference equations are to be set up and solved can consist of a fairly arbitrary connected subset of the points (3.100), subject to conditions that we shall specify shortly.

The problems that we shall approximate are the two-dimensional counterparts to those considered so far in this chapter, and can be written in the form

$$Lu := -\epsilon\nabla^2 u + \mathbf{b}\cdot\nabla u + cu = S \quad \text{on} \quad \Omega \subset \mathbb{R}^2 \quad (3.101a)$$

$$u = u_B \quad \text{on} \quad \partial\Omega_D, \frac{\partial u}{\partial n} = 0 \quad \text{on} \quad \partial\Omega_N. \quad (3.101b)$$

As well as the usual assumptions on $\Omega, \partial\Omega_D$ and $\partial\Omega_N$ as set out in section 2.1, we shall also make assumptions on how these relate to the mesh. The operator L in (3.101a) is the sum of two parts $L^{(x)}$ and $L^{(y)}$, in the x- and y- derivatives respectively,

$$L^{(x)}u := -\epsilon\frac{\partial^2 u}{\partial x^2} + b^{(x)}\frac{\partial u}{\partial x} + \tfrac{1}{2}cu \quad (3.102a)$$

$$L^{(y)}u := -\epsilon\frac{\partial^2 u}{\partial y^2} + b^{(y)}\frac{\partial u}{\partial y} + \tfrac{1}{2}cu; \quad (3.102b)$$

and we shall mainly be concerned with the three-point difference approximations to each of them that have been studied in sections 3.2-3.5. Thus we shall have five-point approximations to $Lu = S$; using a convenient compass point notation, the value at the centre-point U_P will in general be combined with the values U_N, U_S, U_E and U_W at the four neighbouring points.

3.6.1 Maximum principles

We can state our assumptions about the discrete domain Ω^h as follows. It consists of the subset of the mesh points (3.100) at which U is unknown and therefore for which we need to set up a difference approximation to (3.101a); we call these *interior points*, although Ω^h will contain all mesh points that lie in the two open sets Ω and $\partial\Omega_N$. For each such point P we make the following assumptions:

(i) at least one of its neighbours N, S, E, W must be a member of Ω^h. This is the sense in which Ω^h is assumed to be connected, and we shall also assume that the whole line between the two points in Ω^h is contained in Ω;

(ii) at most two of the lines PN, PS, PE, PW may cross the Neumann boundary $\partial\Omega_N$, they must then be at right-angles and enclose the normal to $\partial\Omega_N$ through P. (The lines are considered to be closed sets so PN, say, is still said to cross $\partial\Omega_N$, when $P \in \partial\Omega_N \subset \Omega^h$ and $N \notin \bar{\Omega}$.) When there is only one crossing of $\partial\Omega_N$, including the case when $\partial\Omega_N$ lies along a mesh line, the normal to $\partial\Omega_N$ is defined in terms of the angle, less than a right angle, it makes with the line that is crossed.

In addition we make the overall assumption that

(iii) the set $\partial\Omega_D^h$, consisting of all the points on $\partial\Omega_D$ crossed by mesh lines PN etc. out of all points $P \in \Omega^h$, is not empty. Dirichlet boundary conditions will be imposed at these points.

These conditions are directed towards ensuring that a maximum principle holds over Ω^h after all of the modifications to the basic difference scheme that are necessary near the boundary have been carried out.

We state the maximum principle for more general difference schemes in which more neighbours, such as those in the diagonal directions NE etc., are involved. For this purpose we write the difference scheme at each point $P \in \Omega^h$ as

$$L^h U_P := p_P U_P + \sum_{\substack{Q \in \Omega^h \\ Q \neq P, p_Q \neq 0}} p_Q U_Q + \sum_{\substack{B \in \partial\Omega_D^h \\ p_B \neq 0}} p_B u_B$$

$$= R^h S_P := \sum_{\substack{R \in \Omega^h \\ q_R \neq 0}} q_R S_R. \tag{3.103}$$

The notation is reasonably self-explanatory and consistent with that used in one dimension; note that each of the coefficients appearing in the sums is assumed nonzero, and the complete set of mesh points thus involved in the scheme is often called the *difference stencil*. The set Ω^h is called *stencil-connected* if the sum over Q is not empty, for each $P \in \Omega^h$, and a path of such *stencil links* exists between every pair of points in Ω^h. Then we have the following generalisation of Lemma 3.2.3.

Lemma 3.6.1 *Suppose the set Ω^h of mesh points is stencil-connected in respect of the difference scheme (3.103), whose coefficients satisfy the following inequalities*

$$p_Q < 0 \quad \forall Q, \quad p_B < 0 \quad \forall B, \tag{3.104a}$$

and

$$p_P \geq - \sum_{\substack{Q \in \Omega^h \\ Q \neq P, p_Q \neq 0}} p_Q - \sum_{\substack{B \in \partial\Omega_D^h \\ p_B \neq 0}} p_B > 0. \tag{3.104b}$$

Then, if for some $M \geq 0$ we have $u_B \leq M \ \forall B \in \partial\Omega_D^h$, it follows from $L^h U_P \leq 0 \ \forall P \in \Omega^h$ that

$$U_P \leq M \quad \forall P \in \Omega^h. \tag{3.105}$$

If equality is attained at some point P, then U is constant over Ω^h.

Proof. Let us suppose that the proposition is not true and that $\{U_P\}$ has its maximum greater than M at some point P^*. Then $L^h U_{P^*} \leq 0$ implies through (3.104) that $U_Q = U_{P^*}$ and $u_B = U_{P^*}$ for each of the Q and B appearing in the stencil sums of (3.103). If the second sum is not empty we already have a contradiction; otherwise we repeat the argument for each of the points Q in the first sum until by the stencil connectedness, we have covered all the points Ω^h and, by the assumption that $\partial\Omega_D^h$ is not empty, we do reach a contradiction. The last part of the theorem clearly follows by this latter argument. □

When the coefficients of a difference scheme satisfy the conditions (3.104) for a maximum principle, at every point where $L^h U_P = 0$ the value U_P lies in the range of values attained at its neighbours, that is the values U_Q and u_B occurring in the sums of (3.103). This is an often sought after property and, in a sense, the conditions (3.104) are necessary for it to hold. Equivalently, we can say that (3.104a,b) are necessary for the maximum principle to hold on every subset of Ω^h: if they fail to hold at a point P, we take the subset to consist of just this point with all its neighbours treated as boundary points; then if $p_Q > 0$ and (3.104b) holds, we set $L_h U_P = 0$ with $U_Q = -1$ and zero at the other neighbours to obtain $U_P > 0$; while if the equation is scaled to ensure $p_P > 0$ but (3.104b) does not hold, we set $L_h U_P = 0$ with $U_Q = 1$ for all the neighbours, to obtain $p_P < p_P U_P$ and hence $U_P > 1$. In either case the maximum principle is clearly violated.

3.6.2 Typical difference schemes and error bounds

Let us now consider some specific difference schemes, starting with simple upwinding. In this latter case we apply schemes of the form (3.59) to each of the operators in (3.102), with backward or forward differences for the two convection terms depending on the signs of $b_P^{(x)}$ and $b_P^{(y)}$, the components of b at the point P. We introduce the notation

$$\nabla_h^2 U_P := \frac{1}{(\Delta x)^2}\delta_x^2 U_P + \frac{1}{(\Delta y)^2}\delta_y^2 U_P \tag{3.106}$$

in terms of the operators introduced in section 1.3, and when $b_P^{(x)}$ and $b_P^{(y)}$ are both positive we get

$$-\epsilon\nabla_h^2 U_P + b_P^{(x)} D_{x-} U_P + b_P^{(y)} D_{y-} U_P + c_P U_P = S_P. \tag{3.107}$$

Assuming as usual that $c_P \geq 0$, this clearly satisfies all the hypotheses (3.104) needed for the maximum principle; we will show that these remain satisfied near the boundary if the boundary conditions are imposed appropriately.

We consider each mesh line such as PN in turn: if PN cuts $\partial\Omega_D$ at a point B where $PB/PN = \alpha$, the coefficient of U_N in (3.107) is eliminated by the replacement

$$U_N - U_P \to \alpha^{-1}(u_B - U_P) \tag{3.108}$$

which substitutes the divided difference between the points B and P for that between N and P; if PN cuts $\partial\Omega_N$, a crude approximation to the Neumann boundary condition would be to set $U_N - U_P = 0$ in order to eliminate U_N, but if the normal to $\partial\Omega_N$ through P makes an angle θ with PN, and lies between PN and PE, a better approximation would be

$$\frac{U_N - U_P}{\Delta y}\cos\theta + \frac{U_E - U_P}{\Delta x}\sin\theta = 0, \tag{3.109a}$$

or alternatively

$$\frac{U_N - U_P}{\Delta y}\cos\theta + \frac{U_P - U_W}{\Delta x}\sin\theta = 0. \tag{3.109b}$$

In the Dirichlet case, the change (3.108) clearly leaves unchanged the conditions (3.104) if these are initially satisfied by a scheme (3.103), with the negative coefficient of U_N replaced by one for U_B and that of U_P adjusted appropriately. However, this is not so easily ensured in the Neumann case, where we can assume that $\theta > 0$. If (3.109a) is used to replace U_N by U_E, a negative coefficient of U_E is reduced in magnitude and may go positive. Thus only if the point E is in Ω^h and the angle θ is small enough to ensure that the coefficient of U_E remains negative after the change do we use formula (3.109a) — see Fig 3.3 for a typical situation. Otherwise we use (3.109b) to replace U_N by U_W, which as with the Dirichlet case ensures that the conditions (3.104) remain satisfied. Note that the assumption (ii) above on the mesh ensures that if the normal to $\partial\Omega_N$ through P passes between N and E and so gives $\theta > 0$, then the points W and S must be members of Ω^h. If $E \notin \Omega^h$ as well as N — see Fig 3.3 — we use the analogous formula to (3.109b) with W replaced by S. We summarise what we have just shown in the following lemma.

Lemma 3.6.2 *Suppose that a five-point difference scheme when written in the form (3.103) satisfies the conditions (3.104) for a maximum principle to hold at mesh points well away from the boundary. Then if the assumptions (i), (ii), (iii) given above are satisfied by the discrete domain Ω^h, and the boundary conditions are applied as described above using (3.108) and (3.109), the discrete equations satisfy the conditions (3.104) at all mesh points.*

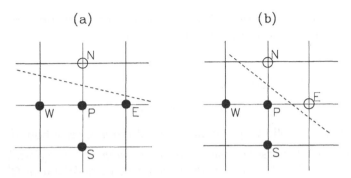

Figure 3.3. *Typical mesh configurations at a Neumann boundary: (a) with one exterior point; (b) with two exterior points*

The schemes that result from the above constructions are not properly centred so that the formal order of the truncation error may be reduced at points near the boundary. However, our main concern here is to maintain the maximum principle, so we will not consider approaches which place their emphasis on reducing the truncation error near the boundary; in any case, this is not always necessary in order to maintain the overall accuracy of the approximation — see Morton and Mayers (1994) for an introductory discussion of this point and further references. Quite often the mesh can be so positioned that there is little loss of accuracy at the boundary; and, if this is not so, we would prefer to adopt the finite volume formulations of Chapter 6 to resolve these competing requirements. So for the remainder of this section we assume that we are either using the upwind scheme (3.107), for which the truncation error is not very good even away from the boundary, or if we are using one of the more sophisticated schemes of section 3.2 or 3.3 then the boundary is well positioned relative to the mesh. Thus we assume we have used the boundary approximations given above and have therefore both maintained the maximum principle and a fairly balanced distribution of truncation errors.

Let us therefore suppose that we have generated a five-point difference scheme $L^h U_P = R^h S_P$, by applying one of the one-dimensional schemes described in sections 3.2 and 3.3 to each of the operators in (3.102), ensuring that the conditions for a maximum principle to apply are satisfied by making use of the one-dimensional Péclet numbers $\beta^{(x)} := |b^{(x)}| \Delta x / \epsilon$ and $\beta^{(y)} := |b^{(y)}| \Delta y / \epsilon$. Note that the values of these parameters are based on the velocity field b at the point P so that the coefficients p_Q, p_B and q_R in (3.103) also depend on P; and, indeed, the direction of upwinding may change completely as $b^{(x)}$ or $b^{(y)}$ is much more likely to change sign than was the case with one-dimensional problems. We also suppose that the boundary conditions are applied as above so as to maintain the maximum principle. Then at each point P we can define the truncation error in the usual way, assuming proper scaling such that $\sum q_R = 1$,

$$T_P := L^h u_P - R^h S_P; \tag{3.110a}$$

thence we obtain the error equation

$$L^h(u_P - U_P) = T_P. \tag{3.110b}$$

This leads to the following standard finite difference error bound.

Theorem 3.6.3 *Suppose the conditions for a maximum principle given in Lemma 3.6.1 are satisfied, and a non-negative mesh function Φ is defined on $\Omega^h \cup \partial \Omega_D^h$ such that*

$$L^h \Phi_P \geq 1 \quad \forall P \in \Omega^h. \tag{3.111}$$

Then one obtains the error bound

$$|U_P - u_P| \leq \Phi_P \max_{Q \in \Omega^h} |T_Q|, \quad \forall P \in \Omega^h. \tag{3.112}$$

Proof. Let T denote the maximum of $|T_Q|$, and apply Lemma 3.6.1 to $U - u - T\Phi$ with $M = 0$, to deduce from

$$L^h(U_P - u_P - T\Phi_P) \leq -T_P - T \leq 0$$

that

$$U_P - u_P - T\Phi_P \leq \max_{B \in \partial \Omega_D^h} (U_B - u_B - T\Phi_B) \leq 0,$$

as we have ensured that $U_B = u_B$ for $B \in \partial \Omega_D^h$. Since the argument can be repeated with $U - u$ replaced by $u - U$, the result follows. $\square$

As already mentioned, this argument can be refined to take account of the larger truncation errors that might occur near nonrectangular boundaries. However, we shall still have the global error expressed in terms of these truncation errors and we shall not attempt, as we did for the one-dimensional schemes analysed in section 3.4, to replace these by *a priori* bounds on the derivatives of the solution in boundary layers etc. The mesh function Φ introduced in the above theorem plays a similar rôle to the comparison functions $z^{(i)}$ used in Lemma 3.4.2 but it can be of simpler form as it does not attempt to reflect the behaviour of the truncation error. For example, suppose the region Ω is contained in the disc $x^2 + y^2 \leq R^2$ and both components of b are non-negative. A typical choice of Φ used in a conventional error analysis for Laplace's equation would be proportional to $R^2 - x^2 - y^2$; but here we would need a coefficient of $(4\epsilon)^{-1}$ in order to satisfy (3.111) and this would give an unacceptable bound. Instead, as in section 3.4, we can take $\Phi = (2R + x + y)/b_\phi$ which satisfies (3.111) if $b^{(x)} + b^{(y)} \geq b_\phi$. This results in a crude error bound which grows linearly from the inflow Dirichlet boundary, below and to the left, but everywhere reflects the maximum truncation error which is likely to occur in boundary layers above and to the right.

Finally, we should note how the construction of Φ or of more sophisticated comparison functions is related to the discrete Green's function for the problem. The error equation (3.110b) can be written as a matrix equation

$$\sum_{(Q)} K_{P,Q}(u_Q - U_Q) = T_P, \qquad (3.113a)$$

in which K is an M-matrix (see Lemma 1.5.1), because of the hypotheses of Lemma 3.6.1 and the assumption that the set Ω^h is stencil-connected. Then the inverse of K is the discrete Green's function G^h, a matrix with all its entries non-negative, and we have

$$u_P - U_P = \sum_{(Q)} G^h_{P,Q} T_Q. \qquad (3.113b)$$

It follows that

$$|u_P - U_P| \le \left(\sum_{(Q)} G^h_{P,Q}\right) \max_Q |T_Q|. \qquad (3.114)$$

Comparing with (3.112), we see that the row sums of the discrete Green's function G^h provide the ultimate comparison functions for estimating the nodal error.

3.6.3 Crosswind diffusion

Examination of the truncation error reveals a new source of concern with these five-point difference schemes. Suppose that both c and S are zero, so that any such scheme can be written in terms of weighted averages of central and upwind differences, as in (3.26),

$$-\epsilon \nabla^2_h U_P + b_P^{(x)} \left[(1 - \alpha^{(x)})D_{x0} + \alpha^{(x)}D_{x-}\right] U_P$$
$$+ b_P^{(y)} \left[(1 - \alpha^{(y)})D_{y0} + \alpha^{(y)}D_{y-}\right] U_P = 0. \qquad (3.115)$$

The leading terms in the truncation error arise from the upwinding and are readily seen to be given by

$$T_P = -\tfrac{1}{2}\left[\alpha^{(x)}b^{(x)}\Delta x \frac{\partial^2 u}{\partial x^2} + \alpha^{(y)}\Delta y \frac{\partial^2 u}{\partial y^2}\right]_P + O(h^2). \qquad (3.116a)$$

Now let us express these in terms of a local coordinate system with ξ in the flow direction and η in the crosswind direction. Supposing that $b^{(x)} = b\cos\theta, b^{(y)} = b\sin\theta$ with $\theta \in [0, \tfrac{1}{2}\pi]$, and taking $\Delta x = \Delta y = h$, we obtain

$$T_P = -\tfrac{1}{2}bh\left[(\alpha^{(x)}\cos^3\theta + \alpha^{(y)}\sin^3\theta)\frac{\partial^2 u}{\partial \xi^2}\right.$$

$$- 2\cos\theta\sin\theta(\alpha^{(x)}\cos\theta - \alpha^{(y)}\sin\theta)\frac{\partial^2 u}{\partial\xi\partial\eta}$$

$$+ \cos\theta\sin\theta(\alpha^{(x)}\sin\theta + \alpha^{(y)}\cos\theta)\frac{\partial^2 u}{\partial\eta^2}\Big] + O(h^2). \tag{3.116b}$$

Taking $\alpha^{(x)} > 0$ and $\alpha^{(y)} > 0$ gives the extra diffusion in the flow direction that damps the spurious oscillations obtained from the central difference scheme; but at the same time it gives an enhanced *crosswind diffusion* expressed by the $\partial^2 u/\partial\eta^2$ term in (3.116b). This is worst when the flow is diagonal to the mesh; it can cause a severe loss of accuracy which cannot be avoided with an upwinded five-point scheme — see Griffiths and Mitchell (1979) for further analysis and numerical examples; the effect when approximating the IAHR/CEGB model problem will be shown in section 4.6.2.

It is natural to try to avoid this problem by use of some diagonal differences, as employed for example in the *skew upstream differencing scheme* (SUDS) of Raithby (1976). This is most clearly represented by using a stencil notation, in which the position of a coefficient on the page represents the neighbour of U_P to which it is applied: for the case when $\frac{1}{2} \leq \tan\theta \leq 2$ it takes the following form

$$\frac{\epsilon}{h^2}\begin{array}{|ccc|} \hline & -1 & \\ -1 & 4 & -1 \\ & -1 & \\ \hline \end{array} U + \frac{1}{2}\frac{b^{(x)}}{h}\begin{array}{|cc|} \hline -1 & 1 \\ -1 & 1 \\ \hline \end{array} U + \frac{1}{2}\frac{b^{(y)}}{h}\begin{array}{|cc|} \hline 1 & 1 \\ -1 & -1 \\ \hline \end{array} U = 0. \tag{3.117}$$

Thus when $\theta = \frac{1}{4}\pi$ it gives pure upwind differencing in the diagonal direction and introduces no crosswind diffusion.

As we shall see in the next two chapters, finite element methods based on bilinear trial functions on a rectangular mesh lead naturally to full nine-point difference stencils. Very similar stencils, however, can be generated by the higher order compact difference methods used by Ciment and Leventhal (1975) for the wave equation; thus a Taylor series expansion readily shows that $D_{ox}f$ is a fourth order approximation to $(1 + \frac{1}{6}\delta_x^2)\partial f/\partial x$, and hence that

$$(1 + \tfrac{1}{6}\delta_y^2)D_{ox}(b^{(x)}u) + (1 + \tfrac{1}{6}\delta_x^2)D_{oy}(b^{(y)}u)$$
$$= (1 + \tfrac{1}{6}\delta_x^2)(1 + \tfrac{1}{6}\delta_y^2)\nabla\cdot(\mathbf{b}u) + O(h^4), \tag{3.118}$$

giving a difference approximation to the convective terms that should be compared with that in equation (4.45) of the next chapter. The analysis of Griffiths and Mitchell (1979) shows how upwinding based on such schemes can then be achieved with minimal crosswind diffusion; the effectiveness of these schemes will be demonstrated in numerical examples presented in Chapters 4 and 5 and we will not pursue the point further here. Finally, however, we should note that the widely used streamline diffusion method

which will be studied in some detail in Chapter 5 was motivated by just
this objective of avoiding crosswind diffusion — see Hughes and Brooks
(1979).

3.6.4 *Adaptive and nonlinear difference schemes*

For steady, linear problems in one or two dimensions, the systems of algebraic equations arising from finite difference or finite element methods
will usually be solved by direct methods, or by 'black-box' iterative methods, including multigrid, especially as the pursuit of a maximum principle
will have yielded a diagonally dominant matrix. However, for nonlinear
problems and for unsteady problems, the continual presence of an approximation to the solution prompts the idea of adapting the choice of
discretisation according to the local character of the solution. Thus where
the solution is smooth a high order scheme can be used with little danger
of spurious oscillations making an appearance; while, where the solution
is changing rapidly through a boundary or interior layer, high formal accuracy can be sacrificed in the interest of maintaining such properties as
monotonicity or positivity where they are warranted by *a priori* knowledge
of the solution. This is a very large topic of great and continuing interest.
Its pursuit in any detail would take us well beyond the scope of the present
book, but its importance necessitates some discussion which we shall defer
to Chapters 6 and 7.

The point that needs to be made here is that, in a real sense, the methods
and the analysis in this and the next two chapters should be regarded as
providing the foundations on which the adaptive methods required for
such challenging problems as solving the unsteady, compressible Navier–
Stokes equations can be built. Many of the sophistications possible in
simple problems, such as for instance exponential fitting, will have to be
abandoned for more complicated problems; yet they always provide a
yardstick by which to judge the more straightforward methods that have
to be adopted.

CHAPTER 4

Finite element methods: Galerkin schemes

Over the last fifteen years, finite element methods have proved to be much more powerful than finite difference methods for convection-diffusion problems, both practically and theoretically; and the more recent finite volume methods are also best treated as variants of finite element methods. Our sub-title to this chapter, and the title to the next, may not accurately reflect the contributions made to the origins of the subject by Petrov, Galerkin, or, indeed, others such as Bubkin; but the term Petrov–Galerkin is now widely used to denote those weighted residual methods in which the weighting functions do not lie in the same space as the approximate solution, and the term Galerkin reserved for those where they do. The former are particularly important for convection-diffusion problems, and in the next chapter we discuss mainly conforming methods, in which the weighting functions still lie in $H^1_{E_0}$. Finite volume methods, dealt with in Chapter 6, use piecewise constant weighting functions and so can often be regarded as nonconforming Petrov–Galerkin methods.

In this chapter, then, we will describe standard finite element methods that use the same space of functions for both approximate solutions and weighting functions, which we refer to as *Galerkin methods*. In the interests of completeness, we shall give quite detailed derivations of such methods based on linear and quadratic basis functions over both triangles and quadrilaterals. Some typical quadrature rules will be given in each case, but more often it will be assumed that exact integration is possible. Finite difference equivalents of the schemes on uniform meshes are also given in order to link up with the schemes of the previous chapter. It is hoped that the chapter will be helpful to readers with little previous experience of finite element methods; the more experienced reader may wish to skip much of this material and move on to the next chapter covering Petrov–Galerkin methods.

In section 4.6 the first detailed numerical comparisons of accuracy are given, both for the Galerkin methods of this chapter and the difference methods of Chapter 3. They cover two one-dimensional model problems

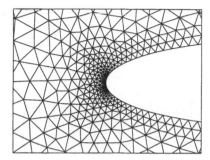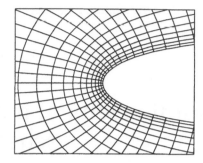

Figure 4.1. *Examples of an unstructured triangular mesh and a body-fitted quadrilateral mesh*

and the first of the IAHR/CEGB problems described in section 1.2.

4.1 Triangular and quadrilateral meshes

A feature of finite element methods that distinguishes them from finite differences is that in their derivation and implementation attention is focused on the computations in an individual element. This is the key to the facility with which irregular and unstructured meshes are dealt with. A *local coordinate system* is set up in each element and all the contributions to the complete system of equations which arise from that element are computed in the local coordinates; the *assembly* of these contributions into the complete system then requires information determining mappings from local to global coordinate systems to be held in an appropriate data structure.

As representative of the meshes used in two dimensions, in this chapter we will consider two cases: a regular triangulation of the region Ω which is completely unstructured — so that the number of triangles meeting at a vertex varies from vertex to vertex — and a subdivision of Ω (or an approximation to Ω) into quadrilaterals which corresponds to a continuous global mapping onto a rectangular mesh. Examples are shown in Fig. 4.1. Of course, not all regions Ω can be subdivided by the second method; even if Ω is divided into blocks which are each dealt with by this means, where the blocks are joined there may be nodes which are common to other than four quadrilaterals. Hence the subdivision into triangles is the more general and flexible technique. This is even more true in three dimensions, and provides a strong motivation for the development of triangular meshes and associated discretisations capable of modelling boundary layer flows as illustrated in Fig. 1.2.

In the first instance we shall assume that all the triangles and quadrilaterals have straight sides; and part of the definition of a *regular triangulation* is that each side is wholly shared by two neighbouring triangles, i.e.

there are no "hanging" nodes. Straight sides limit the accuracy with which curved boundaries can be approximated but we shall later indicate how triangles and quadrilaterals with curved sides can be introduced into the approximations.

It will be clear from what follows that it is quite feasible to have a mesh composed of a combination of triangles and quadrilaterals. This is quite common in practical computer codes for mesh generation and the construction of finite element schemes; and the corresponding combination of tetrahedra, hexahedra and wedges is even more important in three dimensions.

4.1.1 Local mapping for a triangle

We map each triangle onto a unit right-angled triangle by an affine transformation from the global Cartesian coordinates $\mathbf{r} \equiv (x, y)$ to the local coordinates (ξ, η) — see Fig. 4.2. With the notation $\mathbf{r}_\alpha \equiv (x_\alpha, y_\alpha)$, $\alpha = 1, 2, 3$

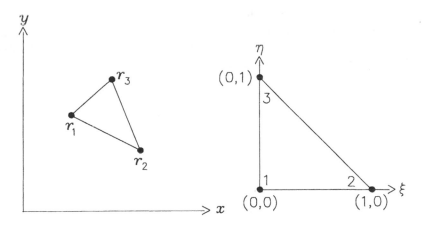

Figure 4.2. *Global and local coordinates for a triangle*

for the vertices of the triangle which are to be mapped onto $(0, 0), (1, 0)$ and $(0, 1)$ respectively in the local coordinate system, it is clear that the mapping is given by

$$\mathbf{r}(\xi, \eta) = (1 - \xi - \eta)\mathbf{r}_1 + \xi\mathbf{r}_2 + \eta\mathbf{r}_3. \tag{4.1}$$

This defines *element basis functions* $N_\alpha(\xi, \eta)$ in terms of which we can write (4.1) as

$$\mathbf{r}(\xi, \eta) = \sum_{\alpha=1}^{3} N_\alpha(\xi, \eta)\mathbf{r}_\alpha \tag{4.2a}$$

where

$$N_1(\xi, \eta) := 1 - \xi - \eta, \quad N_2(\xi, \eta) := \xi, \quad N_3(\xi, \eta) := \eta. \qquad (4.2b)$$

The Jacobian of the transformation is given by

$$J^L = \frac{\partial(x, y)}{\partial(\xi, \eta)} = \begin{bmatrix} x_2 - x_1 & y_2 - y_1 \\ x_3 - x_1 & y_3 - y_1 \end{bmatrix} \qquad (4.3)$$

which is constant over the triangle and gives

$$\det J^L = \det \begin{bmatrix} x_1 & y_1 & 1 \\ x_2 & y_2 & 1 \\ x_3 & y_3 & 1 \end{bmatrix} = 2A_{123}, \qquad (4.4)$$

where A_{123} is the area of the triangle $\triangle(\mathbf{r}_1, \mathbf{r}_2, \mathbf{r}_3) \equiv \triangle_{123}$.

A second part of the assumption that the triangulation is regular is that, if we take the ratio of the triangle diameter to the diameter of the inscribed circle for each of the triangles, this ratio is bounded; moreover, if we parametrize a triangulation by h the diameter of the largest triangle, then these ratios remain bounded as a family of triangulations is generated for which $h \to 0$. Thus there are no arbitrarily long thin triangles, and the Jacobians (4.3) are all nonsingular with $\det J^L \geq \sigma h^2$ for some $\sigma > 0$. This is also important in calculating gradients of finite element approximations V. We have

$$J^L \begin{bmatrix} \partial V / \partial x \\ \partial V / \partial y \end{bmatrix} = \begin{bmatrix} \partial V / \partial \xi \\ \partial V / \partial \eta \end{bmatrix} \qquad (4.5a)$$

and hence

$$\frac{\partial V}{\partial x} = \frac{1}{\det J^L} \left((y_3 - y_1) \frac{\partial V}{\partial \xi} - (y_2 - y_1) \frac{\partial V}{\partial \eta} \right)$$

$$\frac{\partial V}{\partial y} = \frac{1}{\det J^L} \left(-(x_3 - x_1) \frac{\partial V}{\partial \xi} + (x_2 - x_1) \frac{\partial V}{\partial \eta} \right), \qquad (4.5b)$$

from which it is easy to see that

$$(\det J^L)^2 |\nabla V|^2 = |\mathbf{r}_3 - \mathbf{r}_1|^2 \left(\frac{\partial V}{\partial \xi} \right)^2 + |\mathbf{r}_1 - \mathbf{r}_2|^2 \left(\frac{\partial V}{\partial \eta} \right)^2$$

$$+ 2(\mathbf{r}_3 - \mathbf{r}_1) \cdot (\mathbf{r}_1 - \mathbf{r}_2) \frac{\partial V}{\partial \xi} \frac{\partial V}{\partial \eta}. \qquad (4.6)$$

We will make use of these relations in later sections where we calculate stiffness matrices for finite element approximations.

4.1.2 Local mapping for a quadrilateral

In a similar way, each quadrilateral is mapped onto a canonical square as in Fig. 4.3. For convenience in using Gaussian quadrature formulae,

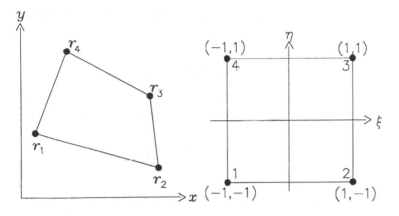

Figure 4.3. *Global and local coordinates for a quadrilateral*

this square is taken as $\{(\xi, \eta) \in [-1,1] \times [-1,1]\}$, with the element basis functions being tensor products of those used in one dimension. Thus we have

$$\mathbf{r}(\xi, \eta) = \sum_{\alpha=1}^{4} N_\alpha(\xi, \eta) \mathbf{r}_\alpha \qquad (4.7a)$$

where

$$N_1(\xi, \eta) := \tfrac{1}{4}(1 - \xi)(1 - \eta) \quad N_2(\xi, \eta) := \tfrac{1}{4}(1 + \xi)(1 - \eta)$$
$$N_3(\xi, \eta) := \tfrac{1}{4}(1 + \xi)(1 + \eta) \quad N_4(\xi, \eta) := \tfrac{1}{4}(1 - \xi)(1 + \eta). \qquad (4.7b)$$

The Jacobian of this bilinear transformation is given by

$$J^B = \frac{1}{4} \begin{bmatrix} (x_3 - x_4)(1 + \eta) & (y_3 - y_4)(1 + \eta) \\ + (x_2 - x_1)(1 - \eta) & + (y_2 - y_1)(1 - \eta) \\ (x_3 - x_2)(1 + \xi) & (y_3 - y_2)(1 + \xi) \\ + (x_4 - x_1)(1 - \xi) & + (y_4 - y_1)(1 - \xi) \end{bmatrix} \qquad (4.8)$$

which therefore varies across the element. It is easily checked that at (x_1, y_1), or $\xi = \eta = -1$, we have $\det J^B = \tfrac{1}{2}A_{412}$ with similar results for the other vertices. Moreover, because the corresponding coefficients of ξ and η in (4.8) are equal, it is clear that $\det J^B$ varies linearly across the element — see section 4.4.1 for explicit forms; hence if the quadrilateral is convex, so that all the triangles like $\triangle_{412}$ have positive area, the Jacobian is nonsingular throughout the element. Note that in the special case of a parallelogram, which corresponds to the midpoints of the diagonals $\tfrac{1}{2}(\mathbf{r}_1 + \mathbf{r}_3)$ and $\tfrac{1}{2}(\mathbf{r}_2 + \mathbf{r}_4)$ being coincident, the Jacobian is constant.

For a general quadrilateral, the nonconstancy of the Jacobian means that gradients in the global coordinates, given by formulae corresponding to

(4.5), involve rational functions of ξ and η which will necessitate the use of numerical quadrature for the evaluation of any integrals involving them.

4.2 Standard error bounds for Galerkin schemes

Let us consider the approximation of the convection-diffusion problem (2.1), written in the weak form given by (2.50) and (2.52): find $u \in H_E^1$ such that

$$B_\gamma(u, v) = l(v) \quad \forall v \in H_{E_0}^1, \tag{4.9a}$$

where

$$B_\gamma(u, v) := \epsilon(A\nabla u, \nabla v) + (\mathbf{b}\cdot\nabla u + cu, v) + \epsilon\gamma \int_{\partial\Omega_N} uv d\Gamma, \tag{4.9b}$$

$$l(v) := (S, v) + \epsilon \int_{\partial\Omega_N} g_B v d\Gamma \tag{4.9c}$$

and all the data and coefficients satisfy the hypotheses of Theorem 2.5.1 establishing the existence and uniqueness of the solution. A form for $B_\gamma(\cdot, \cdot)$ which is often more convenient for finite element or finite volume approximation, and which often occurs in applications, is one in which the convection is in divergence form as in the model problem of (1.18). Integrating by parts the corresponding term in (4.9b) we get the equivalent expression

$$B_\gamma(u, v) = (\epsilon A\nabla u - \mathbf{b}u, \nabla v) + ([c - \nabla\cdot\mathbf{b}]u, v)$$

$$+ \int_{\partial\Omega_N} [\epsilon\gamma + \mathbf{b}\cdot\mathbf{n}]uv d\Gamma. \tag{4.9d}$$

Note how the first term combines the diffusive and convective fluxes in the expression $-\epsilon A\nabla u + \mathbf{b}u$.

4.2.1 Approximation properties of the trial spaces

As a start, we assume that the region Ω is two-dimensional, its boundary $\partial\Omega$ is polygonal and it coincides exactly with its discretisation into triangles or quadrilaterals. Then an approximation space S^h is constructed of functions $V(x, y)$ which are polynomial in form over each element of the discretisation and we assume that they are continuous between elements; this ensures that $S^h \subset H^1(\Omega)$, so defining it as a *conforming* approximation space. If the Dirichlet data u_B on $\partial\Omega_D$ is extended over the region Ω to give a function $\tilde{u}_B \in H_E^1$, we can define the *trial space* S_E^h in which we seek an approximation to u as follows:

$$S_0^h := S^h \cap H_{E_0}^1 \tag{4.10a}$$

$$S_E^h := \tilde{u}_B \oplus S_0^h. \tag{4.10b}$$

Then the Galerkin approximation to (4.9) is given by: find $U \in S_E^h$ such that

$$B_\gamma(U,V) = l(V) \quad \forall V \in S_0^h, \tag{4.11}$$

where for the moment we assume that all the integrals required in evaluating the two sides of the equation are evaluated exactly.

Because of the conforming assumption embodied in (4.10a), we can substitute V for v in (4.9a) and obtain by subtracting (4.11)

$$B_\gamma(u - U, V) = 0 \quad \forall V \in S_0^h \tag{4.12}$$

as the starting point for deriving error bounds. In the absence of convection, $B_\gamma(\cdot, \cdot)$ is symmetric and (4.12) is an orthogonality relation which ensures that U is the best approximation to u from the trial space S_E^h, in the norm corresponding to $B_\gamma(\cdot, \cdot)$; then the approximation properties of the trial space can be called upon to establish precise error bounds in various norms. This can be found in any standard text on the theory of finite element methods, such as Strang and Fix (1973) or Ciarlet (1978). With convection present the most natural norm to use is obtained from the symmetric part of $B_\gamma(\cdot, \cdot)$, that is from $\frac{1}{2}B_\gamma(u,v) + \frac{1}{2}B_\gamma(v,u)$. By making use of Gauss' theorem, as in (2.59), we can write $B_\gamma(v,v)$ as

$$B_\gamma(v,v) = \epsilon(A\nabla v, \nabla v) + ([c - \tfrac{1}{2}\nabla \cdot \mathbf{b}]v, v) + \int_{\partial\Omega_N} [\epsilon\gamma + \tfrac{1}{2}\mathbf{b}\cdot\mathbf{n}]v^2 d\Gamma$$

$$\forall v \in H_{E_0}^1. \tag{4.13}$$

This is positive definite through our assumption that the hypotheses of Theorem 2.5.1 hold, and hence that $c - \frac{1}{2}\nabla \cdot \mathbf{b} \geq 0$ on Ω and $\mathbf{b}\cdot\mathbf{n} \geq 0$ on $\partial\Omega_N$. However, we then have to bound the antisymmetric part of $B_\gamma(\cdot, \cdot)$ in terms of the symmetric part; and roughly speaking this increases the error bound by a factor which has the character of a mesh Péclet number. It will be sufficient to demonstrate this in the theorem below for the case when $\gamma = c = \nabla \cdot \mathbf{b} = 0$.

First we set out the standard approximation properties that the trial space is assumed to have. We state these, not in their most general form, but in terms of the Hilbert spaces $H^s(\Omega)$, the seminorms $|\cdot|_{l,\Omega}$ formed from the integrals over Ω of the squares of all derivative or order l, and the norm $\|\cdot\|_{H^s(\Omega)}$ formed similarly from all derivatives of order up to s.

Approximation properties of S_E^h: we assume that S^h is composed of piecewise polynomial functions and is characterised by an integer $k \geq 2$ and a constant K independent of h, such that for every $w \in H^l(\Omega) \cap H_E^1$ with $l \leq k$ there exists a $W \in S_E^h$ for which we have

$$\|w - W\|_{H^s(\Omega)} \leq Kh^{l-s}|w|_{l,\Omega}, \quad s = 0, 1. \tag{4.14}$$

For both the linear elements on triangles and the bilinear elements on quadrilaterals that we shall consider below we have $k = 2$, so that function values are approximated to $O(h^2)$ and gradients to $O(h)$ if $w \in H^2(\Omega)$. For the corresponding quadratic and biquadratic elements we have $k = 3$, giving an extra order of accuracy in each case if w is smoother, namely in $H^3(\Omega)$. Note that these results depend on the triangulation being regular, with a similar property holding in the case of quadrilaterals.

4.2.2 A general error bound

The following theorem indicates the accuracy that can be achieved with Galerkin methods, and how that depends on the mesh Péclet number.

Theorem 4.2.1 *Suppose that in the problem (4.9) we have $\gamma = 0, c = 0$ and $\nabla \cdot \mathbf{b} = 0$, and that ϵ is chosen so that $A \geq I$. Then the Galerkin approximation given by (4.11), from a trial space satisfying (4.14), satisfies*

$$|(A\nabla(u - U), \nabla(u - U))|^{\frac{1}{2}} \leq \left[\|A\|_{L_\infty}^{\frac{1}{2}} + \|\mathbf{b}\|_{L_\infty} h/\epsilon \right] Kh^{l-1} |u|_{l,\Omega} . \quad (4.15)$$

Proof. For any $W \in S_E^h$, by using (4.12) with $V = U - W$ and by comparing (4.9b) and (4.13), we obtain

$$\epsilon(A\nabla(u - U), \nabla(u - U) \leq B_\gamma(u - U, u - U) = B_\gamma(u - U, u - W)$$
$$= \epsilon(A\nabla(u - U), \nabla(u - W)) + (\mathbf{b}\cdot\nabla(u - U), u - W).$$

The second term on the right is bounded by

$$\|\mathbf{b}\|_{L_\infty(\Omega)} \|\nabla(u - U)\|_{L_2(\Omega)} \|u - W\|_{L_2(\Omega)} ;$$

so if we introduce the notation $\epsilon \|u - U\|_A^2$ for the left-hand side we get

$$\|u - U\|_A^2 \leq \|u - U\|_A \left[\|u - W\|_A + \epsilon^{-1} \|\mathbf{b}\|_{L_\infty(\Omega)} \|u - W\|_{L_2(\Omega)} \right],$$

from which (4.15) follows with the application of (4.14). $\square$

Thus if the solution u is sufficiently smooth the Galerkin approximation gives the full order of approximation, in the H^1 norm, that is achievable with the given approximation space S^h: that is, the order of approximation is optimal as $h \to 0$; but the constant in the error bound (4.15) is magnified by a mesh Péclet number, so the actual optimality of the approximation on each discrete mesh, which is so characteristic of the Galerkin method for self-adjoint problems, is lost in the presence of the convection terms. Error bounds in other norms are derived from H^1 bounds such as that given above, and therefore show the same loss of optimality, as we shall show in section 5.4; and in section 4.5 we shall see that even for simple model problems these bounds are realistic.

In the next chapter we shall show how sometimes optimal approximations can be generated for nonself-adjoint problems; so it is worth noting here how the gradients of such approximations often exhibit *superconvergence* properties. For example, the L_2 norm of the error in the gradient of a bilinear approximation is only $O(h)$ by (4.14); but at element centroids the optimal approximation has an error that is $O(h^2)$, as shown by Zlámal (1978) and Lesaint and Zlámal (1979). This is a common and important phenomenon which will be frequently referred to; for example, the improved gradient approximation can form the basis of an *a posteriori* error indicator as in (1.105).

4.2.3 Inverse estimates for the trial spaces

We shall on occasion, mainly in the next chapter, have to estimate one norm of a finite element approximation in terms of another, typically the L_∞ norm in terms of the L_2 norm. Thus we use the Sobolev space norms $W_p^s(\Omega)$ obtained from the $L_p(\Omega)$ norms of all derivatives of order up to s, so that $W_2^s(\Omega) \equiv H^s(\Omega)$; then on each element e of diameter h we shall require that there is a constant M such that, for $\forall V \in S^h$ or for $V = \nabla W$ with $W \in S^h$, and $1 \le q \le p \le \infty$, we have

$$\|V\|_{W_p^s(e)} \le M h^{-s-2(1/q-1/p)} \|V\|_{L_q(e)}, s = 0, 1. \qquad (4.16a)$$

In particular, we have

$$\|V\|_{L_\infty(e)} \le M h^{-1} \|V\|_{L_2(e)}. \qquad (4.16b)$$

These are readily established for the finite element spaces on regular meshes that we shall describe below.

In order to extend these estimates to the whole mesh, and hence to give global significance to M and h in (4.16), it is necessary to make a quasi-uniformity assumption on the mesh; namely that for some constants K and h, the diameters diam(e) and the diameters of the inscribed circles $\rho(e)$ satisfy

$$C^{-1} \text{diam}(e) \le h \le C\rho(e). \qquad (4.17)$$

Since $s \ge 0$ and $p \ge q$ in (4.16a), giving negative powers of h, these bounds are generally referred to as *inverse estimates*.

4.3 Galerkin methods on a triangular mesh

4.3.1 Linear approximation

A piecewise linear approximation on a triangle requires three parameters for its definition, and two of these need to be shared with a neighbouring triangle in order to ensure continuity and therefore that the approximation is conforming. It follows that $V \in S_0^h$ in the piecewise linear case is parametrized by its values $\{V_j, j = 1, 2, \ldots, N\}$ at all the interior vertices of the triangulation of Ω and of the Neumann boundary $\partial\Omega_N$, together with the zero values at the Dirichlet boundary vertices, $V_j = 0$ for

$j = N + 1, N + 2, \ldots, N + M$. Within each triangle, V is also linear in the local coordinates (ξ, η), and can be expressed in terms of the same nodal basis functions $N_\alpha(\xi, \eta)$ used in the coordinate transformation (4.2); so we have, within the triangle,

$$V(x(\xi, \eta), y(\xi, \eta)) = \sum_{\alpha=1}^{3} N_\alpha(\xi, \eta) V_\alpha, \tag{4.18a}$$

where V_α refers to the local numbering of the vertices of the triangle and, to be completely unambiguous, should perhaps be written $V_{j(\alpha)}$. In terms of global basis functions $\phi_j^L(x, y)$, defined by the properties of being linear on each triangle and such that $\phi_j^L(x_i, y_i) = \delta_{i,j}$, we have the global expansion

$$V(x, y) = \sum_{j=1}^{N} \phi_j^L(x, y) V_j \quad \forall V \in S_0^h. \tag{4.18b}$$

Hence the Galerkin approximation U in S_E^h, which by (4.10b) has an expansion

$$U(x, y) = \tilde{u}_B(x, y) + \sum_{j=1}^{N} \phi_j^L(x, y) U_j, \tag{4.19}$$

is given by

$$B_\gamma(U, \phi_i^L) = l(\phi_i^L) \quad i = 1, 2, \ldots, N \tag{4.20a}$$

which are called the *Galerkin equations*. In terms of the $N \times N$ *global stiffness matrix* K defined by

$$K_{i,j} := B_\gamma(\phi_j^L, \phi_i^L) \quad i, j = 1, 2, \ldots, N, \tag{4.20b}$$

these can be written as

$$K\mathbf{U} = \mathbf{F} \tag{4.20c}$$

where $\mathbf{U}$ is the vector of nodal unknowns, $\mathbf{U} := (U_1, U_2 \ldots, U_N)^T$, and $\mathbf{F}$ is the *load vector* with components given by

$$F_i := l(\phi_i^L) - B_\gamma(\tilde{u}_B, \phi_i^L) \quad i = 1, 2, \ldots, N. \tag{4.20d}$$

The first step in forming these equations is to calculate the *element stiffness matrix* for each triangle, a 3×3 matrix defined by

$$K_{\alpha,\beta}^e = B_\gamma(N_\beta, N_\alpha) \quad \alpha, \beta = 1, 2, 3 \tag{4.21}$$

where in a further abuse of the notation we have introduced the element basis functions as arguments in the global bilinear form. What the entries in this matrix correspond to are the contributions to entries in the global stiffness matrix arising from integration over the particular triangle or element e. For a triangle that does not share an edge with $\partial\Omega_N$, application of the coordinate transformation given by (4.2) and (4.5) to (4.9d) gives

$$
K^e_{\alpha,\beta} = \int_{\Delta^e} \left\{ \left(\epsilon A J^{-1} \begin{bmatrix} \partial_\xi N_\beta \\ \partial_\eta N_\beta \end{bmatrix} - \begin{bmatrix} b^{(x)} \\ b^{(y)} \end{bmatrix} N_\beta \right)^T J^{-1} \begin{bmatrix} \partial_\xi N_\alpha \\ \partial_\eta N_\alpha \end{bmatrix} \right.
$$
$$
\left. + (c - \nabla\cdot b)N_\alpha N_\beta \right\} (\det J)\,d\xi d\eta, \tag{4.22a}
$$

where $b^{(x)}, b^{(y)}$ are the components of b, and we have used the abbreviations $\partial_\xi, \partial_\eta$ for $\partial/\partial\xi$ and $\partial/\partial\eta$ and omitted the superscript L in the Jacobian. The corresponding components of the element load vector are

$$
F^e_\alpha = \int_{\Delta^e} \left\{ (b\tilde{u}_B - \epsilon A\nabla\tilde{u}_B)^T J^{-1}(\partial_\xi N_\alpha, \partial_\eta N_\alpha)^T \right.
$$
$$
\left. + [(\nabla\cdot b - c)\tilde{u}_B + S]N_\alpha \right\}(\det J)\,d\xi d\eta, \tag{4.22b}
$$

where the terms in $\tilde{u}_B$ will usually be zero except in triangles adjacent to the Dirichlet boundary.

In general these rather daunting expressions will have to be evaluated using numerical quadrature. For linear elements on triangles one-point quadrature based on the centroid of the triangle ($\xi = \eta = \frac{1}{3}$) will normally be adequate to maintain the accuracy of the scheme as it integrates linear functions exactly; the result for (4.22a) will be same except for the $c - \nabla\cdot b$ term as assuming that A, b, $c - \nabla\cdot b$ and S are all constant when evaluating (4.21). We proceed by making this assumption, and for further simplicity assume henceforth the common situation in which the matrix A is replaced by the scalar diffusion coefficient a. Then, separating out the four matrices arising from the element basis functions, we obtain

$$
K^e = \epsilon a K_2 - b^{(x)} K_1^{(x)} - b^{(y)} K_1^{(y)} + (c - \nabla\cdot b)K_0, \tag{4.23a}
$$

where

$$
K_2 = \frac{1}{4A_{123}} \begin{bmatrix} |r_2 - r_3|^2 & (r_2 - r_3)\cdot(r_3 - r_1) & (r_2 - r_3)\cdot(r_1 - r_2) \\ \bullet & |r_3 - r_1|^2 & (r_3 - r_1)\cdot(r_1 - r_2) \\ \bullet & \bullet & |r_1 - r_2|^2 \end{bmatrix} \tag{4.23b}
$$

$$
K_1^{(x)} = \frac{1}{6} \begin{bmatrix} y_2 - y_3 \\ y_3 - y_1 \\ y_1 - y_2 \end{bmatrix} \begin{bmatrix} 1 & 1 & 1 \end{bmatrix} \qquad K_1^{(y)} = -\frac{1}{6} \begin{bmatrix} x_2 - x_3 \\ x_3 - x_1 \\ x_1 - x_2 \end{bmatrix} \begin{bmatrix} 1 & 1 & 1 \end{bmatrix} \tag{4.23c}
$$

and

$$
K_0 = \frac{A_{123}}{12} \begin{bmatrix} 2 & 1 & 1 \\ \bullet & 2 & 1 \\ \bullet & \bullet & 2 \end{bmatrix}. \tag{4.23d}
$$

Note that K_2 and K_0 are symmetric matrices. K_0 is called the *mass matrix* and would have had all its entries equal to $\frac{1}{9}A_{123}$ if one-point quadrature had been used in its evaluation; its replacement by a diagonal matrix, by summing the row entries to give $\frac{1}{3}A_{123}$, is a common simplification called *mass lumping*.

In the load vector, the $\tilde{u}_B$ terms will usually be evaluated by approximating $\tilde{u}_B$ in terms of piecewise linear basis functions, so the same calculations can be used as in the stiffness matrix. Thus using a notation mixing (4.20) and (4.21) we write

$$F_\alpha^{(e)} = \tfrac{1}{3}A_{123}S - B_\gamma(\tilde{u}_B, N_\alpha) \qquad (4.24)$$

for the element load vector.

The global stiffness matrix and load vector are then obtained from the element matrices and vectors by what is termed the *assembly process*. In a program implementing the scheme, a *connectivity array* is held giving the global node number for each node of each triangle. If for element e this association gives $\alpha \mapsto i, \beta \mapsto j$, then the component $K_{\alpha,\beta}^e$ is added to the component $K_{i,j}$ of the global matrix and F_α^e is added to F_i. (Note that, with the way we have chosen to deal with the inhomogeneous Dirichlet boundary data, any contributions that would be allocated to a node on the boundary $\overline{\partial\Omega}_D$ are discarded because the $\phi_i^L, \phi_j^L \in S_0^h$ which are used in (4.20b) exclude any such element basis functions.) The assembly process can be represented in algebraic terms by introducing a $N \times 3$ Boolean matrix L^e for each element, which has an entry 1 in the (i, α) position when the association $\alpha \mapsto i$ holds, and $i \in \{1, 2, \ldots, N\}$ so that it does not correspond to a node on $\overline{\partial\Omega}_D$. Then we have

$$K = \sum\nolimits_{(e)} L^e K^e (L^e)^T, \quad \mathbf{F} = \sum\nolimits_{(e)} L^e \mathbf{F}^e \qquad (4.25)$$

to represent the assembly process.

Great simplification occurs on a uniform square mesh with the diagonals drawn from the top left to the bottom right, so that six triangles meet at every vertex. Then each triangle is of the same shape as the canonical triangle, but with horizontal and vertical sides of length h and area of $\frac{1}{2}h^2$; and half of them have the same orientation. With two sides being perpendicular, K_2 reduces to

$$K_2 = \tfrac{1}{2} \begin{bmatrix} 2 & -1 & -1 \\ -1 & 1 & 0 \\ -1 & 0 & 1 \end{bmatrix} \qquad (4.26a)$$

for all the triangles. For those with the same orientation as the canonical triangle, $K_1^{(x)}$ and $K_1^{(y)}$ reduce to the following forms with the positive sign,

$$K_1^{(x)} = \pm \frac{h}{6} \begin{bmatrix} -1 & -1 & -1 \\ 1 & 1 & 1 \\ 0 & 0 & 0 \end{bmatrix} \quad K_1^{(y)} = \pm \frac{h}{6} \begin{bmatrix} -1 & -1 & -1 \\ 0 & 0 & 0 \\ 1 & 1 & 1 \end{bmatrix}, \quad (4.26b)$$

while for the other triangles the negative sign must be taken. The assembly process for a typical node, when all the coefficients are constants, can be demonstrated on the mesh as follows, using Fig. 4.4. The figure shows the six triangles that meet at the node and therefore form the support of the global basis function at the node. On the left we have made entries corresponding to the element stiffness matrices $2K_2$, and on the right to $(6/h)K_1^{(x)}$. Starting with the triangle above and to the right, the central node corresponds to node 1 in the canonical triangle; so the entries in the first rows of the matrices are written in at the positions of the corresponding nodes, i.e., 2 at node 1, -1 at node 2, -1 at node 3 in the case of $2K_2$. Moving in a counter-clockwise direction, the central node corresponds to node 3 in the next triangle; so the entries from the third rows of the matrices are written in, and for $K_1^{(x)}$ the negative sign would be taken, (because from (4.3) we see that J^L is minus the identity matrix) though in this case they are all zero. In this way all the triangles are dealt with, and summing the contributions at each node gives a difference stencil; the contributions from the load vector are accumulated in the same way. For points away from the boundary, so that there is no contribution from $\tilde{u}_B$, the final result can

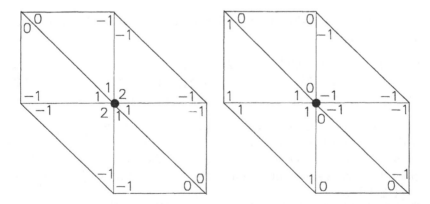

Figure 4.4. *Assembly of the stiffness matrix at a typical node for the linear approximation; on the left are shown the entries of $2K_2$ and on the right those for $(6/h)K_1^{(x)}$*

then be summarised and represented in stencil form as follows,

$$\epsilon a \begin{bmatrix} & -1 & \\ -1 & 4 & -1 \\ & -1 & \end{bmatrix} U + \frac{b^{(x)}h}{6} \begin{bmatrix} -1 & 1 \\ -2 & 0 & 2 \\ -1 & 1 \end{bmatrix} U + \frac{b^{(y)}h}{6} \begin{bmatrix} 1 & 2 \\ -1 & 0 & 1 \\ -2 & -1 \end{bmatrix} U$$

$$+ \frac{(c - \nabla \cdot \mathbf{b})h^2}{12} \begin{bmatrix} 1 & 1 \\ 1 & 6 & 1 \\ 1 & 1 \end{bmatrix} U = Sh^2. \tag{4.27}$$

Thus we have the familiar five-point stencil for the diffusion operator; but two seven-point central difference stencils which reflect the mesh orientation arise for the convection terms, and we have accumulated the full mass matrix for the cu term. The generalisation of this scheme when $a, \mathbf{b}$ etc. are only assumed constant on each element is readily deduced from the assembly process used in Fig. 4.4.

4.3.2 Quadratic approximation

To obtain continuous variation of a quadratic from one triangle to its neighbour there need to be three shared parameters; the usual choice is to add the mid-edge value to the value at the two vertices. This gives six parameters per triangle, which uniquely determines the quadratic variation in the triangle. Using the same local coordinate system (4.1) that was used for the linear elements, we now need six quadratic basis functions in terms of which we can write

$$V(x(\xi, \eta), y(\xi, \eta)) = \sum_{\alpha=1}^{6} N_\alpha^Q(\xi, \eta) V_\alpha. \tag{4.28}$$

With the local numbering of Fig. 4.5 it is easy to deduce that these basis functions are given by

$$\begin{aligned} N_1^Q(\xi, \eta) &:= (1 - \xi - \eta)(1 - 2\xi - 2\eta) & N_4^Q(\xi, \eta) &:= 4\xi\eta \\ N_2^Q(\xi, \eta) &:= \xi(2\xi - 1) & N_5^Q(\xi, \eta) &:= 4\eta(1 - \xi - \eta) \\ N_3^Q(\xi, \eta) &:= \eta(2\eta - 1) & N_6^Q(\xi, \eta) &:= 4\xi(1 - \xi - \eta). \end{aligned} \tag{4.29}$$

Sometimes a hierarchical basis is to be preferred in which N_4^Q, N_5^Q, N_6^Q are added to the linear basis functions given by (4.26b). Then, of course, the expansion (4.28) is replaced by one in which the coefficients do not correspond to nodal values; for example, the coefficient of $N_4^Q(\xi, \eta)$ becomes $V_4 - \frac{1}{2}(V_2 + V_3)$.

The construction of the Galerkin equations $K\mathbf{U} = \mathbf{F}$ through calculation of the element and global stiffness matrices and load vectors takes place in exactly the same way as with the piecewise linear approximation. It is sometimes convenient to replace the global expansion (4.19) by

$$U(x,y) = \tilde{u}_B(x,y) + \sum_{j=1}^{N_1} \phi_j^{QV}(x,y)V_j + \sum_{j=N_1+1}^{N_1+N_2} \phi_j^{QS}(x,y)V_j, \quad (4.30)$$

in order to distinguish the global basis functions ϕ_j^{QV}, which are associated with a vertex and hence usually have support extending over six triangles, from the ϕ_j^{QS} associated with a triangle side or edge, and hence have support over only the two triangles on either side. Other notational changes are obvious and in (4.21) it is necessary only to replace N_α and N_β by N_α^Q and N_β^Q, and let α, β run over the values $1, 2, \ldots, 6$. Because we are using the same coordinate transformation, the relations (4.5) and (4.6) also still hold true. The difference is that instead of $\partial V/\partial \xi = V_2 - V_1, \partial V/\partial \eta = V_3 - V_1$, from (4.28) and (4.29) we now have

$$\frac{\partial V}{\partial \xi} = (4\xi + 4\eta - 3)V_1 + (4\xi - 1)V_2$$
$$+ 4\eta(V_4 - V_5) + (4 - 8\xi - 4\eta)V_6 \quad (4.31a)$$

$$\frac{\partial V}{\partial \eta} = (4\xi + 4\eta - 3)V_1 + (4\eta - 1)V_3$$
$$+ 4\xi(V_4 - V_6) + (4 - 4\xi - 8\eta)V_5. \quad (4.31b)$$

If we again assume that a, $\mathbf{b}$ and $c - \nabla \cdot \mathbf{b}$ are constant over each element, the integrals giving the entries in the element stiffness matrix K_2 therefore involve quadratic functions of ξ and η. A convenient quadrature formula which is exact for any quadratic consists of taking the average of the integrand values at the mid-edge points. By this means we obtain the first of the four 6×6 matrices corresponding to those in (4.23). We need to simplify the notation by writing $\rho_1^2 := |\mathbf{r}_2 - \mathbf{r}_3|^2$ and $\rho_{12} := (\mathbf{r}_2 - \mathbf{r}_3) \cdot (\mathbf{r}_3 - \mathbf{r}_1)$ etc., with $\rho^2 := \rho_1^2 + \rho_2^2 + \rho_3^2$, and then as the first part of

$$K^e = \epsilon a K_2^Q - b^{(x)} K_1^{(x)Q} - b^{(y)} K_1^{(y)Q} + (c - \nabla \cdot \mathbf{b}) K_0^Q \quad (4.32a)$$

we obtain

$$K_2^Q = \frac{1}{12A_{123}} \begin{bmatrix} 3\rho_1^2 & -\rho_{12} & -\rho_{13} & 0 & 4\rho_{13} & 4\rho_{12} \\ \cdot & 3\rho_2^2 & -\rho_{23} & 4\rho_{23} & 0 & 4\rho_{12} \\ \cdot & \cdot & 3\rho_3^2 & 4\rho_{23} & 4\rho_{13} & 0 \\ \cdot & \cdot & \cdot & 4\rho^2 & 8\rho_{12} & 8\rho_{13} \\ \cdot & \cdot & \cdot & \cdot & 4\rho^2 & 8\rho_{23} \\ \cdot & \cdot & \cdot & \cdot & \cdot & 4\rho^2 \end{bmatrix} \quad (4.32b)$$

The convective terms involve cubic functions of ξ and η, which can be integrated exactly by a quadrature rule that combines $\frac{2}{15}$ of the values at the mid-edges with $\frac{1}{20}$ of the values at the vertices and $\frac{9}{20}$ of the value at the centroid. Introducing the diagonal matrices $Y := \text{diag}\{y_2 - y_3, y_3 - y_1, y_1 -$

y_2} and $X := \text{diag}\{x_2 - x_3, x_3 - x_1, x_1 - x_2\}$, the resulting contributions to the stiffness matrix can be computed and written as follows,

$$K_1^{(x)Q} = \frac{1}{30} \begin{bmatrix} 3Y - YE & 3YE - 4Y \\ 6Y - 2YE - 3EY & 4(EY - 2Y - YE) \end{bmatrix} \qquad (4.32c)$$

where

$$E := \begin{bmatrix} 1 & 1 & 1 \\ 1 & 1 & 1 \\ 1 & 1 & 1 \end{bmatrix},$$

with $K_1^{(y)Q}$ having the same form with $-X$ replacing Y. Integrating the mass matrix exactly gives

$$K_0^Q = \frac{A_{123}}{180} \begin{bmatrix} 6 & -1 & -1 & -4 & 0 & 0 \\ \cdot & 6 & -1 & 0 & -4 & 0 \\ \cdot & \cdot & 6 & 0 & 0 & -4 \\ \cdot & \cdot & \cdot & 32 & 16 & 16 \\ \cdot & \cdot & \cdot & \cdot & 32 & 16 \\ \cdot & \cdot & \cdot & \cdot & \cdot & 32 \end{bmatrix}. \qquad (4.32d)$$

When the seven-point quadrature used for the convection terms is also used for the mass matrix there are considerable discrepancies; for example, the first row becomes $(10, 1, 1, -4, -4, -4)$.

Assembly of the element matrices for a uniform square mesh eventually leads to a difference stencil, as in the case of the linear approximation dealt with in the previous section and shown in Fig. 4.4; but there are now four different types of node (vertices, and mid-edge nodes on the horizontal, vertical and diagonal edges) and these give different stencils. However, the geometric factors are the same; namely, $A_{123} = \frac{1}{2}h^2, \rho_1^2 = 2h^2, \rho_2^2 = \rho_3^2 = h^2, \rho_{12} = \rho_{13} = -h^2$ and $\rho_{23} = 0$, with $Y = \text{diag}\{-h, h, 0\}$ and $X = \text{diag}\{h, 0 - h\}$. Substituting these into (4.32b) and (4.32c) we obtain the stiffness matrices

$$K_2^Q = \frac{1}{6} \begin{bmatrix} 6 & 1 & 1 & 0 & -4 & -4 \\ 1 & 3 & 0 & 0 & 0 & -4 \\ 1 & 0 & 3 & 0 & -4 & 0 \\ 0 & 0 & 0 & 16 & -8 & -8 \\ -4 & 0 & -4 & -8 & 16 & 0 \\ -4 & -4 & 0 & -8 & 0 & 16 \end{bmatrix} \qquad (4.33a)$$

and

$$
K_1^{(x)Q} = \frac{h}{30}
\begin{bmatrix}
-2 & 1 & 1 & 1 & -3 & -3 \\
-1 & 2 & -1 & 3 & -1 & 3 \\
0 & 0 & 0 & 0 & 0 & 0 \\
-1 & -1 & 2 & 8 & 8 & 4 \\
1 & 1 & -2 & -8 & -8 & -4 \\
3 & -3 & 0 & -4 & -4 & 0
\end{bmatrix},
\tag{4.33b}
$$

with $K_1^{(y)Q}$ having a similar form. For a constant coefficient problem, the assembly into a difference stencil centred at each of the four types of node is demonstrated in Fig. 4.5. The top two diagrams correspond to the assembly from the six triangles meeting at a vertex, with the assembly from K_2^Q shown on the left and that from $K_1^{(x)Q}$ on the right; they use the first three rows of (4.33a) and (4.33b) to give the entries in the triangles labelled (1),(3),(5) respectively; and then again for triangles (2),(4),(6) but, remembering that the Jacobian $J^L = -I$ in this case, the entries for $K_1^{(x)Q}$ have their signs changed. The next pair of diagrams correspond to the two triangles which make up the support of the global basis function centred at the midpoint of a diagonal; and the last two pairs correspond to the midpoints of a vertical edge and a horizontal edge.

The mass matrix is less interesting than the rest of the stiffness matrix, so in writing out the difference stencils corresponding to these diagrams we omit it by assuming $c - \nabla \cdot b = 0$. Then for a vertex node away from the boundary we obtain in stencil form

$$
\frac{\epsilon a}{3}
\begin{bmatrix}
 & 1 & \\
 & -4 & \\
1 & -4 \quad 12 \quad -4 & 1 \\
 & -4 & \\
 & 1 &
\end{bmatrix}
U +
\frac{b^{(x)}h}{30}
\begin{bmatrix}
1 & 0 & -1 \\
1 & -3 & 3 & -1 \\
2 & -6 & 0 & 6 & -2 \\
1 & -3 & 3 & -1 \\
1 & 0 & -1
\end{bmatrix} U
$$

$$
+ \frac{b^{(y)}h}{30}
\begin{bmatrix}
-1 & -1 & -2 \\
0 & 3 & 6 & -1 \\
1 & -3 & 0 & 3 & -1 \\
1 & -6 & -3 & 0 \\
2 & 1 & 1
\end{bmatrix} U = Sh^2.
\tag{4.34a}
$$

Notice how the stencils for the first order convection terms reflect the diagonal orientation of the mesh and that they give far from obvious difference approximations; the diffusion operator, however, is quite symmetric and consists of a $(\frac{4}{3}, -\frac{1}{3})$ weighted combination of five-point schemes on the full mesh and the $\frac{1}{2}h$ mesh respectively. For a node at the midpoint of a

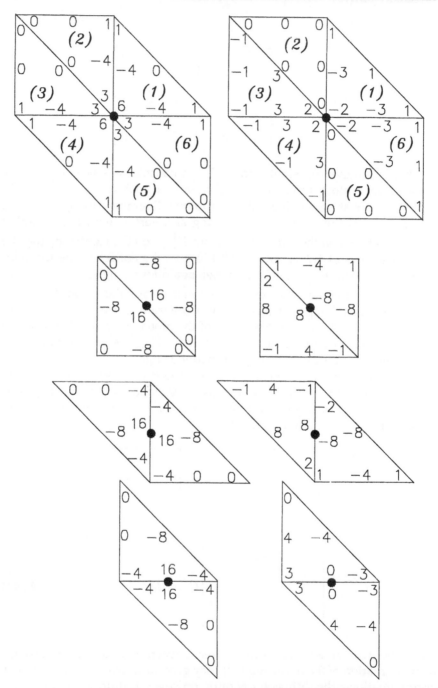

Figure 4.5. *Assembly of the stiffness matrix for the quadratic approximation at the four types of node; on the left are shown the entries for* $6K_2^Q$ *and on the right those for* $(30/h)K_1^{(x)Q}$

diagonal edge we obtain the stencil

$$
\frac{4}{3}\epsilon a
\begin{bmatrix}
 & -1 & \\
-1 & 4 & -1 \\
 & -1 &
\end{bmatrix} U
+ \frac{b^{(x)}h}{30}
\begin{bmatrix}
-3 & 4 & -1 \\
-8 & 0 & 8 \\
1 & -4 & 3
\end{bmatrix} U
$$

$$
+ \frac{b^{(y)}h}{30}
\begin{bmatrix}
3 & 8 & -1 \\
-4 & 0 & 4 \\
1 & -8 & -3
\end{bmatrix} U = Sh^2; \qquad (4.34b)
$$

while for the midpoints of the vertical and horizontal edges we get

$$
\frac{4}{3}\epsilon a
\begin{bmatrix}
 & -1 & \\
-1 & 4 & -1 \\
 & -1 &
\end{bmatrix} U
+ \frac{b^{(x)}h}{30}
\begin{bmatrix}
1 & -4 & 3 \\
-8 & 0 & 8 \\
-3 & 4 & -1
\end{bmatrix} U
$$

$$
+ \frac{b^{(y)}h}{30}
\begin{bmatrix}
4 & 6 \\
-4 & 0 & 4 \\
-6 & -4
\end{bmatrix} U = Sh^2 \qquad (4.34c)
$$

and

$$
\frac{4}{3}\epsilon a
\begin{bmatrix}
 & -1 & \\
-1 & 4 & -1 \\
 & -1 &
\end{bmatrix} U
+ \frac{b^{(x)}h}{30}
\begin{bmatrix}
-4 & 4 \\
-6 & 0 & 6 \\
-4 & 4
\end{bmatrix} U
$$

$$
+ \frac{b^{(y)}h}{30}
\begin{bmatrix}
-1 \\
4 & 8 \\
-3 & 0 & 3 \\
-8 & -4 \\
1
\end{bmatrix} U = Sh^2. \qquad (4.34d)
$$

Note that in these cases, while generally the stencils extend only over a 3×3 square, a 5×3 rectangle is used to approximate the x-derivative across a vertical side and a 3×5 rectangle for the y-derivative across a horizontal side.

We shall see in section 4.5 that, although each of these difference operators has only second order accuracy, their combination when solving the completely assembled system of equations gives third order accuracy; indeed on a uniform mesh one often obtains fourth order accuracy. It is this that accounts for the popularity of these quadratic elements in practical computations. We have written out the stencils in detail because of their importance and to aid comparison with difference methods. In this form they appear rather complicated, but this is a rather unnatural form for them and it should be emphasised that when embodied in a computer program there is little extra complication as compared with linear elements.

4.4 Galerkin methods on quadrilaterals

4.4.1 Isoparametric bilinear approximation

Along each edge of the canonical square in Fig. 4.3, the local to global mapping for a general convex quadrilateral that is given by (4.7) is linear. Thus its specification through the global coordinates r_α of the vertices gives a straight edge to the quadrilateral, and ensures continuity of the mappings used in two neighbouring quadrilaterals. (Note that in three dimensions the generally trilinear mapping reduces to a bilinear mapping on a face; it still provides for continuity but in general does not give a plane face.) Similarly, the use of global vertex values $\{V_j, j = 1, 2, \ldots, N\}$ to parametrize $V \in S_0^h$ will ensure continuity, and hence that S_0^h is conforming, in an approximation that is bilinear in the local coordinates on each element: that is, one that is given by

$$V(x(\xi, \eta), y(\xi, \eta)) = \sum_{\alpha=1}^{4} N_\alpha(\xi, \eta) V_\alpha, \qquad (4.35a)$$

where the element basis functions are those given in (4.7b); and globally,

$$V(x, y) = \sum_{j=1}^{N} \phi_j^B(x, y) V_j \quad \forall V \in S_0^h, \qquad (4.35b)$$

where the support of the global basis function ϕ_j^B extends over up to four quadrilaterals that meet at the vertex (x_j, y_j), in each of which it equals the element basis function corresponding to that vertex. Note that here we have a difference from the approximation on triangles in that the bilinear variation in (ξ, η) does not always lead to bilinear variation in (x, y) over each element. It does so when the Jacobian is constant, which occurs for parallelograms as we have already noted. Hence the approximation is referred to as *isoparametric* bilinear to denote the fact that the same bilinear expansion is used for the global coordinates in terms of the local coordinates as is used for V.

The global expansion for the Galerkin approximation $U \in S_E^h$ has exactly the same form as that given in (4.19) for the piecewise linear approximation on triangles, with the linear global basis functions ϕ_j^L replaced by their bilinear counter parts ϕ_j^B. With this change, all the global equations (4.20a–4.20d) are the same.

The element stiffness matrix on each quadrilateral, however, is now 4×4 and the Jacobian J that has to be substituted into (4.22b) is the nonconstant J^B of (4.8). As it is the inverse that is needed, the integrals in general involve rational functions and numerical quadrature is necessary. However, this is more straightforward than in the case of triangles; just as the element basis functions are tensor products of the one-dimensional basis functions

$\frac{1}{2}(1\pm\xi)$ and $\frac{1}{2}(1\pm\eta)$, so tensor product Gaussian quadrature formulae can be used. Although formally exact only for bilinear polynomials, the use of one-point quadrature at the centroid is usually adequate for the bilinear approximation; we will compare this with more accurate quadrature rules that make use of vertex values.

To confirm statements already made, and for general reference, we give two alternative forms for $\det J^B$. By direct computation of (4.8) at the four vertices, we see that

$$\det J^B = \frac{1}{2}(N_1 A_{412} + N_2 A_{123} + N_3 A_{234} + N_4 A_{341}) \qquad (4.36\text{a})$$

in terms of the element basis functions $N_\alpha \equiv N_\alpha(\xi,\eta)$. If A_Q is the area of the quadrilateral, we have $A_{412} + A_{234} = A_Q = A_{123} + A_{341}$, so that the coefficient of $\xi\eta$ in this expansion is zero and the constant term is $\frac{1}{4}A_Q$. Now consider the linear terms: the coefficient of ξ is $\frac{1}{8}(-A_{412}+A_{123}+A_{234}-A_{341})$ and it is clear from the diagram in Fig. 4.3 that $A_{123} - A_{421} = A_{234} - A_{341}$, with the common value being zero if and only if the opposite sides $(\mathbf{r}_1,\mathbf{r}_2)$ and $(\mathbf{r}_3,\mathbf{r}_4)$ are parallel; denoting this value by $\Delta_{12,34}$ and using a similar notation for the other pair of sides, we have

$$\det J^B = \frac{1}{4}(A_Q + \Delta_{12,34}\xi + \Delta_{23,41}\eta), \qquad (4.36\text{b})$$

which is constant if and only if the quadrilateral is a parallelogram.

From (4.35a) and (4.7b), and corresponding to the columns of J^B, we have in each element

$$\begin{bmatrix} \partial V/\partial\xi \\ \partial V/\partial\eta \end{bmatrix} = \frac{1}{4}\begin{bmatrix} (V_3 - V_4)(1+\eta) + (V_2 - V_1)(1-\eta) \\ (V_3 - V_2)(1+\xi) + (V_4 - V_1)(1-\xi) \end{bmatrix}. \qquad (4.37\text{a})$$

If we abbreviate this to $(V_\xi, V_\eta)^T$, and use the same notation $(x_\xi, x_\eta)^T$ and $(y_\xi, y_\eta)^T$ for the columns of J^B, we can write the global gradient as

$$\begin{bmatrix} \partial V/\partial x \\ \partial V/\partial y \end{bmatrix} = \frac{1}{\det J^B}\begin{bmatrix} y_\eta V_\xi - y_\xi V_\eta \\ x_\xi V_\eta - x_\eta V_\xi \end{bmatrix}. \qquad (4.37\text{b})$$

At the centroid $\xi = \eta = 0$, (4.36b) shows that $\det J^B = \frac{1}{4}A_Q$ and (4.37b) reduces to

$$\begin{bmatrix} \partial V/\partial x \\ \partial V/\partial y \end{bmatrix}_{(0,0)} = \frac{1}{2A_Q}\begin{bmatrix} (V_1 - V_3)(y_2 - y_4) - (V_2 - V_4)(y_1 - y_3) \\ (V_2 - V_4)(x_1 - x_3) - (V_1 - V_3)(x_2 - x_4) \end{bmatrix}; \qquad (4.38\text{a})$$

and at a typical vertex $\alpha = 1$, or $\xi = \eta = -1$, we have

$$\begin{bmatrix} \partial V/\partial x \\ \partial V/\partial y \end{bmatrix}_{(-1,-1)} = \frac{1}{2A_{412}}\begin{bmatrix} (V_2 - V_1)(y_4 - y_1) - (V_4 - V_1)(y_2 - y_1) \\ (V_4 - V_1)(x_2 - x_1) - (V_2 - V_1)(x_4 - x_1) \end{bmatrix} \qquad (4.38\text{b})$$

which is exactly the same as that which would be obtained with a linear transformation over the triangle $\triangle_{412}$.

Let us gather these results to get the element stiffness matrix obtained from centroid quadrature. By allowing the substitution of N_α and N_β for V into (4.38a), we write instead of (4.22a)

$$K_{\alpha,\beta}^e = \int_{\square^e} \left\{ \left(\epsilon a \begin{bmatrix} \partial_x N_\beta \\ \partial_y N_\beta \end{bmatrix} - \begin{bmatrix} b^{(x)} \\ b^{(y)} \end{bmatrix} N_\beta \right)^T \begin{bmatrix} \partial_x N_\alpha \\ \partial_y N_\alpha \end{bmatrix} \right.$$
$$\left. + (c - \nabla \cdot \mathbf{b}) N_\alpha N_\beta \right\} (\det J^B) d\xi d\eta. \tag{4.39}$$

Then with centroid quadrature, noting that $N_\alpha(0,0) = \frac{1}{4} \ \forall \alpha$, we get

$$K_{\alpha,\beta}^e|_{\text{cent}} = A_Q \left\{ \left(\epsilon a \begin{bmatrix} \partial_x N_\beta \\ \partial_y N_\beta \end{bmatrix} - \frac{1}{4} \begin{bmatrix} b^{(x)} \\ b^{(y)} \end{bmatrix} \right)^T \begin{bmatrix} \partial_x N_\alpha \\ \partial_y N_\alpha \end{bmatrix} \right.$$
$$\left. + \frac{1}{16}(c - \nabla \cdot \mathbf{b}) \right\}. \tag{4.40a}$$

With our standard notation (4.23a) for the component matrices, this gives for $4 A_Q K_2^B|_{\text{cent}}$

$$\begin{bmatrix} |\mathbf{r}_2 - \mathbf{r}_4|^2 & (\mathbf{r}_2 - \mathbf{r}_4) \cdot (\mathbf{r}_3 - \mathbf{r}_1) & (\mathbf{r}_2 - \mathbf{r}_4) \cdot (\mathbf{r}_4 - \mathbf{r}_2) & (\mathbf{r}_2 - \mathbf{r}_4) \cdot (\mathbf{r}_1 - \mathbf{r}_3) \\ \bullet & |\mathbf{r}_3 - \mathbf{r}_1|^2 & (\mathbf{r}_3 - \mathbf{r}_1) \cdot (\mathbf{r}_4 - \mathbf{r}_2) & (\mathbf{r}_3 - \mathbf{r}_1) \cdot (\mathbf{r}_1 - \mathbf{r}_3) \\ \bullet & \bullet & |\mathbf{r}_4 - \mathbf{r}_2|^2 & (\mathbf{r}_4 - \mathbf{r}_2) \cdot (\mathbf{r}_1 - \mathbf{r}_3) \\ \bullet & \bullet & \bullet & |\mathbf{r}_1 - \mathbf{r}_3|^2 \end{bmatrix}$$
$$\tag{4.40b}$$

and also

$$K_1^{(x)B}|_{\text{cent}} = \frac{1}{8} \begin{bmatrix} y_2 - y_4 \\ y_3 - y_1 \\ y_4 - y_2 \\ y_1 - y_3 \end{bmatrix} [1,1,1,1]$$

$$K_1^{(y)B}|_{\text{cent}} = -\frac{1}{8} \begin{bmatrix} x_2 - x_4 \\ x_3 - x_1 \\ x_4 - x_2 \\ x_1 - x_3 \end{bmatrix} [1,1,1,1] \tag{4.40c}$$

and $K_0|_{(0,0)}$ has all entries equal to $\frac{1}{16} A_Q$; note that we have not simplified the terms in (4.40b) in order to emphasise the symmetry and highlight the similarity with the corresponding matrices of (4.23) for the case of linear approximation on triangles. However, since only the diagonals are involved in the entries, considerable simplification is always possible and for a rectangle half of the terms are zero and the other half have the same magnitude.

When the integrand in the stiffness matrix (4.39) is evaluated at a vertex, because of what we have already observed for the gradient (4.38b), the entries for K_2 exactly correspond to those in (4.23b) for a triangle apart from a multiplying factor. Thus for the vertex $\xi = \eta = -1$ associated with the triangle $\triangle_{412}$, entries in the third row and column are zero but in the others they are as in (4.23b), though with a leading factor of $(8A_{412})^{-1}$ when account is taken of the value of $\det J^B$. Similarly for $K_1^{(x)}$ and $K_1^{(y)}$, the columns of y and x differences are as given by (4.23c) for triangle $\triangle_{412}$, though with a zero inserted in the third row and a leading factor of $\frac{1}{4}$; but the row obtained from $N_\beta(-1, -1)$ is of course $(1, 0, 0, 0)$ and the contribution to the K_0 matrix has only the entry $\frac{1}{2}A_{412}$ in the top left position. Four point quadrature with the vertices is, like centroid quadrature, exact for bilinears. It will give an approximation to K_2^β which is similar to (4.40b), but considerably more complicated because of the triangle areas occurring in the coefficients. These could be eliminated if the contribution from each vertex were weighted by the corresponding triangle area; but this does not seem justified and would make the computation of $K_1^{(x)}$ and $K_1^{(y)}$ more complicated. For the former the combination of vertex values gives

$$K_1^{(x)B}\Big|_{\text{vert}} = \frac{1}{4}\begin{bmatrix} y_2 - y_4 & y_3 - y_1 & 0 & y_1 - y_3 \\ y_4 - y_1 & y_1 - y_2 & y_4 - y_2 & 0 \\ 0 & y_2 - y_3 & y_2 - y_3 & y_3 - y_4 \\ y_1 - y_2 & 0 & y_3 - y_4 & y_4 - y_1 \end{bmatrix} \qquad (4.41)$$

while $K_1^{(y)}$ is obtained by using the x differences and making a sign change throughout. Clearly $K_0^B\big|_{\text{vert}} = \frac{1}{2}\text{diag}\{A_{412}, A_{123}, A_{234}, A_{341}\}$.

Combining centroid quadrature with vertex quadrature with a $\frac{2}{3} : \frac{1}{3}$ weighting will give a scheme which is exact for full quadratics. But, unlike 2×2 Gaussian quadrature at $\xi, \eta = \pm 1/\sqrt{3}$, they cannot be combined to integrate exactly the products of bilinears which clearly occur in (4.37b) for general quadrilaterals. So the Gaussian quadrature is to be preferred if centroid quadrature is inadequate, though the centroid-vertex combination can be useful for parallelograms. However, our main purpose in giving the details needed for vertex quadrature is that these formulae will be used in the finite volume methods of the next chapter.

We conclude this section by computing and assembling the element stiffness matrices on a uniform square mesh of side h. In each square element, J^B is constant with $\det J^B = \frac{1}{4}h^2$, and from (4.37) we have

$$\begin{bmatrix} \partial V/\partial x \\ \partial V/\partial y \end{bmatrix} = \frac{2}{h}\begin{bmatrix} \partial V/\partial \xi \\ \partial V/\partial \eta \end{bmatrix} = \frac{1}{2h}\begin{bmatrix} (V_3 - V_4)(1 + \eta) + (V_2 - V_1)(1 - \eta) \\ (V_3 - V_2)(1 + \xi) + (V_4 - V_1)(1 - \xi) \end{bmatrix}.$$
$$(4.42)$$

Thus even for the integration to give K_2, centroid quadrature will not give the exact result because of the quadratic terms. This corresponds to the fact that the basis functions are tensor products of the linear one-dimensional basis functions; so when we integrate $(\partial N_\alpha/\partial \xi)(\partial N_\beta/\partial \xi)$ the integral over

η yields the same terms as in the one-dimensional mass matrix, as given in Chapter 1. With exact integration the component matrices are as follows,

$$K_2^B = \tfrac{1}{6} \begin{bmatrix} 4 & -1 & -2 & -1 \\ -1 & 4 & -1 & -2 \\ -2 & -1 & 4 & -1 \\ -1 & -2 & -1 & 4 \end{bmatrix} \tag{4.43a}$$

$$K_1^{(x)B} = \frac{h}{12} \begin{bmatrix} -2 & -2 & -1 & -1 \\ 2 & 2 & 1 & 1 \\ 1 & 1 & 2 & 2 \\ -1 & -1 & -2 & -2 \end{bmatrix} \quad K_1^{(y)B} = \frac{h}{12} \begin{bmatrix} -2 & -1 & -1 & -2 \\ -1 & -2 & -2 & -1 \\ 1 & 2 & 2 & 1 \\ 2 & 1 & 1 & 2 \end{bmatrix} \tag{4.43b}$$

$$K_0^B = \frac{h^2}{36} \begin{bmatrix} 4 & 2 & 1 & 2 \\ 2 & 4 & 2 & 1 \\ 1 & 2 & 4 & 2 \\ 2 & 1 & 2 & 4 \end{bmatrix}. \tag{4.43c}$$

For comparison, note that the mass matrix given by centroid quadrature would have all entries equal to $\tfrac{1}{16}h^2$, and that given by vertex quadrature would be diagonal with entries equal to $\tfrac{1}{4}h^2$.

The assembly process is very simple when all the coefficients are constant, with each row of the matrices being used to give entries in the four elements meeting at a vertex. The result is the following scheme

$$\frac{\epsilon a}{3} \begin{bmatrix} -1 & -1 & -1 \\ -1 & 8 & -1 \\ -1 & -1 & -1 \end{bmatrix} U + \frac{b^{(x)}h}{12} \begin{bmatrix} -1 & 0 & 1 \\ -4 & 0 & 4 \\ -1 & 0 & 1 \end{bmatrix} U$$

$$+ \frac{b^{(y)}h}{12} \begin{bmatrix} 1 & 4 & 1 \\ 0 & 0 & 0 \\ -1 & -4 & -1 \end{bmatrix} U + \frac{(c - \nabla \cdot b)h^2}{36} \begin{bmatrix} 1 & 4 & 1 \\ 4 & 16 & 4 \\ 1 & 4 & 1 \end{bmatrix} U = Sh^2. \tag{4.44}$$

The main points to notice are: the use of all eight neighbours to give the maximum degree of cylindrical symmetry in the approximation of the Laplacian operator; the $\tfrac{1}{6}(1,4,1)$ weighting obtained from the one-dimensional mass matrix which occurs at right angles to the convection differencing; and the tensor product of this mass matrix that is assembled from the K_0^B matrix. This can be written out in terms of the standard one-dimensional difference operators, which highlights the fact that it is derived from tensor product basis functions,

$$-\epsilon a \left[(1 + \tfrac{1}{6}\delta_y^2)\delta_x^2 + (1 + \tfrac{1}{6}\delta_x^2)\delta_y^2 \right] U + b^{(x)}h(1 + \tfrac{1}{6}\delta_y^2)\Delta_{0x}U$$

$$+ b^{(y)}h(1 + \tfrac{1}{6}\delta_x^2)\Delta_{0y}U + (c - \nabla \cdot b)h^2(1 + \tfrac{1}{6}\delta_x^2)(1 + \tfrac{1}{6}\delta_y^2)U = Sh^2. \tag{4.45}$$

There is clearly a very great simplification when mass lumping is used everywhere to remove the mass matrix operators $1 + \tfrac{1}{6}\delta^2$; in effect, this has already been done for the load vector by assuming S is constant.

Note that when a, b and c are variables it is easy to introduce their values into the quadrature schemes leading to the element stiffness matrices, and then to assemble these values to give the Galerkin equation at each vertex. But it is difficult to express the result of this process in a form that can be compared directly with the difference schemes given in Chapter 3. One case in which it might be done is when centroid values of the coefficients are used in each element. Suppose that the $\frac{1}{6}(1,4,1)$ averaging in the y-direction is factored into two operators that might be written

$$m_y^{(1)} m_y^{(2)} \equiv \tfrac{1}{2} \left[\tfrac{1}{3}(1,2,0)_y + \tfrac{1}{3}(0,2,1)_y \right].$$

Then the convective difference in the x-direction given in (4.44) can be written

$$h \mu_x m_y^{(1)} (b^{(x)} \delta_x m_y^{(2)}) U,$$

and this could be approximated by replacing both $m_y^{(1)}$ and $m_y^{(2)}$ by the standard averaging operator μ_y, to give a $\frac{1}{4}(1,2,1)$ weighting.

4.4.2 Biquadratic approximation

This is again a tensor product approximation in the local coordinates, so we start by considering the one-dimensional case. As for the piecewise linear elements described in section 1.4, the unit interval is divided into J subintervals

$$0 = x_0 < x_1 < \ldots < x_{J-1} < x_J = 1$$

with the jth interval $e_j \equiv [x_{j-1}, x_j]$ having length h_j and midpoint $x_{j-\frac{1}{2}}$. There are three element basis functions

$$N_1^Q(\xi) := -\tfrac{1}{2}\xi(1-\xi), \quad N_2^Q(\xi) := 1 - \xi^2, \quad N_3^Q(\xi) := \tfrac{1}{2}\xi(1+\xi), \quad (4.46a)$$

where

$$x = \tfrac{1}{2}(1-\xi)x_{j-1} + \tfrac{1}{2}(1+\xi)x_j$$

is the local coordinate transformation corresponding to (4.7), in terms of which $V \in S^h$ can be written

$$V(x(\xi)) = V_{j-1} N_1^Q(\xi) + V_{j-\frac{1}{2}} N_2^Q(\xi) + V_j N_3^Q(\xi), \quad x \in e_j. \quad (4.46b)$$

Here $V_j = V(x_j)$ and $V_{j-\frac{1}{2}} = V(x_{j-\frac{1}{2}})$. The global expansion becomes

$$V(x) = \sum_{j=0}^{J} V_j \phi_j^Q(x) + \sum_{j=1}^{J} V_{j-\frac{1}{2}} \psi_{j-\frac{1}{2}}^Q(x), \quad (4.47a)$$

where for an interior vertex, $j = 1, 2, \ldots, J - 1$,

$$\phi_j^Q(x) = \begin{cases} N_3^Q(2(x - x_{j-\frac{1}{2}})/h_j) & x \in e_j \\ N_1^Q(2(x - x_{j+\frac{1}{2}})/h_{j+1}) & x \in e_{j+1}, \end{cases} \tag{4.47b}$$

with $\phi_0^Q(x) = N_1^Q(2x/h_1 - 1)$ on e_1 and $\phi_J^Q(x) = N_3^Q(1 - 2(1 - x)/h_J)$ on e_J, and for all elements

$$\psi_{j-\frac{1}{2}}^Q(x) = N_2^Q(2(x - x_{j-\frac{1}{2}}/h_j) \quad x \in e_j. \tag{4.47c}$$

We will suppose for the purpose of this presentation that

$$u(0) = u_L, \quad u'(1) = 0$$

are the boundary conditions for the problem, in nonconservation form,

$$B(u, v) \equiv \epsilon(au', v') + (bu' + cu, v) = (S, v) \quad \forall v \in H_{E_0}^1$$

where $H_{E_0}^1 = \{v \in H^1(0, 1) | v(0) = 0\}$.

In constructing the element stiffness matrix and load vector the third order accurate two-point Gaussian quadrature formula would normally be used in practice; but in order to compare with difference schemes given earlier we will approximate a, b, c and S by element values, denoted by the subscript $j - \frac{1}{2}$, and integrate basis function products exactly. Then we have, in a notation consistent with (4.23a),

$$K^e = \epsilon a_{j-\frac{1}{2}} K_2 - b_{j-\frac{1}{2}} K_1 + c_{j-\frac{1}{2}} K_0, \quad F^e = S_{j-\frac{1}{2}} 1 \tag{4.48a}$$

where elementary integrations give

$$K_2 = \frac{1}{3h_j} \begin{bmatrix} 7 & -8 & 1 \\ -8 & 16 & -8 \\ 1 & -8 & 7 \end{bmatrix}, \quad K_1 = \frac{1}{6} \begin{bmatrix} 3 & -4 & 1 \\ 4 & 0 & -4 \\ -1 & 4 & -3 \end{bmatrix} \tag{4.48b}$$

$$K_0 = \frac{h_j}{30} \begin{bmatrix} 4 & 2 & -1 \\ 2 & 16 & 2 \\ -1 & 2 & 4 \end{bmatrix}, \quad 1 = \frac{h_j}{6} \begin{bmatrix} 1 \\ 4 \\ 1 \end{bmatrix} \tag{4.48c}$$

Note, however, that if the equation is in conservation form and the convective term is written as in (4.9d) to give $-(bu, v')$ in the inner products, the element stiffness matrix K_1 obtained by assuming a constant value of b in the element is changed to

$$K_1^{(\text{cons})} = \frac{1}{6} \begin{bmatrix} -3 & -4 & 1 \\ 4 & 0 & -4 \\ -1 & 4 & 3 \end{bmatrix} \tag{4.48d}$$

The switch of signs in the top left and bottom right corners results after assembly in the addition of a term $(b_{i+\frac{1}{2}} - b_{i-\frac{1}{2}})U_i$ to approximate the extra $b'u$ term in the differential equation. On the other hand, in the numerical example given in section 4.6.2 this is found not to be a good enough approximation: it is much better to consider the convective flux bU as a single quantity with the same form as U when calculating the integrals; then using the form $((bU)', v)$ to compute K_1 from (4.48b) or using (4.48d) makes no difference after assembly.

In the assembly process for the nonconservative case, the first and last rows of the matrices are combined to give the difference scheme associated with a typical inter-element node at x_i,

$$\frac{1}{3}\left[\frac{\epsilon a_{i-\frac{1}{2}}}{x_i - x_{i-1}}\left(U_{i-1} - 8U_{i-\frac{1}{2}} + 7U_i\right)\right.$$

$$\left. + \frac{\epsilon a_{i+\frac{1}{2}}}{x_{i+1} - x_i}\left(7U_i - 8U_{i+\frac{1}{2}} + U_{i+1}\right)\right]$$

$$+ \frac{1}{6}\left[b_{i-\frac{1}{2}}(U_{i-1} - 4U_{i-\frac{1}{2}} + 3U_i) + b_{i+\frac{1}{2}}(-3U_i + 4U_{i+\frac{1}{2}} - U_{i+1})\right]$$

$$+ \frac{1}{30}\left[c_{i-\frac{1}{2}}(x_i - x_{i-1})(-U_{i-1} + 2U_{i-\frac{1}{2}} + 4U_i)\right.$$

$$\left. + c_{i+\frac{1}{2}}(x_{i+1} - x_i)(4U_i + 2U_{i+\frac{1}{2}} - U_{i+1})\right]$$

$$= \frac{1}{6}\left[S_{i-\frac{1}{2}}(x_i - x_{i-1}) + S_{i+\frac{1}{2}}(x_{i+1} - x_i)\right], \quad i = 1, 2, \ldots, J-1. \qquad (4.49a)$$

The left-hand boundary condition is imposed by setting $U_0 = u_L$ and at the right we obtain a similar equation with $i = J$ and all the terms with coefficients $a_{J+\frac{1}{2}}$ etc. set to zero. At the element midpoints no assembly is necessary and the middle rows of the matrices yield the difference scheme, after multiplying through by $\frac{3}{2}$,

$$4\frac{\epsilon a_{i-\frac{1}{2}}}{x_i - x_{i-1}}(-U_{i-1} + 2U_{i-\frac{1}{2}} - U_i) + b_{i-\frac{1}{2}}(U_i - U_{i-1})$$

$$+ \frac{1}{10}c_{i-\frac{1}{2}}(x_i - x_{i-1})(U_{i-1} + 8U_{i-\frac{1}{2}} + U_i) = S_{i-\frac{1}{2}}(x_i - x_{i-1})$$

$$i = 1, 2, \ldots, J. \qquad (4.49b)$$

Apart from the mass matrix which affects the cu term, this is a standard three-point central difference scheme based on the mesh length $\frac{1}{2}h_i$, while the difference operators in (4.49a) make use of differences on both the full mesh and the half mesh. Treated as separate difference schemes they are each of second order accuracy; it is through their combination that one obtains third order accuracy — see the next sub-section.

In two dimensions, the calculational procedures and their results are generalisations of those for the bilinear approximation which were given in detail in the previous section, and make use of the same local to global mapping given by (4.7) with the same relationship (4.37b) between the gradients in the local and global coordinate systems. What is involved has much in common with the generalisation from linear to quadratic approximation on triangles described in section 4.2.3. Thus to avoid unnecessary repetition we will summarise the results and move quickly to the special case of a uniform square mesh.

The global expansion for the Galerkin approximation can be written in terms of three types of basis function, c.f. (4.30),

$$U(x,y) = \tilde{u}_B(x,y) + \sum_{j=1}^{N_1} \phi_j^{BQV}(x,y)U_j + \sum_{j=N_1+1}^{N_1+N_2} \phi_j^{BQS}(x,y)U_j$$

$$+ \sum_{j=N_1+N_2+1}^{N_1+N_2+N_3} \phi_j^{BQC}(x,y)U_j; \tag{4.50}$$

here ϕ_j^{BQV} denotes the biquadratic basis function at a vertex, which normally has support over four neighbouring quadrilaterals, ϕ_j^{BQS} denotes that at the midpoint of a side with support over two quadrilaterals, and ϕ_j^{BQC} denotes that at a centroid of a quadrilateral with support confined to that element.

The element stiffness matrices $K_{\alpha,\beta}^e$ are given by the same expression (4.39) as in the bilinear case but with the biquadratic element basis functions substituted to give 9×9 matrices. Tensor product 2×2 Gaussian quadrature formulae will normally be used for their evaluation, with the gradients given by (4.37b). Little would be gained by attempting to gather the results of this process in explicit formulae corresponding to those in (4.40)–(4.43) for the bilinear case, or (4.48)–(4.49) for the one-dimensional quadratic case. So we jump directly to the case of constant coefficients and a uniform square mesh. We use the difference notation as in (4.45) but also need difference operators on the half mesh for which we use the notation $\delta_{x/2}^2$ etc.; for further simplification we introduce the mass matrix operator, whose form can be deduced from (4.49a),

$$M_x := 1 + \tfrac{1}{5}\delta_{x/2}^2 - \tfrac{1}{10}\delta_x^2. \tag{4.51}$$

Then, at a vertex, multiplying (4.49a) by 3 and specialising to the present situation leads to

$$- \epsilon a[M_y(8\delta_{x/2}^2 - \delta_x^2) + M_x(8\delta_{y/2}^2 - \delta_y^2)]U + b^{(x)}hM_y(4\Delta_{0x/2} - \Delta_{0x})U$$

$$+ b^{(y)}hM_x(4\Delta_{0y/2} - \Delta_{0y})U + (c - \nabla \cdot \mathbf{b})h^2 M_x M_y U = Sh^2 \tag{4.52a}$$

which should be compared to (4.45). At the midpoint of a vertical side we similarly obtain from (4.49b)

$$- \epsilon a[(1 + \tfrac{1}{10}\delta_{y/2}^2)(8\delta_{x/2}^2 - \delta_x^2) + 4M_x\delta_{y/2}^2]U$$

$$+ b^{(x)}h(1 + \tfrac{1}{10}\delta_{y/2}^2)(4\Delta_{0x/2} - \Delta_{0x})U + 2b^{(y)}hM_x\Delta_{0y/2}U$$

$$+ (c - \nabla \cdot \mathbf{b})h^2 M_x(1 + \tfrac{1}{10}\delta_{y/2}^2)U = Sh^2, \tag{4.52b}$$

with the corresponding scheme on a horizontal side; and at a centroid we obtain

$$- 4\epsilon a[(1 + \tfrac{1}{10}\delta_{y/2}^2)\delta_{x/2}^2 + (1 + \tfrac{1}{10}\delta_{x/2}^2)\delta_{y/2}^2]U$$
$$+ 2b^{(x)}h(1 + \tfrac{1}{10}\delta_{y/2}^2)\Delta_{0x/2}U + 2b^{(y)}h(1 + \tfrac{1}{10}\delta_{x/2}^2)\Delta_{0y/2}U$$
$$+ (c - \nabla\cdot\mathbf{b})h^2(1 + \tfrac{1}{10}\delta_{x/2}^2)(1 + \tfrac{1}{10}\delta_{y/2}^2)U = Sh^2. \tag{4.52c}$$

Note that on a mesh with $x = rh, y = sh$ as the vertices, the scheme in (4.52a) is centred at $U_{r,s}$ where this argument is abbreviated to U; that in (4.52b) is centred at $U_{r-\frac{1}{2},s}$ with the corresponding scheme for a horizontal side at $U_{r,s-\frac{1}{2}}$; and that in (4.52c) is centred at $U_{r-\frac{1}{2},s-\frac{1}{2}}$.

4.5 Comparative approximation properties

The starting point for comparing the accuracy of the various schemes given in the previous four sub-sections is the general error bound of section 4.2.2. Since they are all Galerkin schemes this reduces to ascertaining the approximation properties of the four elements in the form given by (4.14). These are derived in standard texts on finite element methods (see e.g. Strang and Fix (1973) or Ciarlet (1978)) and we summarise them here.

The assumed regularity of both the triangular and the quadrilateral meshes is most simply embodied in the bounds that can be placed on the derivatives of the global basis functions. The derivatives with respect to the local coordinates are clearly bounded in all cases, as are the basis functions themselves; so any problem of unboundedness arises from the transformation of the gradients given by (4.5) for a triangle and from inverting the Jacobian (4.8) in the case of a quadrilateral. In the former case $\det J^L = 2A_{123}$ and in the latter $\det J^B \geq \frac{1}{2}\min(A_{412}, A_{123}, A_{234}, A_{341})$. Thus suppose h_e is the diameter of an element in either case; then we shall have

$$\det J \geq ch_e^2$$

for some positive constant c independent of the element, if all the angles in the mesh are uniformly bounded away from 0 and π. Since all the entries in J are $O(h_e)$, we then obtain the required bound for any global basis functions ϕ in the form

$$\|\nabla\phi\|_{L_\infty(e)} \leq K/h_e, \tag{4.53}$$

where K is a constant for the mesh which is determined by this angle bound.

With this regularity condition we can state the required approximation result in terms of seminorms $|\cdot|_{s,e}$, formed from integrals over an element e of the squares of all derivatives of order s. Approximation of a function w is by its *interpolant* w^I, the function in the finite element space S^h whose nodal values equal those of w.

Theorem 4.5.1 *Suppose that S^h contains all polynomials of degree strictly less than k over each element e and its basis functions satisfy the regularity condition (4.53). Then for any function $w \in H^k(e)$ there are constants C_0 and C_1 which are independent of e such that*

$$\left| w - w^I \right|_{s,e} \le C_s h_e^{k-s} |w|_{k,e}, \quad s = 0, 1. \tag{4.54}$$

Proof. This is based on use of the *Bramble–Hilbert lemma* (Bramble and Hilbert 1970) to generalise the classical bound for the remainder in a Taylor expansion for w — see, e.g., Strang and Fix (1973), pp144-6 for details of a theorem stated in these terms. □

Remark In Chapter 3 of Ciarlet (1978) similar theorems are derived directly from the regularity condition given at the beginning of this chapter, namely that the ratios of the diameters of the triangles to the diameters of their inscribed circles are bounded. This condition is equivalent to a minimum angle condition; and in fact the weaker maximum angle condition (that all angles are uniformly bounded from π) is sufficient to obtain such interpolation bounds — see Křížek (1991) and the references therein.

To extend this result to the whole domain Ω, rather than merely to Ω^h, requires us to pay more attention to the boundary $\partial\Omega$ and its approximation. For both the linear elements on triangles and the bilinear elements on quadrilaterals we have $k = 2$ and in the first place need $w \in H^2(\Omega)$, which holds for the solution of our boundary-value problem if the boundary is locally Lipschitz continuous — see Chapter 2 and the references quoted there. Thus if Ω is a polygonal domain and the discretised domain Ω^h is identical to it, we obtain immediately the desired result

$$\left| w - w^I \right|_{s,\Omega} \le C_s h^{k-s} |w|_{k,\Omega}, \quad s = 0, 1 \tag{4.55}$$

where $h = \max(h_e, e \in \Omega^h)$. More generally, suppose the boundary nodes of Ω^h are made to lie on $\partial\Omega$ which is smooth enough between them to ensure that the measure of the "skin" lying between $\partial\Omega^h$ and $\partial\Omega$ is $O(h^2)$. Then if $\Omega^h \subset \Omega$ we can extrapolate w^I and still obtain (4.55); and if, on the other hand, some parts of $\partial\Omega$ lie in the interior of Ω^h we can obtain (4.55) by truncating the integrals, although as we shall see below this is not what will be done in practical applications.

In the case of quadratic elements on triangles and biquadratics on quadrilaterals, we have $k = 3$ and need both a smoother boundary and better approximations to it to maintain the full order of accuracy right up to the boundary. The answer lies in the use of *isoparametric elements*; that is, the same local basis functions are used to define the local to global transformation as are used in the trial space. This was already the case for linear and bilinear elements, but for quadratics and biquadratics it introduces the possibility of elements with curved sides. We need this flexibility only on the boundary and it is sufficient to have only one curved side on an element.

The most convenient approach is a hierarchical one in which the linear or bilinear basis is augmented by just one basis function from the higher order scheme. Thus for a triangle, suppose that the side opposite the vertex r_1 is to be curved. Then we introduce the quadratic basis function corresponding to the midpoint of this side, which is to correspond to a point r_4 on the boundary $\partial\Omega$ between r_2 and r_3. The mapping (4.1) is thus changed, by using $N_4^Q(\xi, \eta)$ of (4.29), to

$$r(\xi, \eta) = (1 - \xi - \eta)r_1 + \xi r_2 + \eta r_3 + 4\xi\eta q_4. \tag{4.56}$$

By substitution of $\xi = \eta = \frac{1}{2}$ we find the parameter $q_4 \equiv (s_4, t_4)$ corresponds to

$$q_4 = r_4 - \frac{1}{2}(r_2 + r_3). \tag{4.57}$$

We also obtain the Jacobian of the transformation as

$$J = \begin{bmatrix} (x_2 - x_1) + 4s_4\eta & (y_2 - y_1) + 4t_4\eta \\ (x_3 - x_1) + 4s_4\xi & (y_3 - y_1) + 4t_4\xi \end{bmatrix} \tag{4.58a}$$

instead of (4.3). Expansion of this and substitution from (4.57) gives

$$\det J = 2A_{123} + 4(2A_{124} - A_{123})\xi + 4(2A_{134} - A_{123})\eta \tag{4.58b}$$

where, for instance, A_{124} denotes the area of the straight-sided triangle with vertices r_1, r_2 and r_4. It is clear that to maintain the nonsingularity of the transformation we need $\det J > 0$ in the triangle and hence

$$A_{124} > \frac{1}{4}A_{123} \text{ and } A_{134} > \frac{1}{4}A_{123}; \tag{4.59}$$

this is also sufficient and it puts only a mild constraint on the placing of the new boundary vertex r_4.

Similarly, in the case of quadrilaterals the mapping given by (4.7) can be modified to make the side between vertices r_2 and r_3 curved, with a new boundary node r_6, by writing

$$r(\xi, \eta) = \sum_{\alpha=1}^{4} N_\alpha^B(\xi, \eta)r_\alpha + \frac{1}{2}(1 + \xi)(1 - \eta^2)q_6, \tag{4.60a}$$

where by setting $\xi = 1, \eta = 0$ we have for $q_6 \equiv (s_6, t_6)$

$$q_6 = r_6 - \frac{1}{2}(r_2 + r_3). \tag{4.60b}$$

If we denote by j_{11}, j_{12}, etc. the entries in the Jacobian (4.8) for the straight-sided quadrilateral, we can write that for the modified quadrilateral as

$$\begin{bmatrix} j_{11} + \frac{1}{2}s_6(1 - \eta^2) & j_{12} + \frac{1}{2}t_6(1 - \eta^2) \\ j_{21} - s_6(1 + \xi) & j_{22} - t_6(1 + \xi) \end{bmatrix}$$

In calculating the determinant, the terms in s_6t_6 cancel; but there will still be cubic terms and the sufficient conditions for the transformation to be nonsingular are quite complicated. However, it is a quite straightforward matter to calculate det J at a number of points and thus obtain necessary conditions. The determinant is clearly unchanged at r_1 and r_4 (i.e., $\xi = -1, \eta = \pm 1$) and at three of the key points we have

$$(\det J)(r_2) = 2A_{126} - \tfrac{1}{2}A_{123}, \ (\det J)(r_3) = 2A_{346} - \tfrac{1}{2}A_{234}$$
$$(\det J)(r_6) = (A_{126} - \tfrac{1}{4}A_{123}) + (A_{346} - \tfrac{1}{4}A_{234}) + \tfrac{1}{4}A_{263}.$$

Thus the conditions $A_{126} > \tfrac{1}{4}A_{123}$ and $A_{346} > \tfrac{1}{4}A_{234}$, which correspond to the conditions (4.59) on the triangle, may have to be strengthened to take account of the triangle Δ_{263} formed at the curved side. If the triangle is convex, the area A_{263} is positive and the condition of the Jacobian is improved; but if the triangle is made concave because of the insertion of the point r_6, the positioning of the point is further restricted.

The approximation of curved boundaries by such as these isoparametric elements, the use of numerical quadrature to evaluate both the load vectors and the entries in the stiffness matrices, and the use of nonconforming methods which we have not yet discussed, are called *variational crimes* by Strang and Fix (1973); this is because they invalidate the assumption $V \in H^1_{E_0}$ that led to the basic relation (4.12). However, finite element methods would not have achieved the great practical successes that they have without the use of these devices; and all the error estimates that we have given can be extended to cover their use. A good general reference is Ciarlet (1978), especially Chapter 4 and the references given there; in particular, it is shown that the use of isoparametric elements requires no more accurate quadrature formulae than do standard elements.

4.6 Numerical comparisons of accuracy

To conclude these sections on Galerkin schemes, we will present the results of some numerical experiments carried out on model problems that were introduced in Chapter 1. These will demonstrate that the error estimates of Theorems 4.2.1 and 4.5.1 are not unduly pessimistic and that at quite modest values of the mesh Péclet number we need to use the Petrov–Galerkin methods that we study in the next chapter. We shall also take the opportunity of comparing accuracies with various finite difference schemes that were described in Chapter 3.

4.6.1 One dimension, constant b

We start in one dimension, where we have just the two schemes based on piecewise linears and piecewise quadratics respectively. Our first model problem,

$$-\epsilon u'' + bu' = S \ \text{on} \ (0,1) \tag{4.61a}$$

$$u(0) = 0, \quad u(1) = 1, \tag{4.61b}$$

particularly the special case $S \equiv 0$, has already been referred to frequently because on a uniform mesh the approximations can be given in a simple explicit form. The piecewise linear Galerkin scheme, as we have already noted, reduces to the central difference scheme considered in section 1.3.1; in terms of the mesh Péclet number $\beta = bh/\epsilon$, the exact and approximate solutions are

$$u(x_j) = \frac{e^{\beta j} - 1}{e^{\beta J} - 1}, \quad U_j = \frac{\lambda_+^j - 1}{\lambda_+^J - 1} \tag{4.62}$$

where $\lambda_+ = (1 + \frac{1}{2}\beta)/(1 - \frac{1}{2}\beta)$. So the piecewise linear approximation gives an oscillatory solution if $\beta > 2$, as shown in Fig. 1.4.

In the case of piecewise quadratics, the scheme given in (4.49) reduces for the present problem to

$$(\delta_x^2 - 8\delta_{x/2}^2)U_i + \beta(4\Delta_{0x/2} - \Delta_{0x})U_i = 0 \tag{4.63a}$$

at an inter-element node, and

$$-4\delta_{x/2}^2 U_{i-\frac{1}{2}} + 2\beta\Delta_{0x/2}U_{i-\frac{1}{2}} = 0 \tag{4.63b}$$

at an element midpoint. The second equation can be used to eliminate the midpoint values, to give after a little algebra the three-point scheme

$$[1 + \tfrac{1}{2}\beta - 4(1 + \tfrac{1}{4}\beta)^2]U_{i-1} + [6 + \tfrac{1}{2}\beta^2]U_i$$
$$+ [1 - \tfrac{1}{2}\beta - 4(1 - \tfrac{1}{4}\beta)^2]U_{i+1} = 0. \tag{4.63c}$$

By consistency, $\lambda = 1$ gives one root of the corresponding characteristic equation while the other is given by

$$\lambda_+^Q = \frac{1 + \tfrac{1}{2}\beta + \tfrac{1}{12}\beta^2}{1 - \tfrac{1}{2}\beta + \tfrac{1}{12}\beta^2}. \tag{4.64}$$

This is in fact the $(2, 2)$ Padé approximant to e^β, so that it has the same fourth order approximation properties as the generalised OCI scheme given in section 3.2.3 for which we obtained the $(4, 0)$ Padé approximant (3.29). Moreover, as with that scheme, λ_+^Q never changes sign and hence the approximation given by substituting λ_+^Q into (4.62) is never oscillatory. The two schemes are therefore strictly comparable, which should be borne in mind in the further tests below. It should be noted, however, that the midpoint values are given by interpolation only if $\beta \leq 4$ and even for lower values may appear to give some oscillation. Also, although the two formulae (4.49a) and (4.49b) depend on the load vectors $S_{i-\frac{1}{2}}$ and $S_{i+\frac{1}{2}}$ through positive coefficients, when the midpoint values are eliminated

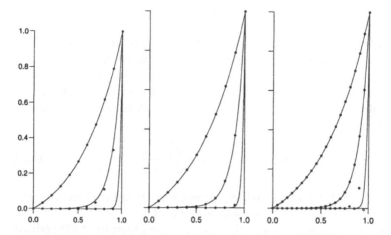

Figure 4.6. *Comparison of the approximations, for the model problem (4.61) with $S \equiv 0$ for $\beta = 0.2, 1.0$ and 5.0, given by the adaptive upwind scheme (3.21) and (3.22) (on the left), the generalised OCI difference scheme (in the centre) and the piecewise quadratic Galerkin scheme (on the right, where open circles denote the element midpoint values)*

one obtains terms that would give

$$\frac{3}{2}\frac{h^2}{\epsilon}\left[(1 + \tfrac{1}{6}\beta)S_{i-\frac{1}{2}} + (1 - \tfrac{1}{6}\beta)S_{i+\frac{1}{2}}\right]$$

as the right-hand side of (4.63c). Thus the conditions (3.32) for a maximum principle to be satisfied hold only for $\beta \leq 6$.

We have already in Fig. 1.4 compared the approximations given by the central difference and an upwind difference scheme with the true solution in the case $S \equiv 0$. Now in Fig. 4.6 we compare those given by the OCI scheme, the piecewise quadratic Galerkin scheme, and the adaptive mix of central and upwind difference schemes given by (3.21) and (3.22), for the same cases $\beta = 0.2, 1.0$ and 5.0. Each scheme gives a monotone sequence of nodal values in every case; but in the last case of $\beta = 5.0$, as expected the midpoint values given by (4.63b) for the Galerkin scheme begin to oscillate, with one going negative; while the simple adaptive scheme becomes a two-point scheme for $\beta \geq 2$, so it does not pick up the right-hand boundary data at all and in this case gives zero everywhere. The OCI scheme is remarkably accurate in all cases; the Galerkin nodal values are comparably accurate except in the last case where they are disappointing and suggest a boundary layer double the actual thickness; and the simple adaptive scheme is quite accurate, even allowing for the complete neglect of the boundary layer in the last case.

Before leaving this simple example it is worth comparing what is given by the general bound of Theorem 4.2.1 in this case with what is given by a specific calculation.

Lemma 4.6.1 *For the problem (4.61) with $S \equiv 0$, approximated by the piecewise linear Galerkin scheme, the general bound of (4.15) gives*

$$\|u' - U'\|_{L_2} \le K\beta(1 + \beta)(b/\epsilon)^{\frac{1}{2}}, \tag{4.65a}$$

while an explicit calculation gives

$$\|u' - U'\|_{L_2} \le \beta(8 + \beta^2)^{-\frac{1}{2}}(b/\epsilon)^{\frac{1}{2}} \tag{4.65b}$$

when terms of order $e^{-b/\epsilon}$ are neglected.

Proof. With $l = 2$ in (4.15) and by neglecting $e^{-b/\epsilon}$ to get

$$u'' \approx (b/\epsilon)^2 e^{-b(1-x)/\epsilon},$$

one obtains $\|u''\|_{L_2}^2 \approx \frac{1}{2}(b/\epsilon)^3$ and hence (4.65a). On the other hand, a direct calculation using (4.62) and also neglecting λ_+^{-J} gives

$$\int (u' - U')^2 dx \approx \frac{1}{2}\frac{b}{\epsilon} + \frac{1}{h}\frac{\lambda_+ - 1}{\lambda_+ + 1} - 2\frac{1 - e^{-\beta}}{h}\frac{\lambda_+ - 1}{\lambda_+ - e^{-\beta}}$$

$$= \frac{b}{\epsilon}\frac{\coth\frac{1}{2}\beta - 2/\beta}{\coth\frac{1}{2}\beta + 2/\beta}. \tag{4.66}$$

Note the occurrence here of the upwinding coefficient (3.34) used in the Allen and Southwell scheme. Using the bounds $0 \le e^{-\beta} \le (1 + \frac{1}{2}\beta)^{-2}$ gives an upper bound for (4.66) when $\beta \ge 2$ of $(b/\epsilon)[\frac{1}{2}\beta/(1 + \frac{1}{2}\beta)]^2$, which in turn is bounded by (4.65b); and for $\beta < 2$ a Taylor expansion of the numerator gives a bound of $\frac{1}{12}(b/\epsilon)\beta^2$, which since $12 \ge 8 + \beta^2$ again gives (4.65b). □

We see from this result that the general bound of Theorem 4.2.1 gives the correct $O(h)$ behaviour for the error in the gradient as $\beta \to 0$; and for moderate values of β it gives the correct dependence on (b/ϵ), although it exaggerates the effect of large β.

4.6.2 A one-dimensional conservation law

As a second model problem in one dimension we introduce a variable velocity field, a Neumann outflow boundary condition and a localised source function:

$$-\epsilon u'' + ((1 + x)^3 u)' = \frac{1}{0.01 + (2x - 1)^2} \quad \text{on } (0, 1) \tag{4.67a}$$

$$u(0) = 0, u'(1) = 0. \tag{4.67b}$$

We will compare the results given by the central difference scheme, the adaptive upwind difference scheme, the generalised OCI scheme and the two Galerkin schemes based on piecewise linears and piecewise quadratics — in each case on a uniform mesh with $h = 0.1$. In the first, a central difference is used for the convection term $(bu)'$ at $x_1, x_2 \ldots, x_{10}$, with the Neumann boundary condition imposed by setting $U_{11} = U_9$; and also for the adaptive upwind scheme, we modify (3.21) to use the conservation form $[(1 - \alpha)\Delta_0 + \alpha\Delta_-](b_i U_i)$ with α based on the value of β_i. However, for the OCI scheme we use the nonconservative form as in (3.1a), namely $bu' + b'u$, or $bu' + cu$ where $b = (1 + x)^3$ and $c = 3(1 + x)^2$, while still setting $U_{11} = U_9$ to apply the Neumann boundary condition; thus the scheme has the form — c.f. (3.7), (3.8) and (3.12) —

$$p_i^- (U_{i-1} - U_i) + p_i^+ (U_{i+1} - U_i) = h^2[q_i^- S_{i-1} - c_{i-1}U_{i-1})$$
$$+ q_i(S_i - c_i U_i) + q_i^+ (S_{i+1} - c_{i+1}U_{i+1})], \qquad (4.68a)$$

where

$$p_i^- = -\epsilon - \tfrac{1}{2}h(3q_i^- b_{i-1} + q_i b_i - q_i^+ b_{i+1})$$
$$p_i^+ = -\epsilon - \tfrac{1}{2}h(q_i^- b_{i-1} - q_i b_i - 3q_i^+ b_{i+1}) \qquad (4.68b)$$

and q_i^-, q_i, q_i^+ are functions of $\beta_i = b_i h / \epsilon$ given in equations (3.26).

In the Galerkin schemes we have generally taken b and S as constant on each interval, equal to their midpoint values $b_{i-\frac{1}{2}}$ and $S_{i-\frac{1}{2}}$, so that the (conservative form) stiffness matrices and load vectors given explicitly in earlier sections of the chapter could be used. However, more accurate quadrature has also been applied and the results compared.

The results in all these cases are gathered together in Fig. 4.7 and plotted against the solution $\tilde{u}_L$ of the reduced equation, which is easily calculated explicitly as

$$\tilde{u}_L(x) = \frac{5}{(1 + x)^3}[\tan^{-1} 10 + \tan^{-1}(20x - 10)]. \qquad (4.69)$$

The top left figure shows the result for the central difference scheme with $\epsilon = 0.1$, which is already somewhat oscillatory; for all other schemes results are also given for $\epsilon = 0.01$, but for the central difference scheme the solution in this case oscillates between -4 and 6.7 and is not worth plotting. By contrast, the adaptive upwind scheme gives results shown in the top right figure which are nonoscillatory and quite accurate; for $\epsilon = 0.01$ the upwind scheme is applied at all points, but for $\epsilon = 0.1$ we have $\beta_i < 2$ for $i = 1, 2$ so that central differences are used at these points. The two central figures are for the piecewise linear Galerkin method, that on the left resulting from midpoint quadrature of S and that on the right from exact integration. The greatly reduced oscillation compared with the central difference scheme is perhaps surprising. It results from the differing treatment of the convection terms: for central differences, the approximation $\Delta_0(bU)_i, \equiv \mu\delta(bU)_i$ in terms of the half-interval averaging

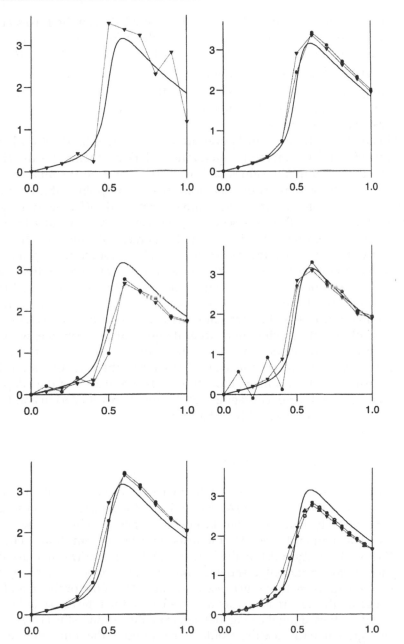

Figure 4.7. *Comparison of approximations to (4.67), with the reduced equation solution (4.69) shown for reference: solid dots correspond to $\epsilon = 0.01$, solid triangles to $\epsilon = 0.1$; from top left, the approximations are for the central difference, adaptive upwind difference, linear Galerkin (midpoint rule), linear Galerkin (exact source terms), generalised OCI, and quadratic Galerkin schemes (with open symbols signifying midpoint values)*

and differencing operators, leaves the coupling between the fluxes at the odd and even points totally dependent on the diffusion; whereas for the Galerkin scheme we obtain $\delta(b_i \mu U_i)$, which introduces a term $\frac{1}{2}(b_{i+\frac{1}{2}} - b_{i-\frac{1}{2}})$ on the diagonal of the stiffness matrix. Note too that the use of the midpoint values of S reduces the contribution from the source term. Correspondingly, use of the nodal values S_i in all the difference schemes overestimates the source term.

Finally, the bottom set of figures are for the generalised OCI scheme, on the left, and the piecewise quadratic Galerkin scheme, on the right; for the latter, midpoint values of S were used and very similar results obtained with the nonconservative form (using midpoint values of b and adding $b'U$ terms) and the conservative form (with bU treated as lying in the trial space). The OCI scheme gives very similar results to those for the much simpler adaptive upwind scheme, with most of the error probably due to an inadequate treatment of the source term peak. The piecewise quadratic Galerkin results, however, are much better than those for the corresponding piecewise linear method, although for $\epsilon = 0.01$ the midpoint values have minor oscillations on the left that are triggered by this source term peak. Further improvements as in the piecewise linear case are obtained by using more accurate quadrature formulae, with the peak and outflow values then accurate to two decimal places, but the midpoint oscillations are enhanced. Thus all of these methods are less than perfect for this problem, though the relative success of the adaptive upwind scheme is an encouraging sign for the Petrov–Galerkin and finite volume methods of the next two chapters.

4.6.3 The first IAHR/CEGB problem

The only two-dimensional problem that we shall use to compare methods at this stage is the first of the IAHR/CEGB problems described in section 1.2. The velocity field is shown in Fig. 1.3 together with the inlet data profile. The standard mesh for the tests reported by Smith and Hutton (1982) has $\Delta x = \Delta y = 0.1$ on the $[-1, 1] \times [0, 1]$ box and we shall use this. It is not always sufficient to show the outlet profile on $y = 0, 0 \leq x \leq 1$ in order to compare methods, because the symmetry of the problem and the mesh can on this line mask errors which show in the interior; so in Fig. 4.8 we show profiles along the diagonal line from the origin to the $(1, 1)$ corner of the region. In each case the profiles correspond to the four values of the diffusion coefficient, $\epsilon = 10^{-6}, 2 \times 10^{-3}, 10^{-2}$ and 10^{-1}, which correspond respectively to average mesh Péclet numbers of $\beta = 10^5, 50, 10$ and 1.0. In Fig. 4.9 a perspective plot of the three-dimensional solution surface is also given for one case.

Many results for these problems have been published in papers already

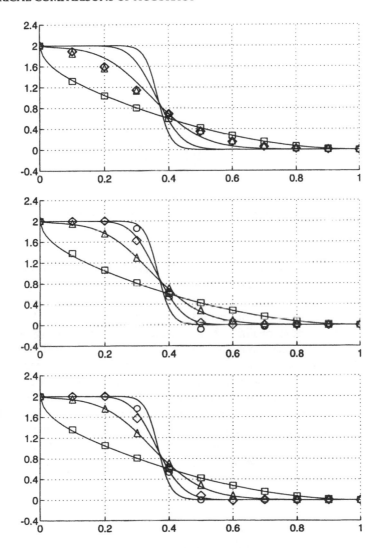

Figure 4.8. *Profiles for the first IAHR/CEGB problem, on the standard test mesh along the diagonal* $(0,0) - (1,1)$*, with differing symbols for approximations at* $\epsilon = 10^{-6}, 0.002, 0.01, 0.1$ *and full lines for exact solutions; from top to bottom, results are shown for the adaptive upwind difference scheme, bilinear Galerkin and biquadratic Galerkin schemes*

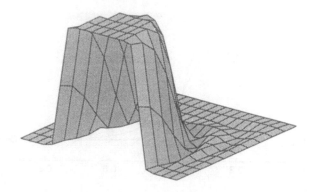

Figure 4.9. *Surface plots of the bilinear Galerkin approximation of the first IAHR/CEGB problem, for $\epsilon = 10^{-6}$*

referred to, and elsewhere. The profiles shown in Fig. 4.8, and compared there with essentially exact solutions obtained on a fine mesh with the methods of the next chapter, are representative of those obtained with the methods that have been described so far. The bilinear Galerkin method in the centre picture of Fig. 4.8 and in the surface plot of Fig. 4.9 shows the oscillations for large mesh Péclet numbers that are typical for standard methods — they are much worse for the central difference scheme. The adaptive upwind scheme in the top picture of Fig. 4.8 shows the thickened profiles typical of simple upwinding; in effect, the upwind switch has forced an upper limit on β in all cases, and this illustrates vividly the effect of the crosswind diffusion that was shown in the analysis of section 3.6.3 to occur in simple five-point schemes. The results for the biquadratic Galerkin method are rather good, since they are on double the mesh size so that they have the same number of free parameters as the other schemes; but they still show some oscillation.

CHAPTER 5

Petrov–Galerkin methods

This chapter and the next two form the heart of the present book. The Petrov–Galerkin methods that we shall consider in this chapter use the same meshes and piecewise polynomial basis functions for the approximations U as in the previous chapter, that is they use the same trial spaces S_E^h. The difference lies in their using test spaces T^h which differ from S_0^h, though having the same dimensions, and these need not be comprised of polynomials. We shall, however, generally assume here that they are conforming so that $T^h \subset H_{E_0}^1$, although eventually we shall consider discontinuous basis functions for T^h. Then a Petrov–Galerkin approximation U is given by

$$B_\gamma(U, V) = l(V) \quad \forall V \in T^h, \tag{5.1}$$

and as a starting point for the error analysis we have the projection property

$$B_\gamma(u - U, V) = 0 \quad \forall V \in T^h, \tag{5.2}$$

corresponding to (4.11) and (4.12). The theoretical problem then is to choose the test space T^h so that (5.2) leads to a useful error bound; moreover, because $B_\gamma(\cdot, \cdot)$ is unsymmetric, there is the associated problem of choosing an appropriate norm in which to measure the error. Both of these aspects of course reflect the practical problem of devising an effective method, and choosing the sense in which it gives a good approximation.

For the most part, in what follows we can work with the simpler problem in which γ and g_B in the boundary condition on $\partial\Omega_N$ are both zero, and the diffusion matrix A is replaced by the scalar a, so that the bilinear form $B_\gamma(\cdot, \cdot)$ and data functional $l(\cdot)$ of (2.50) and (4.9) can be replaced by the $B(\cdot, \cdot)$ of (2.53) and by $(S, \cdot)$. Moreover, we shall usually assume that the convection velocity field is incompressible ($\nabla \cdot \mathbf{b} = 0$) and that $c = 0$. Then we can write for $v, w \in H^1(\Omega)$

$$
\begin{aligned}
B(v, w) &:= \epsilon(a\nabla v, \nabla w) + (\mathbf{b} \cdot \nabla v, w) \\
&= (\epsilon a\nabla v - \mathbf{b}u, \nabla w) + \int_{\partial\Omega} vw(\mathbf{b} \cdot \mathbf{n}) \mathrm{d}\Gamma
\end{aligned}
\tag{5.3a}
$$

as the bilinear form with which we work, in terms of which $u \in H_E^1$ is given by

$$B(u, v) = (S, v) \quad \forall v \in H^1_{E_0}. \tag{5.3b}$$

5.1 Upwinded test functions

Zienkiewicz in Zienkiewicz *et al.* (1975) was among the first to suggest that the analogy in finite element methods to upwinding the finite difference convection operator was provided by upwinding the test function. The idea was taken up vigorously by Mitchell, Griffiths and their associates and developed in a series of papers — Christie *et al.* (1976), Heinrich *et al.* (1977), Griffiths and Mitchell (1979) — the results of which are summarised in the survey article Heinrich and Zienkiewicz (1979).

Consider first of all the simple one-dimensional model problem (4.61) with the piecewise linear trial space; and suppose a conforming test space is constructed by adding to it in a hierarchical manner the quadratic element basis functions $N^Q_2(\xi)$ of (4.46a), i.e., the global basis functions $\psi^Q_{j-\frac{1}{2}}(x)$ of (4.47c). Specifically, if $\phi_j(x)$ is the usual global linear basis function, we define the corresponding global test function by incorporating two such functions to give

$$\psi_j(x) := \phi_j(x) + \alpha\sigma_j(x) \tag{5.4}$$

where

$$\sigma_j(x) = \begin{cases} 3(x - x_{j-1})(x_j - x)/h^2_j & x \in e_j \\ -3(x - x_j)(x_{j+1} - x)/h^2_{j+1} & x \in e_{j+1}. \end{cases}$$

Substitution of this function into the Petrov–Galerkin equations gives no contribution to the diffusion term, because the integral of σ'_j over an element is zero; and because the integral of σ_j over element e_j is $\frac{1}{2}h_j$, the resulting difference scheme on a uniform mesh and with $S \equiv 0$ is

$$-\frac{\epsilon}{h}\delta^2 U_j + b[\frac{1}{2}(1 + \alpha)\Delta_- + \frac{1}{2}(1 - \alpha)\Delta_+]U_j = 0. \tag{5.5}$$

This exactly corresponds to the weighted average of the central and upwind difference schemes given in (3.21); and clearly any function $\sigma_j(x)$ which is zero at x_{j-1}, x_j and x_{j+1} and whose integral over e_j is $\frac{1}{2}h_j$ and over e_{j+1} is $-\frac{1}{2}h_{j+1}$ would give the same result. It is for a nonconstant source function that differences in the choice of $\sigma_j(x)$ would show up. The considerations used in the discussion of section 3.2.2 would apply in this choice and the adaptive upwind scheme of (3.21) and (3.22) could be reproduced by an appropriate choice of $\sigma_j(x)$ and a quadrature rule.

Just as for the difference schemes, choice of the parameter α to satisfy

$\alpha \geq 1 - 2/\beta$ will avoid an oscillatory solution. Moreover, the choice

$$\alpha = \coth \tfrac{1}{2}\beta - 2/\beta \tag{5.6}$$

will reproduce the Allen and Southwell difference operator described in section 3.3.1 on exponentially fitted schemes. This will be the usual choice, based on a local mesh Péclet number. Extension of the scheme to general two-dimensional problems on quadrilateral meshes is straightforward and will be described in later sections where comparisons with other choices of test space will be made.

5.2 Approximate symmetrization

One of the most natural objectives to address in the selection of the test space is the restoration of symmetry in the bilinear form, so as to regain the optimal approximation property that one has when Galerkin methods are applied to symmetric problems. Thus suppose $B_S(\cdot, \cdot)$ is any given symmetric continuous and coercive bilinear form on $H^1_{E_0} \times H^1_{E_0}$, so it defines an inner product and associated norm $\|\cdot\|_{B_S}$. Then by the Riesz representation theorem Riesz and Nagy (1955), for each fixed w the linear functional $B(\cdot, w)$ has a representation $B_S(\cdot, w^*)$ for some $w^* \in H^1_{E_0}$: indeed, one can define a *representer* $R_S : H^1_{E_0} \to H^1_{E_0}$ such that

$$B(v, w) = B_S(v, R_S w) \quad \forall v, w \in H^1_{E_0} \tag{5.7}$$

and it is readily seen that R_S is a linear operator. Now suppose we were able to introduce an 'optimal' test space T^{h^*} such that

$$R_S T^{h^*} = S^h_0. \tag{5.8}$$

Then the Petrov–Galerkin approximation U^*_S obtained with this test space satisfies

$$0 = B(u - U^*_S, W) = B_S(u - U^*_S, R_S W) \quad \forall W \in T^{h^*};$$

but by (5.8) this is equivalent to

$$B_S(u - U^*_S, V) = 0 \quad \forall V \in S^h_0 \tag{5.9a}$$

which is the orthogonality property leading directly to

$$\|u - U^*_S\|_{B_S} = \inf_{Z \in S^h_E} \|u - Z\|_{B_S}. \tag{5.9b}$$

That is, U^*_S is the optimal approximation to u from within the trial space S^h_E and in the sense of the selected norm $\|\cdot\|_{B_S}$.

This approach was proposed and explored in a sequence of papers by Barrett and Morton (Barrett and Morton 1980, 1981, 1984). By using the optimal test space T^{h^*} as a yardstick by which to judge alternative choices, they obtained the following key theorem.

Theorem 5.2.1 *Suppose the test space T^h is conforming and has the same dimension as S_0^h. Let the closeness with which S_0^h can be approximated by $R_S T^h$ be described by the constant $\Delta_S \equiv \Delta_S(h)$ such that*

$$\inf_{W \in T^h} \|V - R_S W\|_{B_S} \leq \Delta_S \|V\|_{B_S} \quad \forall V \in S_0^h. \tag{5.10}$$

Then if $\Delta_S \in [0, 1)$, there exists a unique solution U to the Petrov–Galerkin equations which satisfies the error bound

$$\|u - U\|_{B_S} \leq (1 - \Delta_S^2)^{-\frac{1}{2}} \inf_{Z \in S_E^h} \|u - Z\|_{B_S}. \tag{5.11}$$

Proof. Existence and uniqueness of U follows from showing that the solution U_0 of the problem with zero data is identically zero. Now we have from (5.7) and the definition of U_0

$$0 = B(U_0, W) = B_S(U_0, R_S W) \quad \forall W \in T^h.$$

Hence

$$\|U_0\|_{B_S}^2 \equiv B_S(U_0, U_0) = B_S(U_0, U_0 - R_S W)$$
$$\leq \|U_0\|_{B_S} \|U_0 - R_S W\|_{B_S} \quad \forall W \in T^h;$$

and it follows from (5.10) that

$$\|U_0\|_{B_S}^2 \leq \Delta_S \|U_0\|_{B_S}^2$$

so that $\|U_0\|_{B_S} = 0$ because $\Delta_S < 1$. To obtain (5.11) we make use of the optimal approximation U_S^* given by (5.9) to obtain

$$\|u - U\|_{B_S}^2 = \|u - U_S^*\|_{B_S}^2 + \|U_S^* - U\|_{B_S}^2; \tag{5.12a}$$

and we also have, as for U_0,

$$\|U_S^* - U\|_{B_S}^2 \equiv B_S(U_S^* - U, U_S^* - U) = B_S(u - U, U_S^* - U)$$
$$= B_S(u - U, U_S^* - U - R_S W) \quad \forall W \in T^h$$

and hence

$$\|U_S^* - U\|_{B_S}^2 \leq \Delta_S \|u - U\|_{B_S} \|U_S^* - U\|_{B_S}. \tag{5.12b}$$

Substituting the resulting bound for $\|U_S^* - U\|_{B_S}$ in the right-hand side of (5.12a) yields the desired result (5.11). $\square$

Now there are two natural choices for the symmetric form $B_S(\cdot, \cdot)$. Under the assumption $\nabla \cdot \mathbf{b} = 0$, the convection-diffusion equation can be written in conservation form and the operator factored to give

$$-\nabla \cdot (\epsilon a \nabla u - \mathbf{b} u) \equiv L_1^* L_2 u = S \quad \text{in } \Omega, \tag{5.13a}$$

in terms of the first order operators

$$L_1 v := (\epsilon a)^{\frac{1}{2}} \nabla v, \quad L_2 v := (\epsilon a)^{\frac{1}{2}} \nabla v - [\mathbf{b}/(\epsilon a)^{\frac{1}{2}}] v \tag{5.13b}$$

and where L_1^* is the formal adjoint of L_1. Thus we can introduce two symmetric forms based on (L_1v, L_1w) and (L_2v, L_2w) respectively. A convenient choice is as follows,

$$B_1(v, w) := (a\nabla v, \nabla w) \equiv \epsilon^{-1}(L_1v, L_1w) \qquad (5.14a)$$

$$B_2(v, w) := \epsilon^2(a\nabla v, \nabla w) + ((b^2/a)v, w) \qquad (5.14b)$$

$$\equiv \epsilon(L_2v, L_2w) + \epsilon \int_{\partial\Omega} vw(\mathbf{b}\cdot\mathbf{n})\mathrm{d}\Gamma, \qquad (5.14c)$$

where $b := |\mathbf{b}|$. Clearly $B_1(\cdot, \cdot)$ gives an appropriate norm for diffusion dominated problems, and is in fact closely related to the symmetric part of $B(\cdot, \cdot)$ since in the present case we have

$$\tfrac{1}{2}B(v, w) + \tfrac{1}{2}B(w, v) = \epsilon B_1(v, w) + \tfrac{1}{2}\int_{\partial\Omega} vw(\mathbf{b}\cdot\mathbf{n})\mathrm{d}\Gamma;$$

on the other hand the norm associated with $B_2(\cdot, \cdot)$ tends to the L_2 norm as $\epsilon \to 0$ and might seem more appropriate for convection-dominated problems.

It is important to recognise that the optimal approximations sought with these two norms will have quite different properties. For example, let us consider linear approximations in one dimension with constant a and b. From (5.9a) minimisation with respect to $\|\cdot\|_{B_1}$ gives

$$0 = \int (u - U_1^*)'\phi_j'\mathrm{d}x = \frac{e_j^* - e_{j-1}^*}{x_j - x_{j-1}} - \frac{e_{j+1}^* - e_j^*}{x_{j+1} - x_j} \qquad (5.15)$$

where $e_j^* = U_1^*(x_j) - u(x_j)$. Thus the nodal errors satisfy a maximum principle and, with the application of a Dirichlet boundary condition, must all be zero. Although this nice result does not hold in two dimensions, it is one of the reasons why this norm is attractive even for convection-dominated problems. By contrast, the L_2 norm, which is the appropriate norm for analysing purely hyperbolic problems, leads to a typically oscillatory linear approximation — see Fig. 5.1; such oscillations are easily confused with those produced by, for example, the central difference scheme and therefore require very careful filtering — compare Fig. 1.4 with Fig. 5.1 and see Gresho and Lee (1981).

Before considering specific choices of test space under the two norms, we will give a more explicit expression for the key error parameter Δ_S. Suppose $\{\phi_j, j = 1, 2, \ldots N\}$ are the basis functions for S_0^h and $\{\psi_j, j = 1, 2, \ldots N\}$ those for the test space T^h. Then if W^* minimises the left-hand side of (5.10), we have

$$B_S(R_SW^* - V, R_S\psi_i) = 0 \quad \forall \psi_i \in T^h$$

and hence

$$\|V - R_SW^*\|_{B_S}^2 = \|V\|_{B_S}^2 - \|R_SW^*\|_{B_S}^2 .$$

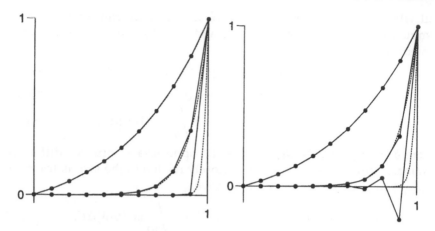

Figure 5.1. *Piecewise linear best fits to the exponential for* $\beta = 0.2, 1.0$ *and* 5.0: *on the left with respect to* $\|\cdot\|_{B_1}$; *and on the right with respect to* $\|\cdot\|_{L_2}$

It follows that we can take

$$\Delta_S^2 = \sup_{V \in S_0^h} \{1 - \|R_S W^*\|_{B_S}^2 / \|V\|_{B_S}^2\}, \tag{5.16}$$

and this is the definition that we use henceforth. Suppose now that we define the following three matrices,

$$A_{i,j} := B_S(R_S \psi_j, R_S \psi_i) \tag{5.17a}$$

$$B_{i,j} := B_S(\phi_j, R_S \psi_i) = B(\phi_j, \psi_i) \tag{5.17b}$$

$$C_{i,j} := B_S(\phi_j, \phi_i) \tag{5.17c}$$

for $i, j = 1, 2, \ldots, N$. Then Δ_S is given by the result of a generalised eigenvalue problem

$$1 - \Delta_S^2 = \inf_{V \in S_0^h} \left\{ \frac{\mathbf{V}^T B^T A^{-1} B \mathbf{V}}{\mathbf{V}^T C \mathbf{V}} \right\}, \tag{5.18}$$

where $\mathbf{V}$ is the vector of components V_j in the expansion $V = \sum V_j \phi_j$; note that if $\mathbf{W}^*$ is similarly the vector in the expansion $W^* = \sum W_j^* \psi_j$, we have $A\mathbf{W}^* = B\mathbf{V}$ and this has been used in obtaining (5.18) from the previous expression for Δ_S^2.

It should also be noted that the proof of existence and uniqueness for the solution of the adjoint to the convection-diffusion problem — using the Lax-Milgram lemma as in Theorem 2.5.1 — establishes the existence of the inverse $R_S^{-1} : H_{E_0}^1 \to H_{E_0}^1$, which satisfies

$$B(v, R_S^{-1} w) = B_S(v, w) \quad \forall v, w \in H_{E_0}^1, \tag{5.19}$$

so complementing the relation (5.7). We shall see below that it is sometimes easier to characterise R_S^{-1} than R_S.

It is worth noting here for future reference the bounds that we have for the operators R_S and R_S^{-1}, and how they depend on earlier assumptions. The lemma given below shows that we may have a bound on R_S that is independent of ϵ, but then that for R_S^{-1} becomes unbounded as $\epsilon \to 0$. This is the case if we choose $B_S(\cdot, \cdot)$ so that Γ_S and γ_S in (5.20b) are independent of ϵ, as for $B_1(\cdot, \cdot)$; but if we scale $B_S(\cdot, \cdot)$ by ϵ^k, then R_S is scaled by ϵ^{-k} and R_S^{-1} by ϵ^k, as for example with the $B_2(\cdot, \cdot)$ of (5.14b).

Lemma 5.2.2 *Suppose that for $\forall v, w \in H_{E_0}^1$ we have*

$$\gamma_\epsilon \|v\|_{H^1}^2 \leq B(v,v) \quad and \quad |B(v,w)| \leq \Gamma \|v\|_{H^1} \|w\|_{H^1}; \tag{5.20a}$$

and

$$\gamma_S \|v\|_{H^1}^2 \leq \|v\|_{B_S}^2 \leq \Gamma_S \|v\|_{H^1}^2, \tag{5.20b}$$

where γ_ϵ, Γ, γ_S and Γ_S are all positive, though $\gamma_\epsilon \to 0$ as $\epsilon \to 0$. Then

$$\|R_S w\|_{H^1} \leq (\Gamma/\gamma_S) \|w\|_{H^1} \tag{5.21a}$$

and

$$\|R_S^{-1} w\|_{H^1} \leq (\Gamma_S/\gamma_\epsilon) \|w\|_{H^1}. \tag{5.21b}$$

Proof. From (5.14) we have $\|R_S w\|_{B_S}^2 = B(R_S w, w)$ and (5.21a) follows from the upper bound for $B(\cdot, \cdot)$; and from (5.19) we have $B_S(R_S^{-1} w, w) = B(R_S^{-1} w, R_S^{-1} w)$, to which we apply the lower bound for $B(\cdot, \cdot)$ to get (5.21b). $\square$

5.2.1 Symmetrization based on $B_1(\cdot, \cdot)$

Denoting by R_1 the Riesz representer in this case, its defining relation (5.7) can be written according to (5.14a) as

$$(\nabla v, a\nabla(\epsilon w - R_1 w) + bw) = 0 \quad \forall v, w \in H_{E_0}^1. \tag{5.22}$$

In one dimension, with Dirichlet boundary conditions at $x = 0, 1$, we can integrate this by parts in a form

$$0 = \int_0^1 v'G\mathrm{d}x = [vG]_0^1 - \int_0^1 vG'\mathrm{d}x = -\int_0^1 vG'\mathrm{d}x$$

and hence deduce that $G = \text{const.}$, i.e.,

$$a(\epsilon w - R_1 w)' + bw = \text{const.} \tag{5.23}$$

with the constant determined by the boundary conditions $(R_1 w)(0) = (R_1 w)(1) = 0$. In this case we therefore have an explicit expression for the representer, $R_1 : H_0^1(0,1) \to H_0^1(0,1)$, namely

$$(R_1 w)(x) = \epsilon w(x) + \int_0^x \frac{bw - C}{a} \mathrm{d}t \quad \forall w \in H_0^1(0,1), \qquad (5.24a)$$

where

$$C = \int_0^1 (bw/a)\mathrm{d}t \Big/ \int_0^1 (1/a)\mathrm{d}t.$$

In particular, for the simple model problem (4.61) in which a and b are constants with $a = 1$, we have

$$(R_1 w)(x) = \epsilon w(x) + b \int_0^x (w - \bar{w})\mathrm{d}t \quad \forall w \in H_0^1(0,1), \qquad (5.24b)$$

where $\bar{w}$ is the average value of w over $(0,1)$. Note that when a homogeneous Neumann boundary condition is applied at $x = 1$, so that there are test functions with $v(1) \neq 0$, the constant C in (5.24a) must be zero and (5.24b) is modified to omit the term $\bar{w}$.

This form can be used in an explicit calculation of the matrices of (5.17) for a given choice of one-dimensional trial space and test space. The resulting error constants in the bound (5.11) will be given in section 5.6 for a number of common choices.

One can also write (5.23) in terms of $z = R_1 w$ as

$$\epsilon a (R_1^{-1} z)' + b R_1^{-1} z = az' + \text{const.}, \qquad (5.25)$$

to deduce from (5.8) that, if the trial space is piecewise linear, then the optimal test space is piecewise exponential in the constant coefficient case. This is the test space derived by Hemker (1977) and discussed in the next section.

In two dimensions, one would have to solve a Poisson equation to obtain $R_1 w$ or a convection-diffusion problem to obtain $R_1^{-1} z$. It is not practicable or worthwhile to do this; but we shall see in section 5.6 that one can nevertheless calculate the error constant in the bound (5.11). One can also make important deductions regarding the form of the optimal test space which we shall use in motivating the streamline diffusion scheme discussed in section 5.5.

5.2.2 Symmetrization based on $B_2(\cdot, \cdot)$

Using R_2 to denote the corresponding Riesz representer, the defining relation (5.7) combined with (5.3) and (5.14b) gives

$$\epsilon(a\nabla v, \nabla w) + (\mathbf{b}\cdot\nabla v, w) = \epsilon^2(a\nabla v, \nabla R_2 w) + ((b^2/a)v, R_2 w).$$

Rewriting as for (5.14c) gives the more useful form

$$(\epsilon a \nabla v - \mathbf{b}v, \nabla w - \epsilon \nabla R_2 w + (\mathbf{b}/a) R_2 w) = \int_{\partial \Omega_N} v(\epsilon R_2 w - w)(\mathbf{b} \cdot \mathbf{n}) \mathrm{d}\Gamma$$

$$\forall v, w \in H^1_{E_0}. \quad (5.26)$$

When this is compared with (5.22) one sees that it is now easier to solve for R_2^{-1} rather than R_2. In particular, for the one-dimensional model problem with $a = 1, b = $ const. and Dirichlet boundary conditions, we have

$$\int_0^1 [(R_2^{-1}z)' - \epsilon z' + bz] e^{bx/\epsilon} [ve^{-bx/\epsilon}]' \mathrm{d}x \quad \forall v, z \in H^1_0(0,1). \quad (5.27a)$$

This can be integrated by parts as in deducing (5.23), but now gives

$$(R_2^{-1}z)' = \epsilon z' - bz + Ce^{-bx/\epsilon}, \quad (5.27b)$$

where the boundary conditions lead to

$$C = (b^2/\epsilon)(1 - e^{-b/\epsilon})^{-1}\bar{z}.$$

Taking $z = \phi_j$ gives the equation for a nonlocal basis function for the optimal test space, which it is convenient to integrate from the right to give

$$\psi_j^*(x) := R_2^{-1}\phi_j(x) = \epsilon \phi_j(x) + b \int_x^1 [\phi_j(t) - \bar{\phi}_j r(t)] \mathrm{d}t, \quad (5.28)$$

where

$$r(t) = (b/\epsilon)e^{-bt/\epsilon}/(1 - e^{-b/\epsilon}).$$

This can be made more local by taking linear combinations of successive basis functions. For piecewise linear ϕ_j on a uniform mesh the nonlocal and localised basis functions are shown in Fig. 5.2: for large values of b/ϵ they are clearly dominated by the piecewise quadratic terms coming from (5.28), and the latter has support over three intervals. Thus a useful set of basis functions consists of

$$\psi_1^*, \psi_2^* - \psi_1^*, \psi_3^* - \psi_2^*, \ldots, \psi_{J-1}^* - \psi_{J-2}^*.$$

If these are matched with the Galerkin basis functions $\phi_1, \phi_2, \ldots, \phi_{J-1}$, their upwind character is clear; moreover, because of their predominantly quadratic form they can be directly compared with those of Heinrich et al. (1977) given in (5.4) and plotted in Fig. 5.3.

Even for this simple model problem these are rather complicated test functions and would seem to be rather awkward to use. However, there is always an alternative formulation of the Petrov–Galerkin equations available when optimal test functions are used and an expression for R_S^{-1} is available, as in this case. Suppose that the problem is formulated to

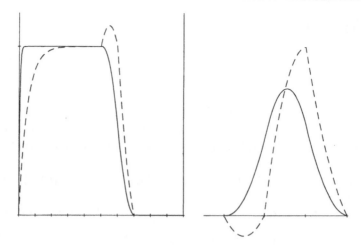

Figure 5.2. *Optimal test functions in the* $\|\cdot\|_{B_2}$ *norm for* $\beta = 2$ *(dashed line) and* $\beta = 20$ *(full line); on the left are shown the nonlocal functions (5.28) and on the right their localised forms*

have homogeneous boundary data. Then the Petrov–Galerkin equations for $U_2^* \in H_{E_0}^1$ are $B(U_2^*, \psi_j^*) = (S, \psi_j^*)$ and can be written as

$$B_2(U_2^*, \phi_j) = (S, R_2^{-1}\phi_j) \quad j = 1, 2, \ldots, J - 1. \tag{5.29a}$$

Suppose further that we can find a modified source function $S_2(x)$ such that

$$(S_2, \phi_j) = (S, R_2^{-1}\phi_j) \quad j = 1, 2, \ldots, J - 1.$$

Then we have to solve

$$B_2(U_2^*, \phi_j) = (S_2, \phi_j) \quad j = 1, 2, \ldots, J - 1 \tag{5.29b}$$

and we have completely symmetrized the original problem. There is the enormous practical advantage here that the algebraic system for the nodal parameters of U_2^* is now symmetric and much easier to solve; in effect, the operator R_2^{-1} has achieved a *pre-conditioning* of the original system based on $B(\cdot, \cdot)$ into a symmetric form. An approximate method based on this approach will be described in section 5.9.

Reinhardt (1982) has developed a very similar formulation based on an extension of the $B_2(\cdot, \cdot)$ symmetrization to the case $c \neq 0$. He has then developed an *a posteriori* error indicator and mesh adaptation strategy capable of detecting and resolving simple boundary layers.

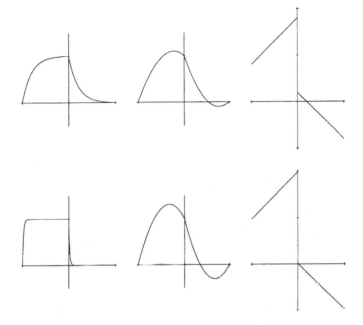

Figure 5.3. *Comparison of one-dimensional test functions for $\beta = 5$ (top) and $\beta = 50$ (bottom); from left to right are shown the Hemker function (5.32), the upwind function (5.4) and the streamline diffusion function (5.41)*

5.3 Hemker test functions

Suppose $G(\mathbf{x}', \mathbf{x})$ is the Green's function for our convection-diffusion problem; thus, as in (2.67), we have

$$B(v, G(\mathbf{x}', \cdot)) = v(\mathbf{x}') \quad \forall v \in H_{E_0}^1. \tag{5.30}$$

Then the error in a Petrov–Galerkin approximation at $\mathbf{x}'$ can be written as

$$
\begin{aligned}
u(\mathbf{x}') - U(\mathbf{x}') &= B(u - U, G(\mathbf{x}', \cdot)) \\
&= B(u - U, G(\mathbf{x}', \cdot) - V) \quad \forall V \in T^h. \tag{5.31}
\end{aligned}
$$

So the pointwise accuracy of U is largely determined by the extent to which this Green's function can be approximated from within the test space. This has been widely used as a means of obtaining or proving *superconvergence* of an approximation at nodal points (that is, having a higher order of approximation than holds generally for the trial space, through properties such as (4.14)) — see Strang and Fix (1973) for references and a general description.

For the one-dimensional model problem of (4.61) this leads to the choice of test space due to Hemker (1977). For each interior node, $\psi_j^H \in H_{E_0}^1$ has support in $[x_{j-1}, x_{j+1}]$, and except at x_j should satisfy $\epsilon \psi'' + b\psi' = 0$.

Normalising to $\psi_j^H(x_j) = 1$ readily gives

$$
\psi_j^H(x) := \begin{cases} (1 - e^{-b(x-x_{j-1})/\epsilon})/(1 - e^{-bh_j/\epsilon}), & x_{j-1} \le x \le x_j \\[2ex] (e^{-b(x-x_j)/\epsilon} - e^{-bh_{j+1}/\epsilon})/(1 - e^{-bh_{j+1}/\epsilon}), & x_j \le x \le x_{j+1}. \end{cases}
$$

$$(5.32)$$

This is shown for a uniform mesh, with $\beta = 5$ and 50, in Fig. 5.3 where it is compared with the upwind test function of (5.4) and that of the streamline diffusion method described in section 5.5. Combining the use of (5.32) with a piecewise linear trial space will give a three-point difference scheme as in (5.5); and it is easy to calculate that the upwind parameter α is given by (5.6) so that the Allen and Southwell operator is always reproduced.

More fundamentally, a simple calculation shows that linear combinations of these basis functions ψ_j^H can exactly reproduce the Green's function $G(x', \cdot)$ for x' corresponding to any interior node; thus, *for any source function S*, it follows from (5.31) that the Petrov–Galerkin approximation obtained with this test space and any conforming trial space S_0^h, of dimension $J - 1$, will be exact at the nodes. In the case of the piecewise linear trial space, this corresponds to the already noted fact that the exactness of U at the nodes is equivalent to its being the best fit to u in the $\|\cdot\|_{B_1}$ norm, coupled with the observation following from (5.25) that span$\{\psi_j^H\}$ is indeed then $R_1^{-1}S_0^h$.

One should pause here to appreciate properly the significance of these Hemker test functions. For the one-dimensional constant coefficient problems $-\epsilon u'' + bu' = S$ with $u(0) = u(1) = 0$, approximated by piecewise linears, they provide the ideal upwinding test space, and the optimal test space in the symmetrization based on the $B_1(\cdot, \cdot)$ inner product. If the integrals occurring in the source terms (S, ψ_j^H) can be evaluated exactly, and all the other calculations carried out exactly, the nodal values U_j will be exact. However, these are difficult integrals to evaluate accurately and Hemker introduced special quadrature formulae for this purpose. When these are used we have a scheme that forms a yardstick against which all alternative schemes for these problems can be measured.

Generalisations to variable coefficient one-dimensional problems are straightforward, but of course will no longer give exact values at the nodes. These will be discussed in section 5.8 and are directly comparable with the difference schemes of Allen and Southwell and of El-Mistikawy and Werle considered in section 3.3. Indeed, it is clear that the local Green's functions introduced for the locally exact difference schemes of section 3.3.2, and plotted in Fig. 3.2, are exactly the Hemker test functions.

As we have seen in section 2.7.2, the Green's functions in two and three dimensions are not wholly exponential, though exponentials in $\mathbf{b}\cdot\mathbf{x}/\epsilon$ largely account for their directional asymmetry. Thus on a rectangular mesh tensor products of Hemker test functions will be used to generate approximations in the numerical experiments of section 5.10.

5.4 Error bounds in L_∞ and L_2 norms

Suppose a Petrov–Galerkin approximation U satisfies the error bound (5.11) in the $\|\cdot\|_{B_S}$ norm. Use of the Green's function $G(\cdot,\cdot)$ in (5.31) shows how errors in the L_∞ norm, or the l_∞ norm over the nodes, can be obtained. We have from that equation

$$
\begin{aligned}
|u(\mathbf{x}') - U(\mathbf{x}')| &= |B(u - U, G(\mathbf{x}',\cdot) - V)| \\
&= |B_S(u - U, R_S(G(\mathbf{x}',\cdot) - V))| \\
&\leq \|u - U\|_{B_S} \|R_S(G(\mathbf{x}',\cdot) - V)\|_{B_S} \quad \forall V \in T^h. \quad (5.33)
\end{aligned}
$$

As long as the second factor here is bounded, the first factor guarantees the expected order of accuracy in the $\|\cdot\|_{B_S}$ or $\|\cdot\|_{H^1}$ norm associated with the trial space and, in addition, retains any improvement in the error constant given by (5.11) through an appropriate choice of test space. Superconvergence, which we shall take to mean improvement over what is to be expected in the L_2 norm, also follows if we can show that the second factor is $O(h^\alpha)$ for $\alpha > 1$. Writing $E := G(\mathbf{x}',\cdot) - V$, from the bounds in Lemma 5.2.2 we have

$$
\|R_S E\|_{B_S} \leq (\Gamma_S^{\frac{1}{2}} \Gamma / \gamma_S) \|E\|_{H^1}, \quad (5.34)
$$

so that the situation is similar to that in a standard self-adjoint problem; in particular, because $G(\mathbf{x}',\cdot)$ has some form of singularity or discontinuity at $\mathbf{x}'$, it will be better approximated by $V \in T^h$ if $\mathbf{x}'$ corresponds to a node. In one dimension we can expect to obtain superconvergence at the nodes by choosing V to approximate piecewise exponentials well; but in more dimensions, as in the self-adjoint case, superconvergence even at the nodes is more difficult to achieve.

Error bounds in the L_2 norm follow what is by now a standard technique due to Aubin (1967) and Nitsche (1968). Define $z \in H^1_{E_0}$ by the following auxiliary problem, using the adjoint bilinear form so as to give

$$
B(v, z) = (u - U, v) \quad \forall v \in H^1_{E_0}. \quad (5.35)
$$

Then setting $v = u - U$ leads to

$$
\begin{aligned}
\|u - U\|_{L_2}^2 &= B(u - U, z) \\
&= B(u - U, z - V) = B_S(u - U, R_S(z - V)) \\
&\leq \|u - U\|_{B_S} \|R_S(z - V)\|_{B_S} \quad \forall V \in T^h. \quad (5.36)
\end{aligned}
$$

As with (5.33) our objective is to obtain the best possible bound on the second factor here. Using definitions of $B_S(\cdot,\cdot)$ and R_S such that the constants γ_S and Γ_S in Lemma 5.2.2 can be taken independent of ϵ, we write the auxiliary problem (5.35) in terms of the symmetric form, giving

$$B_S(R_S z, v) = (u - U, v) \quad \forall v \in H^1_{E_0}. \tag{5.37a}$$

Then it follows that there is a constant K_S, independent of ϵ, for which an *a priori* bound holds,

$$\|R_S z\|_{H^2} \leq K_S \|u - U\|_{L_2}. \tag{5.37b}$$

Now suppose that $R_S T^h$ has an approximation property such that, for $\forall v \in H^1_{E_0} \cap H^2(\Omega)$, we have

$$\inf_{V \in T^h} \|v - R_S V\|_{B_S} \leq K h^\alpha \|v\|_{H^2}. \tag{5.38}$$

Then from (5.36),(5.37) and (5.38) it follows that

$$\begin{aligned}
\|u - U\|^2_{L_2} &\leq K h^\alpha \|R_S z\|_{H^2} \|u - U\|_{B_S} \\
&\leq K K_S h^\alpha \|u - U\|_{L_2} \|u - U\|_{B_S},
\end{aligned}$$

that is,

$$\|u - U\|_{L_2} \leq K K_S h^\alpha \|u - U\|_{B_S}. \tag{5.39}$$

Thus if $R_S T^h$ has the same approximation power as piecewise linear finite elements, we have $\alpha = 1$ and obtain the usual extra order of accuracy in the L_2 norm as compared with that in the $\|\cdot\|_{B_S}$ norm.

It may sometimes be possible to establish (5.38) from the approximate symmetrisation bounds (5.10) and (5.11). Thus there is certainly a function $Z \in S^h_0$ such that

$$\|R_S z - Z\|_{B_S} \leq K' h \|R_S z\|_{H^2};$$

and from (5.10) $\exists V \in T^h$ such that

$$\|Z - R_S V\|_{B_S} \leq \Delta_S \|Z\|_{B_S} \leq \Delta_S (\|Z - R_S z\|_{B_S} + \|R_S z\|_{B_S}).$$

Also, from a Poincaré-Friedrichs inequality $\|R_S z\|_{L_2} \leq K_\Omega \|R_S z\|_{B_S}$ and from (5.37a) with $v = R_S z$, we have $\|R_S z\|_{B_S} \leq K_\Omega \|u - U\|_{L_2}$. Hence by a triangle inequality

$$\begin{aligned}
\|R_S z - R_S V\|_{B_S} &\leq (1 + \Delta_S) \|R_S z - Z\|_{B_S} + \Delta_S \|R_S z\|_{B_S} \\
&\leq [(1 + \Delta_S) K' K_S h + \Delta_S K_\Omega] \|u - U\|_{L_2}, \quad (5.40)
\end{aligned}$$

which gives the sought after result if $\Delta_S = O(h)$.

5.5 Streamline diffusion methods

Having in the earlier sections used the ideas of upwinding, symmetrization and Green's functions to guide the choice of test space in convection-dominated problems, we return to the basic idea of enhancing the diffusion. We have already seen in simple one-dimensional problems that shifting from central to upwind differencing in the convection term is equivalent to adding an extra diffusion term. The idea with the *streamline diffusion method* for multidimensional problems is to add it only in the direction of the convective velocity **b**, thus giving a tensor diffusivity which in general is nondiagonal. In effect, the diffusion term in $B(U, V)$ has added to it a multiple of the term

$$(\mathbf{b}{\cdot}\nabla U, \mathbf{b}{\cdot}\nabla V),$$

which is equivalent to adding a multiple of $\mathbf{b}{\cdot}\nabla V$ to the test function V when calculating the convection term. Similar effects can be achieved by various means, such as the use of upwind quadrature formulae, but the modification of the test functions in this way eventually emerged as the best way of formulating this method — see Hughes and Brooks (1979), Johnson and Nävert (1981), Hughes and Brooks (1982). It can then be regarded as a Petrov–Galerkin method with test functions given by the mapping from the trial functions

$$\psi_j(\mathbf{x}) = M\phi_j(\mathbf{x}) := \phi_j(\mathbf{x}) + \rho\mathbf{b}{\cdot}\nabla\phi_j(\mathbf{x}), \tag{5.41a}$$

with the commonest choice for ρ being

$$\rho|\mathbf{b}|^2 = \max(|\mathbf{b}|\,h - \epsilon, 0), \tag{5.41b}$$

for some mesh parameter h. It is then sometimes called the *SUPG method*, for streamline upwind Petrov–Galerkin. However, it is a nonconforming method because $\psi_j \notin H^1_{E_0}$ and the diffusion term involving $\mathbf{b}{\cdot}\nabla\phi_j$ is evaluated element by element without the usual integration by parts. Thus we define the inner product for the test function (5.41a) as

$$B(U, \psi_j) := \epsilon\left[(a\nabla U, \nabla\phi_j) - \sum_{(e)}\int_{\Omega_e}\nabla{\cdot}(a\nabla U)(\rho\mathbf{b}{\cdot}\nabla\phi_j)\mathrm{d}\Omega\right]$$
$$+ (\mathbf{b}{\cdot}\nabla U, \psi_j), \tag{5.42}$$

where Ω_e denotes an individual element. Note that it is legitimate to use the same notation since (5.42) is consistent with the definition (5.3a) for $B(U, V)$ when $V \in H^1_{E_0}$.

Suppose the pure convection equation $\mathbf{b}{\cdot}\nabla u = f$ were approximated by least squares, that is by determining $U \in H^1_E$ so as to minimise $\|\mathbf{b}{\cdot}\nabla U - f\|^2$. Then the equation system for U would be

$$(\mathbf{b}{\cdot}\nabla U - f, \mathbf{b}{\cdot}\nabla\phi_j) = 0 \quad \forall\phi_j \in S^h_0.$$

This is clearly an alternative way to view the choice of the test function given by (5.41a). Indeed, we see from the lemma below that the use of the mapping M between the trial space and the test space increases the coercivity of $B(\cdot, \cdot)$ as compared with that found for the Galerkin method in Theorem 4.2.1, and this is a key property for the analysis of the streamline diffusion method.

Lemma 5.5.1 *For the mapping $M : H^1 \to L_2$ defined by (5.41a), and the bilinear form $B(\cdot, \cdot)$ defined on $S^h \times M S^h$ by (5.42), we have*

$$B(V, MV) \geq \tfrac{1}{2}\epsilon(a\nabla V, \nabla V) + \tfrac{1}{2}\left\|\rho^{\frac{1}{2}}\mathbf{b}\cdot\nabla V\right\|^2 \quad \forall V \in S_0^h, \qquad (5.43)$$

if the trial space satisfies the inverse estimate

$$\int_{\Omega_e} |\nabla\cdot(a\nabla V)|^2 \, \mathrm{d}\Omega \leq C_S h_e^{-2} \int_{\Omega_e} a \, |\nabla V|^2 \, \mathrm{d}\Omega \qquad (5.44)$$

on each element Ω_e, for some constant C_S, and ρ is chosen so that

$$C_S \rho \, |\mathbf{b}| \leq h_e \quad \text{when} \quad |\mathbf{b}| \, h_e > \epsilon \qquad (5.45)$$

and is zero otherwise.

Proof. Since we have assumed $\nabla\cdot\mathbf{b} = 0$, from Gauss' theorem as in (2.59) we have

$$(\mathbf{b}\cdot\nabla V, MV) = \tfrac{1}{2}\int_{\partial\Omega_N} V^2(\mathbf{b}\cdot\mathbf{n})\mathrm{d}\Gamma + \left\|\rho^{\frac{1}{2}}\mathbf{b}\cdot\nabla V\right\|^2$$

On each element we also have from (5.44)

$$2\int_{\Omega_e} \nabla\cdot(a\nabla V)(\rho\mathbf{b}\cdot\nabla V)\mathrm{d}\Omega \leq \int_{\Omega_e} a \, |\nabla V|^2 \, \mathrm{d}\Omega + C_S h_e^{-2} \int_{\Omega_e} (\rho\mathbf{b}\cdot\nabla V)^2 \mathrm{d}\Omega.$$

Hence

$$B(V, MV) \geq \tfrac{1}{2}\epsilon(a\nabla V, \nabla V) + \left\|\rho^{\frac{1}{2}}\mathbf{b}\cdot\nabla V\right\|^2$$
$$-\tfrac{1}{2}C_S\epsilon\sum_{(e)} h_e^{-2}\rho|_{\max} \int_{\Omega_e} \rho(\mathbf{b}\cdot\nabla V)^2 \mathrm{d}\Omega$$

from which (5.43) follows by using (5.45). □

The inverse estimate (5.44) is a standard property of finite element approximation spaces, closely related to that embodied in (4.14) and deducible from (4.16), except that for convenience we have included some regularity properties of a. If $a \equiv 1$ and S^h consists of piecewise linear elements on triangles, then $\nabla^2 V = 0$ and $C_S = 0$. Otherwise, C_S is mainly determined by the regularity of the mesh through transformations such as (4.5) and (4.37) which give ∇V in terms of V, and similarly for higher derivatives.

For our one-dimensional model problem of (4.61), approximated by piecewise linears, the effect of the gradient term in the streamline diffusion method is to augment the $b\Delta_0 U$ term obtained from the Galerkin equations with a term $-(\rho b^2/h)\delta^2 U$. Thus the Allen and Southwell difference operator will be generated again if we choose ρ as in (5.6) to be given by

$$\rho b = \tfrac{1}{2}h(\coth \tfrac{1}{2}\beta - 2/\beta). \qquad (5.46)$$

The test functions obtained with this choice of the parameter ρ are shown in the comparisons of Fig. 5.3.

A comparison of (5.43) with (5.19) indicates that the mapping M plays a role which is very similar to that of R_S^{-1} in the symmetrization technique. In the error analysis it is bounds of the form (5.43) which are vital and, as we shall see in the examples of the next section, the main importance of the approximate symmetrization theory which is summarised in Theorem 5.2.1 is to provide such a bound for any test space. The advantage of this theory is that it is independent of the mapping M which is used to generate the test functions from the trial functions; its disadvantage is that it is limited to conforming test spaces, although we shall see that it seems to give valid results in the present case. In the next chapter, on finite volume methods, we shall again see the important role played by mappings between trial and test spaces, and how different choices of mapping for the same method can be used to give different error bounds.

5.6 Comparative error estimates

5.6.1 One-dimensional model problem

We begin in one dimension with the model problem (4.61); and since exactness at the nodes corresponds to optimality in the $\|\cdot\|_{B_1}$ norm we concentrate our comparisons on this norm. The operator R_1 is given explicitly in this case by (5.24), which enables all of the matrices A, B and C given in (5.17) and used to define Δ_1 to be calculated. Thus

$$A_{i,j} := B_1(R_1\psi_j, R_1\psi_i)$$

$$= \int_0^1 [\epsilon\psi_j' + b(\psi_j - \bar{\psi}_j)][\epsilon\psi_i' + b(\psi_i - \bar{\psi}_i)]\mathrm{d}x \qquad (5.47a)$$

$$B_{i,j} := B_1(\phi_j, R_1\psi_i) = \int_0^1 \phi_j'[\epsilon\psi_i' + b(\psi_i - \bar{\psi}_i)]\mathrm{d}x \qquad (5.47b)$$

and

$$C_{i,j} := B_1(\phi_j, \phi_i) = \int_0^1 \phi_j'\phi_i'\mathrm{d}x, \quad i,j = 1, 2, \ldots, J-1. \qquad (5.47c)$$

It is worth noting that if a mapping M can be identified such that $\psi_j = M\phi_j$, then $B_{i,j} = B(\phi_j, \psi_i) = B(\phi_j, M\phi_i)$ so that a coercivity relation of the form (5.43) which is needed for an error analysis ensures that the matrix B is nonsingular. Hence the quantity $1 - \Delta_1^2$ can be calculated either as the smallest eigenvalue of the generalised eigenvalue problem

$$B^T A^{-1} B V = \lambda C V, \qquad (5.48a)$$

as in (5.18); or, by setting $V = C^{-1}B^T W$, we can calculate it from

$$BC^{-1}B^T W = \lambda A W, \qquad (5.48b)$$

which is often more convenient. Scotney (1985) has used this to carry out the calculations for a number of test spaces and the piecewise linear trial space on a uniform mesh, and we will reproduce his results later in this section, in Table 5.1.

In fact for all these cases, as was pointed out by Scotney, the error factors depend on just two parameters that have to be calculated for the given test space; and we will present this analysis before considering the results in the table. The support of each ψ_j consists of the two intervals (x_{j-1}, x_j) and (x_j, x_{j+1}) so that all the matrices would be tridiagonal if it were not for the constant $\bar{\psi}_j$ making A a full matrix. Moreover, for all of the cases considered we have $\psi_j(x) + \psi_{j+1}(x) \equiv 1$ for $x \in (x_j, x_{j+1})$, where ψ_j is normalised such that $\bar{\psi}_j = h$; that is, the test space is determined by a single function ψ over the interval $[0, h]$, with $\psi(0) = 0, \psi(h) = 1$ and giving

$$\psi_j(x) = \begin{cases} \psi(x - x_{j-1}) & x \in [x_{j-1}, x_j] \\ 1 - \psi(x - x_j) & x \in [x_j, x_{j+1}]. \end{cases} \tag{5.49}$$

Hence we have

$$\int_0^1 (\psi_j + \psi_{j+1})^2 dx = \int_0^{x_j} \psi_j^2 dx + h + \int_{x_{j+1}}^1 \psi_{j+1}^2 dx = h + \int_0^1 \psi_j^2 dx$$

and thence that

$$2 \int_0^1 \psi_j \psi_{j+1} dx = \int_0^1 (\psi_j + \psi_{j+1})^2 dx - 2 \int_0^1 \psi_j^2 dx$$

$$= h - \int_0^1 \psi_j^2 dx = 2 \int_0^h \psi(1 - \psi) dx.$$

Similarly, we find that

$$2 \int_0^1 \psi_j' \psi_{j+1}' dx = \int_0^1 (\psi_j' + \psi_{j+1}')^2 dx - 2 \int_0^1 (\psi_j')^2 dx$$

$$= - \int_0^1 (\psi_j')^2 dx = -2 \int_0^h (\psi')^2 dx.$$

It follows that, since from (5.47a)

$$A_{j,j} = \epsilon^2 \int_0^1 (\psi_j')^2 dx + b^2 \left[\int_0^1 \psi_j^2 dx - h^2 \right],$$

$$A_{j,j+1} = \epsilon^2 \int_0^1 \psi_j' \psi_{j+1}' dx + b^2 \left[\int_0^1 \psi_j \psi_{j+1} dx - h^2 \right],$$

the matrix A can be written as

$$A = \kappa h C + b^2 h (I - hE), \tag{5.50a}$$

where the matrix C of (5.47c) is just tridiag$(-1, 2, -1)/h$, I is the identity matrix, E is a full matrix of ones and the parameter κ is given by

$$\kappa = \epsilon^2 \int_0^h (\psi')^2 \mathrm{d}x - b^2 \int_0^h \psi(1-\psi) \mathrm{d}x. \qquad (5.50b)$$

The matrix B is tridiagonal, and from (5.47b) we have

$$B_{j,j} = \frac{2\epsilon}{h} + \frac{b}{h} \int_0^h (2\psi - 1)\mathrm{d}x = 2\left(\frac{\epsilon}{h} + \frac{b}{h}\int_0^h \psi \mathrm{d}x\right) - b,$$

$$B_{j,j+1} = -\frac{\epsilon}{h} + \frac{b}{h}\int_0^h (1-\psi)\mathrm{d}x = -\left(\frac{\epsilon}{h} + \frac{b}{h}\int_0^h \psi \mathrm{d}x\right) + b$$

and $B_{j,j-1} = B_{j,j+1} - b$. Thus by introducing the matrix $G = $ tridiag $(-1, 1, 0)/h$ so that $G + G^T = C$, it is clear that $B = $ tridiag $(-q, 2q - b, -q + b) = h(qC - bG^T)$, where $q = -B_{j,j-1}$. Thence, it is shown in Scotney (1985) that $BC^{-1}B^T$ has the same form as A; specifically, we can show that

$$BC^{-1}B^T = \eta C + b^2 h(I - hE) \qquad (5.51a)$$

where

$$\eta = \left[\epsilon + b\int_0^h \psi \mathrm{d}x\right]\left[\epsilon - bh + b\int_0^h \psi \mathrm{d}x\right]. \qquad (5.51b)$$

To do so, we note first that

$$BC^{-1}B^T = h^2(qC - bG^T)C^{-1}(qC - bG)$$
$$= h^2[q(q-b)C + b^2 G^T C^{-1}G],$$

and next that

$$G^{-1}C(G^T)^{-1} = G^{-1}(G + G^T)(G^T)^{-1} = G^{-1} + (G^T)^{-1}$$
$$= h(I + E).$$

Finally,

$$(I + E)(I - hE) = I + (1 - h)E - hE^2$$

and $E^2 = (I - h)E$, so that $(I - hE) = (I + E)^{-1} = hG^T C^{-1}G$. The expressions (5.50) and (5.51) can then be gathered together to give the following result.

Theorem 5.6.1 *Suppose problem (4.61) is approximated on a uniform mesh with a piecewise linear trial space and a conforming test space given by (5.49). Then the error constant in the bound of (5.11) using the $\|\cdot\|_{B_1}$ norm is given by*

$$(1 - \Delta_1^2)^{-\frac{1}{2}} = \begin{cases} \left(\dfrac{b^2 h^2 + 4\kappa h}{b^2 h^2 + 4\eta} \right)^{\frac{1}{2}} & \text{if } J \text{ is even} \quad (5.52a) \\[2em] \left(\dfrac{b^2 h^2 + 4\kappa h \sin^2 \frac{1}{2}\pi(J-1)h}{b^2 h^2 + 4\eta \sin^2 \frac{1}{2}\pi(J-1)h} \right)^{\frac{1}{2}} & \text{if } J \text{ is odd}, \quad (5.52b) \end{cases}$$

where κ and η are given by (5.50b) and (5.51b) respectively.

Proof. From (5.48), (5.50) and (5.51) the quantity $(1 - \Delta_1^2)$ is the smallest eigenvalue λ of the problem

$$[\eta C + b^2 h(I - hE)]\mathbf{W} = \lambda[\kappa h C + b^2 h(I - hE)]\mathbf{W}.$$

Since we have shown that $(I - hE)^{-1} = I + E$, we suppose that μ is the eigenvalue of $(I + E)C$ corresponding to λ, so that

$$\lambda = (b^2 h + \eta\mu)/(b^2 h + \kappa h \mu). \quad (5.53)$$

We shall show that $0 < \mu h \le 4$. Then the smallest eigenvalue λ corresponds to the largest eigenvalue μ if we can show that $\kappa h \ge \eta$; and this will establish that $0 \le \lambda \le 1$. The inequality $\kappa h \ge \eta$ follows from direct computation, with a significant application of the Cauchy-Schwarz inequality to the residual formed by applying the adjoint operator to the test function; that is, we use

$$\left[\int_0^h (\epsilon\psi' + b\psi)\mathrm{d}x \right]^2 \le h \int_0^h (\epsilon\psi' + b\psi)^2 \mathrm{d}x. \quad (5.54)$$

Then from (5.50b) and (5.51b) we write $\kappa h - \eta$ in the form

$$\kappa h - \eta = \int_0^h \left[\epsilon^2 \ (\psi')^2 - b^2\psi(1 - \psi) \right] \mathrm{d}x$$
$$+ bh\left[\epsilon + b\int_0^h \psi\mathrm{d}x \right] - \left[\int_0^h (\epsilon\psi' + b\psi)\mathrm{d}x \right]^2,$$

from which $\kappa h \ge \eta$ follows by application of (5.54).

The eigenvectors of the second difference operator $\tilde{C}$, obtained by extending C to pick up the boundary values at $j = 0$ and $j = J$, can be constructed from $\mathrm{e}^{2\pi\mathrm{i}(mjh)}$ and written as $\{\mathrm{e}^{2\pi\mathrm{i}(mjh)}\}$, with eigenvalues determined from

$$h\tilde{C}\{\mathrm{e}^{2\pi\mathrm{i}(mjh)}\} = 4\sin^2 \pi m h\{\mathrm{e}^{2\pi\mathrm{i}(mjh)}\}. \quad (5.55a)$$

Moreover, hEC is a matrix with two columns of unit entries on the left and on the right and is otherwise zero. Thus, for integer m,

$$hEC\{\mathrm{e}^{2\pi\mathrm{i}(mjh)}\} = \{\mathrm{e}^{2\pi\mathrm{i}m(J-1)h} + \mathrm{e}^{2\pi\mathrm{i}mh}\}$$
$$= \{2\cos 2\pi m h\}. \quad (5.55b)$$

Now suppose that J is odd and let $m = \pm 1, \pm 2, \ldots, \pm \frac{1}{2}(J-1)$, which gives $J - 1$ distinct vectors. Combining the imaginary parts gives $\frac{1}{2}(J - 1)$ eigenvectors $\{\sin 2\pi mjh\}$ of $h(I + E)C$ with eigenvalues $4\sin^2 \pi mh$, because the boundary values are zero so that $\tilde{C} = C$, and (5.55b) gives a zero result. The other $\frac{1}{2}(J - 1)$ eigenvectors with the same eigenvalues are given by $\{1 - \cos 2\pi mjh\}$; again the boundary values are zero so that $\tilde{C} = C$, and from (5.55) we have

$$
\begin{aligned}
h(I + E)C\{1 - \cos 2\pi mjh\} &= 4\sin^2 \pi mh\{-\cos 2\pi mjh\} \\
&\quad + \{2 - 2\cos 2\pi mh\} \\
&= 4\sin^2 \pi mh\{1 - \cos 2\pi mjh\}.
\end{aligned}
$$

When J is even, one obtains similar pairs of eigenvectors with the same eigenvalue for $m = 1, 2, \ldots, \frac{1}{2}J - 1$; and this is supplemented by $\frac{1}{2}\{1 - \cos \pi j\} \equiv (1, 0, 1, \ldots, 1)$ with eigenvalue 4, obtained from setting $m = \frac{1}{2}J$.

These results on the eigenstructure of $(I + E)C$ establish that all of the eigenvalues μ are positive and satisfy $\mu h \leq 4$ as claimed above. Moreover, it is clear from writing $\eta = qh(qh - bh)$ that application of the Cauchy-Schwarz inequality gives $\eta \geq -\frac{1}{4}b^2 h^2$ and hence that

$$
b^2 h + \eta \mu \geq b^2 h(1 - \tfrac{1}{4}\mu h) \geq 0. \tag{5.56}
$$

Thus, from (5.53), $0 \leq \lambda \leq 1$ and substituting the largest value of μ in the two cases of even and odd J gives the desired result (5.52). $\quad\square$

Remark The Cauchy-Schwarz inequality was used twice to obtain bounds on the two parameters κ and η which characterise the test space in this analysis: we have seen from (5.54) that if and only if ψ satisfies the adjoint equation (i.e. it corresponds to the Hemker test function) do we have $\kappa h = \eta$, and hence $\lambda = 1$ and the error constant can have the minimum value of unity; and, at the other extreme, we obtain the equality sign in (5.56), which allows $\lambda = 0$ and the error constant to be unbounded, if and only if $q = \frac{1}{2}b$, which corresponds to $\epsilon = 0$ and no upwinding (i.e. $\psi_j = \phi_j$). We summarise this last result as a corollary to the theorem.

Corollary 5.6.1 *For any (uniformly) conforming upwinded test function, the error constant is bounded as $\beta \to \infty$. Specifically, if we define*

$$
\|\psi\|_{2,h}^2 := \frac{1}{h}\int_0^h \psi^2 \mathrm{d}x, \quad \bar{\psi} := \frac{1}{h}\int_0^h \psi \mathrm{d}x, \quad \bar{\phi} := \frac{1}{h}\int_0^h \phi_j \mathrm{d}x = \tfrac{1}{2},
$$

we have for even J,

$$
(1 - \Delta_1^2)^{-\frac{1}{2}} \to \frac{(\frac{1}{4} + \|\psi\|_{2,h}^2 - \bar{\psi})^{\frac{1}{2}}}{\bar{\psi} - \bar{\phi}} \quad \text{as } \beta \to \infty. \tag{5.57}
$$

Proof. Since $\kappa h \geq \eta$, the case of even J gives the larger constant. By the uniformly conforming assumption we mean that $h^{-1}\int_0^h (\psi')^2 \mathrm{d}x$ is bounded;

Table 5.1. *Error factors in the bound of (5.11), using the* $\|\cdot\|_{B_1}$ *norm, for various choices of test space and the 1D model problem.*

	Error factor $(1 - \Delta_1^2)^{-\frac{1}{2}}$				
β	Galerkin	CGMZ	CGMZ*	HB	HB*
2	1.1547	1.0060	1.0060	1.0924	1.0178
5	1.7559	1.0468	1.0428	1.1509	1.0597
50	14.468	1.2022	1.0945	1.1547	1.1406
500	144.34	1.2344	1.0954	1.1247	1.1532
10^5	28868.	1.2383	1.0954	1.1547	1.1546

this implies that the first term in (5.50b) can be neglected as $\beta \to \infty$, and leads to the numerator in (5.57). For the denominator, we have

$$b^2 + 4\eta/h^2 = b^2 + 4q(q - b) = (2q - b)^2,$$

where $q = b\bar{\psi} + \epsilon/h \to b\bar{\psi}$ as $\beta \to \infty$. This gives the desired result, using the interpretation that an upwinded test function has $\bar{\psi} > \bar{\phi}$. $\quad\square$

Now let us consider the particular choices of test space for which the results obtained by Scotney (1985) are given in Table 5.1; they are all for even J so that the simple form (5.52a) can be used. In the first column are the results with no upwinding, showing that the error constant diverges in direct proportion to the mesh Péclet number β; as indicated in the remark above, the Galerkin choice of $\psi_j = \phi_j$ gives $\eta = \epsilon^2 - \frac{1}{4}b^2h^2$, and with $\kappa h = \epsilon^2 - \frac{1}{6}b^2h^2$ one obtains the error constant as $(1 + \frac{1}{12}\beta^2)^{\frac{1}{2}}$, which gives the results in the table. The second column, headed CGMZ, corresponds to the test functions of (5.4) given in Christie *et al.* (1976) and Heinrich *et al.* (1977) and shown in Fig. 5.3; the parameter α determining the amount of quadratic upwinding is taken here as $\coth \frac{1}{2}\beta - 2/\beta$, so as to reproduce the Allen and Southwell difference scheme. As anticipated by Corollary 5.6.1 the error constant is bounded as $\beta \to \infty$; what is perhaps surprising is how close the bound is to unity. Its value is readily calculated from (5.57) to be $\alpha^{-1}(\frac{1}{3} + \frac{6}{5}\alpha^2)^{\frac{1}{2}}$, which gives the value $(23/15)^{\frac{1}{2}} = 1.238278\ldots$ for $\alpha = 1$. The next column, headed CGMZ*, then gives the best error constant obtainable with this type of test function. It turns out that the optimal choice of the parameter α is given by

$$\alpha_{opt} = \frac{5\beta(\beta^2 + 12)}{36(\beta^2 + 10)};$$

that is, as $\beta \to \infty$ the quadratic modification completely dominates the original linear shape function, instead of being given an equal weight as in the conventional choice of test function. From computations given earlier it is then clear that the error constant tends to $(6/5)^{\frac{1}{2}} = 1.0954\ldots$ as $\beta \to \infty$.

The third set of test functions shown in Fig. 5.3 are those for the streamline diffusion method given by (5.41a), but with the parameter ρ chosen to satisfy (5.46) so that again the Allen and Southwell difference operator is generated. This is discontinuous and therefore nonconforming, so the above error analysis does not apply directly. However, suppose we introduce a conforming test function to approximate that of (5.41a), by writing the latter as $\psi_j = \phi_j + \sigma_j$ and multiplying the piecewise constant σ_j by a mollifier which on each interval is a flat-topped pulse of height $(1 - \delta)^{-1}$ with sloping sides over intervals δh. The resulting values of κ, η and the error constant can be calculated — see Scotney (1985) for details; moreover, δ can be chosen to minimise the error constant, which gives

$$\beta\delta \sim \sqrt{3} \text{ as } \beta \to \infty.$$

Thus the discontinuous test function of the streamline diffusion method is obtained in the limit, and it falls outside the hypotheses of Corollary 5.6.1; with $\rho h = \frac{1}{2}h$ in this limit, the error constant tends to $(4/3)^{\frac{1}{2}} = 1.1547\ldots$. The final column in the table, labelled HB*, gives the error constants for this mollified and optimised streamline diffusion method.

Alternatively, we can analyse the streamline diffusion method by redefining the matrices A and B of (5.47) in an element-by-element manner as we defined the bilinear form $B(\cdot, \cdot)$ in (5.42). The resulting calculations lead to the values given in the column headed HB in Table 5.1, but we defer any consideration of the validity of this procedure until the next section. At this point we merely note that the results for HB and HB* are very similar, with the latter giving slightly better bounds as one would expect, and both being better than CGMZ.

5.6.2 Two-dimensional problems

Let us turn now to two-dimensional problems, where we cannot expect to have an explicit representation of the operator R_1. However, as shown in Morton and Scotney (1985) and Morton et al. (1992), computations to obtain the matrices A, B and C can still be carried out and hence the error constant $(1 - \Delta_1^2)^{-\frac{1}{2}}$ computed from the eigenvalue problem (5.48). The matrix C is of course independent of the test functions, and by (5.7) we can write $B_{i,j} := B_1(\phi_j, R_1\psi_i) = B(\phi_j, \psi_i)$ so that it is only the matrix A that we need be concerned with. We will show that explicit knowledge of R_1 can be avoided by a decomposition of the Hilbert space $[L_2(\Omega)]^2$ into the direct sum of curl-free and divergence-free vectors.

We make further simplifications of the problem (5.3) by setting $a \equiv 1$ and assuming that the boundary $\partial\Omega$ is polygonal. From the definition (5.14a) of $B_1(\cdot, \cdot)$, and that of R_1 in (5.7) we can write

$$
\begin{aligned}
A_{i,j} &:= B_1(R_1\psi_j, R_1\psi_i) = B(R_1\psi_j, \psi_i) \\
&= \epsilon(\nabla(R_1\psi_j), \nabla\psi_i) + (\mathbf{b}\cdot\nabla(R_1\psi_j), \psi_i) \\
&= \epsilon(\nabla(R_1\psi_j), \nabla\psi_i + \epsilon^{-1}\mathbf{b}\psi_i),
\end{aligned}
\tag{5.58}
$$

and it is the vector $\mathbf{b}\psi_i$ that we shall decompose. We first need to recall some definitions and vector identities. The two-dimensional vector and scalar curl are defined by

$$
\mathbf{curl}\,\phi := \left(\frac{\partial\phi}{\partial y}, -\frac{\partial\phi}{\partial x}\right)^T
\tag{5.59a}
$$

$$
\nabla \times \mathbf{v} := \frac{\partial v_2}{\partial x} - \frac{\partial v_1}{\partial y},
\tag{5.59b}
$$

where $\mathbf{v} = (v_1, v_2)^T$, and introducing the function space

$$
H(\mathrm{curl}; \Omega) := \{\mathbf{v} \in [L_2(\Omega)]^2; \nabla \times \mathbf{v} \in L_2(\Omega)\},
\tag{5.59c}
$$

we have the Green's formula

$$
(\mathbf{v}, \mathbf{curl}\,\phi) - (\nabla \times \mathbf{v}, \phi) = -\int_{\partial\Omega} \phi(\mathbf{v}\cdot\mathbf{t})\mathrm{d}\Gamma
$$
$$
\forall \mathbf{v} \in H(\mathrm{curl}; \Omega), \quad \forall\phi \in H^1(\Omega),
\tag{5.59d}
$$

where $\mathbf{t}$ is a unit tangent vector with $\mathbf{t} = (-n_2, n_1)^T$ in terms of the outward normal $\mathbf{n}$. Then we have the following lemma.

Lemma 5.6.2 *For $\mathbf{v} \in [H^1_{E_0}(\Omega)]^2$ there exist unique functions $p \in H^1(\Omega)\backslash\mathbb{R}$ and $q \in H^1_{E_0}(\Omega)$ such that*

$$
\mathbf{v} = \mathbf{curl}\,p + \nabla q
\tag{5.60}
$$

and $\nabla^2 q \in L_2(\Omega), \partial q/\partial n = \mathbf{n}\cdot\mathbf{v}$ on $\partial\Omega_N$.

Proof. See Morton *et al.* (1992) where it is shown that p and q are given by the problems

$$
\nabla^2 q = \nabla\cdot\mathbf{v} \text{ in } \Omega
$$
$$
q = 0 \text{ on } \partial\Omega_D, \ \partial q/\partial n = \mathbf{n}\cdot\mathbf{v} \text{ on } \partial\Omega_N,
$$

and

$$
\nabla^2 p = -\nabla \times \mathbf{v}
$$
$$
\partial p/\partial n = (\nabla q - \mathbf{v})\cdot\mathbf{t} \text{ on } \partial\Omega.
$$

The proof depends on results in Girault and Raviart (1979). $\square$

Using the lemma to define p_i and q_i, we substitute

$$
\epsilon^{-1}\mathbf{b}\psi_i = \mathbf{curl}\,p_i + \nabla q_i
\tag{5.61a}
$$

into (5.58) to obtain the matrix entry

$$A_{i,j} = \epsilon(\nabla(R_1\psi_j), \nabla(\psi_i + q_i)) + \epsilon(\nabla(R_1\psi_j), \mathbf{curl}\, p_i).$$

Since $\nabla\cdot(\mathbf{curl}\, p_i) = 0$ and by the construction $\mathbf{n}\cdot(\mathbf{curl}\, p_i) = 0$ on $\partial\Omega_N$, application of Green's formula shows the last term to be zero. Then we can apply (5.7) to the first term to get

$$A_{i,j} = \epsilon(\nabla(\psi_i + q_i), \nabla\psi_j + \epsilon^{-1}\mathbf{b}\psi_j),$$

and substituting back for ∇q_i gives

$$A_{i,j} = (\epsilon\nabla\psi_i + \mathbf{b}\psi_i, \nabla\psi_j + \epsilon^{-1}\mathbf{b}\psi_j) - (\mathbf{curl}\, p_i, \epsilon\nabla\psi_j + \mathbf{b}\psi_j).$$

This has completely eliminated R_1. As a final step we put the last term into a symmetric form; as above, $(\mathbf{curl}\, p_i, \epsilon\nabla\psi_j) = 0$ and by introducing p_j and q_j we obtain

$$(\mathbf{curl}\, p_i, \mathbf{b}\psi_j) = (\mathbf{curl}\, p_i, \epsilon\,\mathbf{curl}\, p_j) + (\mathbf{curl}\, p_i, \epsilon\nabla q_j)$$
$$= \epsilon(\mathbf{curl}\, p_i, \epsilon\,\mathbf{curl}\, p_j).$$

Hence we have

$$A_{i,j} = \epsilon(\nabla\psi_i + \epsilon^{-1}\mathbf{b}\psi_i, \nabla\psi_j + \epsilon^{-1}\mathbf{b}\psi_j) - \epsilon(\mathbf{curl}\, p_i, \mathbf{curl}\, p_j) \quad (5.61b)$$

as our final form of the matrix A.

In Morton *et al.* (1992) this technique was applied to calculate the error constants for the bilinear Galerkin method and the upwind scheme of Heinrich *et al.* (1977) when applied to the first IAHR/CEGB problem described in section 1.2. We will summarise the results here. For the upwind scheme, which is a generalisation of that given in (5.4), we introduce the local mesh Péclet numbers based on the velocity components and averaged in each rectangular element

$$\beta^{(x)} = \tfrac{1}{4}\epsilon^{-1}(x_i - x_{i-1}) \sum_{l,m=0,1} b^{(x)}(x_{i-l}, y_{j-m})$$

$$\beta^{(y)} = \tfrac{1}{4}\epsilon^{-1}(y_j - y_{j-1}) \sum_{l,m=0,1} b^{(y)}(x_{i-l}, y_{j-m}). \quad (5.62)$$

Corresponding to (5.4) one then introduces for each element the parameters

$$\alpha^{(x)} = \coth\tfrac{1}{2}\beta^{(x)} - 2/\beta^{(x)}, \quad \alpha^{(y)} = \coth\tfrac{1}{2}\beta^{(y)} - 2/\beta^{(y)};$$

and corresponding to the element basis functions for the trial space given by (4.7b) one has the element basis functions for the test space

$$M_\gamma(\xi,\eta) = [\tfrac{1}{2}(1 \mp \xi) \mp \tfrac{3}{4}\alpha^{(x)}(\xi^2 - 1)][\tfrac{1}{2}(1 \mp \eta) \mp \tfrac{3}{4}\alpha^{(y)}(\eta^2 - 1)]$$
$$\gamma = 1, 2, 3, 4. \quad (5.63)$$

Note the important consequence of using element-averaged values for the parameters $\alpha^{(x)}$ and $\alpha^{(y)}$, namely that $\sum_{(\gamma)} M_\gamma(\xi,\eta) \equiv 1$, which ensures the conservation property of the solution u is satisfied by the approximation U.

In order to calculate the matrix entries $A_{i,j}$ from (5.63) we need to calculate the functions p_i from each of the test space basis functions ψ_i, and that also requires calculation of the q_i. From the construction of p and q in Lemma 5.6.2 we see that each satisfies a Poisson equation which we suppose is approximated by the bilinear Galerkin method. Thus let $P_i \in S^h \backslash \mathbb{R}$ be given by

$$\epsilon(\nabla P_i, \nabla \phi) = (\nabla \times (\mathbf{b}\psi_i), \phi) + \int_{\partial \Omega} (\epsilon \nabla Q_i - \mathbf{b}\phi_i) \cdot t \phi \mathrm{d}\Gamma$$

$$\forall \phi \in S^h \backslash \mathbb{R}, \qquad (5.64a)$$

for which we also need $Q_i \in S_0^h$ given by

$$\epsilon(\nabla Q_i, \nabla \phi) + (\nabla \cdot (\mathbf{b}\psi_i), \phi) = \int_{\partial \Omega_N} (\mathbf{n} \cdot \mathbf{b}) \psi_i \phi \mathrm{d}\Gamma$$

$$\forall \phi \in S_0^h. \qquad (5.64b)$$

Note that since both $Q_i \in H_{E_0}^1$ and $\psi_i \in H_{E_0}^1$, the boundary integral in (5.64a) extends only over $\partial \Omega_N$. In Morton et $al.$ (1992) the P_i and Q_i are computed on a sequence of meshes with spacing $\frac{1}{2}h$, $\frac{1}{4}h$, $\frac{1}{8}h$ and $\frac{1}{16}h$ with Richardson extrapolation used to improve the accuracy.

In Table 5.2 we give the error constants obtained on two meshes, with 7×4 elements and 9×5 elements respectively, for various values of the diffusion coefficient ϵ. There is a strong indication that on each fixed mesh

Table 5.2. Error factors $(1 - \Delta_1^2)^{-\frac{1}{2}}$ for two choices of the test space and two meshes used for the first IAHR/CEGB problem.

	7×4 Mesh		9×5 Mesh	
ϵ	Galerkin	HHMZ	Galerkin	HHMZ
0.1	2.776	2.652	3.296	3.188
0.01	10.010	6.888	11.823	8.478
0.0^31	24.186	9.905	38.860	13.370
0.0^51	24.541	9.961	39.988	13.356
0.0^71	24.545	9.962	39.999	13.357

the error constant tends to a limit as $\epsilon \to 0$, that is, as $\beta \to \infty$. This is true for both the Galerkin method and for the upwind method. The error constant for the latter tends to its limit somewhat faster, and the limit is considerably lower; but, in contrast to the situation in one dimension, the limit is very much larger than unity. Moreover, results given in Morton et $al.$ (1992) for a variety of meshes and $\epsilon = 10^{-8}$, which are reproduced in

Table 5.3. *Error factors as in Table 5.2 for* $\epsilon = 10^{-8}$ *and various meshes.*

Mesh	Galerkin	HHMZ
5×3	13.409	5.401
9×5	39.999	13.357
13×7	77.571	21.441
21×11	177.947	38.116
27×14	271.602	50.537

Table 5.3, suggest strongly that there is no limit as the mesh size $h \to 0$ for the upwind scheme, though it appears to diverge rather more slowly than that for the Galerkin method.

The conclusion to be drawn from this example, and supported by other evidence, is that the use of simple tensor products of near optimal one-dimensional test spaces does not give near optimal results for two dimensional problems, especially when the convective velocity field is far from constant. We shall see in section 5.10 and the next chapter that other methods can give much better results.

In this section all the detailed analysis has been in the norm $\|\cdot\|_{B_1}$, because most test spaces that have been proposed have aimed at nodal accuracy. The characterisation of the error constants through the matrices A, B and C of (5.17) holds in any norm consistent with the bilinear form $B(\cdot, \cdot)$, and similar comparative estimates could have been made in the $\|\cdot\|_{B_2}$ for instance, at least in one dimension. It is doubtful whether any extra insight would have been gained, however; and we leave consideration of error estimates in the $\|\cdot\|_{B_2}$ norm until section 5.9 where we consider a method aimed at optimality in that norm. Finally, we note from the error bounds given in section 5.4 that comparisons of the various methods in other norms, such as L_2 or L_∞, are likely to lead to similar comparative estimates of accuracy.

5.7 Error analysis of the streamline diffusion method

5.7.1 One-dimensional model problem

We begin by deriving for the one-dimensional model problem error bounds which are comparable to those derived in the last section for conforming Petrov–Galerkin methods. There is a limit to the comparability, however; the error bounds there were valid for source functions $S \in H^{-1}(\Omega)$ but, because now the test functions ψ lie only in $L_2(\Omega)$ and not $H^1(\Omega)$, we must assume that $S \in L_2(\Omega)$. This is not at all restrictive in practice, of course, and is already weaker than what we assumed for the analysis of difference schemes in Chapter 3.

For the problem $-\epsilon u'' + bu' = S$ on $(0,1)$ with Dirichlet boundary conditions, as given in (4.61), piecewise linear trial functions in $\mathrm{span}\{\phi_j\}$ are matched with test functions given by

$$\psi_i = M\phi_i := \phi_i + \rho b\phi_i' \qquad (5.65a)$$

for some constant ρ yet to be chosen; and the bilinear form is given by

$$B(v, Mw) := \epsilon\left[(v', w') - \rho b\sum_{j=1}^{J}\int_{x_{j-1}}^{x_j} v''w'\,dx\right] + b(v', Mw)$$

$$\forall v \in \bigcap_{j=1}^{J} H^2(x_{j-1}, x_j) \quad \forall w \in H^1(0,1), \qquad (5.65b)$$

where these are special cases of (5.41a) and (5.42) respectively. The streamline diffusion approximation is given by

$$B(U, M\phi_i) = (S, M\phi_i) \quad i = 1, 2, \ldots, J-1; \qquad (5.66a)$$

and, because of our definitions and assumptions, the true solution also satisfies

$$B(u, M\phi) = (S, M\phi) \quad \forall\phi \in S_0^h. \qquad (5.66b)$$

Denoting the piecewise linear interpolant of u by $I^h u$, we introduce the two errors

$$\eta := u - I^h u \in H_0^1(0,1), \quad e := U - I^h u \in S_0^h, \qquad (5.67)$$

and by combining (5.66a) and (5.66b) obtain

$$B(e, Me) = B(\eta, Me). \qquad (5.68)$$

Moreover, in line with the coercivity result (5.43), direct calculation gives

$$B(e, Me) = \epsilon\,\|e'\|^2 + b(e', e + \rho be')$$

$$= (\epsilon + \rho b^2)\,\|e'\|^2. \qquad (5.69)$$

By combining (5.68) and (5.69) we obtain the following result.

Theorem 5.7.1 *For the model problem (4.61) approximated on a uniform mesh with a piecewise linear trial space, a near optimal choice of streamline diffusion test function (5.65a) is given by*

$$\rho = \begin{cases} 0 & \text{if } \beta \leq \pi \\ h/b & \text{if } \beta > \pi. \end{cases} \qquad (5.70)$$

The error is then bounded by

$$\|(U - I^h u)'\| \leq (h/\pi)\,\|u''\|. \qquad (5.71)$$

Proof. From (5.69), (5.68) and (5.65b) we have

$$(\epsilon + \rho b^2)\,\|e'\|^2 = B(e, Me) = B(\eta, Me)$$

$$= \epsilon \left[(\eta', e') - \rho b \sum_{j=1}^{J} \int_{x_{j-1}}^{x_j} u''e' \mathrm{d}x \right]$$
$$+ b(\eta', e + \rho b e')$$
$$= b(\eta', e) - \epsilon \rho b(u'', e'),$$

since e' is constant on each element and hence $(\eta', e') = 0$. Now, by a standard approximation result obtained by a sine series expansion on each interval (see, e.g., Strang and Fix (1973)), we have $\|\eta\| \le (h/\pi)^2 \|u''\|$, as well as $\|\eta'\| \le (h/\pi) \|u''\|$. Integrating by parts to replace (η', e) by $-(\eta, e')$ and using the first of these bounds, we get

$$\|e'\| \le \frac{b(\epsilon \rho + h^2/\pi^2)}{\epsilon + \rho b^2} \|u''\|. \tag{5.72}$$

Now the coefficient of $\|u''\|$ in this bound increases with ρ if $\beta < \pi$ and decreases if $\beta > \pi$. Thus if $\beta < \pi$, the optimal choice is $\rho = 0$ which gives a coefficient $bh^2/\epsilon\pi^2$; this increases with β but is less than h/π. This establishes the first case of (5.70) and (5.71). On the other hand, for ρ considered to be independent of ϵ, the coefficient decreases with β if $\rho b > h/\pi$. Thus by taking $\rho = h/b$ when $\beta > \pi$, the coefficient is bounded by its value at $\beta = \pi$; and since its value then of h/π is independent of ρ, this choice of ρ can be regarded as near optimal. $\square$

We have already noted that $\|\eta'\| \le (h/\pi) \|u''\|$; hence the bound (5.71) for $U - I^h u$ is the same as that which we have for $u - I^h u$. Furthermore, because for this problem the coercivity property (5.69) is in just the $\|\cdot\|_{B_1}$ norm, we can make interesting comparisons with the results of the last section for conforming Petrov–Galerkin methods. Indeed, as in the notation of (5.9) we also have $U_1^* = I^h u$, the analysis leading to (5.12) gives for such a conforming approximation U^C the relation

$$\|(U^C - I^h u)'\| \le \frac{\Delta_1}{(1 - \Delta_1^2)^{\frac{1}{2}}} \|(u - I^h u)'\|. \tag{5.73}$$

The coefficient here is less than unity whenever $\Delta_1^2 < \frac{1}{2}$, then giving a better bound than (5.71) in terms of $\|u''\|$; but, on the other hand, we can say that (5.71) is better than comparable results from the last section whenever the error factors $(1 - \Delta_1^2)^{-\frac{1}{2}}$ calculated there are greater than two.

For a final comment on this simple model problem, consider the mollification of the streamline diffusion test function described and analysed in the last section. We can regard this as replacing $M\phi_i$ by $\tilde{M}\phi_i$ in the Petrov–Galerkin equations (5.66a); but $\tilde{M}$ is chosen so that the left-hand side is unchanged, that is

$$B(V, MW) = B(V, \tilde{M}W) \quad \forall V \in S^h, \ \forall W \in S_0^h.$$

Then if $\tilde{U}$ is the solution produced by $\tilde{M}$, we have

$$B(U - \tilde{U}, MW) = (S, (M - \tilde{M})W) \quad \forall W \in S_0^h;$$

and hence

$$(\epsilon + \rho b^2)\|(U - \tilde{U})'\|^2 = (S, (M - \tilde{M})(U - \tilde{U}))$$
$$\leq \|S\|\|(M - \tilde{M})(U - \tilde{U})\|.$$

If we denote by $m(x)$ the mollifier that is applied to the unit square wave on each interval, with sloping sides over intervals δh, we obtain after a little calculation

$$\|(M - \tilde{M})(U - \tilde{U})\|^2 = \rho^2 b^2 \|(U - \tilde{U})'(1 - m)\|^2$$
$$= \rho^2 b^2 \frac{\delta(2 - 3\delta)}{3(1 - \delta)^2}\|(U - \tilde{U})'\|^2.$$

From this we can deduce the following result.

Theorem 5.7.2 *For the model problem (4.61) with $S \in L_2(0,1)$, the streamline diffusion method with a piecewise linear trial space gives a solution which is the limit of that obtained with a suitably mollified conforming method.*

Proof. For the mollified method described above and for $\delta < \frac{1}{3}$ we obtain

$$\|(U - \tilde{U})'\| \leq \frac{\rho b}{\epsilon + \rho b^2} \delta^{\frac{1}{2}} \|S\| \tag{5.74}$$

for any choice of ρ. We therefore obtain convergence in the $\|\cdot\|_{B_1}$ norm as $\delta \to 0$ and hence in the $\|\cdot\|_\infty$ norm. $\square$

Thus by assuming more smoothness of S so that we only need the test functions to converge in $L_2(0,1)$ instead of $H^1(0,1)$, we establish the relevance of the last two columns in Table 5.1 which we were unable to do in the last section. Moreover, since the error factors given there are always much less than two, converging to $2/\sqrt{3}$ as $\beta \to \infty$, the error bounds obtained by that analysis are somewhat better than we have obtained above in Theorem 5.7.1.

5.7.2 Two-dimensional problems

The real point of the streamline diffusion method, however, is directed to two-dimensional problems. The idea is to eliminate *crosswind diffusion* which, as we see in the numerical comparisons of sections 4.6 and 5.10, bedevil many other methods. However, in recent analysis that we shall make use of, it is found necessary to add a small amount of crosswind diffusion; we will omit this in our initial description of the method and only add it as we begin the error analysis. So let us begin by considering problems with a constant convection field across a uniform square mesh, as in the shear layer problem given in section 1.2 for

$$-\epsilon\nabla^2 u + \mathbf{b}\cdot\nabla u = S \text{ on } (0,1) \times (0,1), \tag{5.75}$$

where $\mathbf{b}$ is constant. For either piecewise linear trial functions on triangles, or bilinears on rectangles aligned with the axes, the element Laplacian terms in (5.42) are zero and we have for this problem

$$B(V, MV) = \epsilon\|\nabla V\|^2 + (\mathbf{b}\cdot\nabla V, V + \rho\mathbf{b}\cdot\nabla V)$$
$$\geq \epsilon\|\nabla V\|^2 + \rho\|\mathbf{b}\cdot\nabla V\|^2, \tag{5.76}$$

because $(\mathbf{b}\cdot\nabla V, V) \geq 0$ by the usual arguments; the factor two improvement here over the general case (5.43) is, of course, because we did not have to use the inverse estimate (5.44) with its associated constraint (5.45) on the choice of ρ.

For higher order trial functions it is also common to omit the element integrals of the second order derivatives in defining or using the streamline diffusion method. Thus for the general problem (5.10) we introduce the special bilinear form on $H^1(\Omega) \times H^1(\Omega)$,

$$\bar{B}_{SD}(v,w) := \epsilon(a\nabla v, \nabla w) + (\mathbf{b}\cdot\nabla v, Mw) \tag{5.77a}$$

where

$$Mw := w + \rho\mathbf{b}\cdot\nabla w; \tag{5.77b}$$

and we define the streamline diffusion approximation $U \in S_E^h$ for $S \in L_2(\Omega)$ as

$$B_{SD}(U,V) = (S, MV) \quad \forall V \in S_0^h. \tag{5.78}$$

It is worth noting that the method is then equivalent to applying the Galerkin method to a problem in which both $B(\cdot,\cdot)$ and S have been modified.

In order to demonstrate the absence of crosswind diffusion, most published analysis makes use of a discrete Green's function

$$B_{SD}(V, G^h) = V(\mathbf{x}') \quad \forall V \in S_0^h, \tag{5.79}$$

where we have suppressed the dependence on the special point $\mathbf{x}'$ in the notation for G^h. Thus from (5.78) and (5.79) together we have that $U(\mathbf{x}') = (S, MG^h)$, and obtaining appropriate bounds on G^h will bound the dependence of U on the data. Moreover, if we introduce some projection $P : H_E^1 \to S_E^h$, we can write

$$Pu(\mathbf{x}') - U(\mathbf{x}') = B_{SD}(Pu - U, G^h)$$
$$= B_{SD}(Pu, G^h) - (S, MG^h), \tag{5.80}$$

which expresses the pointwise error entirely in terms of u and S.

To estimate the terms on the right of (5.80) we return to the model problem (5.76), and its approximation by piecewise linears on triangles.

We also make the standard choice for ρ given by (5.41b), namely $\rho b^2 = \max(bh - \epsilon, 0)$, and assume that ϵ is sufficiently small that $\rho b = h - \epsilon/b$; in addition, there is no loss of generality in assuming $b \geq 1$. Furthermore, let us denote by ∂_ξ and ∂_η differentiation in the direction $\mathbf{b}$ and normal to it, respectively. Then we have from (5.80) and (5.77)

$$
\begin{aligned}
(Pu - U)(\mathbf{x}') &= \epsilon(\nabla Pu, \nabla G^h) + (\mathbf{b}\cdot\nabla Pu, MG^h) \\
&\quad - (-\epsilon\nabla^2 u + \mathbf{b}\cdot\nabla u, MG^h) \\
&= B_{SD}(Pu - u, G^h) + \epsilon\rho(\nabla^2 u, \mathbf{b}\cdot\nabla G^h),
\end{aligned}
\tag{5.81a}
$$

where

$$
B_{SD}(w, G^h) = \epsilon(\partial_\eta w, \partial_\eta G^h) + bh(\partial_\xi w, \partial_\xi G^h) + b(\partial_\xi w, G^h) \tag{5.81b}
$$

and

$$
\epsilon\rho(\nabla^2 u, \mathbf{b}\cdot\nabla G^h) = \epsilon(h - \epsilon/b)(\nabla^2 u, \partial_\xi G^h). \tag{5.81c}
$$

To obtain sharp estimates outside any boundary layers in the problem, and hence to demonstate the accuracy achieved through eliminating crosswind diffusion, the domain Ω is divided into a thin strip Ω' extending upwind from $\mathbf{x}'$ to the boundary, but only a short distance in the downwind direction, and the remainder $\Omega\backslash\Omega'$, as in Fig. 5.4. The aim is to assume u is smooth in the strip, so that both $Pu - u$ and $\nabla^2 u$ are small there, and make the strip as small as possible consistent with showing that G^h is small in the rest of the domain.

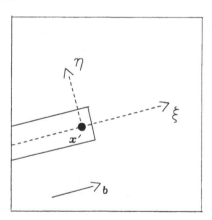

Figure 5.4. *The strip Ω' used in defining the cut-off function $\omega(\mathbf{x})$*

The parameters defining the strip can be obtained by considering a local separation of variables solution to the homogeneous differential equation; this would give a behaviour of the form $\exp(-|\eta|/\epsilon^{\frac{1}{2}})$ in the normal direction, so that for the solution to decay to $O(h^\alpha)$ outside the strip, its width should be of the order of $\epsilon^{\frac{1}{2}}\ln(1/h)$. However, we also need to ensure that the strip is wide enough to span several triangles; so we define

$$\hat{\epsilon} = \max(\epsilon, h^{3/2}) \qquad (5.82a)$$

to replace ϵ in defining the strip thickness. On the other hand, in the downwind direction we have a decay like that in the one-dimensional convection-diffusion problem; so with ϵ increased to bh, the length of the strip in this direction should be of the order $h\ln(1/h)$. Thus it is that we define

$$\Omega' := \{\mathbf{x} \mid (\mathbf{x} - \mathbf{x}') \cdot \mathbf{b} \le 2K'bh\ln(1/h),$$
$$|(\mathbf{x} - \mathbf{x}') \times \mathbf{b}| \le 2K'b\hat{\epsilon}^{\frac{1}{2}}\ln(1/h)\}, \qquad (5.82b)$$

for some $K' > 1$ to be chosen later; with Ω being the unit square, it is clear that meas $\Omega' \le 4K'\hat{\epsilon}^{\frac{1}{2}}\ln(1/h)$.

The first step in the analysis is to introduce a cut-off function which is close to zero over most of $\Omega\backslash\Omega'$. So we define

$$g(s) := \begin{cases} |s| & \text{for } |s| \ge 1 \\ 1 - (2/\pi)\cos\frac{1}{2}\pi s & \text{for } |s| \le 1, \end{cases} \qquad (5.83a)$$

for which we readily see that $g \in C^2(\mathbb{R}), g(t) \ge 1 - (2/\pi)$ and $|g'(t)| \le 1$. Then we define

$$\psi(t) := \left[\int_{-\infty}^{\infty} e^{-g(s)}ds\right]^{-1} \int_t^{\infty} e^{-g(s)}ds. \qquad (5.83b)$$

It is easy to check that ψ is symmetric about $\psi(0) = \frac{1}{2}$ and has the following properties:

$$0 = \psi(\infty) < \psi(t) < \psi(-\infty) = 1; \quad \psi(t) \le e^{-t} \text{ for } t \ge 1;$$
$$0 \le -\psi'(t) \le \psi(t); \text{ and } |\psi''(t)| \le |\psi'(t)|. \qquad (5.84)$$

Now with a new coordinate system (ξ, η) having its origin at $\mathbf{x}'$ and with ξ parallel to $\mathbf{b}$, we let $A = K'h\ln(1/h), B = K'\hat{\epsilon}^{\frac{1}{2}}\ln(1/h)$ and introduce a cut-off function

$$\tilde{\omega}(\mathbf{x}) := \psi\left(\frac{\xi - A}{\sigma_\xi}\right) \cdot \psi\left(\frac{\eta - B}{\sigma_\eta}\right) \cdot \psi\left(\frac{-\eta - B}{\sigma_\eta}\right), \qquad (5.85)$$

where $\sigma_\xi = K''h, \sigma_\eta = K''\hat{\epsilon}^{\frac{1}{2}}$ and K'' is yet to be chosen.

It is also convenient to boost the coercivity condition on $B_{SD}(\cdot, \cdot)$ given by (5.76) by the addition of an L_2 term. This can be done in a standard way by introducing an exponential factor into the first or second argument. For the present purposes it is most convenient to put it in the first argument; so from (5.81b) we have with $\pi(\mathbf{x}) = e^{\alpha\xi}, \alpha > 0$ and $w \in H_{E_0}^1$,

$$B_{SD}(\pi w, w) = \epsilon\left\|\pi^{\frac{1}{2}}\partial_\eta w\right\|^2 + bh\left\|\pi^{\frac{1}{2}}\partial_\xi w\right\|^2$$

$$+ b[\alpha h(\pi w, \partial_\xi w) + (\partial_\xi(\pi w), w)].$$

Integrating the last terms by parts, and using the non-negativity of the boundary terms, we have

$$(\pi w, \partial_\xi w) \geq -(\partial_\xi(\pi w), w) = -(\pi w, \partial_\xi w) - \alpha(\pi w, w)$$

and hence deduce that

$$B_{SD}(\pi w, w) \geq \epsilon \left\| \pi^{\frac{1}{2}} \partial_\eta w \right\|^2 + bh \left\| \pi^{\frac{1}{2}} \partial_\xi w \right\|^2 + \tfrac{1}{2} b\alpha(1 - \alpha h) \left\| \pi^{\frac{1}{2}} w \right\|^2. \quad (5.86)$$

It will be convenient to choose α so that $\tfrac{1}{2}b\alpha = 1$.

In $\Omega \backslash \Omega'$ at least one of the ψ factors in $\tilde{\omega}$ has an argument greater than $(K'/K'')\ln(1/h)$ which we assume is larger than unity. So incorporating the weighting function into the cut-off function to give $\omega = \tilde{\omega}\pi^{-1}$ we have, for some constant C and $\alpha = 2/b$,

$$\omega(\mathbf{x}) := \tilde{\omega}(\mathbf{x})e^{-\alpha\xi} \leq Ch^{K'/K''} \text{ in } \Omega \backslash \Omega'. \quad (5.87a)$$

We also assume that the triangle containing $\mathbf{x}'$ lies in Ω'', a region defined as in (5.82b) with $2K'$ replaced by $\tfrac{1}{2}K'$; and for this region we have

$$\omega(\mathbf{x}) > (1 - h^{K'/2K''})^3 \text{ in } \Omega'' \subset \Omega', \quad (5.87b)$$

because all of the arguments of $\psi(\cdot)$ in (5.85) are less than $-(K'/2K'')\ln (1/h)$ and $\pi(\cdot)$ is exponentially close to unity.

This construction is used to obtain two key bounds for a discrete Green's function, the first over $\Omega \backslash \Omega'$ and the second over the whole domain Ω. The argument is lengthy and is broadly based on that of Johnson *et al.* (1987), as modified and sharpened by Niijima (1990). As indicated at the beginning of the section, it is necessary to enhance the crosswind diffusion by replacing ϵ by $\hat{\epsilon}$ given by (5.82a); we denote the corresponding bilinear form given in (5.81b) by $\hat{B}_{SD}(\cdot, \cdot)$ and the Green's function given as in (5.79) by $\hat{G}^h$. We also assume only Dirichlet boundary conditions. To some extent the powers of h appearing in the bounds over $\Omega \backslash \Omega'$, and the relationship assumed between ϵ and h, are arbitrary; so the following lemma should be regarded as representing typical results.

Lemma 5.7.3 *The discrete Green's function $\hat{G}^h$ defined for the modified form $\hat{B}_{SD}(\cdot, \cdot)$, with all Dirichlet boundary conditions and $\epsilon \leq h$, satisfies*

$$J_1 := \hat{\epsilon} \left\| \partial_\eta \hat{G} \right\|^2_{L_2(\Omega \backslash \Omega')} + bh \left\| \partial_\xi \hat{G}^h \right\|^2_{L_2(\Omega \backslash \Omega')}$$

$$+ \left\| \hat{G}^h \right\|^2_{L_2(\Omega \backslash \Omega')} \leq C_1 h^6 \quad (5.88a)$$

$$J_2 := \hat{\epsilon} \left\| \partial_\eta \hat{G}^h \right\|^2_{L_2(\Omega)} + bh \left\| \partial_\xi \hat{G}^h \right\|^2_{L_2(\Omega)} \leq C_2 h^{-1}\ln(1/h), \quad (5.88b)$$

where C_1 and C_2 are constants independent of h and $\hat{\epsilon}$.

Proof of (5.88a). We manipulate $\widehat{B}_{SD}(\omega^{-1}G^h, G^h)$ as in obtaining (5.86), and get

$$\widehat{B}_{SD}(\omega^{-1}G^h, G^h) = \hat{\epsilon}\left\|\omega^{-\frac{1}{2}}\partial_\eta\widehat{G}^h\right\|^2 + bh\left\|\omega^{-\frac{1}{2}}\partial_\xi\widehat{G}^h\right\|^2$$

$$+ \tfrac{1}{2}b\left\|(\partial_\xi\omega^{-1})^{\frac{1}{2}}\widehat{G}^h\right\|^2$$

$$+ \hat{\epsilon}(\partial_\eta\widehat{G}^h, \widehat{G}^h\partial_\eta\omega^{-1}) + bh(\partial_\xi\widehat{G}^h, \widehat{G}^h\partial_\xi\omega^{-1})$$

$$=: J_\omega + I_\eta + I_\xi, \tag{5.89}$$

where I_η and I_ξ are the last two inner products. Note that

$$\omega\partial_\xi\omega^{-1} = -\omega^{-1}\partial_\xi\omega = \tilde{\omega}\pi^{-1}\partial_\xi(\tilde{\omega}^{-1}\pi) = \alpha - \tilde{\omega}^{-1}\partial_\xi\tilde{\omega}$$

$$= \alpha - \psi'/\sigma_\xi\psi \ge \alpha = 2/b, \tag{5.90}$$

so that the last term in J_ω gives a positive L_2 norm; in fact we have

$$J_\omega \ge \hat{\epsilon}\left\|\omega^{-\frac{1}{2}}\partial_\eta\widehat{G}^h\right\|^2 + bh\left\|\omega^{-\frac{1}{2}}\partial_\xi\widehat{G}^h\right\|^2 + \left\|\omega^{-\frac{1}{2}}\widehat{G}^h\right\|^2. \tag{5.91}$$

Then for each term in J_1 we use (5.87a) to write, for instance,

$$\left\|\partial_\eta\widehat{G}^h\right\|^2_{L_2(\Omega\setminus\Omega')} \le \max_{\Omega\setminus\Omega'}\omega \left\|\omega^{-\frac{1}{2}}\partial_\eta\widehat{G}^h\right\|^2_{L_2(\Omega\setminus\Omega')},$$

to obtain from (5.91)

$$J_1 \le Ch^{K'/K''} J_\omega. \tag{5.92}$$

We shall show below that $J_\omega \le Ch^{-2}$ and hence obtain (5.88a) by taking $K' \ge 8K''$.

Now from the definition of the discrete Green's function as in (5.79), if we introduce a function δ^h which is linear on the triangle containing the point x' but is zero elsewhere, and is defined by $(\delta^h, V) = V(x') \ \forall V \in S_0^h$, we have in fact

$$\widehat{B}_{SD}(V, \widehat{G}^h) = (\delta^h, V) = V(x') \quad \forall V \in S_0^h. \tag{5.93}$$

From inverting the element mass matrix — see (4.23d) — it is clear that $\|\delta^h\|^2 \le Ch^{-2}$; and, moreover, we have from (5.89)

$$J_\omega \le \widehat{B}_{SD}(\omega^{-1}\widehat{G}^h - V, \widehat{G}^h) + (\delta^h, V) - I_\eta - I_\xi \quad \forall V \in S_0^h. \tag{5.94}$$

We choose V to be the interpolant $\chi \in S_0^h$ of $\omega^{-1}\widehat{G}^h$, and estimate the magnitude of each of these terms in order.

Writing $e := \omega^{-1}\widehat{G}^h - \chi$ and using the definition of $\widehat{B}_{SD}(\cdot, \cdot)$, after integrating by parts and using the fact that $\widehat{G}_h$ and therefore e is zero on the boundary, we have

$$\widehat{B}_{SD}(e, \widehat{G}^h) = \hat{\epsilon}(\partial_\eta e, \partial_\eta\widehat{G}^h) + b(h\partial_\xi e - e, \partial_\xi\widehat{G}^h); \tag{5.95}$$

and from approximation theory on each triangle τ we have

$$h\,\|\nabla e\|_\tau + \|e\|_\tau \leq Ch^2 \sum_{|\gamma|=2} \left\| D^\gamma(\omega^{-1}\widehat{G}^h) \right\|_\tau. \tag{5.96}$$

The second derivatives of the cut-off function that occur here have to be estimated fairly carefully. Since $\widehat{G}^h$ is linear, we have

$$\omega \partial_\eta^2(\omega^{-1}\widehat{G}^h) = [2(\omega^{-1}\partial_\eta\omega)^2 - \omega^{-1}\partial_\eta^2\omega]\widehat{G}^h - 2(\omega^{-1}\partial_\eta\omega)\partial_\eta\widehat{G}^h \tag{5.97a}$$

$$\omega \partial_\xi\partial_\eta(\omega^{-1}\widehat{G}^h) = [2(\omega^{-1}\partial_\xi\omega)(\omega^{-1}\partial_\eta\omega) - \omega^{-1}\partial_\xi\partial_\eta\omega]\widehat{G}^h$$
$$- (\omega^{-1}\partial_\xi\omega)\partial_\eta\widehat{G}^h - (\omega^{-1}\partial_\eta\omega)\partial_\xi\widehat{G}^h, \tag{5.97b}$$

with a similar result to the first for the ∂_ξ^2 term. From (5.85) and (5.87a) it is clear that $|\omega^{-1}\partial_\eta\omega| \leq |\psi'/\sigma_\eta\psi| \leq 1/\sigma_\eta$, using (5.84), while $|\omega^{-1}\partial_\eta^2\omega| \leq (2/\sigma_\eta) + (2/\sigma_\eta^2)$; similar bounds hold for the other terms using (5.90), but because of the coefficient of $\widehat{G}^h$ in the definition of J_ω in (5.89), it is useful to retain a factor of $(\omega^{-1}\partial_\xi\omega)$ in the coefficient of $\widehat{G}^h$ in both the $\partial_\xi\partial_\eta$ term and the ∂_ξ^2 term. Finally, in the third term of (5.97b) an inverse estimate is used to bound $|\partial_\eta\widehat{G}^h$ by $|\widehat{G}^h|/h$. Hence we have

$$|\omega\partial_\eta^2(\omega^{-1}\widehat{G}^h)| \leq C[\sigma_\eta^{-1}|\partial_\eta\widehat{G}^h| + \sigma_\eta^{-2}|\widehat{G}^h|] \tag{5.98a}$$

$$|\omega\partial_\xi^2(\omega^{-1}\widehat{G}^h)| \leq C[\sigma_\xi^{-1}|\partial_\eta\widehat{G}^h| + \sigma_\xi^{-3/2}(\omega^{-1}\partial_\xi\omega)^{\frac{1}{2}}|\widehat{G}^h|] \tag{5.98b}$$

$$|\omega\partial_\xi\partial_\eta(\omega^{-1}\widehat{G}^h)| \leq C[\sigma_\eta^{-1}|\partial_\xi\widehat{G}^h| + (h^{-1} + \sigma_\eta^{-1})\sigma_\xi^{-\frac{1}{2}}(\omega^{-1}\partial_\xi\omega)^{\frac{1}{2}}|\widehat{G}^h|]. \tag{5.98c}$$

Collecting all the resultant terms, substituting $\sigma_\xi = K''h, \sigma_\eta = K''\hat{\epsilon}^{\frac{1}{2}}$, and using where necessary the relations $h^{3/2} \leq \hat{\epsilon} \leq h$, we obtain on each triangle τ

$$|\widehat{B}_S\;(e,\widehat{G}^h)_\tau| \leq \max_\tau \omega^{\frac{1}{2}}[\|\nabla e\|_{\tau,D} + h^{-1}\|e\|_\tau]$$

$$\times \left[\hat{\epsilon}\left\|\omega^{-\frac{1}{2}}\partial_\eta\widehat{G}^h\right\|_\tau + bh\left\|\omega^{-\frac{1}{2}}\partial_\xi\widehat{G}^h\right\|_\tau\right]$$

$$\leq \frac{C}{K''}\left[\hat{\epsilon}h\left\|\omega^{-\frac{1}{2}}\partial_\eta\widehat{G}^h\right\|_\tau + h^2\left\|\omega^{-\frac{1}{2}}\partial_\xi\widehat{G}^h\right\|_\tau\right]$$

$$\times \left[\frac{1}{\hat{\epsilon}^{\frac{1}{2}}}\left\|\omega^{-\frac{1}{2}}\partial_\eta\widehat{G}^h\right\|_\tau + \frac{1}{h}\left\|\omega^{-\frac{1}{2}}\partial_\xi\widehat{G}^h\right\|_\tau\right]$$

$$+ \left\|\left(\frac{1}{\hat{\epsilon}} + \frac{1}{h^{3/2}}(\omega^{-1}\partial_\xi\omega)^{\frac{1}{2}}\right)\omega^{-\frac{1}{2}}\widehat{G}^h\right\|_\tau\right], \tag{5.99a}$$

where we have used the fact that the ratio between the largest and smallest values of ω on a triangle is bounded. It is then readily checked that by taking

K'' sufficiently large and summing over the triangles we have

$$\left|\widehat{B}_{SD}(e,\widehat{G}^h)\right| \le \tfrac{1}{4}J_\omega. \tag{5.99b}$$

The I_η and I_ξ terms in (5.94) are estimated in a similar but simpler fashion:

$$\hat{\epsilon}\left|\partial_\eta \widehat{G}^h\right|\left|\widehat{G}^h \partial_\eta(\omega^{-1})\right| \le \frac{\hat{\epsilon}}{K''\hat{\epsilon}^{\frac{1}{2}}}\left|\omega^{-\frac{1}{2}}\partial_\eta\widehat{G}^h\right|\left|\omega^{-\frac{1}{2}}\widehat{G}^h\right|$$

$$bh\left|\partial_\xi\widehat{G}^h\right|\left|\widehat{G}^h\partial_\xi(\omega^{-1})\right| \le \frac{bh}{(K''h)^{\frac{1}{2}}}\left|\omega^{-\frac{1}{2}}\partial_\xi\widehat{G}^h\right|\left|(\partial_\xi\omega^{-1})^{\frac{1}{2}}\widehat{G}^h\right|$$

so that $|I_\eta| + |I_\xi| \le \tfrac{1}{4}J_\omega$ for sufficiently large K''. Also, one of the implications of (5.99) is that

$$\left\|\omega^{\frac{1}{2}}e\right\| \le \hat{\epsilon}^{\frac{1}{2}}\left[\left\|\omega^{\frac{1}{2}}\nabla e\right\| + h^{-1}\left\|\omega^{\frac{1}{2}}e\right\|\right] \le \tfrac{1}{4}J_\omega^{\frac{1}{2}}.$$

Hence

$$|(\delta^h, \chi)| \le |(\delta^h, e)| + \left|(\delta^h, \omega^{-1}\widehat{G}^h)\right|$$

$$\le \left\|\omega^{-\frac{1}{2}}\delta^h\right\|\left[\left\|\omega^{\frac{1}{2}}e\right\| + \left\|\omega^{-\frac{1}{2}}\widehat{G}^h\right\|\right]$$

$$\le \tfrac{1}{4}J_\omega + C\left\|\omega^{-\frac{1}{2}}\delta_h\right\|^2, \tag{5.100}$$

and because of (5.87b) the last term is $O(h^{-2})$. Thus by combining (5.94), (5.99b) and (5.100) we have established that by choosing K'' sufficiently large we have $J_\omega \le Ch^{-2}$, and thence by choice of K' we obtain (5.88a) from (5.92). □

Proof of (5.88b). From the definitions of the Green's function and $\widehat{B}_{SD}(\cdot,\cdot)$,

$$\widehat{G}^h(\mathbf{x}') = \widehat{B}_{SD}(\widehat{G}^h, \widehat{G}^h) = J_2. \tag{5.101}$$

Now suppose that $\mathbf{x}^*$ is a point in $\Omega\backslash\Omega'$ on the streamline directly downwind from the point $\mathbf{x}'$. Then, because $\partial_\xi\widehat{G}_h$ is piecewise constant,

$$|\widehat{G}^h(\mathbf{x}^*) - \widehat{G}^h(\mathbf{x}')| = \left|\int_0^{|(\mathbf{x}^*-\mathbf{x}'|} \partial_\xi\widehat{G}^h(\mathbf{x}' + \xi\mathbf{b}/|\mathbf{b}|)d\xi\right|$$

$$\le \sum_{(\tau)}\frac{(\Delta\xi)_\tau}{\text{meas } \tau}\int_\tau |\partial_\xi\widehat{G}^h|d\Omega,$$

where the sum is over all the triangles τ crossed by the line from $\mathbf{x}'$ to $\mathbf{x}^*$, with $(\Delta\xi)_\tau$ denoting the length of the crossing. From the regularity of the triangulation, $(\Delta\xi)_\tau/(\text{meas } \tau)$ is $O(h^{-1})$; and if $T = \cup\tau$ we have meas $T \leq 2K'h^2\ln(1/h)$ from (5.82). Hence

$$\left|\widehat{G}^h(\mathbf{x}^*) - \widehat{G}^h(\mathbf{x}')\right| \leq Ch^{-1}\int_T \left|\partial_\xi \widehat{G}^h\right|\,\mathrm{d}\Omega$$

$$\leq C[\ln(1/h)]^{\frac{1}{2}}\left\|\partial_\xi \widehat{G}^h\right\|$$

$$\leq \tfrac{1}{2}J_2 + Ch^{-1}\ln(1/h),$$

and thence from (5.101) we have

$$J_2 \leq Ch^{-1}\ln(1/h) + 2|\widehat{G}^h(\mathbf{x}^*)|. \tag{5.102}$$

By an inverse estimate, $|\widehat{G}^h(\mathbf{x}^*)| \leq Ch^{-1}\|\widehat{G}^h\|_{L_2(\Omega\backslash\Omega')} \leq Ch^2$ by (5.88a), and hence we have proved (5.88b). □

It is worth noting here that the result (5.88b) can be somewhat strengthened by introducing a weighting factor as in (5.86) and redeploying the arguments in the above lemma. Let $\pi(\cdot)$ be a linear function of ξ, $a + \alpha\xi$, with a adjusted such that $P(\pi\widehat{G}^h)(\mathbf{x}') = \widehat{G}^h(\mathbf{x}')$, where P is the projector onto S_0^h. Then we have, for $e = P(\pi\widehat{G}^h) - \pi\widehat{G}^h$

$$J_2 = \widehat{G}^h(\mathbf{x}') = \widehat{B}_{SD}(P(\pi\widehat{G}^h), \widehat{G}^h)$$

$$= \widehat{B}_{SD}(\pi\widehat{G}^h, \widehat{G}^h) + \widehat{B}_{SD}(e, \widehat{G}^h). \tag{5.103}$$

It is easy to see that

$$\widehat{B}_{SD}(\pi\widehat{G}^h, \widehat{G}^h) = \hat{\epsilon}\left\|\pi^{\frac{1}{2}}\partial_\eta\widehat{G}^h\right\| + bh\left\|\pi^{\frac{1}{2}}\partial_\xi\widehat{G}^h\right\| + \tfrac{1}{2}b\alpha\|\widehat{G}^h\|^2;$$

and, because $\partial_\xi^2(\pi\widehat{G}^h) = 2\alpha\partial_\xi\widehat{G}^h$ in each triangle, we have

$$|\widehat{B}_{SD}(e, \widehat{G}^h)| \leq C\alpha hJ_2.$$

Combining these bounds in (5.103) shows that $\|\widehat{G}^h\|^2 = O(J_2)$ or, rather,

$$\|\widehat{G}^h\|^2 \leq Ch^{-1}\ln(1/h), \tag{5.104}$$

which we shall use below.

The carefully constructed bounds for the discrete Green's function that have been obtained in Lemma 5.7.3 finally allow the following theorem to be proved; it shows that if the solution of the differential problem is smooth in a narrow strip upwind of a point $\mathbf{x}'$, then at least $O(h)$ accuracy is attained at that point by the streamline diffusion method. Outside of this strip it is necessary only to assume that $\epsilon\nabla^2 u \in L_1(\Omega)$ and $\nabla u \in [L_1(\Omega)]^2$, which as we have seen in Chapter 2 still holds through a typical boundary layer.

Theorem 5.7.4 *Suppose that for the problem (5.76) with Dirichlet boundary conditions we have*

$$\|u\|_{C^2(\Omega')} + \epsilon\left\|\nabla^2 u\right\|_{L_1(\Omega)} + \|\nabla u\|_{L_1(\Omega)} + \|S\|_{L_2(\Omega)} \leq K \tag{5.105}$$

for some constant K; and suppose the approximation $\widehat{U}$ is given by $\widehat{B}_{SD}(\widehat{U}, V) = (S, MV)$ $\forall V \in S_0^h$ with $\epsilon \leq h$ and $\hat{\epsilon} = \max(\epsilon, h^{3/2})$. Then there is a constant $C(K)$, independent of h and ϵ, such that

$$|u(x') - \widehat{U}(x')| \leq C(K) h \hat{\epsilon}^{\frac{1}{4}} \ln(1/h). \tag{5.106}$$

Proof. The starting point for the bound is an expression as in (5.81) for $(Pu - \widehat{U})(x')$, in which we will choose P to be the L_2 projector onto S_E^h; since with $u \in C^2(\widehat{\Omega}')$ by hypothesis we have $(Pu - u)(x') = O(h^2)$, $(Pu - \widehat{U})(x')$ is all we need to bound. As in (5.81) we have after integrating the convection term by parts,

$$
\begin{aligned}
(Pu - \widehat{U})(x') &= \widehat{B}_{SD}(Pu - \widehat{U}, \widehat{G}_h) \\
&= \hat{\epsilon}(\partial_\eta Pu, \partial_\eta \widehat{G}_h) + bh(\partial_\xi Pu, \partial_\xi \widehat{G}_h) - b(Pu, \partial_\xi \widehat{G}^h) - (S, M\widehat{G}_h) \\
&= \hat{\epsilon}(\partial_\eta(Pu - u), \partial_\eta \widehat{G}^h) + bh(\partial_\xi(Pu - u), \partial_\xi \widehat{G}^h) - b(Pu - u, \partial_\xi \widehat{G}^h) \\
&\quad + \epsilon \rho b(\nabla^2 u, \partial_\xi \widehat{G}^h) + (\hat{\epsilon} - \epsilon)(\partial_\eta u, \partial_\eta \widehat{G}^h).
\end{aligned} \tag{5.107a}
$$

All these integrals can be split between those over Ω' and those over $\Omega \backslash \Omega'$; but the last perturbation term, which has arisen because of the enhanced crosswind diffusion that has been used in obtaining $\widehat{U}$, we shall partly integrate by parts to give

$$(\partial_\eta u, \partial_\eta \widehat{G}^h)_{\Omega \backslash \Omega'} + \int_{\partial \Omega'} (\partial_\eta u) \widehat{G}^h n_\eta d\Gamma - (\partial_\eta^2 u, \widehat{G}^h)_{\Omega'} \tag{5.107b}$$

and include the first two terms in the integrals over $\Omega \backslash \Omega'$. Then if we denote by $I_{\Omega'}$ and $I_{\Omega \backslash \Omega'}$ the integrals in (5.107) over Ω' and $\Omega \backslash \Omega'$, in Ω' we have $Pu - u = O(h^2)$, $\nabla(Pu - u) = O(h)$ and $\nabla^2 u = O(1)$, so that

$$
\begin{aligned}
|I_{\Omega'}| &\leq Ch \left[\hat{\epsilon} \left\| \nabla \widehat{G}^h \right\|_{L_1(\Omega')} + h \left\| \partial_\xi \widehat{G}^h \right\|_{L_1(\Omega')} + h^{\frac{1}{2}} \left\| \widehat{G}^h \right\|_{L_1(\Omega')} \right] \\
&\leq Ch(\text{meas } \Omega')^{\frac{1}{2}} \left[\hat{\epsilon} \left\| \partial_\eta \widehat{G}^h \right\|_{L_2(\Omega')} + h \left\| \partial_\xi \widehat{G}^h \right\|_{L_2(\Omega')} + h^{\frac{1}{2}} \left\| \widehat{G}^h \right\|_{L_2(\Omega')} \right] \\
&\leq Ch^{3/2} \hat{\epsilon}^{\frac{1}{4}} (\ln(1/h))^{\frac{1}{2}} \left[\hat{\epsilon} \left\| \partial_\eta \widehat{G}^h \right\|_{L_2(\Omega')}^2 + h \left\| \partial_\xi \widehat{G}^h \right\|_{L_2(\Omega')}^2 + \left\| \widehat{G}^h \right\|_{L_2(\Omega')}^2 \right]^{\frac{1}{2}} \\
&\leq Ch \hat{\epsilon}^{\frac{1}{4}} \ln(1/h),
\end{aligned} \tag{5.108}
$$

from (5.88b) and (5.104).

In $\Omega \backslash \Omega'$ we have from the hypothesis (5.105)

$$
\begin{aligned}
\epsilon \rho b(\nabla^2 u, \partial_\xi \widehat{G}^h)_{\Omega \backslash \Omega'} &\leq \epsilon h \left\| \nabla^2 u \right\|_{L_1(\Omega \backslash \Omega')} \left\| \partial_\xi \widehat{G}^h \right\|_{L_\infty(\Omega \backslash \Omega')} \\
&\leq Ch \left\| \partial_\xi \widehat{G}^h \right\|_{L_\infty(\Omega \backslash \Omega')}.
\end{aligned} \tag{5.109}
$$

We also have on the whole of the unit square Ω that $\|\nabla(Pu - u)\|_{L_1} \leq$

$\|\nabla Pu\|_{L_1} + \|\nabla u\|_{L_1} \leq \|\nabla Pu\|_{L_1} + C$, and $\|\nabla Pu\|_{L_1} \leq \|\nabla Pu\|_{L_2}$; while, by the inverse estimates of section 4.2.3, $\|\nabla Pu\|_{L_2} \leq Ch^{-1}\|Pu\|_{L_2}$ and clearly $\|Pu\|_{L_2} \leq \|u\|_{L_2}$, as well as $\|Pu - u\|_{L_2} \leq \|u\|_{L_2}$. Moreover, by a Sobolev imbedding theorem — see Gilbarg and Trudinger (1983), Theorem 7.10, p155 — the hypothesis (5.105) implies that $\|u\|_{L_2(\Omega')}$ is bounded. Thus we have from (5.107a), (5.109) and (5.107b), again using inverse estimates on each triangle to give $\|\widehat{G}^h\|_{L_\infty} \leq Ch^{-1}\|\widehat{G}^h\|_{L_2}$ with similar results for $\partial_\xi \widehat{G}^h$ and $\partial_\eta \widehat{G}^h$,

$$
\begin{aligned}
|I_{\Omega\setminus\Omega'}| &\leq Ch^{-1}\left[\hat{\epsilon}\left\|\partial_\eta\widehat{G}^h\right\|_{L_\infty(\Omega\setminus\Omega')} + h\left\|\partial_\xi\widehat{G}^h\right\|_{L_\infty(\Omega\setminus\Omega')}\right] \\
&\quad + Ch\left\|\partial_\xi\widehat{G}^h\right\|_{L_\infty(\Omega\setminus\Omega')} \\
&\quad + C\hat{\epsilon}\left[\left\|\partial_\eta\widehat{G}^h\right\|_{L_\infty(\Omega\setminus\Omega')} + \left\|\widehat{G}^h\right\|_{L_\infty(\partial\Omega')}\right] \\
&\leq Ch^{-2}\left[\hat{\epsilon}\left\|\partial_\eta\widehat{G}^h\right\|_{L_2(\Omega\setminus\Omega')}\right. \\
&\quad + \left.h\left\|\partial_\xi\widehat{G}^h\right\|_{L_2(\Omega\setminus\Omega')} + h\left\|\widehat{G}^h\right\|_{L_2(\Omega\setminus\Omega')}\right] \\
&\leq Ch^{3/2} \hspace{4cm} \text{(5.110)}
\end{aligned}
$$

from (5.88a). Since $\hat{\epsilon} > h^2$, (5.108) and (5.110) give the required result.   □

The analysis that we have presented in this section is the most detailed that we have so far undertaken in two dimensions. We have chosen to do so for the streamline diffusion method because the method's key properties are best displayed in two dimensional boundary layers. The fact that we have taken Ω as the unit square is unimportant; but the choice of a constant velocity field is a significant simplification. The effect on such analysis of a variable velocity field will be seen in the methods studied in the next chapter.

5.8 Use of local Green's functions

Extension of the Hemker test functions introduced in section 5.3 to more general one-dimensional problems provides a link between the methods of this chapter and the difference scheme of El-Mistikawy and Werle described in Chapter 3. It also provides a powerful tool for analysing the nodal error of the latter scheme; and the discrete Green's function that is constructed in the course of the analysis gives an explicit example which illustrates some of the more general analysis carried out in the last section. In section 5.8.2 we shall also consider extensions to two-dimensional problems that link up with the cell vertex finite volume methods to be considered in the next chapter.

5.8.1 One-dimensional analysis

We consider the general one-dimensional problem (3.1) for $-\epsilon u'' + b(x)u' + c(x)u = S(x)$; and we make assumptions on the coefficients as given in (3.85), namely $b(x) \geq b_\phi > 0$ and $c(x) \geq 0$, but assume b, c and S are all independent of ϵ. In section 3.3.2 we discussed the use of local Green's functions as the basis for generating various difference schemes; here we consider a particular choice of local Green's functions to construct the test space for a Petrov–Galerkin method.

For simplicity, we assume homogeneous boundary conditions $u(0) = u(1) = 0$, if necessary absorbing boundary data into the source function in the usual way. Then on the usual nonuniform mesh we introduce piecewise constant functions $\bar{b}, \bar{c}$ and $\bar{S}$ by the construction, e.g. in the case of b,

$$\bar{b} := \tfrac{1}{2}\left[b(x_{j-1}) + b(x_j)\right] \text{ on } (x_{j-1}, x_j), \quad j = 1, 2, \ldots, J; \quad (5.111)$$

and as basis functions for the test space T^h define $\{\psi_i(x); i = 1, 2, \ldots, J-1\}$ by

$$-\epsilon\psi_i'' - \bar{b}\psi_i' + \bar{c}\psi_i = 0 \text{ on } (x_{j-1}, x_j), \quad j = 1, 2, \ldots, J \quad (5.112a)$$

$$\psi_i(x_j) = \delta_{i,j}, \quad j = 0, 1, \ldots, J. \quad (5.112b)$$

Note that (5.112a) approximates the adjoint equation so that, apart from a scaling factor, $\psi_i(x)$ approximates $G_i(x_i, x)$ where $G_i(\cdot, \cdot)$ is the Green's function on the interval (x_{i-1}, x_{i+1}) that was introduced in section 3.3.2. With this test space our Petrov–Galerkin approximation U is given by

$$\bar{B}(U, \psi_i) = (\bar{S}, \psi_i) \quad i = 1, 2, \ldots, J-1, \quad (5.113a)$$

where

$$\bar{B}(v, w) := \sum_{j=1}^{J} \int_{x_{j-1}}^{x_j} \left[v'(\epsilon w' + \bar{b}w) + \bar{c}vw\right]dx. \quad (5.113b)$$

This may be used with any conforming trial space $S_0^h \equiv \operatorname{span}\{\phi_j; j = 1, 2, \ldots, J-1\}$; but we assume $\phi_i(x_j) = \delta_{i,j}$ as with the piecewise linears, and this gives a discrete system which is exactly the El-Mistikawy and Werle difference scheme. Indeed, by integrating the first terms in (5.113b) by parts we have, in an obvious notation,

$$\bar{B}(v, \psi) = \sum_{j=1}^{J} \left[v(\epsilon\psi' + \bar{b}\psi)\right]_{x_{j-1}^+}^{x_j^-} \quad \forall v \in H^1(0,1), \ \forall \psi \in T^h; \quad (5.114a)$$

moreover, this can be rearranged to give

$$\bar{B}(v, \psi) = -\sum_{j=1}^{J-1} \left[v(\epsilon\psi' + \bar{b}\psi)\right]_{x_j^-}^{x_j^+} \quad \forall v \in H_0^1(0,1), \ \forall \psi \in T^h. \quad (5.114b)$$

Hence we can write the discrete system (5.113a) as

$$\bar{K}\mathbf{U} = \bar{\mathbf{S}},\qquad\qquad(5.115a)$$

where $\mathbf{U} = (U_1, U_2, \ldots, U_{J-1})^T$ and the entries in the $(J-1) \times (J-1)$ stiffness matrix $\bar{K}$ and the corresponding load vector $\bar{\mathbf{S}}$ are given by

$$\bar{K}_{i,j} = \bar{B}(\phi_j, \psi_i) = -\left[\epsilon\psi_i' + \bar{b}\psi_i\right]_{x_j^-}^{x_j^+},\qquad\qquad(5.115b)$$

$$\bar{S}_i = (\bar{S}, \psi_i).\qquad\qquad(5.115c)$$

It is clear that $\bar{K}$ is tridiagonal and for the off-diagonal elements we have

$$\bar{K}_{i,i-1} = -\epsilon\psi_i'(x_{i-1}^+), \quad \bar{K}_{i,i+1} = \epsilon\psi_i'(x_{i+1}^-).\qquad(5.116a)$$

Moreover, from (5.114) we have

$$
\begin{aligned}
\bar{K}_{i,i-1} + \bar{K}_{i,i} + \bar{K}_{i,i+1} &= \left[\epsilon\psi_i' + \bar{b}\psi_i\right]_{x_{i-1}^+}^{x_i^-} + \left[\epsilon\psi_i' + \bar{b}\psi_i\right]_{x_i^+}^{x_{i+1}^-}\\
&= \left(\int_{x_{i-1}}^{x_i} + \int_{x_i}^{x_{i+1}}\right)(\epsilon\psi_i'' + \bar{b}\psi_i')\,\mathrm{d}x\\
&= \int_{x_{i-1}}^{x_{i+1}} \bar{c}\psi_i\,\mathrm{d}x.
\end{aligned}\qquad(5.116b)
$$

We therefore obtain the following result.

Lemma 5.8.1 *The stiffness matrix for the discrete system (5.115), corresponding to the El-Mistikawy and Werle difference scheme, satisfies*

$$\bar{K}_{i,i} \geq -\bar{K}_{i,i-1} - \bar{K}_{i,i+1} > 0\qquad\qquad(5.117)$$

and is therefore an M-matrix.

Proof. From the definition (5.112) and the conditions $b(x) \geq b_\phi > 0$ and $c(x) \geq 0$, it is clear that each test function ψ_i satisfies a maximum principle and $\psi_i \geq 0$. Thus $\psi_i'(x_{i-1}^+) \geq 0$ and $\psi_i'(x_{i+1}^-) \leq 0$; and the inequality must hold in each case because otherwise ψ_i would be identically zero on the corresponding interval. Substituting into (5.116) establishes (5.117). Then by a standard result (see, for example, Lemmas 1.5.1 and 1.5.2) $\bar{K}$ is nonsingular and its inverse has all its entries non-negative, i.e. it is an M-matrix; indeed, all the entries of $\bar{K}^{-1}$ are positive because $\bar{K}$ is irreducible. □

Let us now consider Green's functions over the whole interval. Corresponding to the continuous Green's function in (5.30) we can introduce a discrete Green's function for any Petrov–Galerkin method, as we did in the last section; specialising to a nodal point x_k and simplifying the notation, we can define in the present case $G_k^h(\cdot) \in T^h$ such that

$$\bar{B}(V, G_k^h) = V(x_k) \quad \forall V \in S_0^h.\qquad\qquad(5.118)$$

In particular, we have

$$\bar{B}(\phi_j, G_k^h) = \delta_{k,j} \tag{5.119}$$

and this holds for $j, k = 1, 2, \ldots, J-1$. So suppose we introduce coefficients $G_{k,i}$, to form a $(J-1) \times (J-1)$ matrix G, such that

$$G_k^h(x) = \sum_{i=1}^{J-1} G_{k,i} \psi_i(x); \tag{5.120a}$$

then, by setting $V = U$ in (5.118), it is clear that

$$U_k = \sum_{i=1}^{J-1} G_{k,i} \bar{B}(U, \psi_i) = \sum_{i=1}^{J-1} \sum_{j=1}^{J-1} G_{k,i} \bar{K}_{i,j} U_j$$

i.e. $\qquad G = \bar{K}^{-1}. \tag{5.120b}$

This, indeed, justifies the claim that G_k^h can be defined by (5.118); but so far this has not used any of the special properties of the present test functions, and (5.120) would hold for any test space giving a nonsingular stiffness matrix. However, in the present case we can show that the discrete Green's function has all the properties of a Green's function for a continuous problem, in the following sense.

Lemma 5.8.2 *The discrete Green's functions defined by (5.118) have the following properties:* $G_k^h \in C[0,1]$,

$$-\epsilon G_k^{h''} - \bar{b} G_k^{h'} + \bar{c} G_k^h = 0 \quad on \quad \bigcup_{j=1}^{J} (x_{j-1}, x_j), \tag{5.121a}$$

$$G_k^h(0) = G_k^h(1) = 0, \tag{5.121b}$$

$$G_k^h(x) \geq 0 \quad \forall x \in [0,1], \tag{5.121c}$$

and

$$-\left[\epsilon G_k^{h'} + \bar{b} G_k^h\right]_{x_j^-}^{x_j^+} = \delta_{k,j}. \tag{5.122}$$

Proof. Properties (5.121) follow from the definition of the test space in (5.112), the non-negativity of the test functions ψ_i and the fact that $K = G^{-1}$ is an M−matrix; and (5.122) follows from combining (5.119) with (5.114), as in deriving (5.115). $\square$

The properties of Lemma 5.8.2 provide the basis for estimating the nodal error in U. From (5.122) and (5.114b) we have

$$u(x_j) - U(x_j) = \bar{B}(u - U, G_j^h);$$

and (5.113) gives

$$\bar{B}(U, G_j^h) = (\bar{S}, G_j^h)$$
$$= (\bar{S} - S, G_j^h) + B(u, G_j^h).$$

Hence we have

$$u(x_j) - U(x_j) = ((\bar{b} - b)u' + (\bar{c} - c)u + (S - \bar{S}), G_j^h). \qquad (5.123)$$

We then need for both u and G_j^h bounds of the form devised in section 2.4 and already used in the error analysis of difference schemes in section 3.4. For the solution u, the most useful bound is given by a minor adaptation of the decomposition given in Theorem 2.4.3; namely, we write with $B(x) := \int_0^x b(s)\mathrm{d}s$

$$u'(x) = (\gamma/\epsilon)\mathrm{e}^{-(1/\epsilon)[B(1) - B(x)]} + z'(x), \qquad (5.124a)$$

where $\gamma = \epsilon u'(1)$ is bounded independently of ϵ, and

$$|z^{(s)}(x)| \leq C\left[1 + \frac{1}{\epsilon^{s-1}}\mathrm{e}^{-(b_\phi/2\epsilon)(1-x)}\right] \qquad s = 0, 1, 2. \qquad (5.124b)$$

Leaving aside for the moment the bounds needed for G_j^h, the $O(h^2)$ nodal accuracy is then dependent on the following approximation result.

Lemma 5.8.3 *Suppose $f \in C^2[0,1]$ with $\bar{f}$ obtained from f as in (5.111), and $g' \in L^1[0,1]$ with $g(0) = g(1) = 0$. Then*

$$\left|\int_0^1 (f - \bar{f})g\mathrm{d}x\right| \leq Ch^2 \int_0^1 |g'(x)|\,\mathrm{d}x. \qquad (5.125)$$

Proof. Writing $F(x) := \int_0^x f(s)\mathrm{d}s$ and similarly for $\bar{F}(x)$, integration by parts gives

$$\int_0^1 (f - \bar{f})g\mathrm{d}x = -\int_0^1 (F - \bar{F})g'\mathrm{d}x.$$

Consideration of the trapezoidal rule establishes that $|F - \bar{F}| \leq Ch^2$, and the result follows. $\square$

The error analysis, which mainly follows that developed in O'Riordan and Stynes (1986) and Stynes and O'Riordan (1986), then needs the following properties of the Green's function.

Lemma 5.8.4 *The discrete Green's functions of Lemma 5.8.2 satisfy the following bounds, where $\bar{B}(x) := \int_0^x \bar{b}(s)\mathrm{d}s$,*

$$0 \leq G_k^h < 1/b_\phi \quad on \ [0,1], \qquad (5.126a)$$
$$G_k^h(x) < (1/b_\phi)\mathrm{e}^{-(1/\epsilon)[B(x) - B(x_k)]} \quad on \ [x_k, 1], \qquad (5.126b)$$

and

$$\int_0^1 |G_k^{h'}|\mathrm{d}x \leq C, \qquad (5.127)$$

where C is independent of ϵ.

Proof. By integrating (5.121a) and (5.122) from $x = 1$ we obtain

$$\epsilon G_k^{h'} + \bar{b} G_k^h = \epsilon G_k^{h'}(1) - \int_x^1 \bar{c} G_k^h ds \quad \text{for} \quad x_k < x \le 1, \qquad (5.128a)$$

$$\epsilon G_k^{h'} + \bar{b} G_k^h = 1 + \epsilon G_k^{h'}(1) - \int_x^1 \bar{c} G_k^h ds \quad \text{for} \quad 0 \le x < x_k. \qquad (5.128b)$$

The first result (5.126a) follows immediately from (5.128) for we have already shown that $G_k^h \ge 0$ and $G_k^{h'}(1) < 0$, so that by (5.128a) G_k^h decreases monotonically on $(x_k, 1]$ from $G_k^h(x_k)$ to $G_k^h(1) = 0$; now suppose $G_k^h(z) \ge 1/b_\phi$ for some $z \in (0, x_k)$, which by (5.128b) would imply that $G_k^{h'}(z) < 0$. Then, by repeating the argument, we would find that $G_k^{h'}(x)$ would continue to grow as x decreased until the boundary condition at $x = 0$ was violated.

By use of an integrating factor, we can readily solve (5.128a) and (5.128b) to obtain

$$G_k^h(x) e^{B(x)/\epsilon} = \begin{cases} \int_x^1 \epsilon^{-1} e^{B(s)/\epsilon} \left[\int_s^1 \bar{c} G_k^h dt - \epsilon G_k^{h'}(1) \right] ds, & x_k \le x \le 1, \\ \int_0^x \epsilon^{-1} e^{B(s)/\epsilon} \left[1 - \int_s^1 \bar{c} G_k^h dt + \epsilon G_k^{h'}(1) \right] ds, & 0 \le x \le x_k. \end{cases}$$
$$(5.129)$$

Continuity at $x = x_k$ gives an equation for $G_k^{h'}(1)$, namely

$$\epsilon G_k^{h'}(1) \int_0^1 \epsilon^{-1} e^{B(s)/\epsilon} ds = \int_0^1 \epsilon^{-1} e^{B(s)/\epsilon} \int_s^1 \bar{c} G_k^h dt ds$$
$$- \int_0^{x_k} \epsilon^{-1} e^{B(s)/\epsilon} ds. \qquad (5.130)$$

We shall repeatedly make use of the obvious bounds, for $y < z$,

$$\frac{1}{\|b\|_\infty} \left[e^{B(z)/\epsilon} - e^{B(y)/\epsilon} \right] \le \int_y^z \frac{1}{\epsilon} e^{B(s)/\epsilon} ds$$
$$\le \frac{1}{b_\phi} \left[e^{B(z)/\epsilon} - e^{B(y)/\epsilon} \right]. \qquad (5.131)$$

From (5.130), because $\bar{c} G_k^h \ge 0$ they are readily seen to lead to

$$0 < -\epsilon G_k^{h'}(1) < (\|b\|_\infty / b_\phi) e^{-(1/\epsilon)[B(1) - B(x_k)]}. \qquad (5.132)$$

Moreover, by bounding the integral from x to 1 in the first case of (5.129) by the integral from 0 to 1 and substituting from (5.130), we obtain the result (5.126b).

For the gradient of the Green's function, the only concern is about possible oscillations in $(0, x_k)$. By introducing $\bar{b}'$ to be interpreted in the sense of distributions, we can integrate (5.121) and (5.122) directly using an integrating factor to get

$$\epsilon G_k^{h'}(x)e^{B(x)/\epsilon} - \epsilon G_k^{h'}(0) = \int_0^x e^{B(s)/\epsilon}\left[(\bar{c} - \bar{b}')G_k^h - \delta(s - x_k)\right]ds$$

$$= \int_0^x e^{B(s)/\epsilon}\bar{c}G_k^h ds - e^{B(x_k)/\epsilon}H_k(x)$$

$$- \sum_{x_j < x} e^{B(x_j)/\epsilon}G_k^h(x_j)\left[\bar{b}\right]_{x_j^-}^{x_j^+} =: M(x), \quad (5.133a)$$

where $H_k(x)$ is the Heaviside function which jumps from zero to one at x_k. Thus we have

$$G_k^{h'}(x) = e^{-B(x)/\epsilon}\left[G_k^{h'}(0) + (1/\epsilon)M(x)\right]. \quad (5.133b)$$

Now each term in $M(x)$ contains an exponential factor with argument less than x and, because each jump in $\bar{b}$ is $O(h)$, the sum of the coefficients of the exponentials is bounded; hence the integral of the second term in (5.133b) is bounded, as is its modulus. Integrating first (5.133b) as it stands, the left-hand side is zero and the coefficient of $G_k^{h'}(0)$ is bounded below by $\epsilon/\|b\|_\infty$. So we deduce that $G_k^{h'}(0) = O(\epsilon^{-1})$. Now we can take the modulus of (5.133b) and integrate to obtain our final result (5.127). □

Combining these last two lemmas yields the following result.

Theorem 5.8.5 *Suppose the coefficients b and c and the source function S all lie in $C^2[0,1]$. Then the Petrov–Galerkin method defined by (5.112)–(5.113), corresponding to the El-Mistikawy and Werle difference scheme, gives uniformly second order accuracy at the nodes.*

Proof. The nodal errors are given by (5.123); and the $O(h^2)$ bound for the term $(S - \bar{S}, G_j^h)$ follows immediately from Lemma 5.8.3 and the bound (5.127). To bound the term $(\bar{c} - c, uG_j^h)$ we also need the decomposition of u given by (5.124), from which we deduce bounds for $\|u\|_{L_\infty}$ and $\|u'\|_{L_1}$ to combine with those for G_j^h given by (5.126a) and (5.127). Finally, the term $(\bar{b} - b, z'G_j^h)$ can be dealt with in the same way because of the bounds for $\|z'\|_{L_\infty}$ and $\|z'\|_{L_1}$ obtained from (5.124b), but the dominant term in $(\bar{b} - b, u'G_j^h)$ must be dealt with more carefully.

What we need to show is that

$$\left|\int_0^1 \frac{1}{\epsilon}(\bar{b} - b)e^{-1/\epsilon[B(1)-B(x)]}G_j^h dx\right| \le Ch^2. \quad (5.134)$$

Integrating by parts and then substituting from (5.129), if the integral is denoted by I we have

$$e^{B(1)/\epsilon}I := \int_0^1 e^{(B-B)/\epsilon}(e^{B/\epsilon}G_j^h)' dx$$

$$= \int_0^{x_j} \frac{1}{\epsilon} e^{B/\epsilon} dx - \int_0^1 \frac{1}{\epsilon} e^{B/\epsilon} \left[\int_x^1 \bar{c} G_j^h dt - \epsilon G_j^{h'}(1) \right].$$

Let us write $B_1(x) := B(1) - B(x) = \int_x^1 b(s) ds$, $\bar{B}_1(x) := \bar{B}(1) - \bar{B}(x)$, introduce

$$Z(x) := \int_x^1 \frac{1}{\epsilon} e^{-B_1/\epsilon} dt, \quad \bar{Z}(x) := -\int_x^1 \frac{1}{\epsilon} e^{-\bar{B}_1/\epsilon} dt \qquad (5.135)$$

and substitute for $\epsilon G_j^{h'}(1)$ from (5.130). Separating out the terms in $\bar{c}$, we obtain $I = I_1 + I_2$, where

$$\begin{aligned}
I_1 &:= Z(0) - Z(x_j) - Z(0) \left[\bar{Z}(0) - \bar{Z}(x_j) \right] / \bar{Z}(0) \\
&= \left[\bar{Z}(x_j) Z(0) - Z(x_j) \bar{Z}(0) \right] / \bar{Z}(0), \qquad (5.136\text{a})
\end{aligned}$$

$$I_2 := \frac{1}{\bar{Z}(0)} \int_0^1 \frac{1}{\epsilon} \left[Z(0) e^{-B_1/\epsilon} - \bar{Z}(0) e^{-\bar{B}_1/\epsilon} \right] \int_x^1 \bar{c} G_j^h dt dx. \quad (5.136\text{b})$$

The second integral is the easier to bound. It is clear that $|B_1(x) - \bar{B}_1(x)| \leq Ch^2$ and hence, since $\min(B_1(x), \bar{B}_1(x)) \geq b_\phi(1 - x)$, we have

$$\begin{aligned}
|e^{-\bar{B}_1(x)/\epsilon} - e^{-B_1(x)/\epsilon}| &\leq e^{-b_\phi(1-x)/\epsilon}(1 - e^{-Ch^2/\epsilon}) \\
&\leq C(h^2/\epsilon) e^{-b_\phi(1-x)/\epsilon} \qquad (5.137)
\end{aligned}$$

It follows that

$$\begin{aligned}
|\bar{Z}(0) - Z(0)| &\leq \int_0^1 C(h^2/\epsilon^2) e^{-b_\phi(1-x)/\epsilon} dx \\
&\leq C(h^2/\epsilon), \qquad (5.138)
\end{aligned}$$

and therefore that

$$\begin{aligned}
|Z(0) e^{-B_1(x)/\epsilon} &- \bar{Z}(0) e^{-\bar{B}_1(x)/\epsilon}| \\
&\leq Z(0) |e^{-\bar{B}_1(x)/\epsilon} - e^{-B_1(x)/\epsilon}| + |Z(0) - \bar{Z}(0)| e^{-\bar{B}_1(x)/\epsilon} \\
&\leq C(h^2/\epsilon) e^{-b_\phi(1-x)/\epsilon}.
\end{aligned}$$

Since $\bar{c}$ and G_j^h are bounded, we have

$$\begin{aligned}
I_2 &\leq Ch^2 \int_0^1 \frac{1}{\epsilon^2} e^{-b_\phi(1-x)/\epsilon}(1 - x) dx \\
&= Ch^2 \left[-\frac{1}{\epsilon b_\phi} e^{-b_\phi/\epsilon} + \int_0^1 \frac{1}{\epsilon b_\phi} e^{-b_\phi(1-x)/\epsilon} \right] \\
&\leq Ch^2, \qquad (5.139)
\end{aligned}$$

after integrating by parts.

More careful estimates of the same type are needed for the first integral.

Firstly, we note that

$$
\begin{aligned}
Z(x_j) &= \int_{x_j}^1 \frac{1}{b(t)} \frac{b(t)}{\epsilon} e^{-B_1(t)/\epsilon} \mathrm{d}t \\
&= \frac{1}{b(1)} - \frac{1}{b(x_j)} e^{-B_1(x_j)/\epsilon} - \int_{x_j}^1 \left(\frac{1}{b}\right)' e^{-B_1/\epsilon} \mathrm{d}t \\
&= \frac{1}{b(1)} - \frac{1}{b(x_j)} e^{-B_1(x_j)/\epsilon} + O(\epsilon), \qquad\qquad (5.140\text{a})
\end{aligned}
$$

and hence also

$$
Z(0) = \frac{1}{b(1)} + O(\epsilon). \qquad\qquad (5.140\text{b})
$$

Then from (5.138), and a similar bound for $Z(x_j) - \bar{Z}(x_j)$, we have

$$
\begin{aligned}
Z(0)I_1 &= \left[Z(0) - \bar{Z}(0)\right] Z(x_j) - \left[Z(x_j) - \bar{Z}(x_j)\right] Z(0) \\
&= \frac{1}{b(1)} \int_0^{x_j} \Phi(t)\mathrm{d}t - \frac{1}{b(x_j)} e^{-B_1(x_j)/\epsilon} \int_0^1 \Phi(t)\mathrm{d}t + O(h^2),
\end{aligned}
$$

where $\Phi(t)$ is the integrand of $\left[Z(x) - \bar{Z}(x)\right]$ given in (5.135). But

$$
\left| \frac{1}{b(1)} - \frac{1}{b(x_j)} \right| e^{-B_1(x_j)/\epsilon} \int_0^1 \Phi(t)\mathrm{d}t \le Che^{-b_\phi h/\epsilon}(h^2/\epsilon) = O(h^2),
$$

and, by the properties of the trapezoidal rule,

$$
|e^{-B_1(x_j)/\epsilon} - e^{-B_1(x_j)/\epsilon}| \le e^{-B_1(x_j)/\epsilon} \left[C(h^2/\epsilon)(1 - x_j) \right] = O(h^2).
$$

Hence, by writing for $t \le x_j$ and with $t' = t + 1 - x_j$,

$$
\int_t^1 b(s)\mathrm{d}s = B_1(x_j) + \int_t^{x_j} b(s)\mathrm{d}s = B_1(x_j) + \int_{t'}^1 b(s' - 1 + x_j)\mathrm{d}s',
$$

we have

$$
\begin{aligned}
b(1)Z(0)I_1 &= e^{-B_1(x_j)/\epsilon} \left[\int_0^{x_j} \frac{1}{\epsilon} \left[e^{-[B_1(x) - B_1(x_j)]/\epsilon} - e^{-[B_1(x) - B_1(x_j)]/\epsilon} \right] \mathrm{d}t \right. \\
&\qquad\qquad \left. - \int_0^1 \Phi(t)\mathrm{d}t \right] + O(h^2) \\
&= e^{-B_1(x_j)/\epsilon} \left[\int_{1-x_j}^1 \left[\Phi_j(t) - \Phi(t) \right] \mathrm{d}t - \int_0^{1-x_j} \Phi(t)\mathrm{d}t \right] + O(h^2), \quad (5.141)
\end{aligned}
$$

where $\Phi_j(t)$ is the same as $\Phi(t)$ with $b(\cdot)$ replaced by $b(\cdot - 1 + x_j)$. Moreover, for the last term we have from (5.137)

$$
e^{-B(x_j)/\epsilon} \int_0^{1-x_j} \Phi(t)\mathrm{d}t \le e^{-b_\phi(1-x_j)/\epsilon} C(h^2/\epsilon) e^{-B_\phi x_j/\epsilon} = O(h^2);
$$

and the final required result will follow from establishing

$$|\Phi(t) - \Phi_j(t)| \leq C e^{b_\phi(1-t)/\epsilon} \left[h^2(1 - x_j)/\epsilon^2 + h^4/\epsilon^3 \right]. \qquad (5.142)$$

A careful analysis of the trapezoidal rule, as carried out in O'Riordan and Stynes (1986), shows that

$$\left| [B_1(t) - \bar{B}_1(t)] - [B_j(t) - \bar{B}_j(t)] \right| \leq C h^2(1 - x_j), \qquad (5.143)$$

where $B_j(t)$ is the same as $B_1(t)$ with $b(\cdot)$ replaced by $b(\cdot - 1 + x_j)$. We also have for $x \geq 0$ that $x - x^2 \leq 1 - e^{-x} \leq x$, and hence

$$e^{-B_1/\epsilon} - e^{-\bar{B}_1/\epsilon} = e^{-\min(B_1, \bar{B}_1)/\epsilon} \left[(B_1 - \bar{B}_1)/\epsilon + O(h^4/\epsilon^2) \right],$$

with a similar expression when B_1 is replaced by B_j; and by again taking the minimum of the arguments in the exponentials when taking the difference and using $|B_1 - B_j| \leq C(1 - x_j)$, we have

$$|\Phi(t) - \Phi_j(t)| \leq e^{-b_\phi(1-t)/\epsilon} \big[|(B_1 - \bar{B}_1) - (B_j - \bar{B}_j)| /\epsilon^2$$
$$+ C h^2(1 - x_j)/\epsilon^2 + O(h^4/\epsilon^3) \big].$$

Using the fact that $e^{-b_\phi h/\epsilon}(h^2/\epsilon^2) \leq C$ and applying (5.143) gives (5.142) and hence the desired result. $\square$

We have given this result in some detail because it is the best result available for any of the widely used methods. It means that in any boundary layer, however steep and containing however many or however few mesh points, the values at the mesh points are always uniformly second order accurate. As we saw in section 3.4.3, the uniform bound can only be obtained by the use of exponentials in some way; and the second order accuracy requires their careful approximation, as with the use of the trapezoidal rule, and a consistent treatment of the source function as in the Petrov–Galerkin formulation.

From the explicit form of the Green's function given in (2.28)–(2.30) it is straightforward to evaluate the entries in the stiffness matrix and load vector given by (5.115)–(5.116). For many purposes it will be adequate to approximate this by a local Hemker test function, that is to neglect the effect of the term in $\bar{c}$. Then on the jth interval the element basis functions for the test space are given, as in (5.32), by

$$\frac{1 - e^{-\beta_j s}}{1 - e^{-\beta_j}} \quad \text{and} \quad \frac{e^{-\beta_j s} - e^{-\beta_j}}{1 - e^{-\beta_j}} \quad \text{for } 0 \leq s \leq 1, \qquad (5.144)$$

where $\bar{\beta}_j = \frac{1}{2}(b_{j-1} + b_j)h_j/\epsilon$. Notice that these basis functions sum to unity identically on the interval. The gradients of the first at $s = 0$ and the second at $s = 1$ are, respectively,

$$\bar{\beta}_j/(1 - e^{-\beta_j}) \quad \text{and} \quad - \bar{\beta}_j e^{-\beta_j}/(1 - e^{-\beta_j});$$

hence by (5.116a) this gives, after multiplying by ds/dx,

$$\bar{K}_{i,i-1} = -\frac{\epsilon\bar{\beta}_i/h_i}{1-e^{-\bar{\beta}_i}}, \quad \bar{K}_{i,i+1} = -\frac{\epsilon(\bar{\beta}_{i+1}/h_{i+1})e^{-\bar{\beta}_{i+1}}}{1-e^{-\bar{\beta}_{i+1}}}, \qquad (5.145a)$$

which is in agreement with the coefficients given for the El-Mistikawy and Werle scheme on a uniform mesh in (3.60a). Similarly, by integrating the test basis functions in (5.144), we obtain from (5.115c)

$$\bar{S}_i = \tfrac{1}{2}h_i(S_{i-1}+S_i)\left[\frac{1}{1-e^{-\bar{\beta}_i}} - \frac{1}{\bar{\beta}_i}\right]$$

$$+ \tfrac{1}{2}h_{i+1}(S_i+S_{i+1})\left[\frac{1}{\bar{\beta}_{i+1}} - \frac{e^{-\bar{\beta}_{i+1}}}{1-e^{-\bar{\beta}_{i+1}}}\right], \qquad (5.145b)$$

in similar agreement with (3.60b). As discussed in Chapter 3, modifications to allow for the term cu in the differential equation can then be made by including it in the source function.

5.8.2 Extensions to two dimensions

It is clear that a direct extension of all of these ideas into an analysis in two dimensions is not practically feasible. As we have seen in section 2.7.2 even the free-space Green's function has the form of a modified Bessel function; and to apply boundary conditions on the boundary of the support of each global basis function is quite impractical. On a rectangular mesh we can of course use tensor product test functions to derive a numerical method, each factor constructed from a one-dimensional Green's function using the appropriate component of the velocity vector **b**. Results obtained with such a scheme will be presented in section 5.10; they are not particularly noteworthy in that instance. Several such schemes have also been compared and analysed by Hegarty *et al.* (1993); they have confirmed that on a rectangular mesh the Petrov–Galerkin method based on a bilinear trial space and tensor product exponential test space gives significantly better accuracy than the Galerkin method that uses the exponentials for both spaces. Thus this is probably the best that one can achieve by using essentially one-dimensional ideas.

A much more interesting extension, however, has been proposed recently by Craig and Perella (1995). Strictly speaking, this should not be categorised as a scheme using the local Green's function, but it does generalise (5.112) by using test functions that satisfy the adjoint equation on each element. Thus suppose we are to approximate $-\epsilon\nabla^2 u + \nabla\cdot(\mathbf{b}u) = S$ on a general mesh, of triangles, quadrilaterals or a mix of the two, and for simplicity we assume homogeneous Dirichlet boundary conditions are applied on all of $\partial\Omega$. We let Ω_e denote an arbitrary element of the mesh with boundary $\partial\Omega_e$; and we suppose $\{\phi_j\}$ are the global basis functions of S_0^h, each associated with a node r_j and having support Σ_j with boundary $\partial\Sigma_j$. Now for each ϕ_i introduce a test function $\psi_i \in C(\Omega)$, with the properties

$$-\epsilon\nabla^2\psi_i - \mathbf{b}\cdot\nabla\psi_i = 0 \quad \forall\Omega_e, \tag{5.146a}$$

$$\psi_i(\mathbf{r}_i) = 1, \tag{5.146b}$$

$$\psi_i = 0 \quad \text{on } \Omega_e \not\subset \Sigma_i. \tag{5.146c}$$

If U is the Petrov–Galerkin approximation obtained with $T^h = \text{span}\{\psi_i\}$, we have

$$0 = B(u - U, V) \equiv (\epsilon\nabla(u - U) - \mathbf{b}(u - U), \nabla V) \quad \forall V \in T^h$$

$$= \sum_{(e)}\left\{\int_{\partial\Omega_e} \epsilon(u - U)\mathbf{n}\cdot\nabla V d\Gamma - \int_{\Omega_e} (u - U)(\epsilon\nabla^2 V + \mathbf{b}\cdot\nabla V)d\Omega,\right.$$

$$\tag{5.147a}$$

and all the element integrals are zero by (5.146a). Thus there is a projection of the error onto the element boundaries that is set to zero. This is the appropriate generalisation of the one-dimensional result that the nodal errors are set to zero, and is used by Craig and Perella to obtain error estimates in a similar form to those in (5.11).

Let $\mathcal{D}$ be the space of functions defined on the mesh boundaries, with values given by the jump in the normal derivative across each $\partial\Omega_e$ of functions $V \in T^h$; and let $\mathcal{C}$ be the space of functions defined on the mesh boundaries by restriction of the trial space. We introduce a mesh inner product $(\cdot, \cdot)_h$, and associated norm $\|\cdot\|_h$, given by

$$(u, v)_h := h\sum_{(e)}\int_{\partial\Omega_e} uv d\Gamma,$$

where the maximum element diameter h is introduced so that $\|\cdot\|_h$ does not diverge as $h \to 0$; one can show that this inner product is well defined on $H^1(\Omega) \times H^1(\Omega)$. Then (5.147a) can be written

$$(u - U, d)_h = 0 \quad \forall d \in \mathcal{D}, \tag{5.147b}$$

and we have the following result.

Theorem 5.8.6 *Suppose the test space T^h defined by (5.146) is used to obtain a Petrov–Galerkin approximation U, and with the spaces of functions on the mesh $\mathcal{D}$ and $\mathcal{C}$ defined as above let Δ_T be defined by*

$$\Delta_T := \inf_{d\in\mathcal{D}} \sup_{c\in\mathcal{C}} \frac{\|c - d\|_h}{\|c\|_h}. \tag{5.148a}$$

Then we have

$$\|u - U\|_h \leq (1 - \Delta_T^2)^{-\frac{1}{2}} \inf_{c\in\mathcal{C}} \|u - c\|_h. \tag{5.148b}$$

Proof. Let c^* be the projection into C of the restriction of u to the mesh boundaries, so we have

$$(u - c^*, c)_h = 0 \quad \forall c \in C.$$

Then since $c^* - U \in C$, where U here is restricted to the mesh boundaries, we have

$$\|u - U\|_h^2 = \|u - c^*\|_h^2 + \|c^* - U\|_h^2. \tag{5.149}$$

Also, and by using (5.147b), it follows that

$$\|c^* - U\|_h^2 = (c^* - U, c^* - U)_h = (u - U, c^* - U)_h$$
$$= (u - U, c^* - U - d)_h \quad \forall d \in \mathcal{D}.$$

From (5.148a), d can be chosen so that $\|c^* - U - d\|_h \leq \Delta_T \|c^* - U\|_h$, so that

$$\|c^* - U\|_h^2 \leq \|u - U\|_h \|c^* - U - d\|_h \leq \Delta_T \|u - U\|_h \|c^* - U\|_h.$$

Substitution into (5.149) gives

$$\|u - U\|_h^2 \leq \|u - c^*\|_h^2 + \Delta_T^2 \|u - U\|_h^2$$

and hence the result (5.148b). $\square$

We have implicitly assumed above that $\Delta_T < 1$, which has been demonstrated in Craig and Perella (1995) for various cases by explicit calculation. On a rectangular mesh and with constant b, the obvious choice for T^h is composed of tensor product Hemker test functions; but these authors have also considered alternative separable solutions of the adjoint equation and carried out numerical tests on problems that include the IAHR/CEGB problems. There are clearly various possibilities for similar test functions on triangular meshes, since each ψ_i need only be continuous across the edges that meet at r_i; and choices between these possibilities would undoubtedly be illuminated by an extension of the error analysis to cover the use of element averages of b in (5.146), as was done in the previous section for the one-dimensional case.

5.9 A mixed norm method

Virtually all of the methods that we have so far described have been aimed at pointwise or nodal accuracy; as a consequence the detailed analyses of the last two sections have been based on discrete Green's functions. The exception has been the idea of using approximate symmetrization based on the $B_2(\cdot, \cdot)$ bilinear form, as discussed in section 5.2.2. In the limit $\epsilon \to 0$ the sought-after approximation is in that case an L_2 best fit to the exact solution. Although no generally practicable method has so far been developed from this approach, the very different behaviour that it produces on model problems, as shown in the next section, illuminates our understanding of discretisation errors for all convection-diffusion problems. Moreover, the practical value of the symmetrization that is applied, and its connection with pre-conditioning, is widely appreciated in the closely related Lagrange–Galerkin methods used for unsteady problems, as we shall see in Chapter 7.

The starting point of the method is to rewrite the convection-diffusion problem (5.3), with $a \equiv 1$ and $\nabla \cdot \mathbf{b} = 0$, in terms of

$$B_2(v, w) := \epsilon^2(\nabla v, \nabla w) + (\mathbf{b}v, \mathbf{b}w) \qquad (5.150a)$$

by using the relation $B(v, w) = B_2(v, R_2 w)$ valid for $v, w \in H^1_{E_0}$. Thus if we extend the Dirichlet boundary data u_B to give $\tilde{u}_B \in H^1_E$, so that $u = u_0 + \tilde{u}_B$ with $u_0 \in H^1_{E_0}$, we can write

$$B_2(u, w) = B_2(\tilde{u}_B, w) + B(u_0, R_2^{-1}w),$$

i.e.

$$B_2(u, w) = (S, R_2^{-1}w) + [B_2(\tilde{u}_B, w) - B(\tilde{u}_B, R_2^{-1}w]$$
$$\forall w \in H^1_{E_0}. \qquad (5.150b)$$

Unfortunately, neither the Riesz representer R_2 nor its inverse R_2^{-1} are known in general, so the effect of the source function and the boundary data may be difficult to calculate. On the other hand, when a trial space S^h_0 is used to give an approximation, no corresponding test space is needed: because $B_2(\cdot, \cdot)$ is symmetric, the optimal approximation U^* would be generated from exactly evaluating the right-hand side of (5.150b) for every W in S^h_0, which we might denote as (S_2, W) for some function S_2; and an approximation U to U^* is obtained from approximating these quantities. Indeed, if the approximation is equivalent to replacing S_2 by S_2^{approx}, it is clear that we have

$$\|U^* - U\|_{B_2} = (S_2 - S_2^{approx}, U^* - U)/\|U^* - U\|_{B_2}$$
$$\leq \|(\mathbf{b}/b^2)(S_2 - S_2^{approx})\|_{L_2}. \qquad (5.151)$$

We will describe below how these approximations can be generated in both one and two dimensions.

5.9.1 In one dimension

Here we can generalise the explicit expression for R_2^{-1} given in (5.28) to the problem

$$-\epsilon u'' + (bu)' = S \quad \text{on } (0, 1) \qquad (5.152a)$$
$$u(0) = u_L, \quad u(1) = u_R, \qquad (5.152b)$$

in conservation law form. Having a nonconstant b in one dimension corresponds to dropping the assumption $\nabla \cdot b = 0$ made in the formulation of (5.150); then, as used in the original papers of Barrett and Morton (1980, 1981, 1982, 1984), we define for $v, w \in H^1(0, 1)$

$$B_2(v, w) := \epsilon^2(v', w') + ([b^2 + \epsilon b']v, w). \tag{5.153}$$

The defining relation for the Riesz representer is valid only on $H_0^1(0, 1)$, as in (5.7), and in the present case generalises (5.27b) to

$$(R_2^{-1}z)' = \epsilon z' - bz + Ce^{-B/\epsilon} \quad \forall z \in H_0^1(0, 1) \tag{5.154a}$$

where

$$B(x) = \int_0^x b(t)dt \quad \text{and} \quad C = \left[\int_0^1 bz\,dt\right] \Big/ \left[\int_0^1 e^{-B/\epsilon}dt\right]. \tag{5.154b}$$

It is useful to introduce the weighting function

$$r(x) := e^{-B(x)/\epsilon} \Big/ \left[\int_0^1 e^{-B/\epsilon}dt\right], \tag{5.155a}$$

in terms of which we then have

$$R_2^{-1}z = \epsilon z + \int_x^1 \left[bz - (\overline{bz})r\right] dt. \tag{5.155b}$$

Here $(\overline{bz})$ denotes the usual average of bz over $(0, 1)$; but it is also useful to introduce the weighted average

$$\hat{f} := \int_0^1 rf\,dt, \tag{5.156a}$$

in terms of which a little manipulation readily gives

$$(S, R_2^{-1}z) = \epsilon(S, z) + (T_S - \hat{T}_S, bz), \tag{5.156b}$$

where we have written $T_S(x) := \int_0^x S(t)dt$ for the indefinite integral. To deal with the inhomogeneous Dirichlet data, we can use the linear extension of the data $\tilde{u}_B(x) = u_L + (u_R - u_L)x$ and obtain, by substituting $(b\tilde{u}_B)'$ for S in (5.156b),

$$B_2(\tilde{u}_B, z) - B(\tilde{u}_B, R_2^{-1}z)$$
$$= ([b^2 + \epsilon b']\tilde{u}_B, z) - \epsilon((b\tilde{u}_B)', z) - (b\tilde{u}_B - (\widehat{b\tilde{u}_B}), bz)$$
$$= [\epsilon(u_L - u_R) + (\widehat{b\tilde{u}_B})](b, z).$$

Because br and xbr can both be integrated exactly, this expression can be further simplified so that the exact symmetrized form of the equation (5.152) can be written

$$B_2(u, w) = \epsilon(S, w) + (T_S - \hat{T}_S, bw)$$

$$+ \epsilon \left[\int_0^1 e^{-B/\epsilon} dt \right]^{-1} \left[u_L - u_R e^{-B(1)/\epsilon} \right] (b, w)$$

$$\forall w \in H_0^1(0, 1). \tag{5.157}$$

Note that, since $\int_0^1 e^{-B/\epsilon} dt = O(\epsilon)$, the influence of the inlet data u_L is felt throughout the domain by means of the last term here; in some sense this is how convection is embodied in this formulation.

This suggests that it might be worthwhile to solve the pure convection problem as an initial value problem and use this as a guide to approximating the full convection-diffusion problem. Thus suppose we define $\tilde{u}_L$ by

$$(b\tilde{u}_L)' = S \text{ with } \tilde{u}_L(0) = u_L, \tag{5.158a}$$

and set

$$v := b(u - \tilde{u}_L) - \epsilon u'. \tag{5.158b}$$

Substitution into the original equation (5.152a) shows that v is constant, and this constant can be taken to be $-\epsilon u'(0)$. We then have

$$B_2(u, w) - \epsilon B(u, w) = ([b^2 + \epsilon b']u - \epsilon(bu)', w)$$
$$= (bu - \epsilon u', bw) = (v + b\tilde{u}_L, bw),$$

which gives the alternative formulation

$$B_2(u, w) = \epsilon(S, w) + (v + b\tilde{u}_L, bw), \quad v = -\epsilon u'(0). \tag{5.159}$$

This implicit form corresponds to a very useful formulation in two dimensions. Here we could have made more explicit use of $\tilde{u}_L$ by writing $u_0 = u - \tilde{u}_L$ to get $-\epsilon u_0'' + (bu_0)' = \epsilon \tilde{u}_L''$, and then applying the formula (5.157) for u_0 by replacing S, u_L and u_R by $\epsilon u_0'', 0$ and $u_R - \tilde{u}_L(1)$ respectively.

Suppose now we construct approximations using a trial space S^h. The optimal approximation U^* in the $\|\cdot\|_{B_2}$ norm is given from (5.157) by

$$B_2(U^*, W) = \epsilon(S, W) + (T_S - \widehat{T}_S, bW)$$

$$+ \epsilon \left[\int_0^1 e^{-B/\epsilon} dt \right]^{-1} \left[u_L - u_R e^{-B(1)/\epsilon} \right] (b, W)$$

$$\forall W \in S_0^h, \tag{5.160}$$

and various approximations to this can be obtained by approximating the integrals on the right in various ways. The u_R term is the most obvious candidate for complete omission, and $\int_0^1 e^{-B/\epsilon} dt$ is well approximated by $\epsilon/b(0)$. Moreover, we can replace T_S by $b\tilde{u}_L - b(0)\tilde{u}_L(0)$ by using (5.158a);

and hence we have, after integration by parts,

$$\widehat{T}_S = \int_0^1 r\left[b\tilde{u}_L - b(0)\tilde{u}_L(0)\right] \mathrm{d}x$$

$$\approx b(0)\left[\tilde{u}_L(0) + \int_0^1 \tilde{u}_L' e^{-B/\epsilon} \mathrm{d}x\right] - b(0)\tilde{u}_L(0)$$

$$\approx \epsilon\tilde{u}_L'(0).$$

Substituting all of these into (5.160) we obtain for an approximation U

$$B_2(U, W) = \epsilon(S, W) + (b\tilde{u}_L - \epsilon\tilde{u}_L'(0), bW) \quad \forall W \in S_0^h, \qquad (5.161)$$

which is of exactly the same form as (5.159).

Moreover, working directly from (5.159) without any approximation, we can generate a sequence of approximations U^m to U^* by setting

$$B_2(U^{m+1}, W) = \epsilon(S, W) + (b\tilde{u}_L - \epsilon(U^m)'(0), bW) \quad \forall W \in S_0^h \qquad (5.162a)$$

and starting with U^0 given by (5.161); more practically, we write

$$B_2(U^{m+1} - U^m, W) = -\epsilon(b, W)[(U^m)'(0) - (U^{m-1})'(0)]$$
$$\forall W \in S_0^h. \qquad (5.162b)$$

Approximations to model problems obtained by this means can be found in the references given above, and an example will be given in the next section.

5.9.2 In two dimensions

We assume that the flow field is incompressible and that it and the source term are such that we can define $\tilde{u}_- \in H^1(\Omega)$ by

$$\mathbf{b}\cdot\nabla\tilde{u}_- = S \text{ in } \Omega, \quad \tilde{u}_- = u_B \text{ on } \partial\Omega_-, \qquad (5.163)$$

that is by integrating the source term along the flow field starting at the inflow boundary. Then we define the flux vector $\mathbf{v}$ by

$$\mathbf{v} := \mathbf{b}(u - \tilde{u}_-) - \epsilon\nabla u. \qquad (5.164a)$$

Substitution into the differential equation gives $\nabla\cdot\mathbf{v} = 0$, so that we can introduce a stream function ψ; with the definition of **curl** (and also of $\nabla\times$) introduced in (5.59), and with the vector normal to the flow field $\mathbf{b}^\perp := (-b_2, b_1)^T$, we have

$$\mathbf{v} = \mathbf{curl}\,\psi \quad \text{and} \quad \mathbf{b}\cdot\mathbf{v} = \mathbf{b}^\perp\cdot\nabla\psi. \qquad (5.164b)$$

Hence we can write

$$B_2(u, w) - \epsilon B(u, w) = (bu, bw) - \epsilon(\mathbf{b}\cdot\nabla u, w)$$
$$= (b\tilde{u}_-, bw) + (\mathbf{b}^\perp\cdot\nabla\psi, w),$$

i.e.

$$B_2(u, w) = \epsilon(S, w) + (b\tilde{u}_-, bw) + (b^\perp \cdot \nabla \psi, w) \quad \forall W \in H^1_{E_0}. \quad (5.165a)$$

To obtain the matching equation for ψ, to which we apply Neumann boundary conditions everywhere, we have

$$\nabla^2 \psi = -\nabla \times \mathbf{v} = -\nabla \times (b(u - \tilde{u}_-)) = \nabla \cdot (b^\perp(u - \tilde{u}_-)).$$

Hence

$$\int_{\partial\Omega} \frac{\partial \psi}{\partial n} w d\Gamma - (\nabla \psi, \nabla w)$$

$$= \int_{\partial\Omega} (u - \tilde{u}_-)(b^\perp \cdot \mathbf{n}) w d\Gamma - (u - \tilde{u}_-, b^\perp \cdot \nabla w);$$

but if the tangent to the boundary is denoted by $\mathbf{t} := (-n_2, n_1)^T$, we have $b^\perp \cdot \mathbf{n} = -b \cdot \mathbf{t}$ and $\mathbf{v} \cdot \mathbf{t} = -\partial \psi / \partial n$, so that (5.164a) gives

$$(u - \tilde{u}_-)(b^\perp \cdot \mathbf{n}) = \frac{\partial \psi}{\partial n} - \epsilon \frac{\partial u}{\partial t}.$$

Thus we obtain

$$(\nabla \psi, \nabla w) = \epsilon \int_{\partial\Omega} \frac{\partial u}{\partial t} w d\Gamma + (u - \tilde{u}_-, b^\perp \cdot \nabla w) \quad \forall w \in H^1(\Omega) \quad (5.165b)$$

to define ψ up to an arbitrary constant. Note the much greater complication, compared with the second equation of (5.159), occasioned by having to solve $\nabla \cdot \mathbf{v} = 0$ rather than $v' = 0$.

Suppose the same piecewise bilinear trial space S^h is used to approximate u and ψ by U and Ψ. We assume that the ordinary differential equations corresponding to (5.163) can be integrated approximately to give an approximation $\tilde{U}_-$ to $\tilde{u}_-$ that can be evaluated at the 2×2 Gauss points that we shall use to obtain the equations for U and Ψ; in the next chapter, where we shall also consider pure convection problems, we shall show how $\tilde{U}_- \in S^h$ can be computed very easily. Then (5.165a) and (5.165b) are approximated alternately, starting with $U^0 = \tilde{U}_-$, to give

$$(\nabla \Psi^{m+1}, \nabla W) = \epsilon \int_{\partial\Omega} \frac{\partial U^m}{\partial t} W d\Gamma + (U^m - \tilde{U}_-, b^\perp \cdot \nabla W)$$

$$\forall W \in S^h$$

$$B_2(U^{m+1}, W) = \epsilon(S, W) + (b\tilde{U}_-, bW) + (b^\perp \cdot \nabla \Psi^{m+1}, W)$$

$$\forall W \in S^h_0. \quad (5.166)$$

Results obtained by Morton and Scotney (1985) with this method will be shown in the next section, where it will be seen quite clearly that U is close to the best approximation to u in the $\|\cdot\|_{B_2}$ norm, displaying the expected L_2 oscillations in a sharp boundary layer.

It is worth noting that it is easy to enhance the predominantly L_2 norm in $\|\cdot\|_{B_2}$ by adding a multiple of the crosswind gradients. Thus suppose we define the bilinear form

$$B_\sigma(v, w) := B_2(v, w) + \sigma^2(\mathbf{b}^\perp \cdot \nabla v, \mathbf{b}^\perp \cdot \nabla w); \qquad (5.167a)$$

we have from (5.164) for substitution into it

$$\epsilon \mathbf{b}^\perp \cdot \nabla u = -\mathbf{b}^\perp \cdot \mathbf{v} = \mathbf{b} \cdot \nabla \psi. \qquad (5.167b)$$

Hence U^{m+1} can be obtained from (5.166) by replacing $B_2(U^{m+1}, W)$ by $B_\sigma(U^{m+1}, W)$ on the left and adding $(\sigma^2/\epsilon)(\mathbf{b} \cdot \nabla \Psi^{m+1}, \mathbf{b}^\perp \cdot \nabla W)$ on the right.

5.10 Numerical comparisons

We use the same pair of problems as in section 4.6, but add the more demanding second IAHR/CEGB problem. In each case, results for only a limited number of representative methods will be presented.

5.10.1 The one-dimensional conservation law

The problem given by (4.67) is typical of those for which the methods of this chapter were designed and we give results obtained with Hemker test functions, with the streamline diffusion method in the form (5.41), and with the mixed norm method of the previous section. These are shown in Fig. 5.5 and should be compared with those in Fig. 4.7. We have, however, compared results at $h = 0.1$ and $h = 0.05$, and chosen the more extreme case $\epsilon = 0.001$ to compare with $\epsilon = 0.1$.

In all cases, the conservation law form of the equations is preserved and the exact integral of the source function over each interval is used in a one-point quadrature; cell-averaged mesh Péclet numbers are used as required in the test function integrals and on the right in (5.161). As was to be expected, all three methods are extremely accurate for the $\epsilon = 0.001$ case. With the greater diffusion of $\epsilon = 0.1$ the Hemker method, which represents all those methods based on local Green's functions, is still very accurate with the maximum errors at 0.4 and 0.5 being due to the simple quadrature; but the simpler streamline diffusion method gives comparable accuracy. The mixed norm method, which is of limited applicability, gives the highest accuracy; and there is evidence at the peak of its least squares approximation property.

5.10.2 The IAHR/CEGB problems

In Fig. 5.6 results obtained for the first problem with tensor product Hemker test functions and with the streamline diffusion method are

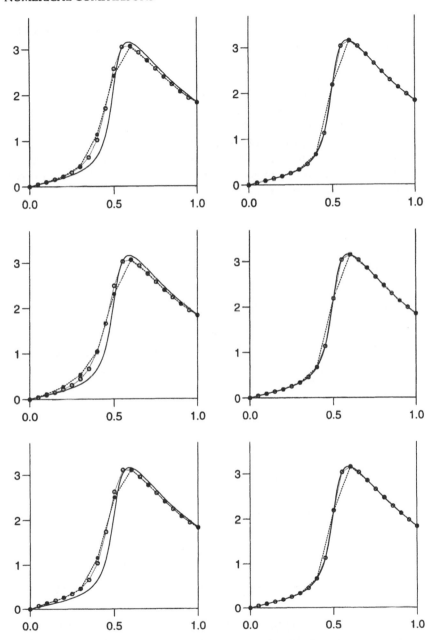

Figure 5.5. *Comparison of approximations to model problem (4.67), with the reduced equation solution shown for reference: figures on the left are for $\epsilon = 0.1$ and on the right for $\epsilon = 0.001$; solid dots joined by a dashed line correspond to $h = 0.1$, open circles and a dotted line to $h = 0.05$; from the top, the approximations are for the Hemker test space, the streamline diffusion method and the mixed norm method*

plotted in the format used for Fig. 4.8. In the former case, the profiles remain monotone but there is some smearing in the highest Péclet number case; this is considerably improved on a halved mesh, but then there is some overshoot. Thus, though the worst effects of crosswind diffusion are avoided by the Petrov–Galerkin formulation, the limitations of a tensor product form for the test functions are evident. On the other hand, the streamline diffusion method which makes more use of the flow direction, and was designed to eliminate crosswind diffusion, shows both overshoots and smearing in the two highest Péclet number cases on the standard mesh; on a halved mesh these are much improved but still persist in the highest Péclet number case.

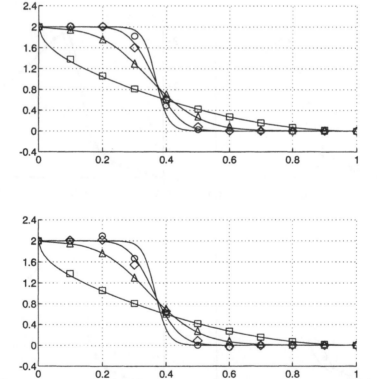

Figure 5.6. *Profiles for the first IAHR/CEGB problem at $\epsilon = 10^{-6}, 0.002, 0.01$ and 0.1, as in Fig. 4.8; results at the top are for tensor product Hemker test functions, and at the bottom for the streamline diffusion method*

The much more demanding second problem that is described in section 1.2 shows up greater differences. In this problem, with the same flow field but no inlet data, the boundary condition of $u = 100$ on the right creates a boundary layer that develops from top to bottom.

Results are given in Fig. 5.7 in the form of profiles, for the same two methods on the same mesh and for the same set of ϵ values, at three stations $y = 0.9, y = 0.5$ and $y = 0$. The tensor product form of the Hemker functions shows up well here, giving monotone profiles which show the thickening of the profile where it can be captured by the mesh. The streamline diffusion scheme, however, gives quite severe oscillations in the outlet profile for $\epsilon = 0.002$ and 10^{-6}. Addition of extra crosswind diffusion of the form (5.82a) used in the error analysis of section 5.7.2 is quite effective in reducing these oscillations, and $\hat{\epsilon} = \frac{1}{4}h^{3/2}$ eliminates them completely.

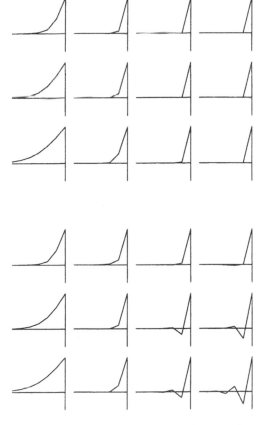

Figure 5.7. *Profiles for the second IAHR/CEGB problem with (from left to right)* $\epsilon = 0.1, 0.01, 0.002$ *and* 10^{-6}*, and (from top to bottom) at* $y = 0.0, 0.5$ *and* 0*; the top set of results are for the tensor product Hemker test function and the bottom set for the streamline diffusion method*

In Morton and Scotney (1985) similar oscillatory profiles are shown with these two values of ϵ for the Petrov–Galerkin method based on the test function of (5.4), as well as for the mixed norm method; but for the latter method they show that this correctly reflects the least squares approximation property sought through the $B_2(\cdot, \cdot)$ symmetrization — c.f. Fig. 5.1.

CHAPTER 6

Finite volume methods for steady problems

6.1 Introduction

The term *finite volume method* appears to have been coined in the early 1970's for discretisations of the full potential equation of gas dynamics $\nabla \cdot (\rho \nabla \phi) = 0$, where $\rho = \rho(|\nabla \phi|)$ is given by Bernoulli's equation. However, the essential idea of using an integral formulation, integrating the equation over a mesh region and applying Gauss' theorem before carrying out the discretisation, was used in Varga's classical book (Varga 1962) to derive finite difference schemes for elliptic equations on irregular two-dimensional meshes — for early references see also MacNeal (1953), McDonald (1971) and Jameson and Caughey (1977). Similarly, for unsteady hyperbolic conservation laws of the form $u_t + f_x = 0$, the basis of the very influential *Godunov scheme* (Godunov 1959) is an integral formulation; and one could give a like interpretation of the widely used difference scheme, which is referred to as the *Preissmann box scheme* (Preissmann 1961) by hydraulic engineers who apply it to the unsteady St. Venant equations — see also Wendroff (1960) and Thomée (1962). The same difference scheme is also often called the *Keller box scheme* because of that author's application of the method to parabolic equations written as first order systems, and thence applied to the modelling of viscous boundary layers (Keller and Cebeci (1971, 1972)). We shall therefore refer to as finite volume methods all those discretisations which start from integrating the differential equations over a set of *control volumes*; in the terminology of finite element methods, this includes all schemes that use piecewise constant test functions. In these early references the control volumes are usually rectangular and occasionally quadrilateral, extending to hexahedral volumes in three dimensions. More recently, the same ideas have been applied to unstructured triangular and tetrahedral meshes. We shall concentrate on the quadrilateral case but will briefly discuss the use of triangles.

With quadrilateral control volumes, the main distinction between the various finite volume methods lies in the positioning of the mesh points or

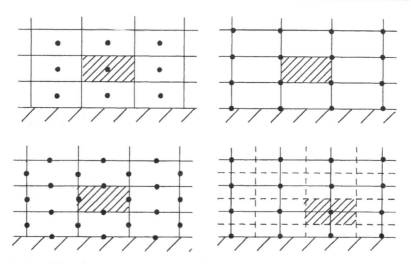

Figure 6.1. *Positioning of the discrete unknowns in alternative finite volume schemes; reading from the top left we have the cell centre, cell vertex, cell edge and vertex-centred schemes*

nodal parameters used in the discretisation. There are three main choices: we call the schemes typified by those in Varga (1962) and Jameson (1979) *cell centre* schemes, because the unknowns are associated with the centres of the cells (or control volumes) over which conservation is applied; the box scheme is a forerunner of the *cell vertex* schemes in which unknowns are associated with the vertices of the cell; and we might call the Godunov scheme a *cell edge* scheme because the unknowns are associated with the edges of the cell, though this is a less commonly used term — see Fig. 6.1 illustrating the three cases. There are, however, other variants that come about by considering the alternative arrangements that might be used near the boundary. We have tacitly assumed in the above categorisation that we have a *primary mesh* that is *body-fitted*, i.e. the domain boundary normally coincides with a mesh line, and conservation is applied to each cell of this primary mesh. But in the *vertex-centred* schemes of Dick (1988) and Hall (1991) conservation is applied on each cell of a dual mesh that is centred on a vertex of the primary mesh, as in the bottom right-hand diagram of Fig. 6.1; then conservation is applied to half cells at the boundary. There are clearly other possible alternatives to be derived in this way, and for a system of equations the treatment of either the equations or the unknowns may be different as in so-called staggered schemes. However, the four cases of Fig. 6.1 will be sufficient for our purposes.

Let us write a general scalar conservation law in two dimensions as

$$\frac{\partial f}{\partial x} + \frac{\partial g}{\partial y} = S, \quad \text{or} \quad \nabla \cdot (f, g) = S; \tag{6.1}$$

for a first order equation, as in pure convection, we have $f = f(\mathbf{x}, u), g = g(\mathbf{x}, u)$ for the scalar unknown u, while for a second order equation, as in convection-diffusion, we have $f = f(\mathbf{x}, u, \nabla u)$, $g = g(\mathbf{x}, u, \nabla u)$. We have adopted this general notation for two reasons: firstly, it embodies the emphasis that finite volume schemes place on the *fluxes* f and g in a conservation law; and, secondly, it enables the methods to be applied directly to nonlinear problems, with systems of equations, where finite volume methods have had widest use. Indeed, in compressible fluid dynamics — either for the first order inviscid Euler equations or the second order viscous Navier–Stokes equations — they are by far the most generally used methods. Application of Gauss' theorem to a control volume Ω_α gives

$$\int_{\Omega_\alpha} \nabla \cdot (f, g) \mathrm{d}\Omega = \int_{\partial\Omega_\alpha} (f, g) \cdot \mathbf{n} \mathrm{d}\Gamma = \int_{\partial\Omega_\alpha} [f \mathrm{d}y - g \mathrm{d}x] \qquad (6.2)$$

and it is the boundary integral that is to be approximated. In convection-diffusion, we have $(f, g) \cdot \mathbf{n} = (\mathbf{b} \cdot \mathbf{n})u - \epsilon \partial u / \partial n$. Clearly the convection term is most readily approximated in the cell vertex and cell edge schemes; while the diffusion term is most easily approximated in the cell centre and vertex-centred schemes.

The cell centre schemes are probably most usefully regarded from a finite difference viewpoint. The $\partial u / \partial n$ is approximated by a divided difference across the edge and $(\mathbf{b} \cdot \mathbf{n})u$ can be either centrally-averaged or upwind-averaged. On the other hand, the cell vertex and cell edge schemes and perhaps also the vertex-centred schemes are best thought of as nonconforming finite element methods. In many of these cases, then, one can regard both the trial and test spaces as prescribed so there is little room for manoeuvre in trying to obtain accurate nonoscillatory solutions for a wide range of mesh Péclet numbers, as was achieved in the last chapter with many of the Petrov-Galerkin schemes. Thus finite volume methods usually make some use of *artificial viscosity* or *artificial dissipation* terms, both second order and fourth order.

The broader nonlinear viewpoint and the use of artificial dissipation terms (either explicitly or implicitly) provide the most distinctive new features of the methods to be discussed in this chapter. They can be summarised under two main design objectives:

(i) artificial dissipation is used to provide coercivity (as in the streamline diffusion method), but mainly in respect of high frequency solution components, while low frequency components are approximated well and are to be largely unaffected by the added terms;

(ii) the local form of the scheme is *adaptive* and depends on properties of the solution, so that even for a linear problem the method is nonlinear. Thus the solution is sought iteratively, there is a close link with the methods used for unsteady problems, and many of the design criteria that have been developed in that area (such as *monotonicity preservation, total variation diminishing* or *TVD*) are brought into play.

Of all the various finite volume methods, that which we most wish to draw to the reader's attention is the cell vertex method. The box scheme as used by Preissmann for the St. Venant equations is a familiar special case; but Keller's box scheme, as used for the parabolic system formed by boundary layer equations, is quite different in its treatment of the diffusive terms from the schemes we shall describe. We apply the term cell vertex only to those methods that consistently calculate both convective and diffusive fluxes on control volume surfaces that pass through the vertices of the mesh. In one dimension the control volume is an interval between two mesh points; and if the gradients either side of a mesh point are used to define the diffusive flux there, a four-point difference scheme results. This is quite unlike any other scheme we have so far used for our one-dimensional model problems.

Such cell vertex methods have many attractive properties; but they have one outstanding difficulty, namely, that there is no natural correspondence between the unknowns, which are associated with the vertices, and the discrete equations, which are associated with the cells. Because of both the good properties and distinctive character of the methods and the special problem posed by their analysis, the cell vertex methods are the only finite volume methods that are analysed in any detail in this chapter, in sections 6.4 and 6.5.

6.2 Difference-based methods

We can start in one dimension and on a uniform mesh, because many of the methods have not been developed beyond independent application in the two coordinate directions on a square mesh, as in section 3.6. Let us consider the problem

$$(-\epsilon u' + b(x)u)' = S(x) \text{ on } (0, 1); \tag{6.3a}$$

$$u(0) = u_L, \quad u(1) = u_R. \tag{6.3b}$$

Around the node x_i we suppose we have a control volume $(x_{i-\frac{1}{2}}, x_{i+\frac{1}{2}})$, over which we integrate the equation to derive an approximation of the form

$$F_{i+\frac{1}{2}} - F_{i-\frac{1}{2}} = S_i \Delta x, \tag{6.4}$$

where $\Delta x = x_{i+\frac{1}{2}} - x_{i-\frac{1}{2}}$. We can suppose that S_i is the average of $S(x)$ over the control volume, but we leave this aside as well as how the boundary conditions are to be applied, while concentrating on the specification of the fluxes $\{F_{i-\frac{1}{2}}\}$. Let us regard (6.3) and (6.4) as the limiting steady solution of an unsteady problem as $t \to \infty$; since $\epsilon > 0$, it is clear that the simplest appropriate unsteady problem is obtained by adding a term $\partial u/\partial t$ on the left of (6.3) and some corresponding time difference such as $(U_i^{n+1} - U_i^n)/\Delta t$ on the left of (6.4). Then the steady state results from a balance between the contribution of the source term over a time interval Δt and the difference of the fluxes over the same interval. It is this that motivates the idea that, if $b_{i-\frac{1}{2}} \equiv b(x_{i-\frac{1}{2}}) > 0$, then the flux $F_{i-\frac{1}{2}}$ should depend more on the values of U to the left of $x_{i-\frac{1}{2}}$ than on those to the right; for example, simple upwinding would use only U_{i-1} for the convective flux.

6.2.1 Convective fluxes

Some common linear schemes that use at most the three values U_i, U_{i-1} and U_{i-2} to calculate $F_{i-\frac{1}{2}}$ when $b_{i-\frac{1}{2}} > 0$ have the following form for the convective flux, while all using $-\epsilon(U_i - U_{i-1})/\Delta x$ for the diffusive flux:

(simple upwind)	$U_{i-1}b_{i-\frac{1}{2}}$	(6.5a)
(central difference)	$\frac{1}{2}(U_i + U_{i-1})b_{i-\frac{1}{2}}$	(6.5b)
(second order upwind)	$\frac{1}{2}(3U_{i-1} - U_{i-2})b_{i-\frac{1}{2}}$	(6.5c)
(Lax–Wendroff)	$\frac{1}{2}[(1 - \nu_{i-\frac{1}{2}})U_i + (1 + \nu_{i-\frac{1}{2}})U_{i-1}]b_{i-\frac{1}{2}}$	(6.5d)
(QUICK)	$\frac{1}{8}(3U_i + 6U_{i-1} - U_{i-2})b_{i-\frac{1}{2}}.$	(6.5e)

The first can be regarded as using extrapolation by a constant function to give a value for $U_{i-\frac{1}{2}}$, the second and third use a linear function to interpolate and extrapolate respectively, and the last uses quadratic interpolation. Introduced by Leonard (1979a), this QUICK scheme (denoting Quadratic Upstream Interpolation for Convective Kinematics) uses the quadratic interpolation consistently for both the convective and diffusive terms in the flux. The scheme therefore has much to commend it for smooth flows, and we shall have much more to say about its stablemate QUICKEST in the next chapter on unsteady problems. Indeed, QUICKEST is related to the only one of the above schemes (6.5d), based on the two-step Lax–Wendroff scheme, which predicts $U_{i-\frac{1}{2}}$ half a time step ahead by use of a CFL parameter $\nu_{i-\frac{1}{2}} := b_{i-\frac{1}{2}}\Delta t/\Delta x$.

However, when these convective fluxes are incorporated in a difference scheme only the simple upwind form gives a scheme satisfying the conditions for a maximum principle for all mesh Péclet numbers β. As we know, the central difference scheme fails when $\beta > 2$, and it is easy to see that QUICK fails in a similar manner when $\beta > \frac{8}{3}$. Moreover, both QUICK and the second order upwind scheme have positive coefficients for U_{i-2} for all mesh Péclet numbers and so are always liable to give oscillatory results; and the Lax–Wendroff scheme also never satisfies a maximum principle. How can they then be of use if we want 'wiggle-free' solutions? The answer lies in using an iterative approach and making the method depend on the solution being obtained.

6.2.2 *Time-stepping to steady state*

As already indicated, this viewpoint of using an iteration motivated by time-stepping will increasingly influence the description and design of methods in this chapter, leading naturally to the next chapter on unsteady problems; the crucial difference, however, is that here we are concerned only with solution efficiency and not with time-accuracy. We shall therefore use a notation in which the time step Δt is replaced by an iteration parameter ω, which may depend on the solution, the mesh point and the iteration step, and we introduce the schemes in a fairly general context ready for later use.

(i) *Convergence criteria.* Thus, in the notation of section 3.6, we suppose that we wish to solve the difference scheme

$$L^h U_P = R^h S_P \tag{6.6}$$

at a set of mesh points $P \in \Omega^h$. Simple *Richardson relaxation* gives the iteration

$$U_P^{n+1} = U_P^n - \omega(L^h U_P^n - R^h S_P). \tag{6.7}$$

For example, for the simple upwind scheme (6.5a) applied to (6.3) with $b(x) \geq 0$ we have

$$U_i^{n+1} = \left[1 - \omega\left(\frac{2\epsilon}{(\Delta x)^2} + \frac{b_{i+\frac{1}{2}}}{\Delta x}\right)\right] U_i^n + \omega\left(\frac{\epsilon}{(\Delta x)^2} + \frac{b_{i-\frac{1}{2}}}{\Delta x}\right) U_{i-1}^n$$
$$+ \omega \frac{\epsilon}{(\Delta x)^2} U_{i+1}^n + \omega S_i, \tag{6.8a}$$

which clearly converges (see Lemmas 1.5.1 and 1.5.2) when

$$\omega\left(\frac{2\epsilon}{(\Delta x)^2} + \frac{b_{i+\frac{1}{2}}}{\Delta x}\right) \leq 1. \tag{6.8b}$$

This corresponds to the practical stability condition of (1.62) with $\omega = \Delta t$; for large mesh Péclet numbers, fastest convergence is obtained when $\omega \equiv \omega_i$ is chosen to give equality in (6.8b), becoming a direct marching method from the inflow data on the left when ϵ becomes zero and the update is applied in the order of increasing i. Similarly, the central difference scheme converges when all the coefficients in the update are non-negative, that is when

$$\omega\left(\frac{2\epsilon}{(\Delta x)^2} + \frac{b_{i+\frac{1}{2}} - b_{i-\frac{1}{2}}}{2\Delta x}\right) \leq 1 \quad \text{and} \quad \frac{b_{i+\frac{1}{2}}}{2\Delta x} \leq \frac{\epsilon}{(\Delta x)^2}, \tag{6.9}$$

which reduces to the practical stability conditions of (1.48) when b is constant, and includes the severe mesh Péclet number restriction which precludes the use of this scheme entirely when $\epsilon = 0$.

We have used here the most demanding criterion for convergence[*], namely that the matrix norm subordinate to the maximum vector norm be less than or equal to unity; it follows that at each iteration we then have $\|U^{n+1} - U^\infty\|_\infty \leq \|U^n - U^\infty\|_\infty$, with the boundary conditions ensuring actual convergence. At the other extreme, on a fixed finite mesh the necessary and sufficient condition for convergence (in any norm) is that all the eigenvalues of the iteration matrix G lie inside the unit circle, i.e. the spectral radius satisfies $\rho(G) < 1$. This is in fact too weak for our purposes: for example, the upwind scheme (6.8a) with $\epsilon = 0$ has a lower triangular iteration matrix with spectral radius less than one if $0 < \omega b_{i+\frac{1}{2}}/\Delta x < 2 \ \forall i$, while the familiar CFL condition gives half this range; what happens between these conditions is that there is an initial large (polynomial) growth in error before the exponential decay obtained from $\rho(G) < 1$ sets in, and the peak of the growth gets higher and higher if the mesh is refined (so that the order of the matrix gets larger) — see Trefethen (1992) and Reddy and Trefethen (1994) for a full discussion of the phenomenon, and also Elman and Chernesky (1993) for a detailed analysis of a one-dimensional convection-diffusion problem.

For any matrix G we have

$$\rho(G) \leq \|G\|_2 \leq (\|G\|_1 \|G\|_\infty)^{\frac{1}{2}}, \tag{6.10}$$

and for a typical difference scheme the maximum row sum equals the maximum column sum so that $\|G\|_1 = \|G\|_\infty$. For symmetric matrices $\rho(G) = \|G\|_2$, but the dominant convection terms that we are dealing with means that we are always very far from this position. As we are presently dealing with scalar problems, Parseval's relation ensures that when Fourier analysis can be used to calculate the practical stability condition this will coincide with the condition $\|G\|_2 \leq 1$; and when an energy method is used this gives the same condition directly. Moreover, Fourier analysis shows how each mode behaves and, in particular, draws attention to any constraints which arise from the high frequency modes that are not properly representative of the exact solution.

Thus we adopt the following *practical convergence criteria* for the linear iteration matrix G,

$$(convergence \ criteria) \ \ \|G\|_2 \leq 1 \ \text{and} \ \rho(G) < 1. \tag{6.11}$$

When the coefficients are all non-negative in (6.8a), and similarly for the

---

[*] It is worth noting that this criterion is closely related to the property of a time-stepping method to be *monotone*, which is also one of the most demanding of those commonly in use — see LeVeque (1992, pp 169–172).

central difference case, it is easy to see that $\|G\|_2 = \|G\|_\infty = 1$ and that this is attained with the lowest frequency Fourier mode or where the solution is locally constant; then $\rho(G) < 1$ because of the boundary conditions. When these conditions are not satisfied, both $\|G\|_2$ and $\|G\|_\infty$ in each case will be larger than unity; maximum amplification will be attained with an oscillatory vector, but the form of the vector in the two cases will be different and the norms unequal. Thus the conditions (6.8b) and (6.9) are necessary as well as sufficient for the satisfaction of our convergence criteria (6.11) in the upwind and central difference case, respectively.

With the criteria (6.11), however, we can also consider relaxation with difference schemes based on the fluxes (6.5c)–(6.5e) which never satisfy maximum principles. For example, if we simplify the Lax–Wendroff case by setting $\epsilon = 0, b = $ constant and $\nu = b\omega/\Delta x$, we obtain the familiar formula for the amplification factor of a Fourier mode e^{ikx}

$$\kappa(k) = 1 - i\nu \sin k\Delta x - 2\nu^2 \sin^2 \tfrac{1}{2}k\Delta x, \qquad (6.12a)$$

$$\text{so that} \quad |\kappa(k)|^2 = 1 - 4\nu^2(1 - \nu^2) \sin^4 \tfrac{1}{2}k\Delta x, \qquad (6.12b)$$

from which we see that $\|G\|_2 = 1$ if $\nu^2 \le 1$ and this is attained as $k\Delta x \to 0$; for $\nu^2 < 1$ all other modes are damped, albeit rather slowly. Notice, moreover, that this is a considerably better performance than that obtained with the central difference scheme; in that case, when $\epsilon = 0$ we have $|\kappa|^2 = 1 + \nu^2 \sin^2 k\Delta x$ and all modes are amplified, with the most oscillatory having the greatest amplification. The advantage shows in the practical stability condition for the Lax–Wendroff scheme when applied to convection-diffusion, which has the form, compared with (6.8b) and (6.9),

$$(2\nu/\beta) + \nu^2 \le 1, \qquad (6.13)$$

where $\beta = b\Delta x/\epsilon$ and we have again put $\omega = \Delta t$. Thus, as with the upwind scheme, there is no limit placed on the value of β, but only a restriction on $\omega = \Delta t$ which becomes more severe as β is decreased; but in contrast with the upwind scheme, as it does not satisfy a maximum principle the Lax–Wendroff scheme will exhibit damped oscillations streaming upwind of a source function or boundary layer.

Stability analysis of time-stepping schemes which use a four-point stencil, as would be the case with the fluxes (6.5c) and (6.5e), is much more difficult to carry out, though some such schemes will be studied in the next chapter. The only simple case is (6.5c) with $\epsilon = 0$ as then only the values U_i, U_{i-1} and U_{i-2} are involved; a straightforward analysis shows that in that case we never have practical stability for any value of ν.

(ii) *Iteration schemes.* Let us now consider more general iteration schemes than the Richardson relaxation of (6.7). There are three main lines of development: the combination of algebraic residuals $L^h U^n - R^h S$ at more than the single point P; multiple level time-stepping; and the introduction of some implicitness by having an operator operating on U_P^{n+1}. We shall make only a brief comment here on the first, because it will be a major topic in the discussion of cell vertex methods in the next three sections. We merely note that the Lax–Wendroff scheme just discussed is closely related to a scheme obtained in this way. To see this, suppose we use the upwind flux (6.5a) and combine the resulting residuals $L^h U_i^n - S_i$ and $L^h U_{i+1}^n - S_{i+1}$. To simplify notation we write $\mu = \omega\epsilon/(\Delta x)^2, \beta_- = b_{i-\frac{1}{2}}\Delta x/\epsilon, \beta_0 = b_{i+\frac{1}{2}}\Delta x/\epsilon, \beta_+ = b_{i+\frac{3}{2}}\Delta x/\epsilon$, in terms of which the relaxation scheme (6.8a) using a single residual can be written

$$U_i^{n+1} = U_i^n - \mu\left[(\beta_0 U_i^n - \Delta_+ U_i^n) - (\beta_- U_{i-1}^n - \Delta_+ U_{i-1}^n) - \frac{\omega}{\mu}S_i\right];$$

but if instead the residuals associated with x_i and x_{i+1} are combined with the weights $\frac{1}{2}(1 \pm \nu)$ we obtain

$$U_i^{n+1} = U_i^n - \tfrac{1}{2}\mu\left[(1 - \nu)(\beta_0 U_{i+1}^n - \Delta_+ U_{i+1}^n) + 2\nu(\beta_0 U_i^n - \Delta_+ U_i^n)\right.$$

$$\left. -(1 + \nu)(\beta_- U_{i-1}^n - \Delta_+ U_{i-1}^n) - \frac{\omega}{\mu}((1 - \nu)S_{i+1} + (1 + \nu)S_i)\right]. \qquad (6.14)$$

Although the diffusion term and the source term are treated differently, it is readily seen that when these are set to zero and b is taken as constant, this reduces to exactly the same scheme as that obtained with the Lax–Wendroff flux (6.5d) when simple relaxation is used. Note that at convergence it is the quantity in square brackets in (6.14) which is set to zero; whether the individual residuals which have been averaged to obtain this quantity are driven to zero depends on the boundary conditions that are imposed.

As regards multiple level time-stepping, by far the most commonly used schemes are based on *Runge–Kutta methods*, although the leap-frog or midpoint rule which can be written $U^{n+1} = U^{n-1} - 2\omega(L^h U^n - R^h S)$ is sometimes advocated. Both can handle the iteration of the central difference scheme, based on the flux (6.5b), even when $\epsilon = 0$, that is, for arbitrarily high mesh Péclet numbers; but the leap-frog scheme needs to incorporate the Du Fort–Frankel treatment of the diffusion terms.

Very general Runge–Kutta methods have been derived for systems of ordinary differential equations, but if we limit our consideration to explicit methods of at most four stages, and denote the nodal residual $L^h U - R^h S$ by $N(U)$, we obtain schemes of the general form

$$U^{n+1} = U^n - \omega \sum_{r=1}^{4} b_r K_r \qquad (6.15a)$$

$$K_r = N(U^n - \omega \sum_{s=1}^{r-1} a_{rs} K_s), \qquad (6.15b)$$

where $K_1 = N(U^n)$ and the coefficients $\{a_{rs}, b_r\}$ can be chosen to give schemes with appropriate properties. Using the same approach as with simple relaxation, the most useful guidance in this choice is obtained from a linearisation and Fourier analysis in space. Thus we replace $N(U)$ by $\widehat{N}\widehat{U}$, where $\widehat{N} \equiv \widehat{N}(k\Delta x)$ is a complex number, sometimes called the *Fourier footprint* and otherwise the *symbol* of N, with the imaginary part representing the convection and the real part the diffusion plus any artificial damping; the descriptive term footprint derives from the plot of $\widehat{N}$ on an Argand diagram as $k\Delta x$ varies from $-\pi$ to π. The amplification, or damping, of such a mode by the scheme (6.15) is then given by

$$1 - \omega\left[b_1\widehat{N} + b_2\widehat{N}(1 - a_{21}\omega\widehat{N}) + b_3\widehat{N}(1 - a_{31}\omega\widehat{N} + \ldots) + \ldots\right] \quad (6.16)$$

in a form corresponding to that in (6.12). Clearly this is a polynomial in $\omega\widehat{N}$ of degree equal to the number of stages in the formula. Standard notation is to denote it by $\lambda(z)$ with $z = -\omega\widehat{N}$ so that $\lambda(z)$ approximates e^z; and it is usual to plot contours of $|\lambda(z)|$ in the complex z-plane for $|\lambda| \leq 1$ in order to show the properties of the scheme, with $|\lambda| = 1$ giving the stability limit. To obtain the maximum stable value of ω for a given combination of Runge–Kutta scheme and discretisation L^h, one has to compare the curve or footprint given by $z = -\omega\widehat{N}(\xi)$ for $-\pi \leq \xi \leq \pi$ with the contour $|\lambda| = 1$; the largest value of ω for which the footprint of the discretisation scheme lies inside the stability limit curve for the Runge–Kutta scheme gives the required result. It is easily checked that this is equivalent to finding practical stability limits for an unsteady computation, as in the examples of (1.48) and (1.64).

A simple one-parameter family of three-stage formulae (see Eberle, Rizzi and Hirschel 1992, pp 137–139) is given by $a_{21} = 1, a_{31} = 1-\theta, a_{32} = \theta; b_1 = 1 - \theta, b_2 = 0, b_3 = \theta$; that is,

$$K_1 = N(U^n), \quad K_2 = N(U^n - \omega K_1)$$
$$K_3 = N(U^n - \omega[(1-\theta)K_1 + \theta K_2])$$
$$U^{n+1} = U^n - \omega[(1-\theta)K_1 + \theta K_3]. \quad (6.17a)$$

This gives, after simplification and with $z = -\omega\widehat{N}$,

$$\lambda(z) = 1 + z + \theta z^2 + \theta^2 z^3 \quad (6.17b)$$

and contours of $|\lambda(z)|$ are plotted in Fig 6.2 for $\theta = \frac{1}{2}$ and $\theta = 1$. It can be seen that for $\theta = \frac{1}{2}$ the stability region extends to $z = \pm 2i$; for the central difference scheme applied to pure convection, where $\widehat{N} = i(b/\Delta x)\sin k\Delta x$, this means we have stability for $\nu = b\omega/\Delta x \leq 2$. Increasing θ to $\theta = 1$ greatly increases the damping on the imaginary z-axis, it clearly being four times larger for low frequencies, but reduces the stability limit to

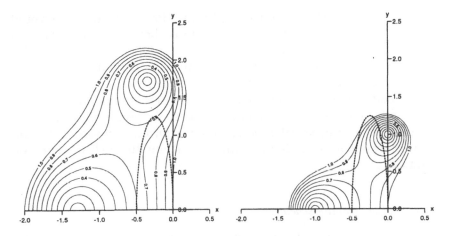

Figure 6.2. *Stability limits and contours of* $|\lambda(z)|$ *for the Runge–Kutta formula (6.17): on the left for* $\theta = \frac{1}{2}$ *and on the right for* $\theta = 1$; *the Fourier footprint for the central difference scheme with* $\nu = 1.25$, $\mu = 0.125$ *is shown in each case as a dark dashed line*

$\sqrt{((1+\sqrt{5})/2)} \approx 1.27$. In Fig. 6.2 we have also plotted the Fourier footprint for $\nu = 1.25$ after diffusion corresponding to $\mu \equiv \epsilon\omega/(\Delta x)^2 = \frac{1}{8}$ has been added to the scheme.

The classical (and most widely used for ordinary differential equations) four-stage Runge–Kutta scheme, which is fourth order accurate because $\lambda(z)$ matches the first five terms in the Taylor expansion of e^z, is given by

$$K_1 = N(U^n), \qquad\qquad K_2 = N(U^n - \tfrac{1}{2}\omega K_1)$$
$$K_3 = N(U^n - \tfrac{1}{2}\omega K_2), \quad K_4 = N(U^n - \omega K_3)$$
$$U^{n+1} = U^n - \tfrac{1}{6}\omega\left(K_1 + 2K_2 + 2K_3 + K_4\right). \qquad (6.18)$$

It is stable for $|\mathrm{Im}\,z| \leq 2\sqrt{2} \approx 2.83$, but of course because of its high accuracy it has very little damping on the imaginary axis until $|\mathrm{Im}\,z| \geq 1$.

The last of the three developments from simple relaxation that we should mention is variously described as using implicit methods, residual smoothing, or pre-conditioning; these are distinct but closely related techniques. *Implicit methods* for solving unsteady problems will be discussed at greater length in the next chapter. Their objective is to allow the use of larger time steps, but for our present purposes they are of little use since by introducing the term $L^h U^{n+1}$ they require the inversion of this operator, which is after all our present objective. Their link to the other techniques comes from the possibility of using an alternative discretisation $\tilde{L}^h$ here which is more easily invertible, leading thence to the use of defect correction or difference correction techniques. *Residual smoothing* is the term used to describe the replacement of the residual $N(U)$ by $\bar{N}(U)$, where $M^h\bar{N}(U) = N(U)$ and

M^h is an easily invertible discrete elliptic operator; a typical choice would be $M^h = (1 - \epsilon_x \delta_x^2)(1 - \epsilon_y \delta_y^2)(1 - \epsilon_z \delta_z^2)$ in three dimensions so that only a sequence of tridiagonal equations has to be solved. Finally, *preconditioning* is a general term used for operating on an algebraic system to improve its condition before applying an iteration process to it, and so might be said to include both the use of implicit schemes and residual smoothing. Indeed, it is clear that there is much in common between the preconditioned conjugate gradient or subspace iteration methods referred to in section 1.5 and the use of residual smoothing in combination with Runge–Kutta iteration methods.

6.2.3 *Solution-adaptive schemes*

Various techniques for adapting a difference scheme according to the solution have been developed for both unsteady hyperbolic conservation laws and also for the steady convection-diffusion problems with which we are presently concerned. There is much in common between the two situations; yet there seems to have been relatively little interaction between the two fields. We shall treat the unsteady case in the next chapter, but because the problems are more severe there, and in order to emphasise the common features, we shall also use here much of the terminology and notation which has become standard in that case.

(i) *Adaptive choice of fluxes.* In both cases there are two main issues: firstly, how should other quantities such as fluxes be estimated from given discrete data, represented by $\{U_i\}$; and secondly, how does the choice for this process affect the solution obtained from the difference scheme. We have seen in the present case that to approximate the solution of (6.3) we first need to construct fluxes $\{F_{i-\frac{1}{2}}\}$, with examples for the convective flux being given in (6.5). In effect we need to infer a value for $U_{i-\frac{1}{2}}$ from, say, U_i, U_{i-1} and U_{i-2}. Choosing U_{i-1} is always 'safe' but can be inaccurate, choosing $\frac{1}{2}(U_i + U_{i-1})$ is sometimes safe and generally more accurate, while choosing $\frac{1}{8}(3U_i + 6U_{i-1} - U_{i-2})$ is never safe but is most accurate. By the term 'safe' here we mean that a maximum principle holds so that a reliable error estimate is available *for any data.* The key question then is whether we can deduce from the solution when it is safe to use the more accurate methods. Normally this is done by switching between these choices according to the data: thus if the three data values U_i, U_{i-1} and U_{i-2} are judged to be sufficiently smooth, we use the quadratic interpolation formula; if not, this formula is modified to be closer to a safe alternative. Several alternative criteria and strategies have been proposed; we shall consider their general justification and a few examples.

Consider first some simple approximation arguments. Suppose we are given exact values u_i, u_{i-1} and u_{i-2} on a uniform mesh for an unknown function u of which we have no *a priori* knowledge, except that it is differentiable to some order. The approximation u_{i-1} to $u_{i-\frac{1}{2}}$ will have an error of $-\frac{1}{2}hu'(\xi_1)$, the approximation $\frac{1}{2}(u_i + u_{i-1})$ an error of $\frac{1}{8}h^2u''(\xi_2)$, and the approximation $\frac{1}{8}(3u_i + 6u_{i-1} - u_{i-2})$ an error of $\frac{1}{16}h^3u'''(\xi_3)$. The derivatives here are assumed to exist but are at unknown points and have unknown values. The first two, however, can be estimated from the data; so, if we find that $\frac{1}{8}|\delta^2 u_{i-1}|$ is greater than $\frac{1}{2}|\Delta_+ u_{i-1}|$ we have some reason to use the lowest order approximation u_{i-1}, although the justification is not strong since we always have $|\delta^2 u_{i-1}| \le 2\max(|\Delta_+ u_{i-1}|, |\Delta_- u_{i-1}|)$. On the other hand, if we make further assumptions such as that any high derivatives are likely to occur in boundary layers of exponential form, we know that ratios of successive derivatives are roughly constant. This implies that if $|\delta^2 u_{i-1}| < 2|\Delta_+ u_{i-1}|$ we might go on to deduce that $h^3|u'''| < 2h^2|u''|$, and hence we should use the quadratic approximation; thus, by combining these two cases, we see that these approximation arguments seldom justify the use of the intermediate linear approximation.

Moreover, we do in fact have *a priori* knowledge of u if it is the solution of (6.3), for example, it satisfies local maximum and minimum principles, so that if $S \le 0$ in (x_{i-1}, x_i), then $u_{i-\frac{1}{2}} \le \max(u_i, u_{i-1})$, and $S \ge 0$ implies $u_{i-\frac{1}{2}} \ge \min(u_i, u_{i-1})$. If these are to be satisfied they clearly place limits on the use of quadratic interpolation.

This sort of application of a maximum principle is probably the commonest argument used in the design of adaptive difference schemes for both steady and unsteady problems. The most widely used measure of solution variation, on which the choice of local difference scheme is based, is the ratio of successive differences; so for the approximate solution $\{U_i\}$ we define

$$r_i := \frac{\Delta_- U_i}{\Delta_+ U_i} = \frac{U_i - U_{i-1}}{U_{i+1} - U_i}, \qquad (6.19)$$

in terms of which the ratio used above can be written $\delta^2 U_{i-1}/\Delta_+ U_{i-1} = (1 - r_{i-1})$. Also the quadratically interpolated $U_{i-\frac{1}{2}}$ of (6.5e) can be written as

$$U_{i-\frac{1}{2}} = \left[1 + \tfrac{1}{8}(3 + r_{i-1})\Delta_+\right] U_{i-1}; \qquad (6.20)$$

this clearly lies between U_i and U_{i-1} if and only if we have $-3 \le r_{i-1} \le 5$, a condition on $1 - r_{i-1}$ which therefore exactly matches that suggested above based on $|\delta^2 U_{i-1}| \le 4|\Delta_+ U_{i-1}|$. However, to have $r_{i-1} < 0$ means that U_{i-1} does not lie between U_{i-2} and U_i; and if S is small it would suggest that the data values U_{i-2}, U_{i-1}, U_i were not accurate enough to warrant their use in quadratic interpolation. Thus one might arrive at the simple adaptive scheme: if $0 \le r_{i-1} \le 5$, use quadratic interpolation for $U_{i-\frac{1}{2}}$; otherwise set $U_{i-\frac{1}{2}} = U_{i-1}$. This can be given a simple functional form if we normalise the three data values by defining $\widehat{U}_j := (U_j - U_{i-2})/(U_i - U_{i-2})$. Then the upwind choice is $\widehat{U}_{i-1} \equiv r_{i-1}/(1 + r_{i-1})$ and quadratic interpolation becomes $\widehat{U}_{i-\frac{1}{2}} = \frac{3}{8}(1 + 3r_{i-1})/(1 + r_{i-1})$. These two curves, and the switching points between them at $r_{i-1} = 0$ and 5, can be seen in Fig. 6.3.

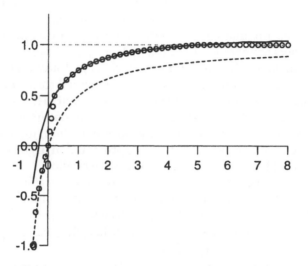

Figure 6.3. *The normalised convective flux* $\widehat{U}_{i-\frac{1}{2}}$ *plotted against the solution gradient ratio* r_{i-1}; *the curves shown are for the upwind scheme (dashed line), the QUICK scheme (full line), and the solution-adaptive SMART scheme (open circles)*

The SMART scheme (for Sharp and Monotonic Algorithm for Realistic Transport) introduced by Gaskell and Lau (1988) is a solution-adaptive method which is based on the QUICK scheme (Leonard 1979a) in a manner very similar to that outlined above. For the iterative solution of the equations it is desirable to have the flux depend continuously on the data; so in the SMART scheme there is a smooth transition from the upwind method to QUICK over the intervals $0 \leq r_{i-1} \leq \frac{1}{5}$ and $5 \leq r_{i-1} < \infty$. This scheme is also represented in Fig 6.3 as a plot of $\widehat{U}_{i-\frac{1}{2}}$ against r_{i-1}. This figure should be compared with Fig 7.5 in section 7.2.4 where flux-limiters for similar unsteady schemes are plotted against r.

(ii) *Algorithm properties.* Let us now consider the second key issue — how the choice of convective flux affects the overall algorithm. Continuing to confine the discussion to one dimension and to the alternative fluxes given in (6.5), we suppose that the quadratically interpolated flux (6.5e) is the normal choice for $F_{i-\frac{1}{2}}$, and that only if r_{i-1} lies outside some given range is another choice made. The result is that the QUICK difference scheme is applied at the point x_i if both r_i and r_{i-1} lie in the given range; for constant

b, a uniform mesh and $\beta = bh/\epsilon$, this gives

$$\tfrac{1}{8}\beta U_{i-2} - (1 + \tfrac{7}{8}\beta)U_{i-1} + (2 + \tfrac{3}{8}\beta)U_i - (1 - \tfrac{3}{8}\beta)U_{i+1} = \frac{h^2}{\epsilon}S_i,$$

i.e. $\qquad - \delta^2 U_i + \beta(\Delta_0 - \tfrac{1}{8}\Delta_-\delta^2)U_i = \frac{h^2}{\epsilon}S_i,$ $\qquad\qquad$ (6.21)

as the approximation to (6.3).

Apart from the situation when r_{i-1} lies outside the given range, an alternative flux will also have to be used at the left boundary; whatever the choice, because each flux is used consistently, one always has the conservation property obtained from summing (6.4) over each interval,

$$F_{J-\frac{1}{2}} - F_{\frac{1}{2}} = h \sum_{i=1}^{J-1} S_i. \qquad\qquad (6.22)$$

In order to relate the present approach to that based on using artificial dissipation, we note that when the normal quadratic flux $F_{i-\frac{1}{2}}$ is changed to the simple upwind, central difference or second order upwind fluxes of (6.5a–c), this is equivalent to adding the following terms to (6.21),

(simple upwind) $\qquad\qquad \tfrac{1}{8}\beta(-U_{i-2} - 2U_{i-1} + 3U_i)$ $\qquad$ (6.23a)

(central difference) $\qquad\quad \tfrac{1}{8}\beta(-U_{i-2} + 2U_{i-1} - U_i)$ $\qquad\quad$ (6.23b)

(second order upwind) $\qquad \tfrac{3}{8}\beta(U_{i-2} - 2U_{i-1} + U_i).$ $\qquad\qquad$ (6.23c)

At the same time, of course, the identical term is subtracted from the equation at the point x_{i-1}. A switch to the central difference flux therefore enhances the diffusion terms for the equation at x_i, but reduces it at x_{i-1}; use of the second order upwind flux has the opposite effect; and a simple upwind switch at one point adds first order terms to the equations, while the same change to both of the fluxes in a scheme clearly enhances the diffusion.

Whatever switching strategy is used, the result will be at most a four-point scheme, which we can write as either a three-point or a two-point scheme by using relationships given by (6.19),

$$U_{i-2} = U_{i-1} - r_{i-1}(U_i - U_{i-1}), \quad U_{i+1} = U_i + (1/r_i)(U_i - U_{i-1}). \quad (6.24)$$

The two-point form will be unique but we will use the less restrictive three-point form; thus we write the nodal residual, on a uniform mesh and with constant b, as

$$N_i(U) = \frac{1}{h}\left[C_{i-\frac{1}{2}}(U_i - U_{i-1}) - D_{i+\frac{1}{2}}(U_{i+1} - U_i)\right] - S_i, \qquad (6.25)$$

where $C_{i-\frac{1}{2}}$ and $D_{i-\frac{1}{2}}$ are functions of r_{i-1}, and $C_{i-\frac{1}{2}}$ is defined for $i = 1, 2, \ldots, J - 1$ while $D_{i-\frac{1}{2}}$ is defined for $i = 2, 3, \ldots, J$. When simple upwinding is used for both fluxes, we can take $C_{i-\frac{1}{2}} = b + \epsilon/h$ and

$D_{i+\frac{1}{2}} = \epsilon/h$, both of which are always positive and are independent of r_{i-1}. This corresponds to the scheme always satisfying a maximum principle; in general we have the following result.

Lemma 6.2.1 *Suppose a solution of $N_i(U) = 0, i = 1, 2, \ldots, J-1$ with N_i given by (6.25), is such that $C_{i-\frac{1}{2}} \geq 0, D_{i+\frac{1}{2}} \geq 0$ and $C_{i-\frac{1}{2}} + D_{i+\frac{1}{2}} > 0$. Then the solution satisfies a maximum principle, as in the nonadaptive case covered by Lemma 3.2.3.*

Proof. Exactly the same arguments as in Lemma 3.2.3 can be used. □

The unfortunate difference from the nonadaptive case is that the hypotheses in the lemma cannot be used to guarantee the existence of a solution to the equations because these are now nonlinear. We can, however, make use of techniques developed in the study of hyperbolic conservation laws to obtain a result for homogeneous equations. Following Harten (1983) and Tadmor (1988), we define the *total variation* of a mesh function U by

$$TV(U) := \sum_{j=1}^{J} |U_j - U_{j-1}|, \qquad (6.26)$$

and consider when the simple relaxation process

$$U_i^{n+1} = U_i^n - \omega_i^n N_i(U^n) \quad i = 1, 2, \ldots, J-1, \qquad (6.27)$$

where N_i is given by (6.25) with $S_i = 0$, is *TVD (total variation diminishing)*. Clearly we have both

$$U_i^{n+1} = \left[1 - (\omega_i^n/h)(C_{i-\frac{1}{2}}^n + D_{i+\frac{1}{2}}^n)\right] U_i^n$$
$$+ (\omega_i^n/h)\left[C_{i-\frac{1}{2}}^n U_{i-1}^n + D_{i+\frac{1}{2}}^n U_{i+1}^n\right] \qquad (6.28a)$$

and

$$\Delta_- U_i^{n+1} = \left[1 - (\omega_i^n/h)C_{i-\frac{1}{2}}^n - (\omega_{i-1}^n/h)D_{i-\frac{1}{2}}^n\right] \Delta_- U_i^n$$
$$+ (\omega_i^n/h)D_{i+\frac{1}{2}}^n \Delta_- U_{i+1}^n + (\omega_{i-1}^n/h)C_{i-\frac{3}{2}}^n \Delta_- U_{i-1}^n, \qquad (6.28b)$$

where the notation $C_{i-\frac{1}{2}}^n$ expresses the dependence of $C_{i-\frac{1}{2}}$ on U^n through r_{i-1}, and the $C_{i-\frac{1}{2}}$ and $D_{i-\frac{1}{2}}$ outside the subscript ranges needed to define N_i are defined as zero. Thence we have the following result.

Lemma 6.2.2 *The iteration (6.27) with $U_0^n = u_L, U_J^n = u_R$ has a convergent subsequence to a monotone solution of the homogeneous problem, if the adaptive strategy and the choice of $\omega_i^n > 0$ is such as to guarantee*

$$C_{i-\frac{1}{2}}^n \geq 0, \quad D_{i-\frac{1}{2}}^n \geq 0$$

and

$$\omega_i^n \leq h\left[\max(C_{i-\frac{1}{2}}^n, C_{i+\frac{1}{2}}^n) + \max(D_{i-\frac{1}{2}}^n + D_{i+\frac{1}{2}}^n)\right] \quad \forall i, n. \qquad (6.29)$$

Proof. By the hypothesis (6.29), all the coefficients of both (6.28a) and (6.28b) are non-negative. Hence, by taking absolute values and summing we have

$$\left\|U^{n+1}\right\|_{l_1} \leq \left\|U^n\right\|_{l_1} \text{ and } TV(U^{n+1}) \leq TV(U^n).$$

The sequence $\{U^n\}$ can therefore be regarded as a sequence of piecewise linear functions whose variation and L_1 and L_∞ norms are all bounded, so that Helly's theorem can be applied to deduce the existence of a convergent subsequence. (See LeVeque (1992) or Godlewski and Raviart (1991) for a discussion of such compactness arguments and their use in the study of approximating hyperbolic conservation laws.)   □

We conclude this section by considering adaptive strategies that attempt to make maximum use of the QUICK scheme; we could equally well consider mixing central and upwind differencing. In writing (6.21) in the form (6.25) we need use (6.24) only to eliminate U_{i-2}; but this results in defining $C_{i-\frac{1}{2}} = \frac{1}{8}b(6 - r_{i-1}) + \epsilon/h$ and $D_{i+\frac{1}{2}} = -\frac{3}{8}b + \epsilon/h$, which will fail the conditions (6.29) whenever $\beta > \frac{8}{3}$, whatever the value of the difference ratios. Hence, instead, we write for the QUICK scheme

$$C_{i-\frac{1}{2}} = \frac{1}{8}b(6 - r_{i-1} + 3/r_i) + \epsilon/h, \quad D_{i+\frac{1}{2}} = \epsilon/h, \quad (6.30)$$

which enables (6.29) to be satisfied if $0 < r_i \leq 6 \; \forall i$; note that these are close to the conditions under which the SMART scheme described earlier uses the QUICK convective flux.

We can simplify further by introducing the normalised convective fluxes used in plotting Fig. 6.3; thence, by eliminating U_{i-2} through (6.24), a general four-point convective flux difference can be written

$$b(U_{i+\frac{1}{2}} - U_{i-\frac{1}{2}}) = b\left\{\widehat{U}_{i+\frac{1}{2}}\Delta_+U_i \right.$$
$$\left. + \left[\widehat{U}_{i+\frac{1}{2}} + r_{i-1} - (1 + r_{i-1})\widehat{U}_{i-\frac{1}{2}}\right]\Delta_-U_i\right\}. \quad (6.31)$$

Then we obtain the following result for the SMART scheme.

Theorem 6.2.3 *Suppose an adaptive strategy is determined through defining $\widehat{U}_{i-\frac{1}{2}}(r_{i-1})$ for $i = 2, 3, \ldots, J - 1$ by the SMART scheme depicted in Fig. 6.3, and we set $U_{\frac{1}{2}} = U_0$. Then parameters ω_i^n can be chosen to ensure that the simple relaxation (6.27) has a convergent subsequence to a monotone solution of the homogeneous problem.*

Proof. In (6.31) we replace Δ_+U_i by $(\Delta_-U_i)/r_i$, so we can set $D_{i+\frac{1}{2}} = \epsilon/h$ and, for $i \geq 2$,

$$C_{i-\frac{1}{2}} = (1 + 1/r_i)\widehat{U}_{i+\frac{1}{2}} + r_{i-1} - (1 + r_{i-1})\widehat{U}_{i-\frac{1}{2}} + \epsilon/h.$$

For the SMART scheme we have $\widehat{U}_{i+\frac{1}{2}} = r_i/(1+r_i)$ for $r_i \leq 0$ and $\widehat{U}_{i+\frac{1}{2}} > r_i/(1+r_i)$ for $r_i > 0$, so that $(1+1/r_i)\widehat{U}_{i+\frac{1}{2}} \geq 1$; similarly, $\widehat{U}_{i-\frac{1}{2}}$ lies between $r_{i-1}/(1+r_{i-1})$ and 1, so that $(1+r_{i-1})(1-\widehat{U}_{i-\frac{1}{2}}) \geq 0$ and hence $C_{i-\frac{1}{2}} \geq \epsilon/h$. Moreover, by considering $r_i \to 0$ and $r_{i-1} \to \infty$, where upwinding comes into play, we see that $C_{i-\frac{1}{2}}$ is bounded; and when $i = 1, C_{i-\frac{1}{2}}$ reduces to $(1 + 1/r_i)\widehat{U}_{i+\frac{1}{2}} + \epsilon/h$. Thus (6.29) can be satisfied by appropriate choice of ω_i^n and the result follows from Lemma 6.2.2. □

In practice the iteration process (6.27) may be rather slow, and in their numerical experiments Gaskell and Lau (1988) solve the linearised equations $N_i(U^{n+1}) = 0$ directly, with $C_{i-\frac{1}{2}}^n$ and $D_{i+\frac{1}{2}}^n$ calculated from U^n. In these tests the authors apply a two-dimensional extension of the algorithm described above to the convection of a step profile and a box profile at various angles across the mesh, as in the test problem posed by Raithby (1976) and described in section 1.2 on model problems; their results amply demonstrate the valuable monotonicity-preserving properties that one can readily deduce for these special problems by arguments similar to those employed above in one dimension.

6.2.4 A central/upwind difference example

At this point it may be helpful to illustrate the techniques described in the last two sections with a simple example of the form given in (6.3). As a challenge to the time-stepping approach we will mainly use central differences, but mindful of the advice given in the paper by Gresho and Lee (1981) with its famously provocative title, we shall try to grade the mesh to eliminate oscillations. This could already make the scheme partially adaptive; but we will also use solution-adaptive switches to the simple upwind scheme, of the form described in the previous section. In Chapters 4 and 5 we used the example given in (4.67), and in Fig. 4.7 results were given for the central difference scheme on a uniform mesh with $h = 0.1$ for $\epsilon = 0.1$; for smaller values of ϵ the results were too oscillatory to be presented. So we attempt here to obtain results for $\epsilon = 0.01$.

Using the notation for the mesh of section 1.4 and that for defining the flux in (6.5a), we have the following definitions for the flux $F_{i-\frac{1}{2}}$, the difference operator L^h and the nodal residual N_i,

$$F_{i-\frac{1}{2}}(U) := b_{i-\frac{1}{2}}U_{i-\frac{1}{2}} - (\epsilon/h_i)(U_i - U_{i-1}) \tag{6.32a}$$

$$L^h U_i := \left[F_{i+\frac{1}{2}}(U) - F_{i-\frac{1}{2}}(U) \right] / \tfrac{1}{2}(h_i + h_{i+1}) \tag{6.32b}$$

$$N_i(U) := L^h U_i - S_i, \tag{6.32c}$$

where S_i is the average of S over the control volume and in the first instance we will take $U_{i-\frac{1}{2}} = \tfrac{1}{2}(U_i + U_{i-1})$; application of the Neumann condition $u'(1) = 0$ will give a special form for $F_{J+\frac{1}{2}}$. The solution we seek is denoted by U^∞ and given by

$$N_i(U^\infty) = 0, \; i = 1, 2, \ldots, J; \; U_0^\infty = u_L. \tag{6.33}$$

Successive approximations U^n are obtained from a Runge–Kutta scheme, which we take to be that given in (6.17) with $\theta = 1$ and locally selected parameters ω_i. We need to choose these parameters so that the nodal errors $E_i^n := U_i^n - U_i^\infty$ are driven to zero as quickly as possible; but convergence is monitored by the l_2 norm of the nodal residuals $\|N(U^n)\|_{l_2}$.

Suppose we carry out a 'local' Fourier analysis assuming a locally constant h and b. Then we have, with $\xi = kh$,

$$N(E) \to \widehat{N}\widehat{E} = \left[\mathrm{i}(b/h)\sin\xi + 4(\epsilon/h^2)\sin^2\tfrac{1}{2}\xi \right]\widehat{E}. \tag{6.34}$$

From this analysis, which corresponds to a practical stability analysis of the time-stepping scheme, we need to choose ω such that for all wave modes, $|\xi| \le \pi$, the complex variable $z = -\omega\widehat{N}$ lies in the stability region of the Runge–Kutta scheme. For large mesh Péclet numbers the choice will be dominated by the behaviour of $|\lambda(z)|$ for z on the imaginary axis; referring to the right-hand diagram of Fig. 6.2 we deduce that we need $\omega b/h < 1.25$. Since in our example we have $1 \le b \le 8$, and our main concern is to have freedom in the choice of h, we set $\omega_i = \frac{1}{2}\omega(h_i + h_{i+1})$ and ensure that $\omega < 0.15$.

The usual Neumann boundary condition for a central difference scheme, and that used to calculate the results of Fig.4.7, would be to set $U_{J+1} = U_{J-1}$ and $h_{J+1} = h_J$ in the expression for $F_{J+\frac{1}{2}}$. However, for high mesh Péclet numbers this leaves the values of U at odd and even mesh points virtually uncoupled, and the smoothing effect of the boundary condition is lost; the resulting oscillations near the right-hand boundary are evident in the top left graph of Fig.4.7 and the results are much worse at $\epsilon = 0.01$. We therefore use the boundary condition $U_{J+1} = U_J$, which is equivalent to taking the upwind convected flux $b_{J+\frac{1}{2}}U_J$ at the extrapolated cell boundary $x_J + \frac{1}{2}h_J$, and eliminates the diffusive flux to give

$$F_{J+\frac{1}{2}}(U) = b_{J+\frac{1}{2}}U_J. \tag{6.35}$$

The other source of oscillation with the central difference scheme in this example is the sharp source term peak at $x = \frac{1}{2}$. We therefore define a mesh that equidistributes the source and combine this with a uniform mesh. The results shown on the left of Fig.6.4 were obtained with a $\frac{1}{4} : \frac{3}{4}$ mix of these meshes for $J = 20$; Runge–Kutta iteration with $\omega = 0.1$ gave converged results with $\|N\|_{l_2} < 0.01$ after 78 steps. The results are clearly very good apart from the continued presence of oscillations on the right and at the peak.

In a second calculation on the same mesh, we use the difference ratio r_i given by (6.19), or with divided differences replacing undivided differences, to switch between the convective fluxes (6.5a) and (6.5b). Convergence of the Runge–Kutta scheme is then much more erratic; but with $U_{i-\frac{1}{2}}$ determined by the value of r_{i-1} and with smooth transitions over the intervals $[0, \frac{1}{5}]$ and $[5, \infty)$ as described above for the SMART scheme, a reduced parameter of $\omega = 0.075$ gave convergence to plotting accuracy with $\|N\|_{l_2} \leq 0.01$ after 99 steps. The result is shown on the right of Fig.6.4. There is sufficient reduction in the oscillations on the right to give a monotone solution, but it is not dramatically better than with the nonadaptive scheme.

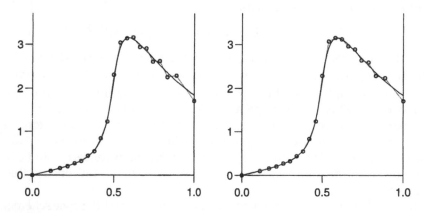

Figure 6.4. *Approximations to the model problem (4.67) for $\epsilon = 0.01$ as in Fig. 5.5 but with a graded mesh; on the left, for a purely central difference scheme and, on the right, for a solution-adaptive scheme using central and upwind $U_{i-\frac{1}{2}}$*

6.3 Finite element-based methods

We shall confine our discussion to the use of linear or multilinear approximations, with the emphasis on two-dimensional quadrilateral meshes. As indicated in the Introduction (section 6.1), all schemes will refer to a primary mesh that fits the boundary of the domain; and most will impose conservation on this mesh. Extensions to three dimensions will be quite natural for all schemes considered, but we shall not discuss specifically three-dimensional issues such as the presence of singular mesh lines and the effect of having nonplanar surfaces to the control volumes.

As we shall see in the next chapter, in a Galerkin formulation the use of multilinear trial functions can give schemes similar to the QUICK difference scheme of the previous section; but using piecewise constant test functions in the present finite volume framework limits comparisons to simple upwind and central difference schemes. These will sometimes emerge as special cases (e.g. in simple geometry and with simple quadrature rules); so some of the following finite volume methods can be regarded as natural generalisations of these difference schemes to two dimensions and nonuniform meshes.

The starting point of a finite volume approximation on a quadrilateral cell Ω_α to the convection-diffusion equation $-\epsilon\nabla^2 u + \nabla\cdot(\mathbf{b}u) = S$ is the integral form

$$\int_{\partial\Omega_\alpha} \left[-\epsilon\frac{\partial u}{\partial n} + (\mathbf{b}\cdot\mathbf{n})u\right] \mathrm{d}\Gamma = \int_{\Omega_\alpha} S\mathrm{d}\Omega, \qquad (6.36)$$

as in (6.2). For both the cell centre and cell vertex schemes which we shall concentrate on, the control volumes Ω_α are the cells of the primary mesh; it is only the approximation of the two sets of integrals that differ, and these are based on the same metric quantities — the measure V_α of the cell and $(\Delta x, \Delta y)$ for each edge, with the latter generalising into three components of surface area in three dimensions.

6.3.1 Cell centre schemes

In these schemes the parameters specifying the approximation U are associated with nodes which are at the centres of the cells, and are often interpreted as cell averages. Although it would be possible to derive the resulting discretisations by supposing that U has an isoparametric bilinear form on the dual mesh which has the primary cell centres as its vertices, this is not the usual practice: firstly, on a general mesh it would require storing or calculating much extra mesh data; and, secondly, it is hardly justified if the U parameters are interpreted as cell averages rather than point values. It is because of this last point that the convective flux through an edge is usually approximated by a simple average between the values in the cells either side. However, there is another viewpoint that places more emphasis on the accurate calculation of these edge fluxes. Indeed, some researchers consider the property of the edges being midway between the nodes to be sufficiently important that it is imposed everywhere on the

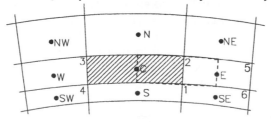

Figure 6.5. *Local mesh labelling for a cell centre scheme*

mesh; this is accomplished by generating the node positions first and then constructing the cells from the orthogonal bisectors of the lines between them. Note that the nodes are then not in the centres of the cells, in general; however, the resulting schemes have the same form and we shall continue to describe them as cell centre schemes.

In the notation of Fig. 6.5, and in order to reflect the preferred averaging in nonlinear problems, the convective flux through the edge between $r_1 \equiv (x_1, y_1)$ and $r_2 \equiv (x_2, y_2)$ is given by

$$\frac{1}{2}\left[(b^{(x)}U)_E + (b^{(x)}U)_C\right](y_2 - y_1)$$

$$- \frac{1}{2}\left[(b^{(y)}U)_E + (b^{(y)}U)_C\right](x_2 - x_1). \tag{6.37}$$

As a result when all four edge contributions are combined, and divided by the measure of the cell V_C, there will be no dependence on U_C and the equivalent of a four-point central difference approximation is obtained. Had an attempt to obtain greater accuracy been made, by having the averaging reflect the nonuniformity of the mesh but ignoring the direction of b, the resulting coefficient of U_C would have been negative for some mesh geometries; this would have made the system of equations even less well conditioned, and in a manner more difficult to control, than the central difference system.

To compute the diffusive flux we use the common device of writing $\partial u/\partial x = \text{div}(u, 0)$ and applying Gauss' theorem to an appropriate secondary control volume. With this volume chosen as that marked by dashed lines in Fig. 6.5, it has two types of mid-edge point — the cell centres, C and E, and cell vertices, r_1 and r_2. The corresponding contributions to $u\Delta y$ from C and r_2 are, respectively,

$$U_C\left[\tfrac{1}{2}(y_1 - y_2) + \tfrac{1}{2}(y_4 - y_3))\right], \tag{6.38a}$$

and

$$\tfrac{1}{4}(U_C + U_E + U_{NE} + U_N)\left[\tfrac{1}{2}(y_2 - y_5) + \tfrac{1}{2}(y_3 - y_2)\right]; \tag{6.38b}$$

and the measure of the cell is taken as $\frac{1}{2}(V_C + V_E)$. It is clear that on a rectangular mesh aligned with the (x, y)-axes, terms such as (6.38b) will all be zero and $y_1 - y_2 = y_4 - y_3 = -\Delta y_C$. Since also $\Delta y_E = \Delta y_W = \Delta y_C$ and $x_1 - x_2 = x_4 - x_3 = 0$ in that case, together with all similar relations, the diffusive flux through the (r_1, r_2) edge reduces to $-\epsilon(U_E - U_C)2\Delta y_C/(\Delta x_E + \Delta x_C)$ and the approximation to $-\epsilon V_C \nabla^2 u$ becomes

$$-\epsilon V_C \nabla^2 u \approx -2\epsilon \Delta y_C \left[\frac{U_E - U_C}{\Delta x_E + \Delta x_C} + \frac{U_W - U_C}{\Delta x_W + \Delta x_C}\right]$$

$$- 2\epsilon \Delta x_C \left[\frac{U_N - U_C}{\Delta y_N + \Delta y_C} + \frac{U_S - U_C}{\Delta y_S + \Delta y_C}\right] \tag{6.39}$$

Note that this is not exactly the same as the standard difference scheme on a nonuniform mesh, reflecting again the fact that the cell edges are not necessarily midway between the corresponding cell centres.

When all these integrals around the boundary of the cell have been computed, and the result has been divided by V_C, we call the result the nodal residual $N_C(U)$.

Boundary conditions, even those of Dirichlet type, will as we see from Fig. 6.1 entail the use of fictitious points outside the domain. Thus if the domain boundary lay along the (r_1, r_2) edge, the fluxes needed to compute the equation associated with an interior value U_C would be defined as above, and U_E would be eliminated either by setting $\frac{1}{2}(U_E + U_C)$ equal to a given boundary value or using the above formulae for $\partial u/\partial n$ to apply a non-Dirichlet condition. Note that even for a pure convection problem a boundary condition is needed at all boundaries; at an outflow boundary along the (r_1, r_2) edge one might set $U_E = U_C$, but see the comments in the next section.

6.3.2 Spurious modes and artificial dissipation

In finite volume methods, difficulties in imposing or proving coercivity properties are often attributed to the presence of *spurious solution modes*, or inadequate control of them. This term is rather loosely applied to mesh functions which lie in the null space, or almost in the null space, of the discrete operator L^h but are not smooth enough to represent the true solution; and such an assessment may ignore the application of boundary conditions, because these may control the mode only rather weakly in large domains or where there are nonlinear effects generating the modes.

Thus for the cell centre schemes just described, the averaging across a cell edge that is embodied in (6.37) means that there are three important spurious mode patterns for the convection operator; their forms are simplest when b is constant, and can then be described diagrammatically as follows,

$$
\begin{array}{ccc}
+ - + \qquad & + - + \qquad & + + + \\
- + - \qquad & + - + \qquad & - - - \\
+ - + \qquad & + - + \qquad & + + +.
\end{array}
\qquad (6.40)
$$

That on the left is the well-known *chequerboard mode* and we call the other two *washboard modes*, one of vertical form and one horizontal. The first is essentially two-dimensional but the latter pair are familiar from one-dimensional central differencing. In the absence of any diffusion, and ignoring the imposition of boundary conditions, arbitrary magnitude errors of the forms in (6.40) can occur in the solution of the cell centre equations. Moreover, the technique we have described for imposing Dirichlet boundary conditions does nothing to control these error modes.

The key effects of the washboard mode can be seen in our standard one-dimensional model problem: with the usual boundary conditions $U_0 = 0, U_J = 1$, we recall that the difference approximation to the convection-diffusion problem with mesh Péclet number β is given in (1.33) as $U_j = (\lambda_+^j - 1)/(\lambda_+^J - 1)$, with the central difference oscillations occurring for $\beta > 2$ because we have $\lambda_+ = (1 + \frac{1}{2}\beta)/(1 - \frac{1}{2}\beta)$; with boundary conditions $\frac{1}{2}(U_0 + U_1) = 0, \frac{1}{2}(U_{J-1} + U_J) = 1$, however, the oscillations are enhanced by roughly $\frac{1}{2}\beta$ when $\beta \gg 1$, as the solution becomes $[\lambda_+^j(1 - \frac{1}{2}\beta) - 1]/(\lambda_+^{J-1} - 1)$. In view of the terminology introduced above, though, it should be noted that λ_+^j is not a truly spurious mode of the convection-diffusion equation since it describes the true solution as $h \to 0$; but as $\epsilon \to 0$ it becomes the spurious washboard mode of the convection equation.

Artificial dissipation terms are normally used to control these spurious modes. It is clear that the true diffusion, modelled by the terms in (6.38) or (6.39), has a controlling effect on all of the spurious modes (6.40); but to enhance this directly can be justified only where the mesh is too coarse to describe the true solution, such as in boundary or interior layers; that is, one is again led to the need for an adaptive method. However, one still needs to control the spurious modes in regions where the solution is smooth and one wishes to retain the basically second order accuracy of the central difference scheme. The use of fourth order as well as second order artificial dissipation for these purposes was pioneered by Jameson in the context of nondissipative compressible flow calculations (Jameson *et al.* 1981). There are several alternative forms that these terms might take, with the guiding design principle that they should be such as to have a maximal effect on the spurious modes and a minimal effect on smooth solutions, and some of the possibilities will be discussed in this and the next chapter.

Again, something can be learned from our standard one-dimensional problem on a uniform mesh. In the interior the natural choice of term for the fourth order dissipation is $h^{-4}\delta^4 U_i$, and it is usual to set the coefficient to be $O(h^3)$ so as to have a minimal effect on the accuracy for smooth solutions. Then, if we set $b = 1$ and write $\epsilon^{(2)} = \epsilon^*/h$, where ϵ^* combines the true diffusion ϵ with the second order artificial dissipation, the complete scheme reduces to

$$-\epsilon^{(2)}\delta^2 U_i + \Delta_0 U_i + \epsilon^{(4)}\delta^4 U_i = hS_i, \tag{6.41a}$$

which has the characteristic polynomial

$$(\lambda - 1)\left[\epsilon^{(4)}\lambda^3 - (3\epsilon^{(4)} + \epsilon^{(2)} - \tfrac{1}{2})\lambda^2 + (3\epsilon^{(4)} + \epsilon^{(2)} + \tfrac{1}{2})\lambda - \epsilon^{(4)}\right]. \tag{6.41b}$$

When both $\epsilon^{(2)}$ and $\epsilon^{(4)}$ are small, one root is close to $(\epsilon^{(2)} - \tfrac{1}{2})/\epsilon^{(4)}$ and gives an oscillatory mode which dies away rapidly from the right-hand boundary, while another is close to $\epsilon^{(4)}/(\epsilon^{(2)} + \tfrac{1}{2})$, giving a mode which dies away rapidly from the left-hand boundary. The really troublesome mode, because being close to -1 it oscillates and spreads over much of the domain, corresponds to the washboard mode and is roughly $(\epsilon^{(2)} + \frac{1}{2})/(\epsilon^{(2)} - \frac{1}{2})$. From an expansion we have

$$\lambda_{\text{washboard}} \sim -(1 + 4\epsilon^{(2)} + 16\epsilon^{(4)} + \ldots), \tag{6.42}$$

which suggests that it is actually better controlled by the fourth order dissipation than the second, while the accuracy of the smooth component of solution is then affected only by $O(h^3)$ rather than $O(h)$. However, this does not apply to our simple model problem with a boundary layer on the right, as depicted in Figs. 1.4 and 4.6: we know from the exponentially-fitted schemes of Chapters 3 and 5 that in this case a judicious increase of $\epsilon^{(2)}$ can give very accurate monotone solutions, and adding too much merely thickens the boundary layer; but by adding increased amounts of $\epsilon^{(4)}$ one finds that the solution is never made monotone or at all accurate. So we conclude that, while fourth order dissipation is useful for eliminating spurious oscillatory modes from an otherwise smooth solution, second order terms are more useful in sharp boundary layers, where the oscillations are more in the nature of poor approximations to steep exponentials than true spurious modes.

Extra boundary conditions are needed when the fourth order operator of (6.41a) is used in the whole domain; because two of the roots of (6.41b) always lie outside the unit circle, it is natural to choose two conditions at either boundary and this is most convenient for the difference scheme. Thus with (6.41a) applied for $i = 1, 2, \ldots, J - 1$ and a Dirichlet condition applied to U_0, a further condition is needed to eliminate U_{-1}. If U^I is any solution of the inhomogeneous equations (6.41a), and if the modes corresponding to roots outside the unit circle have decayed to zero near the left-hand boundary, the general solution there can be written in the form

$$U_j \approx U_j^I + A_0 + A_L \lambda_L^j, \tag{6.43a}$$

where λ_L is the root of (6.41b) lying inside the unit circle. A boundary condition $U_0 - U_{-1} = 0$, using the first difference, would lead to

$$U_0^I + A_0 + A_L = u(0), \quad A_L(\lambda_L^{-1} - 1) = U_0^I - U_{-1}^I;$$

this could clearly result in having $A_L = O(h)$, so giving an $O(h)$ error at this boundary. It is therefore preferable to use a second difference boundary condition $\delta^2 U_0 = 0$ which ensures that $A_L = O(h^2)$, or even the third difference $\Delta_+ \delta^2 U_0 = 0$.

At the right-hand end, when the roots are complex, the general solution can be written

$$U_j \approx U_j^I + A_0 + |\lambda_R|^{j-J} [A_R \cos(j - J)\theta_R + B_R \sin(j - J)\theta_R], \tag{6.43b}$$

where $|\lambda_R| > 1$. By the same reasoning as above, one uses at least second order differencing; so that, with Dirichlet conditions specified at each end, the complete set of boundary conditions

$$U_0 = u(0), \ U_J = u(1), \ \delta^2 U_0 = 0, \ \delta^2 U_J = 0 \qquad (6.44)$$

is used to combine with scheme (6.41a).

It should be noted that the same arguments suggest an improved outflow boundary condition for the cell centre scheme. At the end of the last section, a boundary condition was given for pure convection which for our model problem would give $U_J = U_{J-1}$; it is clear that $\delta^2 U_{J-1} = 0$ would be much better, and this is borne out by practical experience. If fourth order dissipation is also being used, this condition is combined with $\delta^2 U_J = 0$ to give

$$(U_{J-1} - U_{J-2}) + \epsilon^{(4)} \delta^2 U_{J-2} = h S_{J-1}$$

as the equation corresponding to (6.41a) at the last interior point.

On a nonuniform mesh and in multi-dimensions, there are various alternative forms that can be used for the fourth order dissipation terms. In using such high order differences at all we have departed dramatically from the theme followed in the book so far, of using only compact finite difference schemes, and corresponding finite element schemes. However, we do not here require the fourth order terms to be accurate divided differences and we can use the flexibility implicit in their being $O(h^3)$; all that is important is that they have minimal effect on smooth modes, which is commonly interpreted as insisting that they are transparent to linear variations of U. This is achieved most simply by using the divided differences that are used to calculate the diffusion terms; thus for the cell centre scheme on a rectangular mesh, and the notation of (6.39), we can write

$$\delta_x^2 U_C := 2\Delta x_C \left[\frac{U_E - U_C}{\Delta x_E + \Delta x_C} + \frac{U_W - U_C}{\Delta x_W + \Delta x_C} \right], \qquad (6.45a)$$

with a similar expression for $\delta_y^2 U_C$, and hence obtain for the fourth order terms in the x-direction

$$\epsilon^{(4)} V_C \left[\frac{\delta_x^2 U_E - \delta_x^2 U_C}{\Delta x_E + \Delta x_C} + \frac{\delta_x^2 U_W - \delta_x^2 U_C}{\Delta x_W + \Delta x_C} \right]. \qquad (6.45b)$$

In section 6.3.8 we shall show how similar terms can be generated for general quadrilateral and triangular meshes. There is one further decision to be made, however, on whether the fourth order terms should be of the form $\delta_x^4 + \delta_y^4$ or of the form $(\delta_x^2 + \delta_y^2)^2$. Both choices are used in practice, but in the sections on the cell vertex method we shall give reasons why the second alternative might be preferred.

For the reasons that have been given, in cell centre schemes spurious modes are usually controlled by a combination of fourth order and second order dissipation with a solution-dependent switch between them. In compressible flow calculations this switch and the magnitude of the second order terms are usually based on the magnitude of $|\delta^2 p|/(4+\delta^2)p$, where p is the pressure, because the second order terms are of particular importance at shocks and other discontinuities, and pressure is the most reliable monitor function. There is clearly a good deal of common ground here with the solution-adaptive schemes discussed in section 6.2.3, and we will now show how close this is in the case of second order dissipation and a commonly used alternative switching parameter, namely $|\Delta_+ p - \Delta_- p|/(|\Delta_+ p| + |\Delta_- p|)$.

To maintain conservation the switch must be done at the level of the dissipative flux. So, reverting to the 1D scalar case dealt with in section 6.2.4, we combine the notation of (6.41a) with that of (6.5) and (6.19), and write the convective flux with second order artificial dissipation as

$$U_{i-\frac{1}{2}} = \tfrac{1}{2}(U_i + U_{i-1}) - \epsilon^{(2)}_{i-\frac{1}{2}}(U_i - U_{i-1}). \qquad (6.46a)$$

Then, using the difference ratios of (6.19), the above switching parameter takes the form $|1-r|/(1+|r|)$, so it is one when r is negative, is zero when $r = 1$, and tends to one as r tends to zero and infinity. So suppose we take

$$\epsilon^{(2)}_{i-\frac{1}{2}} = \tfrac{1}{2} \max \left(\frac{|1-r_i|}{1+|r_i|}, \frac{|1-r_{i-1}|}{1+|r_{i-1}|} \right) \qquad (6.46b)$$

Then the scheme switches to the upwind scheme for extreme values of r in very much the same way as the SMART scheme shown in Fig. 6.3, or the simple example of section 6.2.4; where the flow is smooth, $\epsilon^{(2)} = O(h)$ so that only an $O(h^2)$ departure from the central difference scheme occurs.

More discussion of these adaptive strategies will be given in Chapter 7 on Unsteady Problems.

6.3.3 Runge–Kutta time-stepping

An outline of these methods has already been given in section 6.2.2, with a general explicit four-stage scheme presented in (6.15) as well as some specific examples. The use of Runge–Kutta methods for time-stepping to steady state owes much to the work of Jameson, who has made extensive use and study of them in connection with cell centre methods for the Euler equations of compressible fluid flow — see, e.g. Jameson et al. (1981). A typical four-stage scheme, instead of the classical choice of (6.18), has $a_{21} = \tfrac{1}{4}, a_{32} = \tfrac{1}{3}, a_{43} = \tfrac{1}{2}, b_4 = 1$ and all other coefficients in the general scheme of (6.15) set to zero. Shu (1988) gives a whole family of similar schemes that maintain the TVD property. They use the form (6.25) and keep the coefficients $C_{i-\frac{1}{2}}$ and $D_{i+\frac{1}{2}}$ fixed at their (non-negative) values, $C^{(0)}_{i-\frac{1}{2}}$ and $D^{(0)}_{i+\frac{1}{2}}$, calculated with U^n at the beginning of the Runge–Kutta step. Thus one has, with $S_i \equiv 0$,

$$U_i^{(k)} = U_i^{(0)} - \alpha_k(\omega/h)[C_{i-\frac{1}{2}}^{(0)} \Delta_- U_i^{(k-1)} - D_{i+\frac{1}{2}}^{(0)} \Delta_+ U_i^{(k-1)}] \quad (6.47a)$$

for $k = 1, 2, \dots, m$ with $U_i^{(0)} = U_i^n$ and $U_i^{(m)} = U_i^{n+1}$. With the choice of coefficients

$$\alpha_k = k/m(m-k+1), \quad k = 1, 2, \dots, m, \quad (6.47b)$$

Shu shows that the scheme maintains its TVD property if the CFL condition $\omega(C_{i-\frac{1}{2}}^{(0)} + D_{i-\frac{1}{2}}^{(0)}) \leq mh$ is satisfied in each interval.

More recently, Jameson has studied the use of Runge–Kutta methods in conjunction with multigrid methods where their damping properties are particularly important. For this purpose he uses a simple five-stage scheme with carefully adjusted parameters, given by

$$K_1 = N_C(U^n), \; K_2 = N_C(U^n - \tfrac{1}{4}\omega K_1), \; K_3 = N_C(U^n - \tfrac{1}{6}\omega K_2)$$
$$K_4 = N_C(U^n - \tfrac{3}{8}\omega K_3), \; K_5 = N_C(U^n - \tfrac{1}{2}\omega K_4)$$
$$U_C^{n+1} = U_C^n - \omega K_5, \quad (6.48)$$

where $N_C(U^n)$ is the nodal residual for the cell with centre C. This gives very good damping of high frequency modes while allowing CFL numbers for the central difference scheme up to $\nu \leq 4$, see Liu and Jameson (1990).

It will be noticed that both this five-stage scheme and the earlier four-stage scheme can be efficiently implemented for very large systems of equations, because they use only the nodal residuals calculated at the immediately preceding stage. In addition, as already pointed out, they are designed for their damping and large allowable CFL numbers, rather than alternative criteria such as time-accuracy and maintenance of the TVD property that are emphasised for unsteady problems by, for example, van der Houwen (1977) and Shu (1988). Further damping is also obtained by applying residual smoothing techniques, as already mentioned in section 6.2.2. In two dimensions, and on a rectangular mesh, this would typically yield smoothed residuals $\bar{N}$ to replace N given by solving the pair of tridiagonal systems in

$$(1 - \epsilon_x \delta_x^2)(1 - \epsilon_y \delta_y^2)\bar{N}(U) = N(U). \quad (6.49)$$

Application to the residuals used in such as (6.48) would usually be at each stage, but in a boundary layer may only be applied in a direction where the mesh is very refined.

However, all of these techniques — Runge–Kutta, residual smoothing, multigrid — are all concerned with solving the discrete equations rapidly and efficiently. Our main concern is with the quality of solution obtained at the end of this process, and the fact that efficient iterative solvers allow solution-adaptive schemes to be used. So we shall not pursue the topic of solvers any further here.

6.3.4 Cell vertex methods

As indicated in Fig. 6.1, the unknowns for these methods are associated with the vertices rather than the centres of the primary mesh cells over which conservation is to be imposed. It is therefore useful sometimes to

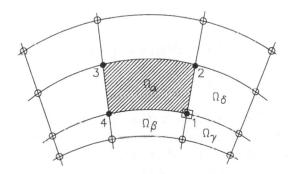

Figure 6.6. *Local mesh labelling for the cell vertex methods*

regard the solution U to be of isoparametric bilinear form in each quadrilateral cell, as described in Chapter 4. However, this is not sufficient to determine how the flux integrals in (6.36) should be approximated, because $\partial U/\partial n$ would then be discontinuous across each edge; we need to use a gradient recovery procedure for this purpose, examples of which will be given in the next section, and the result of which we denote by $\nabla^h U$. To cover both this situation and nonlinear problems, we use the more general form (6.2) and apply the trapezoidal rule directly to the flux components f and g; in effect, we project f and g onto the space of isoparametric bilinear functions before carrying out the integration. Then with the labelling of Fig. 6.6, and writing F_i for $f(U_i, \nabla^h U_i)$ etc., we have

$$\int_{\partial \Omega_\alpha} [f \mathrm{d}y - g \mathrm{d}x] \approx \tfrac{1}{2} \left[(F_1 + F_2)(y_2 - y_1) - (G_1 + G_2)(x_2 - x_1) \right.$$
$$+ (F_2 + F_3)(y_3 - y_2) - (G_2 + G_3)(x_3 - x_2)$$
$$\left. + \ldots \right].$$

The sums in this expression collapse to involve only the diagonal differences; so by introducing the notation $\delta x_{ij} = x_i - x_j, \delta y_{ij} = y_i - y_j$, we define the properly scaled cell residual as

$$R_\alpha(U) := \frac{1}{2V_\alpha} \left[(F_1 - F_3)\delta y_{24} + (F_2 - F_4)\delta y_{31} \right.$$
$$\left. - (G_1 - G_3)\delta x_{24} - (G_2 - G_4)\delta x_{31} \right] - S_\alpha, \qquad (6.50)$$

where V_α is the measure of the cell Ω_α, and S_α is the average of any source function over the cell.

With this definition, $R_\alpha(u)$ gives the truncation error of the scheme, where u is the exact solution of the problem. This error consists of two parts, that arising from the quadrature rule on a generally nonuniform mesh and that due to the approximation of the fluxes at the vertices:

$$R_\alpha(u) = T_\alpha^{(1)} + T_\alpha^{(2)},$$

where

$$T_\alpha^{(1)} = \frac{1}{V_\alpha}\left[\tfrac{1}{2}[(f_1 - f_3)\delta y_{24} + \ldots] - \int_{\partial\Omega_\alpha}[f\,dy - g\,dx]\right] \qquad (6.51a)$$

$$T_\alpha^{(2)} = \frac{1}{2V_\alpha}[(\Delta f_1 - \Delta f_3)\delta y_{24} + \ldots], \qquad (6.51b)$$

and where $f_1 \equiv f(u_1, \nabla u_1), \Delta f_1 \equiv f(u_1, \nabla^h u_1) - f(u_1, \nabla u_1)$ etc. Considering $T_\alpha^{(1)}$ first, it is clear that approximating each edge integral by the trapezoidal rule gives an error which is $O(h^3)$ and proportional to a third derivative of the normal flux in the tangential direction; after division by V_α the whole error is $O(h^2)$ only if there is cancellation between the errors on opposite edges. As pointed out by Morton and Paisley (1989) in their comparison of cell vertex and cell centre methods, this cancellation occurs if each quadrilateral is within $O(h)$ of being a parallelogram; and the method therefore shows greater resilience to mesh distortions than the cell centre scheme, where only an $O(h^2)$ distortion is allowed. Following the work of Süli (1992) on linear convection problems we can state this result more precisely as follows. We write the fluxes as $\mathbf{F} \equiv (f, g)$ where each component, treated as a composite function of (x, y), lies in $H^s(\Omega)$, i.e. $\mathbf{F} \in \mathbf{H}^s(\Omega)$; and we denote by$I^h$ the interpolation projector onto the trial space of isoparametric bilinear functions on each quadrilateral and by Π^h the L_2 projector onto the test space of piecewise constants.

Lemma 6.3.1 *Suppose the bounded polygonal region Ω is covered by a structured quadrilateral mesh satisfying the standard regularity assumptions; and that, for every quadrilateral Ω_α, the distance between the midpoints P_α, Q_α of its two diagonals satisfies* $\mathrm{dist}(P_\alpha, Q_\alpha) \leq C'V_\alpha$, *where C' is independent of α and h. Then for all $\mathbf{F} \in \mathbf{H}^s(\Omega)$*

$$\left\|\Pi^h(\nabla\cdot\mathbf{F} - \nabla\cdot I^h\mathbf{F})\right\|_{L_2(\Omega)} \leq Ch^{r-1}|\mathbf{F}|_{r,\Omega}, \quad 1 \leq r \leq \min(s, 3), \quad (6.52)$$

where the constant C is independent of h.

Proof. See Süli (1992), especially Theorem 4; for a simpler proof on a square mesh see Balland and Süli (1994). $\square$

Estimation of the second term $T_\alpha^{(2)}$ depends very much on the character of the error committed at each vertex; but again one can see that if this is a smooth function of the position of the vertex there will be some degree of cancellation which will depend on the mesh properties. Thus, since it is clear that $V_\alpha = \delta x_{13} \delta y_{24} - \delta x_{24} \delta y_{13}$, if we can show that $\Delta f_1 - \Delta f_3 = \delta x_{13}(Ch^p + o(h^p))$ etc., we will find that $T_\alpha^{(2)} = O(h^p)$.

We will return to these estimates in later sections, but to realise the accuracy that they imply we need to face the key difficulty posed by the cell vertex approach: an approximation U must be found that sets the $R_\alpha(U)$ to zero, or close to zero, despite the fact that there is no natural correspondence between these cell equations and the unknowns that are associated with the nodes or vertices. Sometimes such a correspondence can be set up: suppose, for example, a pure convection problem is posed on a region which is topologically equivalent to a rectangle with $n_x \times n_y$ cells and the velocity b is such that inflow Dirichlet boundary conditions are imposed on the left-hand and bottom boundary and none on the other boundaries; then each vertex that corresponds to an unknown parameter U_i can be associated with the cell below it and to the left — its upwind cell. This is a case where one can carry through the analysis and realise the full accuracy implied by the truncation errors — see Süli (1992). However, suppose the flow diverges and inflow boundary conditions are imposed only on one side; or suppose it converges and they are imposed on three sides; or suppose diffusion is added and almost any combination of boundary conditions may be applied. Then the number of unknowns does not always match the number of available equations; we call this the *counting problem*, and in its presence it is clear that the core method has to be elaborated upon.

This is a difficulty quite unlike any we have met before either with finite difference or finite element methods; hitherto, we have always started by devising a difference stencil or test function associated with each nodal unknown. To tackle the problem, there are two natural alternative courses of action: we can combine the residuals R_α for the cells surrounding a node to give a nodal residual N_j; or we can average the nodal values U_j in a cell to give cell averages U_α. In each case there is a further step to be taken with attendant disadvantages: in the first case, this may still not be enough to determine U_j uniquely and the error analysis has been further complicated; and, in the second case, a procedure still has to be devised to calculate the full set of nodal values $\{U_j\}$ from the boundary conditions and the set of cell averages $\{U_\alpha\}$. This latter step is notoriously ill-conditioned, as the chequerboard mode is in the null space of the node-to-cell mapping. Hence we shall pursue the first course, using residual weightings, or *distribution matrices* as they are called in the general nonscalar case, and artificial dissipation in cases of non-uniqueness. A further argument in favour of this course is that there is an obvious link between the piecewise constant test functions of the cell vertex method and the large β limit of the Hemker test functions shown in Fig. 5.3. The link is not clear cut because the limit is only in the L_2 norm and not the H^1 norm; but the scheme of Craig and Perella (1995) described in section 5.8.2 partly bridges the gap.

6.3.5 Recovery of gradients

The post-processing of finite element approximations, for example to obtain stresses from displacements, has long been a standard procedure; and, indeed, the Gauss points in a bilinear element that Zlámal (1978) and Lesaint and Zlámal (1979) showed were points of superconvergence for the gradients are often called stress points. So some of the techniques we shall require are already well known; but what we need here are gradient recovery operators to use in the construction of the discretisation, rather than to apply to its results. Thus we suppose that U has an isoparametric bilinear form on each quadrilateral, and from this we in general need to construct an approximate gradient $\nabla^h U$ at each vertex to substitute into the calculation of the fluxes in (6.50). However, for our linear convection-diffusion problem, our real need is to approximate the integral of $\partial u/\partial n$ along an edge so this might be approached directly.

We shall describe three basic techniques. Each is based on the assumption that the bilinear U is an approximation, to a smooth unknown function u, which is particularly accurate at the nodes. The justification for this viewpoint is that, for pure convection, the cell vertex approximation is exact at the nodes in one dimension, and in two dimensions we have $O(h^2)$ error bounds — see section 6.5 — though it should be noted that these are often in a cell-averaged norm.

The sampling regions used in the three gradient recovery techniques, called methods A, B and C in the terminology of Mackenzie (1991), are shown in Fig. 6.7. In method A, the same technique as for the cell centre scheme is used to integrate $\mathrm{div}(U,0)$ and $\mathrm{div}(0,U)$ over an appropriate region. A direct application of (6.50) to $\mathrm{div}(U,0)$ and $\mathrm{div}(x,0)$ gives

$$\text{A}: \quad \partial_x^h U_1 = \frac{(U_2 - U_6)\delta y_{48} + (U_4 - U_8)\delta y_{62}}{\delta x_{26}\delta y_{48} + \delta x_{48}\delta y_{62}}, \tag{6.53}$$

with a similar formula holding for $\partial_y^h U_1$. From the formulae given in section 4.4, it is easily checked that (6.53) is exactly equivalent to calculating the derivative $\partial_x U_1$ at the vertex in each quadrilateral, and taking a weighted average using the areas of the triangles A_{124} etc. Unfortunately, this technique which we may also call 'divergence-averaging' is only first order accurate on general irregular meshes.

In method B, which is the generally preferred technique, use is made of the superconvergent gradients at the centres of each cell. For Ω_α, this is obtained directly from substituting $\mathrm{div}(U,0)$ and $\mathrm{div}(0,U)$ into (6.50); and by the formulae of section 4.4 it equals ∇U evaluated at the centroid of the

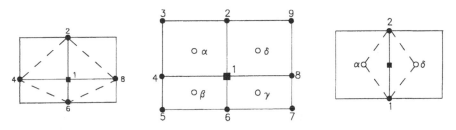

Figure 6.7. *Sampling regions for three gradient recovery techniques; from left to right, methods A, B and C*

cell. A bilinear interpolation between these centroids can then be used to give the value at the vertex. On very curved meshes of high aspect ratios, as might be met in a boundary layer, this might lead to extrapolation rather than interpolation. Thus a simpler weighted average using opposing cells may be preferred, namely

$$\text{B}: \quad \nabla^h U_1 = \frac{V_\gamma \nabla U_\alpha + V_\delta \nabla U_\beta + V_\alpha \nabla U_\gamma + V_\beta \nabla U_\delta}{V_\alpha + V_\beta + V_\gamma + V_\delta}. \tag{6.54}$$

The pairing of opposing cells in this way is exactly equivalent in one dimension to fitting a quadratic between three values and taking its gradient at the middle point. The whole procedure, which we may call 'centroid averaging', clearly fits in very well with the calculation of the cell residuals.

The third method, which can only be used for problems with linear diffusion, gives directly the flux integral along the (r_1, r_2)-edge from cell Ω_α to cell Ω_δ,

$$\text{C}: \quad \int_{r_1}^{r_2} \frac{\partial u}{\partial n} d\Gamma \approx$$
$$\frac{(U_1 - U_2)(\delta y_{21} \delta y_{\delta\alpha} + \delta x_{21} \delta x_{\delta\alpha}) + (U_\delta - U_\alpha)((\delta y_{21})^2 + (\delta x_{21})^2)}{\delta x_{12} \delta y_{\delta\alpha} + \delta x_{\delta\alpha} \delta y_{21}}, \tag{6.55}$$

where $U_\alpha, x_\alpha, y_\alpha$, etc. are cell averages. On a rectangular mesh, where $\delta x_{21} = \delta y_{\delta\alpha} = 0$, this clearly becomes very simple and obvious, but only first order accurate.

To complete the specification of the basic cell vertex methods we need to consider the imposition of boundary conditions, including the recovery of gradients there, and this will lead to consideration of the counting problems. We start with the one-dimensional problem (6.3). On a general mesh, $0 = x_0 < x_1 < \ldots < x_{J-1} < x_J = 1$, with $h_j = x_j - x_{j-1}$, we have residuals defined by integrating (6.3) over a cell and substituting U for u

to get

$$R_{i-\frac{1}{2}} := \frac{1}{h_i} \left[-\epsilon(U_i' - U_{i-1}') + (b_i U_i - b_{i-1} U_{i-1}) - \int_{x_{i-1}}^{x_i} S(x)\mathrm{d}x \right], \quad (6.56)$$

where U_i' is given by the one-dimensional form of (6.53) or (6.54), as (6.55) gives the same as (6.53). This is defined for the interior cells $i = 2, 3, \ldots, J - 1$, and setting each to zero gives $J - 2$ four-point difference equations for the $J - 1$ unknowns $U_1, \ldots, U_{J-1}$ when we set $U_0 = u_L$ and $U_J = u_R$. To define the residual in either end interval requires the gradient to be recovered at the end point. In Mackenzie and Morton (1992) it is argued that only at inflow boundaries should the gradient be extrapolated to the boundary and the resulting residual set to zero; at outflow boundaries with a Dirichlet boundary condition the residual should be ignored for the purpose of solving the equations. Thus under our usual assumption that $b(x) > 0$ everywhere we extrapolate to get

$$U_0' = 2D_- U_1 - U_1'. \quad (6.57)$$

Then setting $R_{\frac{1}{2}} = 0$, as well as the residuals at the interior cells, solves the counting problem. However, one may argue that it is desirable that conservation be applied to all the cells; and, indeed, it can be extended to $R_{J-\frac{1}{2}}$ by using $R_{J-\frac{1}{2}} = 0$ to define U_J'.

In two dimensions, similar principles apply to extrapolating the gradient at other than those outflow points at which a Dirichlet boundary condition is imposed. At a node on a smooth part of the boundary, a tangential derivative is obtained from the differences along the boundary on either side; and extrapolation along the interior edge through the node gives a derivative in a second direction, which can be combined with the tangential derivative to give the gradient vector. At a corner, extrapolation can be carried out in both directions. Simple lower order approximations to these procedures can, however, be obtained by applying method A to just the interior cells that meet at the boundary node.

We defer consideration of other situations until the next section, and in section 6.4 we gather together the principal properties of this scheme. However, in order to highlight the difference of this four-point scheme from the variety of basically three-point schemes that have been our main concern so far, let us consider the standard constant coefficient model problem on a uniform mesh with $u_L = 0, u_R = 1$ and $S = 0$. With $\beta = bh/\epsilon$, the scheme becomes

$$\Delta_-(-\Delta_0 U_i + \beta U_i) = 0, \quad (6.58a)$$

with the corresponding characteristic equation

$$(\lambda - 1)(-\tfrac{1}{2}\lambda^2 + \beta\lambda + \tfrac{1}{2}) = 0, \quad (6.58b)$$

whose roots are 1 and $\mu_\pm = \beta \pm (\beta^2 + 1)^{\frac{1}{2}}$. With the boundary condition (6.57) a short calculation gives the solution to the problem as

$$U_j = \frac{\mu_+^j - 1 + C(\mu_-^j - 1)}{\mu_+^J - 1 + C(\mu_-^J - 1)}, \tag{6.58c}$$

where $C = (\mu_-/\mu_+)[(\mu_+ - 1)/(1 - \mu_-)]^3$. Clearly $\mu_+ > 1$ and μ_+ matches the first three terms in the Taylor expansion of e^β, while μ_- is negative but decaying; as a result, the solution (6.58c) is both second order accurate and is also monotonic increasing for all β, the latter result being proved in section 6.4. The maximum error $\|U - u\|_\infty$ is plotted against h in Fig. 6.8, where its behaviour is compared with that of the upwind and central difference schemes; note that in the latter case the results are for odd numbers of intervals — the errors are worse for even numbers, as discussed in section 6.2.4. The cell vertex scheme closely matches the central difference scheme when $\beta < 1$ and is always better than the upwind scheme when $\beta > 1$.

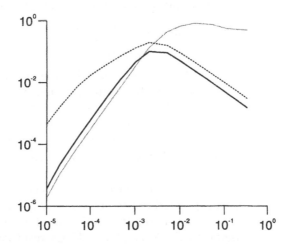

Figure 6.8. *The maximum error, plotted against h in a log-log plot, for the cell vertex (full line), central difference (dotted line) and upwind (dashed line) schemes for the model problem (4.61) with $S = 0$ and $b/\epsilon = 10^3$*

In Fig. 6.9 we also show the results obtained for the problem (4.67), which should be compared with those given in Figs. 4.7 and 5.5. They are as good as any of the very good results obtained with the Petrov–Galerkin methods of Chapter 5.

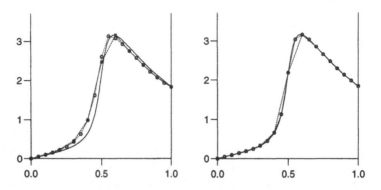

Figure 6.9. *Cell vertex approximations to model problem (4.67); on the left for* $\epsilon = 0.1$, *and on the right for* $\epsilon = 0.001$, *as for Fig.5.5*

6.3.6 Distribution matrices and artificial dissipation

The terminology and techniques we shall use here are largely those used by Crumpton *et al.* (1993) for the compressible Navier-Stokes equations. On a two-dimensional quadrilateral mesh they lead to a set of nodal residuals, defined at each vertex corresponding to an unknown U_i, given by

$$N_i(U) := \frac{\sum_{(\alpha)} V_\alpha (D_{\alpha,i} R_\alpha + A_{\alpha,i})}{\sum_{(\alpha)} V_\alpha}, \qquad (6.59)$$

where the summation is over the cells surrounding the vertex, four in the interior but appropriately modified for boundary nodes — see Fig. 6.6 or the central diagram of Fig.6.7 for a typical layout. To see how these distribution matrices $D_{\alpha,i}$ and artificial dissipation terms $A_{\alpha,i}$ solve the counting problems for the cell vertex method we can start in one dimension with problem (6.3) in which $b(x)$ is allowed to change sign in the interval $(0,1)$.

Suppose first of all that $b(0) > 0$ but $b(1) < 0$, and that both $u(0)$ and $u(1)$ are specified. Then the extrapolated gradient boundary condition (6.57) is applied at both ends of the interval to give J residuals of the form (6.56) to determine the $J-1$ unknowns $U_1, U_2, \ldots, U_{J-1}$. All the residuals should be used in order to maintain overall conservation, and let us suppose that for just one interior node s we have some cell-averaged $b_{s-\frac{1}{2}} > 0$ but $b_{s+\frac{1}{2}} < 0$; then the natural solution is to combine the corresponding residuals and set $R_{s-\frac{1}{2}} + R_{s+\frac{1}{2}} = 0$. Generally, this corresponds to choosing distribution matrices (or residual weights) given by

$$D_{i-\frac{1}{2},i} = 1 + \text{sign } b_{i-\frac{1}{2}}, \quad D_{i+\frac{1}{2},i} = 1 - \text{sign } b_{i+\frac{1}{2}}. \qquad (6.60)$$

The choice of a cell-averaged $b_{i-\frac{1}{2}}$ is based on convenience for the nonlinear situation, where one would use $\Delta_- F_i / \Delta_- U_i$, using just the convective flux;

for the present problem any cell-based value of b will be satisfactory. Note that, in the purely convective case, this algorithm results in $U_1, U_2 \ldots, U_{s-1}$ being computed successively from $R_{\frac{1}{2}} = 0, R_{\frac{3}{2}} = 0, \ldots, R_{s-\frac{3}{2}} = 0$; and $U_{J-1}, U_{J-2} \ldots, U_{s+1}$ are similarly computed successively from the right. Then $R_{s-\frac{1}{2}} + R_{s+\frac{1}{2}} = 0$ is used to determine U_s. Note also that in the general convection-diffusion case, the nodal values cannot be computed successively in this simple way, and instead a symmetric Gauss–Seidel iteration might be used — nonlinear examples will be given in section 6.6. However, in the final equation we do not have to recover U_s' which, as we saw in the asymptotic analysis of Chapter 2, would be poorly defined.

Consider now the complementary situation, where $b_{i-\frac{1}{2}} < 0$ for $i \leq s$ and $b_{i+\frac{1}{2}} > 0$ for $i \geq s$. If Dirichlet conditions specify both $u(0)$ and $u(1)$, the procedure so far outlined is not to define residuals $R_{\frac{1}{2}}$ and $R_{J-\frac{1}{2}}$ in the end intervals, so there are one too few equations; moreover, Dirichlet conditions do not need to be imposed at these outflow boundary points, and if they are not, and the residuals are then used, there is still an equation deficiency. This deficiency clearly cannot be remedied by using distribution matrices to form $J - 1$ linear combinations of the $J - 2$ independent residuals. The resolution of the difficulty is to recognise that the solution is not uniquely determined from the residuals, and to use artificial dissipation to select a sufficiently smooth approximation. This, after all, reflects the ill-posedness of the purely convective differential equation with no boundary conditions; in nonlinear problems, it corresponds to the need to impose an *entropy condition* to define a unique weak solution — see Tadmor (1984) for a discussion of the link with artificial viscosity. Thus, for the problem with Dirichlet data, we set $N_i(U) = 0$ for $i = 1, 2, \ldots, J - 1$ and substitute into the residual definition (6.59) the distribution matrices (6.60) and an artificial dissipation corresponding to the fourth order form (6.45) used for the cell centre scheme.

The most commonly used distribution matrices in two dimensions are those introduced by Ni (1981) and based on matching the update $U_i^{n+1} = U_i^n - \Delta t\, N_i(U)$ to the Lax–Wendroff scheme for solving the unsteady problem. Thus consider the Taylor series for solving $u_t + f_x + g_y = 0$,

$$
\begin{aligned}
u^{n+1} &\approx u^n + \Delta t\, u_t^n + \tfrac{1}{2}(\Delta t)^2 u_{tt}^n \\
&= u^n - \Delta t[(f_x + g_y) + \tfrac{1}{2}\Delta t(f_{xt} + g_{yt})]^n \\
&= u^n - \Delta t[(f_x + g_y) + \tfrac{1}{2}\Delta t((Au_t)_x + (Bu_t)_y)]^n, \quad (6.61)
\end{aligned}
$$

where $A := \partial f/\partial u, B := \partial g/\partial u$, which for a system of equations are Jacobian matrices. The $O(\Delta t)$ term here can be replaced by an average of the cell residuals for cells surrounding a node; it is important that these are volume-weighted (Hall 1985) so that, in effect, $(f_x + g_y)$ is integrated over the whole composite region. The $O((\Delta t)^2)$ terms are also in divergence form and can be approximated by integration around a diamond formed by the diagonals of cells surrounding the node, as in Morton (1988); in Fig. 6.6 this diamond starts with the line $(\mathbf{r}_2, \mathbf{r}_4)$ across Ω_α, and now using the midpoint rule with $-(f_x + g_y)$ replacing u_t, gives a term in the line integral $[A(f_x + g_y)]_\alpha \delta y_{42} - [B(f_x + g_y)]_\alpha \delta x_{42}$. These accumulated line integrals have to be divided by the measure of the diamond, taken to be $\frac{1}{2}(V_\alpha + V_\beta + V_\gamma + V_\delta)$. Taken together all these terms amount to an update from cell Ω_α to node 1 of

$$(U^{n+1} - U^n)_{\alpha,1} = -\Delta t \frac{[V_\alpha R_\alpha - \Delta t(\delta y_{42} A_\alpha - \delta x_{42} B_\alpha) R_\alpha]}{V_\alpha + V_\beta + V_\gamma + V_\delta}. \tag{6.62}$$

Motivated in this way, we can now write down all the distribution matrices, and do so by listing the contributions that R_α makes to the updates at all the vertices of the cell Ω_α, which is how the algorithm is generally programmed. We also distinguish the Δt multiplying the whole update and that in the distribution matrix, make each a local parameter, and introduce the global CFL parameter ν_C. Then we have

$$\left.\begin{aligned}
D_{\alpha,1} &:= I - \nu_C \Delta t_\alpha (\delta y_{42} A_\alpha - \delta x_{42} B_\alpha)/V_\alpha \\
D_{\alpha,2} &:= I - \nu_C \Delta t_\alpha (\delta y_{13} A_\alpha - \delta x_{13} B_\alpha)/V_\alpha \\
D_{\alpha,3} &:= I - \nu_C \Delta t_\alpha (\delta y_{24} A_\alpha - \delta x_{24} B_\alpha)/V_\alpha \\
D_{\alpha,4} &:= I - \nu_C \Delta t_\alpha (\delta y_{31} A_\alpha - \delta x_{31} B_\alpha)/V_\alpha.
\end{aligned}\right\} \tag{6.63a}$$

Note that $D_{\alpha,1} + D_{\alpha,2} + D_{\alpha,3} + D_{\alpha,4} = 4I$, an important property in ensuring that the final solution satisfies a global conservation principle. When used to give an update

$$U_i^{n+1} = U_i^n - \nu_N \Delta t_i N_i(U^n), \tag{6.63b}$$

as in Crumpton et al. (1993), the local time steps Δt_i and Δt_α are defined to be the maximum allowed by local stability criteria; then overall convergence requires that $\nu_N \leq \nu_C$ and $\nu_N \nu_C < 1$. Our main interest here is in the flexibility that this gives to the choice of ν_C in the distribution matrices.

We have derived and given these formulae in their general matrix form. Let us now specialise our considerations to the scalar convection-diffusion problem in the integral form (6.36). Then A_α and B_α are the velocity components $b_\alpha^{(x)}$ and $b_\alpha^{(y)}$, and if we assume the diffusive stability condition of the form $\epsilon \Delta t[(\Delta x)^{-2} + (\Delta y)^{-2}] \leq \frac{1}{2}$ is easily met, we have

$$\Delta t_\alpha = V_\alpha [(b_\alpha^{(x)})^2 \max((\delta y_{42})^2, (\delta y_{13})^2)$$
$$+ (b_\alpha^{(y)})^2 \max((\delta x_{42})^2, (\delta x_{13})^2)]^{-\frac{1}{2}}. \tag{6.64}$$

On a rectangular mesh with $b^{(x)}$ and $b^{(y)}$ of constant sign, (6.63a) therefore reduces to

$$D_{\alpha,j} = 1 \pm \nu_C^{(x)} \pm \nu_C^{(y)}, \tag{6.65}$$

where $\nu_C^{(x)} = \nu_C(b^{(x)}/\Delta x)[(b^{(x)}/\Delta x)^2 + (b^{(y)}/\Delta y)^2]^{-\frac{1}{2}}$, with a similar expression for $\nu_C^{(y)}$. Thus in one dimension one obtains the very effective upwind choice (6.60) by setting $\nu_C = 1$. On the other hand by increasing ν_C one approaches for the purely convection problem the system of equations obtained by minimising the L_2 norm of $(bU)' - S$; note that the option $\nu_C > 1$ would not be allowed by a standard Lax–Wendroff iteration.

In general, particularly for systems of equations, boundary conditions can be incorporated into the set of nodal equations as follows. The nodal residual for a boundary point is assembled as in (6.59) by including dummy cells exterior to the domain, each having the same measure V_α as its interior neighbour but with its residual R_α set to zero. Then any Dirichlet boundary conditions are combined with these nodal residual equations; of course, in the scalar case, this amounts to complete replacement of the nodal residual equation. Note the similarity with the way in which finite element stiffness matrices are assembled.

In two dimensions, no value of ν_C gives a completely upwind scheme for this choice of distribution matrices; for $\nu_C = 1$ for example, flow in the x-direction gives an upwind scheme averaged in the y-direction, and flow in a diagonal direction gives a weight $\frac{1}{4}(1 + \sqrt{2})$ to the upwind cell, weights of $\frac{1}{4}$ on the cells either side of this and a negative weight $\frac{1}{4}(1 - \sqrt{2})$ to the downwind cell. A completely upwind scheme is, however, obtained by replacing (6.65) by

$$D_{\alpha,j} = (1 \pm \nu_C^{(x)})(1 \pm \nu_C^{(y)}) \qquad (6.66)$$

and forcing $\nu_C^{(x)} = \nu_C^{(y)} = 1$, while an attractive scheme involving at most two cells is obtained from (6.66) with $\nu_C^{(x)}$ and $\nu_C^{(y)}$ defined as in (6.65) and ν_C chosen so that $\max(|\nu_C^{(x)}|, |\nu_C^{(y)}|) = 1$. As we shall see in the analysis of section 6.5, the upwind choice makes it much easier to ensure that the system of equations $\{N_i(U) = 0\}$ plus the boundary conditions imply that most of the cell residuals R_α are also set to zero. The addition of fourth order dissipation, as in the one-dimensional example and generalised for the cell centre scheme, is used to resolve the problems of divergent flow when too few boundary conditions are imposed to completely determine the solution. We shall give more details of the appropriate multi-dimensional fourth order dissipation in section 6.3.8.

Solution of the nodal residual equations can normally be achieved by simple Richardson iteration, which with the choice (6.65) corresponds to Lax–Wendroff time-stepping; but this can be very slow, especially if ν_C is chosen greater than one. Moreover, a Fourier analysis readily establishes that Richardson iteration is never convergent for any Δt if the upwind choice possible with (6.66) is made; on a coarse mesh convergence may occur for this choice because of the influence of the boundary conditions, but as the mesh is refined unacceptable error growth occurs. Fortunately, a

Gauss-Seidel successive iteration approach, marching in roughly a down-stream direction is very effective in both cases; in the completely upwind case it can of course give a direct solution in one sweep for the purely convective problem. Illustrations of these techniques, and the use of both second order and fourth order dissipation, will be given in section 6.6 in the context of two simple nonlinear problems.

6.3.7 Vertex-centred schemes

We include in this group of schemes, with the relative positioning of the unknowns and the control volumes illustrated in the bottom right dia-gram of Fig. 6.1, the vertex centroid scheme of Hall (1991), the vertex-based scheme of Dick (1988), the vertex-centred scheme of Crumpton and Shaw (1994) and the upwinded control volume scheme of Moore and Moore (1983). The last scheme, when used with no upwinding, and the Crumpton–Shaw scheme are very close to the Galerkin scheme based on isoparametric bilinear approximations that was described in section 4.4.1; and, when upwinded, the Moores' method corresponds closely to the Petrov–Galerkin methods of section 5.2.1. Hall's scheme is closest to the cell vertex scheme in that, for the pure convection case, it uses the nodal residual (6.59) with the distribution matrices (6.63), taken with $\nu_C = 0$ if Runge–Kutta time-stepping is used. Dick's scheme, on the other hand, is closest to the solution-adaptive difference schemes of section 6.2.3, and forms natural generalisations of them into multi-dimensions.

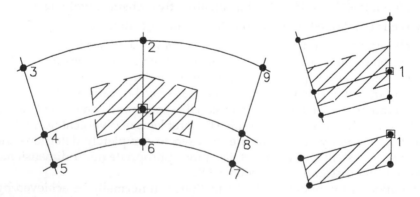

Figure 6.10. *Local mesh labelling for the vertex-centred schemes; and, on the right, typical upwinded control volumes*

Thus the basic vertex-centred scheme is obtained from the integral form (6.36) with the integrals carried out over the boundary and interior of the shaded cell shown in Fig. 6.10, and approximated in various ways. They are evaluated by looping over each element of the primary mesh and distributing the resulting contributions to make up the equations associated with each of the four vertices; that is, the process is similar in character both to the assembly stage of a finite element scheme and the use of distribution matrices in the cell vertex finite volume scheme. In the local primary cell coordinate system, the flux integrals are carried out over the lines from the origin to the mid-sides, and the source function integrals approximated over each quadrant of the element. On a uniform rectangular mesh with constant velocity field b, the integrals are evaluated exactly by trapezoidal rules on half-mesh intervals, and the resultant nine-point scheme takes the form

$$\frac{\Delta y}{\Delta x}(1 + \tfrac{1}{8}\delta_y^2)\delta_x(-\epsilon\delta_x + b^{(x)}\Delta x \mu_x)U$$

$$+ \frac{\Delta x}{\Delta y}(1 + \tfrac{1}{8}\delta_x^2)\delta_y(-\epsilon\delta_y + b^{(y)}\Delta y \mu_y)U = \bar{S}\Delta x \Delta y, \qquad (6.67)$$

where $\bar{S}$ denotes the cell-averages of the source term. Comparing with the corresponding Galerkin finite element scheme which gave (4.48), we see that the only difference is in the mass matrix operators $(1 + \tfrac{1}{8}\delta_y^2)$ and $(1 + \tfrac{1}{8}\delta_w^2)$. If on the other hand the integrals are approximated by using only mid-edge values, these operators are omitted and the scheme becomes virtually identical with the cell centre scheme, except for the positioning of the nodes relative to the boundary; while if, as in the vertex-centroid scheme, the convective fluxes are the averages over those calculated for the cell vertex scheme, these terms would acquire the mass matrix operators $\mu_y^2 \equiv 1 + \tfrac{1}{4}\delta_y^2$ and $\mu_x^2 \equiv 1 + \tfrac{1}{4}\delta_x^2$.

These differences between the various schemes are not very significant, but the associated means by which upwinding is introduced are much more so. In the scheme of Moore and Moore (1983) upwinding is accomplished by shifting the control volume according to the direction and magnitude of b. Various alternatives are discussed by Benson (1991), where the close relationship with Petrov-Galerkin methods based on Hemker test functions is demonstrated. Briefly, except for small values of the mesh Péclet number $|b| h/\epsilon$, each control volume is shifted to one of eight positions: if both components of b relative to the local axes are large, the control volume becomes the upwind primary mesh cell; if one component is small, it moves so as to straddle the upwind edge — see the right-hand diagrams of Fig. 6.10. For the convective terms, these switches have a similar effect to some choices of distribution matrices in the cell vertex method, and are illustrated in Fig. 6.15 where the two choices are compared; however, it should be emphasised that for these schemes the diffusion terms are evaluated in the interior of each element, so the corresponding nine-point central difference stencil in (6.67) is only modified in respect of the mass matrix and convective operators.

Carefully crafted artificial viscosity terms are used in the schemes of

Hall (1991). On the other hand, in the schemes of Dick (1988) which have mainly been used for the Euler equations of inviscid fluid flow, use is made of the approximate Riemann solvers which we shall briefly describe in section 7.3: for our present simple convection problem, this reduces to using upwind differences based on just the three points with coordinate vectors r_1, r_4 and r_6 in Fig. 6.10 if $b(r_1)$ has positive components along the two corresponding edges.

6.3.8 Unstructured triangular meshes

In the last few years there has been increasing attention paid to the use of finite volume methods on the general unstructured triangular meshes used with finite elements, and described in section 4.1. The motivation, as with finite elements, has been the ease with which complicated geometric regions can be discretised, and this argument becomes even stronger in three dimensions; there has also been a rapid increase in the efficiency of methods for generating such meshes, for example around the exterior of a complete aircraft — see Jameson et al. (1986), Weatherill and Hassan (1994) and Marcum and Weatherill (1994). On a primary mesh constructed of triangles one can devise the same variety of finite volume schemes as we have devised above for a quadrilateral mesh. In particular, if a Delauney mesh is constructed it is natural to use the dual Voronoi regions as control volumes in a vertex-centred scheme. However, we shall describe only a cell vertex scheme which directly generalises those described in sections 6.3.4 – 6.3.6.

Integrating the fluxes around the edges of a triangle to obtain the cell residual, one finds a similar collapse of terms to that which led to (6.50): thus for the triangle $\triangle(r_1, r_2, r_3)$ shown in Fig. 6.11, the terms $\frac{1}{2}F_1\delta y_{21}$ and $\frac{1}{2}F_1\delta y_{13}$ combine to give $\frac{1}{2}F_1\delta y_{23}$. To reflect this general phenomenon, we introduce the vector notation $\vec{F}_i$ for the fluxes (f, g) at node r_i and $\vec{dS}_{\alpha,i}$ for the scaled *inward* normal of the side opposite node r_i in the triangle Ω_α, giving $\vec{dS}_{\alpha,1} = (\delta y_{23}, -\delta x_{23})$ in the present case. Hence we have

$$R_\alpha(U) := \frac{1}{2V_\alpha} \sum \vec{F}_i \cdot \vec{dS}_{\alpha,i} - S_\alpha, \qquad (6.68)$$

where the summation is over the vertices of the triangle Ω_α and S_α is the average of the source function. The collapse of terms is due to the identity $\sum \vec{dS}_{\alpha,i} = 0$, and is generally valid for polygonal cells if use is made of subsidiary triangles, and in more dimensions. Thus (6.50) can also be written in the form (6.68) with $\vec{dS}_{\alpha,1} = -\vec{dS}_{\alpha,3} = (\delta y_{24}, -\delta x_{24})$; and (6.68) can also be used in three dimensions with $\vec{F} = (f, g, h)$, appropriate definitions of $\vec{dS}_{\alpha,i}$ and the $\frac{1}{2}$ replaced by $\frac{1}{3}$ for tetrahedra.

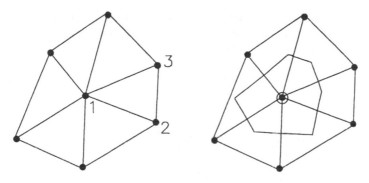

Figure 6.11. *Cell vertex and vertex-centred schemes on a triangular mesh*

Recovery of the gradients at a vertex, or the average of the normal derivative along an edge, can be accomplished by exactly the same means as described in section 6.3.5. The gradient at a centroid is no longer second order accurate, but second order accurate recovery at a vertex is generally possible on a mesh where six triangles meet at each vertex — see Levine (1985) and Zlámal (1994) for details.

There are roughly twice as many triangles as there are vertices in a mesh, so it is quite clear that all the cell residuals cannot be set to zero. The solution is to use distribution matrices to form nodal residuals exactly as in (6.59); and the same range of choice is available, from the identity, through upwinding and generalised Lax–Wendroff, to least squares. Using the same notation as in (6.68), and writing $\vec{A}_\alpha$ for the vector (A_α, B_α) of average Jacobian fluxes over the cell Ω_α, we can rewrite the distribution matrices of (6.63) as

$$D_{\alpha,i} := I + \nu_C \Delta t_\alpha (\vec{A}_\alpha \cdot \vec{dS}_{\alpha,i})/V_\alpha; \qquad (6.69)$$

and this form then covers the case of triangles and tetrahedra if any geometric factors are absorbed in ν_C.

Understanding of what are the best choices of distribution matrices and artificial dissipation terms to substitute into the nodal residual expression (6.59) awaits further analysis. However, we note two points here: linearity-preserving artificial viscosities can be constructed using techniques similar to those used in section 6.3.2; and Fourier analysis of the residual carried out in section 6.5 will show that the chequerboard spurious node is no longer the serious problem on triangular meshes that it is on quadrilateral meshes.

A good example of the forms used for fourth order artificial dissipation is that given by Holmes and Connell (1989), which can equally well be used on quadrilateral or triangular meshes, or in three dimensions. At each node a pseudo-Laplacian is constructed from edge differences in such a way that it is transparent to linear functions, and with weights close to unity; this leads to the following formula

$$\delta_h^2 U_i := \sum_j [1 + \lambda_x(x_j - x_i) + \lambda_y(y_j - y_i)](U_j - U_i) \qquad (6.70a)$$

where

$$\lambda_x = \frac{I_{xy}R_y - I_{yy}R_x}{I_{xx}I_{yy} - I_{xy}^2}, \qquad \lambda_y = \frac{I_{xy}R_x - I_{xx}R_y}{I_{xx}I_{yy} - I_{xy}^2},$$

$$R_x = \sum_j (x_j - x_i), \qquad R_y = \sum_j (y_j - y_i),$$

$$I_{xx} = \sum_j (x_j - x_i)^2, I_{yy} = \sum_j (y_j - y_i)^2, I_{xy} = \sum_j (x_j - x_i)(y_j - y_i),$$

and some 'clipping' of the coefficients in (6.70a) may be necessary. Then the artificial viscosity term $A_{\alpha,i}$ in (6.59), or the corresponding term for a triangular mesh, can be defined as

$$A_{\alpha,i} := \tau_\alpha^{(2)}(U_i - \bar{U}_\alpha) - \tau_\alpha^{(4)}(\delta_h^2 U_i - \overline{\delta_h^2 U_\alpha}), \qquad (6.70b)$$

where the overbar denotes averaging over the nodes of the cell Ω_α, and where $\tau_\alpha^{(2)}$ and $\tau_\alpha^{(4)}$ will usually be given a local scaling as discussed in section 6.3.2.

This form (6.70a) is fairly complicated, and if the mesh is such that edges can be paired to form straight lines, or nearly straight lines, then the simpler divided difference forms as in (6.45) can be used.

6.4 Cell vertex analysis in one dimension

We consider the problem (6.3) on a general mesh and assume $b(x) > 0$. Gathering together all the formulae of (6.53),(6.54),(6.56) and (6.57), we arrive at the following system of $J - 1$ equations in the $J - 1$ interior unknowns $U_1, U_2 \ldots, U_{J-1}$:

$$-\epsilon(U_i' - U_{i-1}') + (b_i U_i - b_{i-1} U_{i-1})$$
$$= \int_{x_{i-1}}^{x_i} S(x)\mathrm{d}x, \quad i = 1, 2, \ldots, J - 1 \qquad (6.71a)$$

where

$$U_0' = 2D_- U_1 - U_1', \qquad (6.71b)$$
$$U_i' = \alpha_i D_+ U_i + (1 - \alpha_i)D_- U_i, \quad i = 1, 2, \ldots, J - 1, \qquad (6.71c)$$

and $\alpha_i := h_{i+1}/(h_i + h_{i+1})$ for method A, $\alpha_i := h_i/(h_i + h_{i+1})$ for method B; on a uniform mesh, both methods give the four-point scheme that was first studied by Gushchin and Shchennikov (1974). We have already given the solution for the homogeneous, constant coefficient and uniform mesh problem in (6.58c) and presented the errors in Fig. 6.8. In this section we use a variety of approaches to deduce properties of the cell vertex method for more general cases and, for example, to explain the very good numerical results shown in Fig. 6.9. The results to be derived in 6.4.1 are based on those in Mackenzie and Morton (1992) and those in 6.4.2 on the analysis in Morton and Stynes (1994).

6.4.1 Monotonicity properties, maximum principles and supraconvergence

Key properties of the Green's function follow from those of the solution to the homogeneous differential equation with $u_L = 0$; in particular, integrating the equation once gives the relation

$$\epsilon u'(x) = \epsilon u'(0) + b(x) \int_0^x u'(s)\mathrm{d}s, \tag{6.72}$$

from which it is clear that u is always monotone. Now we consider the circumstances under which this is true for the cell vertex scheme (6.71).

It is tempting to write the discrete counterpart to (6.72) as

$$\epsilon U_i' = \epsilon U_0' + b_i U_i, \tag{6.73}$$

but to relate U_i to $\{U_j'\}$ it is necessary to invert the mapping between these gradients and the differences $\{D_- U_j\}$ defined by (6.71b,c). Inversion of such an averaging process gives an oscillatory result which is not very helpful for the present purposes. Also, by solving for just U_0' and U_1' we obtain

$$U_0' = (1 - \tfrac{1}{2}\beta_1)D_- U_1, \quad U_1' = (1 + \tfrac{1}{2}\beta_1)D_- U_1, \tag{6.74}$$

where $\beta_1 = b_1 h_1/\epsilon$; so U_0' and U_1' have opposite signs if $\beta_1 > 2$. Although this situation could be avoided by using the lower accuracy boundary condition $U_0' = D_- U_1$, it is preferable to work directly with the differences $\{D_- U_j\}$.

Substitution of (6.71c) into (6.71a) with $S \equiv 0$ leads to the recurrence relation

$$\alpha_i D_- U_{i+1} = [\epsilon^{-1}\Delta_-(b_i U_i) - (1 - \alpha_i - \alpha_{i-1})D_- U_i]$$
$$+ (1 - \alpha_{i-1})D_- U_{i-1} \tag{6.75a}$$

for $i = 2, 3, \ldots, J - 1$, and use of (6.71b) or (6.74) gives

$$\alpha_1 D_- U_2 = (\alpha_1 + \tfrac{1}{2}\beta_1)D_- U_1. \tag{6.75b}$$

Hence we obtain the following result.

Lemma 6.4.1 *Suppose $b(x) \geq 0$ and $b'(x) \geq 0$. Then the approximation to the problem $-\epsilon u'' + (bu)' = 0$ with $u(0) = 0, u(1) = 1$ given by (6.71) is monotone* *increasing if, for method A, we have for $i = 2, 3, \ldots, J - 1$*

$$b_i(h_{i-1} + h_i)(h_i + h_{i+1}) \geq \epsilon(h_{i-1} - h_{i+1}) \tag{6.76a}$$

or, for method B, we have

$$b_i(h_{i-1} + h_i)(h_i + h_{i+1}) \geq \epsilon(h_{i+1} - h_{i-1}). \tag{6.76b}$$

Proof. From (6.75b), D_-U_2 and D_-U_1 have the same sign; and then from the recurrence relation (6.75a) all D_-U_i have the same sign, which by the boundary conditions must be positive, if the first term on the right has the same sign as D_-U_i. Rewriting, we require

$$[b_i h_i - \epsilon(1 - \alpha_i - \alpha_{i-1})](D_-U_i)^2 + (b_i - b_{i-1})U_{i-1}D_-U_i \geq 0.$$

Since $b' \geq 0$ and $U_{i-1} \geq 0$, it is sufficient to have the square-bracketed expression positive, which corresponds to (6.76a,b). $\square$

Corollary 6.4.1a *In the case b is constant and the mesh uniform, methods A and B are identical and the approximation, which is given by (6.58c), is always monotone increasing.*

Corollary 6.4.1b *For method B, if the mesh is nonincreasing into the boundary layer on the right, condition (6.76b) is always satisfied and the approximation monotone increasing.*

To obtain monotone solutions for all mesh Péclet numbers without the assumption $b'(x) \geq 0$, some modification of the boundary condition (6.71b) appears to be necessary. We then have the following result.

Lemma 6.4.2 *For general $b(x) \geq 0$, suppose we ensure $U_0'D_-U_1 \geq 0$ by, for instance, replacing (6.71b) by*

$$U_0' = \begin{cases} 2D_-U_i - U_i' & \text{if } \beta_1 \leq 2 \\ D_-U_i & \text{if } \beta_1 > 2. \end{cases} \tag{6.77}$$

Then the approximations to $-\epsilon u'' + (bu)' = 0, u(0) = 0, u(1) = 1$, are monotone increasing if the following conditions are satisfied for $i = 2, 3, \ldots, J - 1$:

(method A) $b_{i-1}b_i(h_{i-1} + h_i)(h_i + h_{i+1}) \geq$
$$\epsilon[b_{i-1}(h_{i-1} + h_i) - \min(b_{i-1}, b_i)(h_i + h_{i+1})] \tag{6.78a}$$

(method B) $b_{i-1}b_i h_i(h_{i-1} + h_i)(h_i + h_{i+1}) \geq$
$$\epsilon[b_{i-1}h_{i+1}(h_{i-1} + h_i) - \min(b_{i-1}, b_i)h_{i-1}(h_i + h_{i+1})] \tag{6.78b}$$

Proof. The proof proceeds as in Lemma 6.4.1, except that we substitute for U_{i-1} from the relation (6.73) when $b_{i-1} > b_i \geq 0$, to obtain

$$(b_i - b_{i-1})U_{i-1} = \epsilon\left(\frac{b_i - b_{i-1}}{b_{i-1}}\right)[\alpha_{i-1}D_-U_i + (1 - \alpha_{i-1})D_-U_{i-1} - U_0'].$$

Then (6.75a) is replaced by

$$\epsilon \alpha_i D_- U_{i+1} = [b_i h_i - \epsilon(1 - \alpha_i - \alpha_{i-1} b_i / b_{i-1})] D_- U_i$$
$$+ \epsilon(1 - \alpha_{i-1})(b_i / b_{i-1}) D_- U_{i-1} + \epsilon(1 - b_i / b_{i-1}) U_0'.$$

It is therefore sufficient to have the first term in the brackets on the right non-negative and this condition, when combined with that for the case $b_i \geq b_{i-1}$, gives the conditions in (6.78). $\square$

These two lemmas together show that both method A and method B give monotone increasing solutions to the homogeneous problem for all Péclet numbers, apart from when very exceptional meshes are used. Thus suppose $|h_{i+1} - h_i| \leq \alpha h$ for some $\alpha = \alpha(h) \to 0$ as $h \to 0$: then, while the left-hand sides of (6.78a,b) approximately equal $4b^2 h^2$ and $4b^2 h^3$ respectively, the right-hand sides are approximately $2\epsilon b h(2\alpha + |b'| h)$ and $2\epsilon b h^2(2\alpha + |b'| h)$; so the conditions for monotone solutions are satisfied if $(bh/\epsilon) > \alpha + \frac{1}{2}(|b'|/b)h$.

However, the establishment of maximum principles for these schemes is not possible for small mesh Péclet numbers; for example, with constant b on a uniform mesh the scheme reduces for $i > 1$ to

$$\frac{\epsilon}{h}[-\tfrac{1}{2} U_{i-2} - (\beta - \tfrac{1}{2}) U_{i-1} + (\beta + \tfrac{1}{2}) U_i - \tfrac{1}{2} U_{i+1}] = h S_{i-\frac{1}{2}}, \qquad (6.79)$$

so that we need $\beta \geq \frac{1}{2}$ to obtain the desired signs for the coefficients, as in Lemma 3.2.3. Generally the required conditions are as follows.

Lemma 6.4.3 *The schemes given by (6.71) satisfy a maximum principle under the following conditions for $i = 2, 3, \ldots J - 1$:*

$$\text{(method A)} \quad b_{i-1}(h_i + h_{i+1}) \geq \epsilon \qquad\qquad\qquad (6.80a)$$
$$\text{(method B)} \quad b_{i-1} h_{i-1}(h_i + h_{i+1}) \geq \epsilon(h_{i+1} + h_i - h_{i-1}) \quad (6.80b)$$

(assuming the mild condition $h_{i+1} \leq h_i + h_{i-1}$).

Proof. Simplification of the conditions ensuring that the coefficient of U_{i-1} in the scheme is nonpositive gives (6.80a,b). For method A the coefficient of U_i is clearly positive and that for method B is positive if $b_i h_{i+1}(h_{i-1} + h_i) \geq \epsilon(h_{i+1} - h_i - h_{i-1})$, so there is no further condition if the latter term is negative. It is easily checked that for both methods the three-point scheme obtained for $i = 1$ has the desired signs. $\square$

The limited validity of the maximum principle prevents our proving convergence in the maximum norm by this means. Nevertheless we can obtain some error bounds from the truncation error, and in particular a result as $\epsilon \to 0$ on a fixed mesh, which will generalise and complement the error bounds shown in Fig. 6.8. As in (6.51), and denoting by $I^h u$ the linear interpolant of u on the mesh, the truncation error of (6.71) is given by

$$R_{i-\frac{1}{2}}(u) := \frac{1}{h_i}\left[\Delta_-[-\epsilon(I^h u)_i' + b_i u_i] - \int_{x_{i-1}}^{x_i} S(x)\mathrm{d}x\right]$$

$$= \frac{\epsilon}{h_i}\Delta_-[u'(x_i) - (I^h u)_i']. \tag{6.81}$$

Note that, in this one-dimensional case and as specifically identified in the notation of (6.51), the error here is due only to the approximation of the (diffusive) flux at the nodes. Note, too, that the difference form of the error is typical of situations where the global accuracy can be of higher order than that given by the maximum norm of the truncation error — a phenomenon called *supraconvergence* and analysed by Manteuffel and White (1986) and Kreiss *et al.* (1986); we shall make use of this below.

Taylor series expansions readily give for methods A and B the gradient recovery errors

$$[(I^h u)_i' - u'(x_i)]^{(A)} = \tfrac{1}{2}(h_{i+1} - h_i)u''(x_i) + O(h^2) \tag{6.82a}$$

$$[(I^h u)_i' - u'(x_i)]^{(B)} = \tfrac{1}{6}h_i h_{i+1}[u'''(x_i)$$
$$\qquad\qquad + \tfrac{1}{4}(h_{i+1} - h_i)u^{iv}(x_i)] + O(h^4) \tag{6.82b}$$

for $i \geq 1$, and expressions of the same order for $i = 0$. Using the *a priori* bounds of section 2.4 and the techniques of section 3.4, the maximum principle can give various bounds of which the following is typical. It should be noted, however, especially in view of the extra difficulties encountered above when $b'(x) < 0$, that in those earlier chapters the standard equation form was taken as $-\epsilon u'' + bu' + cu = S$, with the assumption $c \geq 0$ to ensure a maximum principle, corresponding in our present conservation form case to assuming $b'(x) \geq 0$.

We consider calculations on a fixed mesh, denoted by M^h, and define

$$\|\tau(M^h, \epsilon)\|_\infty = \max_{1,2,\dots,J-1} \frac{\epsilon}{h_j}|\Delta_-[u'(x_j) - (I^h u)_j']| \tag{6.83}$$

with superscripts denoting method A or B, as in (6.82). Then we have the following.

Theorem 6.4.4 *Suppose the conditions for a maximum principle are satisfied, i.e. (6.80a) or (6.80b), and that $b + b'x \geq b_\phi > 0$. Then, for either method A or method B,*

$$\|U - I^h u\|_\infty \leq \frac{1}{b_\phi}\|\tau(M^h, \epsilon)\|_\infty . \tag{6.84a}$$

Suppose, moreover, that the conditions of Theorem 2.4.3 are satisfied giving a priori bounds on the derivatives of u. Then there is a constant C, independent of ϵ but dependent on M^h, such that

$$\|U - I^h u\|_\infty \leq C\epsilon. \tag{6.84b}$$

Proof. We use the comparison function x, for which the operator L^h of (6.71) gives $L^h x = b + b'x \geq b_\phi$. Thence, as in Lemma 3.4.2, we deduce

$$\left| (U - I^h u)_j \right| \leq (jh/b_\phi) \left\| \tau(M^h, \epsilon) \right\|_\infty$$

from which (6.84a) follows. The bounds for u' given by Theorem 2.4.3 show that the largest value, at x_{J-1}, is bounded independently of ϵ; and, since u is similarly bounded, we deduce from (6.83) that $\|\tau\|_\infty \leq C\epsilon b_\phi$ for some constant C independent of ϵ, which gives (6.84b). $\quad\square$

It is instructive to carry out the above computation in a little more detail for the case of a uniform mesh. Then, as shown by Theorem 2.4.3, the dominant behaviour of the truncation error can be obtained by replacing $u(x_j)$ by $e^{-\beta(J-j)}$, where $\beta = b(1)h/\epsilon$. This gives

$$u'(x_{J-1}) - (I^h u)'_{J-1} = e^{-\beta}h^{-1}[\beta - \tfrac{1}{2}(e^\beta - e^{-\beta})] = e^{-\beta}h^{-1}(\beta - \sinh\beta),$$

with the same expression multiplied by $e^{-\beta}$ at x_{J-2}. Hence we obtain

$$\|\tau\|_\infty = \epsilon h^{-2} e^{-\beta}(1 - e^{-\beta})(\sinh\beta - \beta). \tag{6.85}$$

It is easy to see that the expression in β here increases monotonically from its value of 0.00503 at $\beta = \tfrac{1}{2}$, the smallest value of β for which the maximum principle applies, to the value $\tfrac{1}{2}$ as $\beta \to \infty$. If we write ϵh^{-2} as $b(1)/h\beta$ and keep $b(1)/h$ fixed, then the value of $\|\tau\|_\infty$ rises to a maximum at around $\beta = 3$ before decaying like $1/\beta$; so the behaviour for $\beta > \tfrac{1}{2}$ depicted in Fig. 6.8 is quite typical for the cell vertex method on a uniform mesh.

However, other methods of analysis are needed to cover the case of small β, and hence convergence as $h \to 0$. One such result is given by Mackenzie and Morton (1992) for constant b but a general nonincreasing mesh. Denoting the nodal error $U_i - u_i$ by e_i as usual and the corresponding gradient error $U'_i - u'(x_i)$ by g_i, the combination of (6.71) with (6.81) gives

$$b_i e_i - \epsilon[U'_i - (I^h u)'_i] + \epsilon[u'(x_i) - (I^h u)'_i] = \text{const.}, K \tag{6.86a}$$

$$\text{i.e.} \qquad\qquad b_i e_i - \epsilon g_i = K. \tag{6.86b}$$

Moreover, the mapping in (6.71) giving U'_i in terms of the differences of U can be inverted to give a relation that we write as $D_+ U_i = (M^{-1}U')_i$ for $i = 0, 1, ..., J - 1$, and which we can also apply to e_i. Hence we have from (6.86a)

$$(M^{-1}be)_i - \epsilon D_+ e_i = K + \epsilon(M^{-1}T)_i, \tag{6.87}$$

where expressions for the gradient recovery errors $T_i = (I^h u)'_i - u'(x_i)$ are given in (6.82). These show that on a general mesh the truncation errors $h_i^{-1}\Delta_- T_i$ that are used in the maximum principle estimate of Theorem 6.4.4 are only $O(h)$, while the T_i themselves are $O(h^2)$. One advantage of the equations (6.87) is that the $(M^{-1}T)_i$ are also $O(h^2)$; and another is that bounds on $|e_i|$ can be obtained by estimating the discrete Green's function corresponding to the system, for any value of ϵ. Thus we obtain the following result.

Theorem 6.4.5 *Suppose that $u \in C^4(0,1)$ is the solution to problem (6.3) with constant b, and this is approximated by (6.71) with method B. Then if the mesh is nonincreasing, $h_{i+1} \geq h_i$, we have*

$$\|U - I^h u\|_\infty \leq 2 \|M^{-1} T\|_\infty = O(h^2).$$ (6.88)

Proof. See Mackenzie and Morton (1992). □

The supraconvergence implied by this result is confirmed by numerical experiments on random meshes and meshes refined in the boundary layer that are reported in the same paper — see also García-Archilla and Mackenzie (1991). These experiments also show method A to be significantly less accurate than method B, and to have a lower convergence rate; but it also exhibits supraconvergence since the truncation error (6.83) in that case is only $O(h)$ even when the ratio between successive mesh lengths is $1 + O(h^2)$.

6.4.2 Error estimates in discrete energy norms

We begin with method A and note that we can use the discrete equation (6.71a) in the last cell to define U'_J. Then (6.86b) could be extended and written as

$$\Delta_-(b_i e_i) = \epsilon \Delta_- g_i \quad i = 1, 2, \ldots, J.$$ (6.89)

Also, in the notation of section 1.3, we can write $U'_i = D_0 U_i$ so we have

$$g_i = D_0 e_i + T_i \quad i = 1, 2, \ldots, J - 1,$$ (6.90)

with the boundary condition (6.70b) on the left defining g_0 and the use of the last cell equation giving g_J. Substituting (6.90) into (6.89) offers a variety of opportunities to bound $|e_i|$ in terms of some norm of the gradient recovery errors T_i. Well-known discrete summation by parts formulae will often be helpful in the manipulations. The key identities which we will list here for future reference are as follows:

$$\sum_{j=r}^{s} U_j \Delta_- V_j + \sum_{j=r}^{s} V_j \Delta_+ U_j = U_{s+1} V_s - U_r V_{r-1}$$ (6.91a)

$$\sum_{j=r}^{s} U_j \Delta_0 V_j + \sum_{j=r}^{s} V_j \Delta_0 U_j = \tfrac{1}{2}(U_{s+1} V_s + U_s V_{s+1} - U_r V_{r-1} - U_{r-1} V_r)$$

(6.91b)

$$\sum_{j=r}^{s} U_j \delta V_j + \sum_{j=r}^{s} V_{j+\frac{1}{2}} \delta U_{j+\frac{1}{2}} = U_{s+1} V_{s+\frac{1}{2}} - U_r V_{r-\frac{1}{2}}$$ (6.91c)

$$\sum_{j=r}^{s} U_j \mu V_j + (U_{s+1} V_{s+\frac{1}{2}} - U_r V_{r-\frac{1}{2}}) = \sum_{j=r}^{s} V_{j+\frac{1}{2}} \mu U_{j+\frac{1}{2}}. \tag{6.91d}$$

They are all easily verified by direct computation.

For simplicity here we will assume b is constant but retain the general mesh; we will also replace the boundary condition (6.71b) by $D_+U_0 = 0$, which is consistent with the accuracy of method A on a general mesh, and we will not use the last cell to extend (6.89) to J. Now suppose we multiply each side of (6.89) by $\mu e_{i-\frac{1}{2}} \equiv \frac{1}{2}(e_i + e_{i-1})$ and sum from 1 to $J-1$; the sum on the left collapses to $\frac{1}{2}b(e_{J-1}^2 - e_0^2)$ and we apply (6.91a) to the sum on the right to obtain

$$\frac{1}{2}be_{J-1}^2 = \epsilon \sum_{1}^{J-1} (\Delta_- g_i) \mu e_{i-\frac{1}{2}} = \frac{1}{2}\epsilon(e_{J-1}g_{J-1} - e_1 g_0) - \epsilon \sum_{1}^{J-1} g_i \Delta_+ \mu e_{i-\frac{1}{2}}.$$

Since $\Delta_+ \mu e_{i-\frac{1}{2}} = \Delta_0 e_i$, this relation can be written

$$\frac{1}{2}be_{J-1}^2 + \epsilon \left[\frac{1}{2}g_0 \Delta_+ e_0 + \sum_{1}^{J-1} g_i \Delta_0 e_i + \frac{1}{2}g_{J-1}\Delta_- e_J \right] = 0. \tag{6.92}$$

Substitution for g_i from (6.90), together with $g_0 = D_+ e_0$, suggests defining the inner product

$$\langle U, V \rangle_h := \frac{1}{2}h_1 U_0 V_0 + \sum_{1}^{J-1} \frac{1}{2}(h_i + h_{i+1})U_i V_i + \frac{1}{2}h_J U_J V_J \tag{6.93}$$

with the norm $\|U\|_h^2 \equiv \langle U, U \rangle_h$. Then if we define

$$De := (D_+ e_0, D_0 e_1, D_0 e_2, \ldots, D_0 e_{J-1}, D_- e_J), \tag{6.94}$$

and also formally set $g_J = g_{J-1}$, we can write (6.92) in the compact form

$$\frac{1}{2}be_{J-1}^2 + \epsilon \langle De, g \rangle_h = 0. \tag{6.95}$$

This leads to the following error bound.

Theorem 6.4.6 *Suppose method A with boundary condition $D_+U_0 = 0$ is applied to the problem of (6.3) with b constant. Then if*

$$b(h_{J-1} + h_J) > \frac{1}{4}\epsilon, \tag{6.96}$$

we have

$$|U_i - u(x_i)|^2 \le (2/\gamma^2) \min(x_i, 1 - x_i) \|T\|_h^2, \tag{6.97}$$

where $T_J \equiv T_{J-1}$,

$$\gamma = \max_{\alpha^2} \left[\min \left(1 - \frac{h_J \alpha^2}{2(h_{J-1} + h_J)}, \beta_J - \frac{1}{2\alpha^2} \right) \right], \text{ and } \beta_J = bh_J/\epsilon.$$

Proof. On substituting for g in (6.95) from (6.90) only the terms in e_{J-1}^2 prevent our obtaining the simple relation $\|De\|_h^2 + \langle De, T \rangle_h = 0$, from which an unconditional error bound would follow. Let us therefore collect these terms, to get

$$\tfrac{1}{2}be_{J-1}^2 + \tfrac{1}{2}\epsilon(D_0 e_{J-1} + T_{J-1})(2e_J - e_{J-2} - e_{J-1})$$
$$= \epsilon T_{J-1}[\tfrac{1}{2}(h_{J-1} + h_J)D_0 e_{J-1} + \tfrac{1}{2}h_J D_- e_J]$$
$$+ \tfrac{1}{2}b(\Delta_- e_J)^2 + \epsilon D_0 e_{J-1}(\Delta_0 e_{J-1} + \Delta_- e_J). \qquad (6.98)$$

This is all in the required form except for the cross-product in the last term for which we have, for any $\alpha > 0$,

$$(D_0 e_{J-1})(\Delta_- e_J) \le \frac{\alpha^2}{2} \frac{h_J}{h_{J-1} + h_J} \tfrac{1}{2}(h_{J-1} + h_J)(D_0 e_{J-1})^2$$
$$+ \frac{1}{2\alpha^2} \tfrac{1}{2} h_J (D_- e_J)^2;$$

and hence the contributions from (6.98) to $\epsilon \|De\|_h^2$ are bounded below by

$$\epsilon \left[\left(1 - \frac{h_J \alpha^2}{2(h_{J-1} + h_J)}\right) \tfrac{1}{2}(h_{J-1} + h_J)(D_0 e_{J-1})^2 \right.$$
$$\left. + \left(\beta_J - \frac{1}{2\alpha^2}\right) \tfrac{1}{2} h_J (D_- e_J)^2 \right].$$

Taking the minimum of these coefficients and maximising the result over α^2 gives the γ in the statement of the theorem, and clearly $0 < \gamma < 1$ if (6.96) is satisfied. Hence we obtain the desired result, when (6.98) is combined with the rest of the terms in (6.95), namely

$$0 \ge \gamma \|De\|_h^2 + \langle DE, T \rangle_h, \qquad (6.99)$$

from which it follows that $\|De\|_h \le (1/\gamma) \|T\|_h$.

Finally, for i even we have

$$|e_i|^2 = |(e_2 - e_0) + \cdots + (e_i - e_{i-2})|^2$$
$$\le \left[\frac{(2\Delta_0 e_1)^2}{h_1 + h_2} + \cdots + \frac{(2\Delta_0 e_{i-1})^2}{h_{i-1} + h_i}\right][(h_1 + h_2) + \cdots + (h_{i-1} + h_i)]$$
$$\le 2x_i \|De\|_h^2.$$

When i is odd the same bound is obtained by starting with $(e_1 - e_0)$; and similar bounds in terms of $(1 - x_i)$ can be obtained by starting from the right-hand end. This then gives the final result (6.97). $\square$

Remark The arguments used above are close to those used by Mackenzie and Morton (1992) and, although they made use of the last cell equation and the above analysis did not, the results as regards the right-hand end of the interval are exactly the same; by imposing the boundary condition

(6.71b) on the left, however, in general their argument requires a smaller value of γ.

Turning now to method B, with variable b and on a general mesh, we introduce notation to enable us to treat the method as a nonconforming Petrov-Galerkin scheme. The trial space is the usual continuous piecewise linear space S_E^h; but the test space T^h is composed of the piecewise constant functions on the intervals (x_{j-1}, x_j) for $j = 1, 2, \ldots, J-1$. Then we define $B_i^h(U)$ as the left-hand side of (6.71a), with the gradient definitions of (6.71b,c) and the method B choice of α_i,

$$B_i^h(U) := -\epsilon(U_i' - U_{i-1}') + (b_i U_i - b_{i-1} U_{i-1}), \qquad (6.100a)$$

from which we define for $V \in S_E^h, W \in T^h$ the bilinear form

$$B^h(V, W) := \sum_{j=1}^{J-1} W_j B_j^h(V). \qquad (6.100b)$$

We can also define $B^h(\cdot, \cdot)$ on $C^1[0, 1] \cap T^h$ as

$$B^h(v, W) = \sum_{j=1}^{J-1} W_j[-\epsilon(v'(x_j) - v'(x_{j-1})) - (b_i v_i - b_{i-1} v_{i-1})]. \quad (6.101)$$

These definitions are consistent with $B(v, w)$ in (5.3) as well as with each other since $C^1[0, 1] \cap S_E^h$ consists of the linear function $u_L(1 - x) + u_R x$ for which they give the same result. Hence we have both $B^h(U, V) = (S, V)$ and $B^h(u, V) = (S, V)$ for $V \in T^h$, and hence our familiar error projection property

$$B^h(u - U, W) = 0 \quad \forall W \in T^h. \qquad (6.102)$$

Now the idea of the error analysis in Morton and Stynes (1994) is to construct a mapping $M^h : S_0^h \to T^h$ such that $B^h(V, M^h V)$ is coercive with respect to some discrete norm. Since we have, from (6.102),

$$B^h(U - I^h u, M^h(U - I^h u)) = B^h(u - I^h u, M^h(U - I^h u)), \quad (6.103)$$

applying the coercivity result on the left and a Cauchy–Schwarz inequality on the right eventually gives an error bound in terms of the interpolation error $u - I^h u$.

The mapping used in the above paper was the upwind mapping

$$(M^h V)(x) = V_j \text{ on } (x_{j-1}, x_j), \quad j = 1, 2, \ldots, J-1, \qquad (6.104)$$

for which the coercivity result is as follows.

Lemma 6.4.7 Suppose $b(x) \geq b_\phi > 0$ and $b'(x) \geq b_\psi \geq 0$ for $x \in (0, 1)$; and suppose also that the mesh is nonincreasing, $h_{j+1} \leq h_j \, \forall j$, with $h_2 \geq \frac{1}{4} h_1$. Then

(6.100), (6.104) satisfy

$$B^h(V, M^h V) \geq \tfrac{1}{2} \sum_{j=1}^{J} [b_{j-1}(\Delta_- V_j)^2 + (\Delta_- b_j)V_j^2] \tag{6.105a}$$

$$\geq \tfrac{1}{2} b_\phi \sum_{j=1}^{J} (\Delta_- V_j)^2 + \tfrac{1}{2} b_\psi \sum_{j=1}^{J} h_j V_j^2 \quad \forall V \in S_0^h. \tag{6.105b}$$

Proof. We have from (6.100) and applying summation by parts formula (6.91a),

$$B^h(V, M^h V) = \sum_{j=1}^{J-1} V_j \Delta_- (-\epsilon V_j' + b_j V_j)$$

$$= \epsilon V_1 V_0' - \sum_{j=1}^{J-1} (-\epsilon V_j' + b_j V_j)\Delta_+ V_j$$

$$= \epsilon \sum_{j=0}^{J-1} h_{j+1} V_j' D_+ V_j - \sum_{j=1}^{J-1} b_j V_j \Delta_+ V_j, \tag{6.106}$$

because $V_0 = V_J = 0$. After substitution for V_j' we show that the first sum is non-negative; in particular we have $V_0' = [(2h_1 + h_2)D_- V_1 - h_1 D_+ V_1]/(h_1 + h_2)$, so that the sum can be written

$$\frac{h_1}{h_1 + h_2}(D_- V_1)[2h_1 + h_2)D_- V_1 - h_1 D_- V_2]$$

$$+ \frac{h_2}{h_1 + h_2}(D_- V_2)(h_2 D_- V_1 + h_1 D_- V_2) + \cdots$$

$$= \sum_{j=1}^{J} c_j (D_- V_j)^2 + \sum_{j=1}^{J-1} d_j (D_- V_j)(D_- V_{j+1}) =: S_1,$$

where

$$c_1 = \frac{h_1(2h_1 + h_2)}{h_1 + h_2}, \quad c_j = \frac{h_{j-1}h_j}{h_{j-1} + h_j}, \quad \text{for } j = 2, \ldots, J$$

$$d_1 = h_2 - h_1, \quad d_j = \frac{h_{j+1}^2}{h_j + h_{j+1}}, \quad \text{for } j = 2, \ldots, J-1.$$

Because of the mesh grading, for $j = 2, \ldots, J-1$ we have

$$d_j \leq \tfrac{1}{2} h_{j+1} \leq c_{j+1} \quad \text{and} \quad d_j \leq \tfrac{1}{2} h_j \leq c_j. \tag{6.107}$$

Consequently, and by applying a Cauchy–Schwarz inequality to the products in S_1, we have

$$S_1 \geq c_1 (D_- V_1)^2 + d_1 (D_- V_1)(D_- V_2) + (c_2 - \tfrac{1}{2}d_2)(D_- V_2)^2$$

$$+ \sum_{j=3}^{J-1}(c_j - \tfrac{1}{2}d_j - \tfrac{1}{2}d_{j+1})(D_-V_j)^2 + (c_J - \tfrac{1}{2}d_{J-1})(D_-V_J)^2$$

$$\geq c_1(D_-V_1)^2 + d_1(D_-V_1)(D_-V_2) + \tfrac{1}{2}c_2(D_-V_2)^2.$$

Substituting for the coefficients, this quadratic is non-negative if

$$(h_2 - h_1)^2 \leq 2\frac{h_1^2 h_2(2h_1 + h_2)}{(h_1 + h_2)^2}$$

i.e. $\qquad s^4 - 4s^2 - 4s + 1 \leq 0,$

where $s = h_2/h_1$; it is easily checked that this is satisfied for $1 \geq s \geq 0.207434\ldots$. Hence $S_1 \geq 0$ for $h_2 \geq \tfrac{1}{4}h_1$.

The convective sums can be extended to $j = J$; so from (6.106) and the identity

$$-b_j V_j \Delta_+ V_j = b_j(\Delta_+ V_j)^2 - V_{j+1}[\Delta_+(b_j V_j) - V_{j+1}\Delta_- b_{j+1}],$$

we have, since $V_0 = V_J = 0$,

$$\sum_{j=1}^{J-1} V_j \Delta_-(b_j V_j) = \tfrac{1}{2} \sum_{j=1}^{J}[b_{j-1}(\Delta_- V_j)^2 + (\Delta_- b_j)V_j^2]$$

$$\geq \tfrac{1}{2}b_\phi \sum_{j=1}^{J}(\Delta_- V_j)^2 + \tfrac{1}{2}b_\psi \sum_{j=1}^{J} h_j V_j^2$$

which establishes (6.105). $\square$

Suppose we define the (problem-dependent) discrete energy norm from (6.105a) as

$$\|V\|_h^2 := \sum_{j=1}^{J}[b_{j-1}(\Delta_- V_j)^2 + (\Delta_- b_j)V_j^2], \quad \forall V \in S_0^h. \qquad (6.108)$$

Then we have the following error bound.

Theorem 6.4.8 *Suppose the mesh is nonincreasing and $h_2 \geq \tfrac{1}{4}h_1$; and suppose that $b(x) \geq b_\phi > 0, b'(x) \geq b_\psi \geq 0$. Then for method B of (6.71) we have*

$$\|U - I^h u\|_h^2 \leq (4\epsilon^2/b_\phi) \sum_{j=0}^{J-1} |(I^h u)_j' - u'(x_j)|^2 \equiv \frac{4\epsilon^2}{b_\phi}\|T\|_{l_2}^2, \qquad (6.109)$$

giving the nodal errors in terms of gradient recovery errors.

Proof. Applying Lemma 6.4.7 to (6.103), where M^h is given by (6.104), gives

$$\|U - I^h u\|_h^2 \leq 2B^h(u - I^h u, M^h(U - I^h u))$$

$$= 2\epsilon \sum_{j=1}^{J-1} (U_j - u_j) \Delta_- [(I^h u)'_j - u'(x_j)]$$

$$= -2\epsilon \sum_{j=0}^{J-1} [(I^h u)'_j - u'(x_j)] \Delta_+ (U_j - u_j)$$

$$\leq (2\epsilon/b_\phi^{\frac{1}{2}}) \|U - I^h u\|_h \left[\sum_{j=0}^{J-1} |(I^h u')_j - u'(x_j)|^2 \right]^{\frac{1}{2}},$$

where we have summed by parts and applied a Cauchy–Schwarz inequality. □

Two corollaries make these bounds more explicit, firstly in the case that the boundary layer is well resolved and, secondly, when at the other extreme it is completely missed.

Corollary 6.4.8a *Suppose $b \in C^2[0,1]$ and $S \in C^2[0,1]$, and assume that*

$$h_k \leq \epsilon \ \text{where} \ k = \max\{j : x_j \leq 1 - (3\epsilon/b_\phi)\ln(1/\epsilon)\}.$$

Then

$$\|U - I^h u\|_h^2 \leq Ch^3\epsilon^2 + C'(h_k/\epsilon)^3, \tag{6.110}$$

where C and C' are independent of ϵ and h.

Proof. We use the *a priori* bounds of Theorem 2.4.1 and substitute into the Taylor expansion for $T_j = (I^h u)'_j - u'(x_j)$, as in (6.82b) but with a $u'''(\eta)$ remainder term where $\eta < x_{j+1}$. Then we see first, with the above definition of k, that $|T_j| \leq Ch_j h_{j+1}$ for $j < k$ because $|u'''(\eta)|$ is bounded. More generally,

$$\sum_{j=0}^{J-1} |T_j|^2 \leq C \left[h^3 \sum_{j=0}^{k-1} h_{j+1} + \epsilon^{-6} \sum_{j=k}^{J-1} h_j^2 h_{j+1}^2 e^{-2b_\phi(1-x_{j+1})/\epsilon} \right]$$

$$\leq Ch^3 + C'h_k^3\epsilon^{-6} \sum_{j=k}^{J-1} h_{j+1} e^{-2b_\phi(1-x_j)/\epsilon},$$

because $x_{j+1} - x_j = h_{j+1} \leq \epsilon$ for $j \geq k$. But then the last sum is a Riemann sum for the corresponding integral, with the integrand evaluated on the left of each interval where it is smallest. Hence the sum is $O(\epsilon)$ and we have

$$\sum_{j=0}^{J-1} |T_j|^2 \leq Ch^3 + C'h_k^3\epsilon^{-5},$$

from which (6.109) gives the required result.   □

Corollary 6.4.8b *Suppose* $b \in C^2[0,1]$ *and* $S \in C^2[0,1]$, *and assume that*

$$h_J \geq (\epsilon/b_\phi)\ln(1/\epsilon).$$

Then

$$\|U - I^h u\|_h^2 \leq C(\epsilon/h_J)^2, \tag{6.111}$$

where C *is independent of* ϵ *and* h.

Proof. From the *a priori* bounds on u and u' of Theorem 2.4.1 and the hypothesis on h_J,

$$|u'(x_j)| \leq C \quad \text{and} \quad |(I^h u)'_j| \leq C/h_{j+1} \quad \forall j < J.$$

Define k as in Corollary 6.4.8a and note that

$$x_{J-3} \leq 1 - 3h_J \leq 1 - 3(\epsilon/b_\phi)\ln(1/\epsilon)$$

so that $k \geq J - 3$. Hence

$$\sum_{j=0}^{J-1} |T_j|^2 \leq Ch^3 + \sum_{j=k}^{J-1} |(I^h u)'_j - u'(x_j)|^2$$

$$\leq Ch^3 + C' \sum_{j=k}^{J-1} (1 + h_{j+1}^{-2}) \leq Ch_J^{-2}$$

from which (6.109) gives the required result. □

To relate the energy norm used in this theorem and its corollaries to standard discrete norms, we can use the following discrete Poincaré–Friedrichs inequalities.

Lemma 6.4.9 *For every* $V \in S_0^h$, *and for any mesh, we have*

$$\sum_{j=1}^{J-1} V_j^2 \leq \tfrac{1}{8} J^2 \sum_{j=1}^{J} (\Delta_- V_j)^2 \tag{6.112a}$$

and

$$\sum_{j=1}^{J-1} h_j V_j^2 \leq \tfrac{1}{4} \sum_{j=1}^{J} h_j (D_- V_j)^2. \tag{6.112b}$$

Proof. When J is odd, $J = 2m + 1$, we have

$$\sum_{j=1}^{m} V_j^2 = \sum_{j=1}^{m} \left(\sum_{i=1}^{j} \Delta_- V_i \right)^2 \leq \left[\sum_{i=1}^{m} (\Delta_- V_i)^2 \right] \sum_{j=1}^{m} j$$

$$= \tfrac{1}{2} m(m+1) \sum_{i=1}^{m} (\Delta_- V_i)^2.$$

Similarly,

$$\sum_{j=m+1}^{J-1} V_j^2 \leq \tfrac{1}{2}m(m+1) \sum_{i=m+1}^{J-1} (\varDelta_+ V_i)^2.$$

Adding the two results with $m(m+1) = \tfrac{1}{4}(J^2 - 1)$ gives (6.112a). When J is even, $J = 2m$, we obtain in the same way

$$\sum_{j=1}^{m-1} V_j^2 + \tfrac{1}{2}V_m^2 \leq \tfrac{1}{2}m^2 \sum_{j=1}^{m}(\varDelta_- V_j)^2,$$

with a similar bound from the right; again we obtain (6.112a).

For (6.112b), we choose m such that $x_m \leq \tfrac{1}{2}, x_{m+1} > \tfrac{1}{2}$ and obtain

$$\sum_{j=1}^{m} h_j V_j^2 = \sum_{j=1}^{m} h_j \left(\sum_{i=1}^{j} h_i D_- V_i \right)^2 \leq \left[\sum_{i=1}^{m} h_i (D_- V_i)^2 \right] \sum_{j=1}^{m} h_j \left(\sum_{i=1}^{j} h_i \right)$$

$$\leq \tfrac{1}{4} \sum_{i=1}^{m} h_i (D_- V_i)^2.$$

Adding this to a similar bound obtained at the other end of the interval gives (6.112b). □

One might anticipate that the application of these inequalities to the energy norm might allow some relaxation in the conditions on b that are hypothesised in Theorem 6.4.8, by enabling the two sums in (6.108) to be combined. However, neither is very helpful in this respect: if $\varDelta_- b_j \geq -\delta$ for some $\delta > 0$, then the use of (6.112a) gives

$$\|V\|_h^2 \geq [b_\phi - \tfrac{1}{8}J^2\delta] \sum_{j=1}^{J-1}(\varDelta_- V_j)^2$$

which allows only very small values of δ; similarly, if we allow b_ψ to be negative and use (6.112b) we obtain

$$\|V\|_h^2 \geq \left[\min_j (b_{j-1}h_j) - \tfrac{1}{4}|b_\psi| \right] \sum_{j=1}^{J-1} h_j(D_- V_j)^2,$$

which again allows only very small negative values of b_ψ.

We can summarise the results of this section by referring to the identity (6.103): for, not only are the bounds of Theorem 6.4.8 obtained with the particular upwind mapping given by (6.104), but also those of Theorem 6.4.5 can be cast in a similar form, with the mapping corresponding to replacing V_j in (6.104) by $\tfrac{1}{2}(V_j + V_{j-1})$. This naturally poses the question that we might parametrise possible mappings and optimise the bounds obtained thereby. In the next section, more general mappings will be chosen for some two-dimensional problems which will take this programme a little further.

6.5 Cell vertex analysis in two dimensions

The one-dimensional analysis given by Mackenzie and Morton (1992) was prompted by the very good results they obtained for the two-dimensional IAHR/CEGB problems described in section 1.2. These results are shown here in Figs. 6.12 and 6.13, and should be compared with those obtained with various Petrov–Galerkin schemes and shown in Figs. 5.6 and 5.7 in section 5.10. The profiles for the first problem, shown in Fig. 6.12, are remarkable for their monotonicity and accuracy for all mesh Péclet numbers, the largest discrepancy occurring at the smallest value $\beta = 1$; there are some small interior oscillations, however, which are due to the chequerboard mode and could be removed with a small amount of fourth order dissipation. The results for the second, more demanding, problem shown in Fig. 6.13 are even better: the developing boundary layer is everywhere monotonic, both where it can be properly represented on the mesh and where the mesh is too coarse to show it at all; and the more detailed

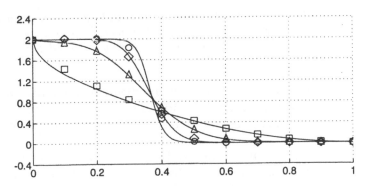

Figure 6.12. *Profiles for the first IAHR/CEGB problem obtained with the cell vertex method, as for Fig. 5.6*

profiles shown in the lower part of the figure exemplify the high nodal accuracy which is typical of the method.

It would be gratifying to demonstrate these features in an analysis in two dimensions, but so far this has not been fully accomplished. This is the reason for the detailed analysis of these properties for one-dimensional problems presented in the last section, although one cannot of course simulate tangential boundary layers in one dimension, just as there is then no phenomenon of crosswind diffusion. There are two principal difficulties in generalising the analysis to two dimensions, and they coincide with the practical difficulties met with the method. First, there is the presence of

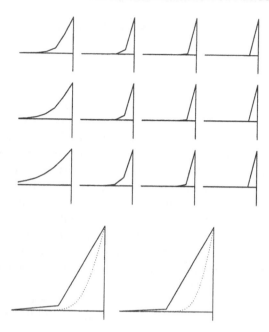

Figure 6.13. *Boundary layer profiles for the second IAHR/CEGB problem obtained with the cell vertex method at the top, as for Fig. 5.7; at the bottom, the profiles compare a fine grid solution with that on the standard grid for $y = 0.0$ and $y = 0.5$ with $\epsilon = 2 \times 10^{-3}$*

undamped spurious modes, particularly the chequerboard mode which is essentially two-dimensional; and, secondly, there is much greater difficulty in matching the number of cell residual equations to the number of nodal unknowns.

In this section we will therefore first discuss these difficulties and what analysis there is to show how they should be overcome. Then we shall present, in section 6.5.4, a detailed analysis for a model problem which is quite close to the IAHR/CEGB problems and where these difficulties have either been eliminated or overcome. All of the problems we shall consider here are of the form

$$\nabla \cdot (-\epsilon \nabla u + \mathbf{b} u) = S \quad \text{on } \Omega$$

$$u = u_B \text{ on } \partial \Omega_D, \quad \mathbf{n} \cdot \nabla u = 0 \quad \text{on } \partial \Omega_N \qquad (6.113)$$

in our standard notation, with Ω a bounded region in $\mathbb{R}^2$.

6.5.1 Distribution matrices and resolution of the counting problem

For an algorithm to be effective for all mesh Péclet numbers, or for its analysis to be similarly valid, it must be able to deal with a pure convection problem. So we will also consider problems of the form

$$\nabla \cdot (\mathbf{b}u) = S \quad \text{on } \Omega, \quad u = u_B \text{ on } \partial\Omega_D, \tag{6.114}$$

where $\partial\Omega_D \supset \partial\Omega_-$, and $\partial\Omega_-$ is the inflow boundary. In Fig. 6.14 we sketch three typical configurations of flow fields on a square mesh which must be considered. If U is given at all mesh points in the closure of $\partial\Omega_D \equiv \partial\Omega_-$ in each case, we wish to determine its value at all other mesh points by setting

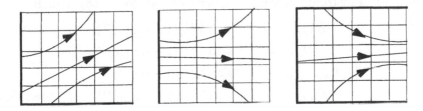

Figure 6.14. *Three typical convection problems; the thickened boundary denotes $\partial\Omega_D$*

to zero some set of cell residuals. Suppose we have a $J \times J$ mesh. In the first case, in which $b^{(x)} > 0$ and $b^{(y)} > 0$, of the $(J + 1)^2$ nodal values of U the boundary conditions determine $2J + 1$, leaving exactly J^2 to be determined by the residuals on the J^2 cells; this is the simple, straightforward case. In the second case, of diverging flow, there are only $J + 1$ boundary conditions so the cell residuals under-determine the unknowns; and in the third case, of converging flow, there are $3J + 1$ boundary conditions so that the cell residuals over-determine the unknowns. This is the counting problem facing the simple cell vertex method that has already been discussed in section 6.3.4.

To cope with all of these cases, and corresponding cases when the diffusion is added, the cell vertex method makes use of distribution matrices and artificial dissipation. These have been described already in sections 6.3.6 and 6.3.8; we gather together here the complete formulation of the method for (6.113) and (6.114) on a general rectangular mesh $\{x_0, x_1, \dots, x_J\} \times \{y_0, y_1, \dots, y_K\}$, where $\Delta x_i := x_i - x_{i-1}, \Delta y_j := y_j - y_{j-1}$. We use method B to recover the gradients and define the fluxes, in terms of the usual forward and backward divided differences, as

$$F_{ij} := -\epsilon \frac{\Delta x_i D_{x+} U_{ij} + \Delta x_{i+1} D_{x-} U_{ij}}{\Delta x_i + \Delta x_{i+1}} + b_{ij}^{(x)} U_{ij},$$

$$G_{ij} := -\epsilon \frac{\Delta y_i D_{y+} U_{ij} + \Delta y_{i+1} D_{y-} U_{ij}}{\Delta y_i + \Delta y_{i+1}} + b_{ij}^{(y)} U_{ij}. \tag{6.115a}$$

Normal gradients required on boundaries are extrapolated as in (6.71b). Then the residual for the cell $\Omega_\alpha \equiv [x_{i-1}, x_i] \times [y_{j-1}, y_j]$, with $\alpha \equiv (i -$

$\frac{1}{2}, j - \frac{1}{2})$ and $V_\alpha \equiv \Delta x_i \Delta y_j$, is from (6.50)

$$R_\alpha(U) := \frac{1}{2} \left[\frac{F_{i,j-1} - F_{i-1,j} + F_{ij} - F_{i-1,j-1}}{\Delta x_i} \right.$$
$$\left. - \frac{G_{i,j-1} - G_{i-1,j} + G_{ij} - G_{i-1,j-1}}{\Delta y_i} \right] - S_\alpha, \qquad (6.115b)$$

where S_α is an average of $S(x, y)$ over the cell. Similarly, let b_α be an average value of b over the cell, with which we will use the upwind definition of distribution matrices given by (6.60) and (6.66), namely

$$D_{\alpha,i-1,j} = (1 - \text{sign } b_\alpha^{(x)})(1 + \text{sign } b_\alpha^{(y)}),$$
$$D_{\alpha,i,j} = (1 + \text{sign } b_\alpha^{(x)})(1 + \text{sign } b_\alpha^{(y)}),$$
$$D_{\alpha,i-1,j-1} = (1 - \text{sign } b_\alpha^{(x)})(1 - \text{sign } b_\alpha^{(y)}),$$
$$D_{\alpha,i,j-1} = (1 + \text{sign } b_\alpha^{(x)})(1 - \text{sign } b_\alpha^{(y)}); \qquad (6.115c)$$

note that these are laid out in (6.115c) so as to correspond positionally with the distribution of R_α to the four corners of the cell, and that for each particular cell three are zero and one equals 4.0. So, finally, we define the nodal residual as in (6.59) by

$$N_{ij}(U) := \frac{\sum_{\alpha=(i \pm \frac{1}{2}, j \pm \frac{1}{2})} V_\alpha (D_{\alpha,i,j} R_\alpha + A_{\alpha,i,j})}{(\Delta x_i + \Delta x_{i+1})(\Delta y_j + \Delta y_{j+1})}, \qquad (6.115d)$$

where it is assumed a fourth order artificial dissipation which is transparent to linear functions is included. Using the general form given by (6.70b), but with the simpler differences of (6.45a), this can be constructed from nodal pseudo-Laplacians as

$$A_{\alpha,i,j} = \tau_\alpha^{(4)} (\delta_h^2 U_{ij} - \overline{\delta_h^2 U_\alpha}),$$
$$\delta_h^2 U_{ij} = \frac{1}{2} [(\Delta x_i + \Delta x_{i-1})(D_{x+} - D_{x-})$$
$$+ (\Delta y_j + \Delta y_{j+1})(D_{y+} - D_{y-})]U_{ij}; \qquad (6.115e)$$

recall that $\overline{\delta_h^2 U_\alpha}$ denotes the average of $\delta_h^2 U$ at the four vertices of the cell.

The nodal residual (6.115d) needs to be defined at every node where U_{ij} is not determined by the Dirichlet boundary conditions; note that, as with finite element methods, Neumann boundary conditions in (6.113) are not imposed directly but $\partial u / \partial n = 0$ is merely used to replace the normal diffusive fluxes in (6.115a). We recall that the general procedure for applying (6.115d) at boundary nodes is to extend the mesh to have dummy cells exterior to the domain, each with the same measure V_α as its interior neighbour but with R_α and $A_{\alpha,i,j}$ set to zero. So let us consider the left boundary, $x = x_0$: if $b_{0j}^{(x)} > 0$, corresponding to an inflow boundary point, Dirichlet boundary conditions determine U_{0j}; if $b_{0j}^{(x)} < 0$, corresponding to an outflow point, an equation is certainly needed for U_{0j} if $\epsilon = 0$ and may be needed when $\epsilon > 0$; and if $b_{0j}^{(x)} = 0$, corresponding to a tangential boundary point, an equation may or may not be needed in either case.

Generally, the choice of distribution matrices contained in (6.115c) will provide nodal equations when needed, but there are two particular circumstances that are worth noting. For any method of defining b_α, every cell residual is assigned to just one of its vertices, and suppose $b_{0j}^{(x)} \leq 0$. Then if U_{0j} is determined by a Dirichlet boundary condition, giving a boundary layer in general, either of the cell residuals corresponding to $\alpha = (\frac{1}{2}, j + \frac{1}{2})$ or $\alpha = (\frac{1}{2}, j - \frac{1}{2})$ may be assigned to N_{0j} and would therefore not be used in the total system of nodal equations; while if $b_{\frac{1}{2}, j+\frac{1}{2}}^{(y)} > 0$ and $b_{\frac{1}{2}, j-\frac{1}{2}}^{(y)} < 0$, corresponding to a diverging flow, neither of the corresponding residuals would be assigned to N_{0j}. It is in this latter case that the artificial viscosity is essential to obtaining a nonsingular system of nodal equations; in this connection it should be noted that in calculating $\delta_h^2 U_{ij}$ for (6.115e) at a boundary, the normal second differences are set to zero but the tangential differences retained. On the other hand the complementary situation of converging flow, in which $b_{\frac{1}{2}, j+\frac{1}{2}}^{(y)} < 0$ and $b_{\frac{1}{2}, j-\frac{1}{2}}^{(y)} > 0$ so that N_{0j} obtains contributions from both cells, is dealt with very satisfactorily and requires no artificial dissipation.

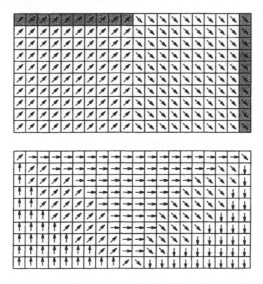

Figure 6.15. *Allocation of cell residuals for the IAHR/CEGB problems; the top diagram shows the allocation of cell residuals to nodes for the cell vertex method, with the unused cells being shaded; the bottom diagram shows a typical allocation for the Moores' method, with an arrow to an edge indicating a shared residual*

The flow field for the IAHR/CEGB problems, shown in Fig. 1.3 illustrates some of these points. The top diagram in Fig. 6.15 shows that in this case each nodal unknown is assigned just one cell residual for its determination, and no artificial dissipation was used in the computations shown in Fig. 6.12 and Fig. 6.13; the residuals for the shaded cells are not needed but can be set to zero, to maintain overall conservation, by using them to define the normal gradients at the nodes to which they are assigned, as in (6.89). For comparison, the typical allocation of cells to nodes that is used in the Moores' method is also shown in Fig.6.15; the effect of the allocation near the right-hand boundary in severe oscillation in second IAHR/CEGB problem, similar to that obtained with the streamline diffusion method and shown in Fig.5.7.

6.5.2 Spurious modes, least squares and artificial dissipation

This leaves us with divergent flow problems typified by the central diagram in Fig. 6.14 as needing some further attention. It is the lack of boundary conditions here which allows the spurious chequerboard mode more scope to grow and pollute the solution. Considerable discussion was devoted in section 6.3.2 to the control of the washboard modes that afflict cell centre schemes, with both the mode and its control described for the model one-dimensional problem; but the chequerboard mode is essentially two-dimensional. Moreover, it is of particular concern for the cell vertex scheme because it lies in the null space of the uniform mesh diffusion operator, as well as in the null space of the constant b convection operator; so, unlike the situation with the cell centre scheme, it is not controlled by increased diffusion. Fourth order dissipation of the form (6.115e) is effective, however, because it is based on a three-point diffusion operator.

Fourier analysis is a useful tool for characterising the mode and considering its control. On a uniform square mesh with the constant $\mathbf{b}$, the dispersion relation for the convective equation $\nabla \cdot (\mathbf{b}u) = 0$ is $b^{(x)}k_x + b^{(y)}k_y = 0$; the cell vertex method for comparison yields

$$h\widehat{R} \equiv ib^{(x)} \cos \tfrac{1}{2}k_y h \sin \tfrac{1}{2}k_x h + ib^{(y)} \cos \tfrac{1}{2}k_x h \sin \tfrac{1}{2}k_y h = 0. \qquad (6.116a)$$

The contours of the amplitude of $h\widehat{R}$ in the $(k_x h, k_y h)$ plane for $b^{(x)}$: $b^{(y)} = 2 : 1$ are plotted in Fig. 6.16. From the left-hand diagram one sees clearly how the $|\widehat{R}| = 0$ contour, which passes through the origin with the correct slope, deviates to run into the $(\pm\pi, \pm\pi)$ corners which correspond to the chequerboard mode. Then in the right-hand diagram contours of the corresponding nodal residual $|h\widehat{N}|$ are plotted after a small amount of fourth order dissipation is added: without dissipation $\widehat{N} = \widehat{R}$, but from (6.115e) we see that hN_{ij} contains dissipation of the form $\epsilon^{(4)}(1 - \mu_x^2 \mu_y^2)(\delta_x^2 + \delta_y^2)U_{ij} = -\tfrac{1}{4}\epsilon^{(4)}(\delta_x^2 + \delta_y^2 + \tfrac{1}{4}\delta_x^2\delta_y^2)(\delta_x^2 + \delta_y^2)U_{ij}$ while the contribution from hR_α is shifted upwind; hence we have

$$h\widehat{N} = h\widehat{R}e^{-\frac{1}{2}(k_x+k_y)hi} - 4\epsilon^{(4)}(s_x^2 + s_y^2 + s_x^2 + s_y^2)(s_x^2 + s_y^2), \quad (6.116b)$$

where $s_x = \sin\frac{1}{2}k_x h, s_y = \sin\frac{1}{2}k_y h$. From the figure we see that the contours near the origin are little changed, while the chequerboard mode is clearly eliminated.

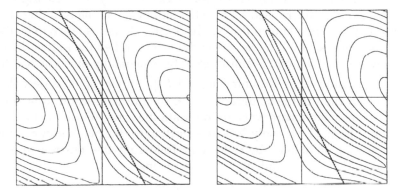

Figure 6.16. *Contours of* $|h\widehat{R}| \equiv |h\widehat{N}|$ *without dissipation on the left, and of* $|h\widehat{N}|$ *with* $\epsilon^{(4)} = 0.06$ *on the right*

There are several possible ways of avoiding the chequerboard mode altogether. One is to redefine the cell residual by using other than the trapezoidal rule to integrate along each edge; however, any compact scheme will remove the mode only at the cost of losing accuracy. Another alternative is to use a triangular mesh. The details of how the elimination is then effected, even on a square mesh divided into triangles by drawing one set of diagonals, depends on how the distribution matrices are defined and may depend on the flow direction. Again, on a uniform mesh Fourier analysis may be used to study the phenomenon but we shall not pursue it further here.

Of course, on a bounded region with Dirichlet inflow boundary conditions the chequerboard mode cannot satisfy all the discrete equations. But the cell residual equations for the diverging flow in the central picture of Fig. 6.14 are singular because they form a rectangular system; and Field (1994) in a detailed analysis of this situation has shown that all the solutions in the null space of the matrix for a typical 10×10 mesh exhibit a modified chequerboard pattern. The critical case, which is intermediate between those shown in the centre and on the right of Fig. 6.14, is where flow is parallel to the mesh lines, leading to characteristic boundaries. The analysis in Morton and Stynes (1994) shows how the normal coercivity properties, that one can demonstrate by the methods used in Lemma 6.4.7, break down for parallel flow. They go on to show how the technique proposed in Morton (1991a) to split certain cell residuals, in order to create the correct number of equations, restores coercivity so that second order accuracy can be established. We shall refer to this splitting technique again in section 6.6, but the use of artificial dissipation is not only more flexible and practical, it can also be given strong theoretical justification by relating it to a constrained least squares formulation.

In his thesis, Field (1994) has explored various ways of selecting U to satisfy the under-determined set of cell residual equations resulting from diverging flow. For a known flow u, the 'best' solution is taken to be that which minimises $\|u - U\|_{l_2}$, i.e. the nodal errors. When the inflow data is a sine wave, this solution is almost exactly matched by one which is the smoothest in the sense of minimising $\left\|\delta_h^2 U\right\|_{l_2}^2$, where δ_h^2 is as described above taken right to the boundary, where the normal second differences are set to zero; for square wave inflow data the error, which is of course now much larger, is still only 30% greater than for the best solution. Now the interesting point made by Field is the following: the constrained minimisation of $\left\|\delta_h^2 U\right\|_{l_2}$ is equivalent to minimising

$$\|R_\alpha(U)\|_{L_2}^2 + \epsilon^{(4)} \left\|\delta_h^2 U\right\|_{l_2}^2$$

as $\epsilon^{(4)} \to 0$; and the resulting normal equations are very close in form to those provided by the nodal residuals $N_{ij}(U) = 0$ when Lax–Wendroff distribution matrices are used and the fourth order dissipation is of the form $(\delta_h^2)^2 U$. This new interpretation and justification for fourth order artificial dissipation also provides useful guidance for its form and the magnitude of $\epsilon^{(4)}$. It can, almost equivalently, be regarded as a stabilization technique for the singular system of cell residual equations.

6.5.3 Coercivity for the continuous problem

Before embarking on an error analysis of the cell vertex method using the techniques of Lemma 6.4.7, let us consider the corresponding analysis for the differential problem (6.113). First we absorb the boundary data into the source term in the usual way so that we consider the problem, find $w \in H^1_{E_0}$ such that

$$B(w, v) \equiv \epsilon(\nabla w, \nabla v) + (\nabla \cdot (\mathbf{b}w), v) = (S, v) \quad \forall v \in H^1_{E_0}. \tag{6.117}$$

Now let us choose $v = gw$, where g is yet to be specified, to obtain firstly

$$(\nabla \cdot (\mathbf{b}w), gw) = ((\nabla \cdot \mathbf{b})w, gw) + (w\mathbf{b} \cdot \nabla w, g)$$

$$= ((\nabla \cdot \mathbf{b})w, gw) + \tfrac{1}{2}(\mathbf{b} \cdot \nabla w^2, g)$$

$$= ((\nabla\cdot\mathbf{b})gw, w) + \tfrac{1}{2}\int_{\partial\Omega_N} gw^2(\mathbf{b}\cdot\mathbf{n})\mathrm{d}\Gamma - \tfrac{1}{2}(w^2, \nabla\cdot(\mathbf{b}g))$$

$$= \tfrac{1}{2}([(\nabla\cdot\mathbf{b})g - \mathbf{b}\cdot\nabla g]w, w) + \tfrac{1}{2}\int_{\partial\Omega_N} gw^2(\mathbf{b}\cdot\mathbf{n})\mathrm{d}\Gamma. \quad (6.118a)$$

Secondly, we have

$$(\nabla w, \nabla(gw)) = \tfrac{1}{2}(\nabla w^2, \nabla g) + (\nabla w, g\nabla w). \qquad (6.118b)$$

Together these lead to the following theorem.

Theorem 6.5.1 *The solution w of (6.117) satisfies the bound*

$$\epsilon\left\|g^{\frac{1}{2}}\nabla w\right\|^2 + \tfrac{1}{4}\|qw\|^2 + \tfrac{1}{2}\int_{\partial\Omega_N} gw^2(\mathbf{b}\cdot\mathbf{n})\mathrm{d}\Gamma \le \|(g/q)S\|^2 \qquad (6.119)$$

for any pair of positive functions g and q satisfying the relation

$$\epsilon(\nabla g, \nabla v^2) + ((\nabla\cdot\mathbf{b})g - \mathbf{b}\cdot\nabla g, v^2) = (q^2, v^2) \quad \forall v \in H^1_{E_0}. \qquad (6.120)$$

Proof. We have only to combine (6.118a,b) and set $v = w$ in (6.120) to obtain

$$(S, gw) = B(w, gw) = \epsilon(g\nabla w, \nabla w) + \tfrac{1}{2}(q^2w, w) + \tfrac{1}{2}\int_{\partial\Omega_N} gw^2(\mathbf{b}\cdot\mathbf{n})\mathrm{d}\Gamma.$$

But $|(S, gw)| = |(g/q)S, qw| \le \tfrac{1}{4}\|qw\|^2 + \|(g/q)S\|^2$ and the result follows.
$\square$

We shall use such a multiplicative mapping $w \longmapsto gw$ in the next section, and in doing so we shall need to carry out the discrete analogues of all the manipulations that led to (6.118a) and (6.118b). A wide choice for g is possible, but we shall use some approximation to $g = \mathrm{e}^{-B}$, where $\nabla B = \mathbf{b}$, for which $-\mathbf{b}\cdot\nabla g = |\mathbf{b}|^2 g$.

It is worth noting two relationships between this class of mappings and methods that have been discussed in earlier chapters. Firstly, when $\nabla\cdot\mathbf{b} = 0$ the operator defining g is just the adjoint of that in the original problem (6.117); so there is a close link to using a local Green's function to define the test space or using a mapping between trial and test spaces to symmetrize the bilinear forms, both of which were described in Chapter 5. Secondly, there is a clear relationship with the streamline diffusion methods also discussed in that chapter, with the function g being replaced by the operator $\mathbf{b}\cdot\nabla$.

6.5.4 Error bounds in a discrete energy norm

The methods we shall use here are based on a discrete Gårding inequality which is a development of that originally formulated by Süli (1992) for the pure convection problem and developed in Morton and Stynes (1994),

Morton (1994), and Mackenzie *et al.* (1994). For simplicity in the manipulations we shall consider only a uniform square mesh of size h, take Ω
to be the unit square $(0,1) \times (0,1)$, and impose homogeneous Dirichlet
boundary conditions on all of $\partial\Omega$. Further assumptions will also be made
later regarding the flow field so that, as with the IAHR/CEGB problems,
no artificial dissipation is needed and the cell vertex method reduces to
setting a certain set of cell residuals to zero.

The cell equation on $\Omega_{i-\frac{1}{2},j-\frac{1}{2}}$, with characteristic function $\chi_{i-\frac{1}{2},j-\frac{1}{2}}$, can
then be written in finite difference notation as

$$B_c(U, \chi_{i-\frac{1}{2},j-\frac{1}{2}}) + B_d(U, \chi_{i-\frac{1}{2},j-\frac{1}{2}}) = (S, \chi_{i-\frac{1}{2},j-\frac{1}{2}}), \qquad (6.121a)$$

where

$$B_c(U, \chi_{i-\frac{1}{2},j-\frac{1}{2}}) \equiv h[\mu_y \delta_x(b^{(x)}U) + \mu_x \delta_y(b^{(y)}U)]_{i-\frac{1}{2},j-\frac{1}{2}} \qquad (6.121b)$$

$$B_d(U, \chi_{i-\frac{1}{2},j-\frac{1}{2}}) \equiv -\epsilon[\mu_y \delta_x(\mu_x \delta_x U) + \mu_x \delta_y(\mu_y \delta_y U)]_{i-\frac{1}{2},j-\frac{1}{2}}. \qquad (6.121c)$$

We suppose that both components of **b** are positive, specifically that $\mathbf{b} \in (C^1(\bar\Omega))^2$ and there are positive constants α_1, α_2 such that

$$b^{(x)} \geq \alpha_1 > 0, \quad b^{(y)} \geq \alpha_2 > 0. \qquad (6.122a)$$

Then, with the Dirichlet boundary conditions, (6.121) is applied only for
$i, j = 1, 2, \ldots, J-1$, i.e. the test space is given by

$$T^h = \text{span}\{\chi_{i-\frac{1}{2},j-\frac{1}{2}}; i, j = 1, 2, \ldots, J-1\}. \qquad (6.122b)$$

Gradients in the normal direction therefore need to be recovered only on
the bottom and left-hand boundaries, where we use extrapolation,

$$\mu_x \delta_x U_{0,j} := 2\delta_x U_{\frac{1}{2},j} - \mu_x \delta_x U_{i,j}, \quad \mu_y \delta_y U_{i,0} = 2\delta_y U_{i,\frac{1}{2}} - \mu_y \delta_y U_{i,1}. \qquad (6.122c)$$

We use I^h to denote the interpolation projector onto the trial space S_0^h, of
bilinear functions which vanish on $\partial\Omega$, and Π^h to denote projection onto
T^h. Then clearly $B_c(U, V) \equiv (\nabla \cdot I^h(\mathbf{b}U), V) \; \forall V \in T^h$; and $(\Pi^h W)_{i-\frac{1}{2},j-\frac{1}{2}}$
$\equiv \mu_x \mu_y W_{i-\frac{1}{2},j-\frac{1}{2}} \; \forall W \in S^h$, while generally $(\Pi^h w)_{i-\frac{1}{2},j-\frac{1}{2}} = \langle w \rangle_{i-\frac{1}{2},j-\frac{1}{2}}$
the average of w over the cell $\Omega_{i-\frac{1}{2},j-\frac{1}{2}}$. We can therefore write our first
obvious coercivity result, for $\gamma > 0$, as

$$B_c(W, \gamma^2 \Pi^h \nabla \cdot I^h(\mathbf{b}W)) = \gamma^2 \left| \nabla \cdot I^h(\mathbf{b}W) \right|^2_{l_2(\Omega^h)}, \quad \forall W \in S_0^h, \qquad (6.123)$$

where

$$|v|_{l_2(\Omega^h)} := \left\{ \sum_{i,j=1}^{J-1} h^2 \langle v \rangle^2_{i-\frac{1}{2},j-\frac{1}{2}} \right\}^{\frac{1}{2}}, \qquad (6.124)$$

which is a norm on S_0^h. On the other hand, for the diffusion form $B_d(W, V)$
we have sums of terms like $p_{i-\frac{1}{2}} \delta_x q_{i-\frac{1}{2}}$ for which the Cauchy-Schwarz

inequality gives

$$\left| p_{i-\frac{1}{2}} \delta_x q_{i-\frac{1}{2}} \right| \leq \frac{1}{\sqrt{2}} \left[p_{i-\frac{1}{2}}^2 + \tfrac{1}{2}(q_i - q_{i-1})^2 \right] \leq \frac{1}{\sqrt{2}} \left[p_{i-\frac{1}{2}}^2 + q_i^2 + q_{i-1}^2 \right];$$

hence it is easy to deduce that, with the useful abbreviations for the average normal gradients across the mesh edges $\bar{W}_x := h^{-1} \mu_y \mu_x \delta_x W$ and $\bar{W}_y := h^{-1} \mu_x \mu_y \delta_y W$,

$$\left| B_d(W, \gamma^2 \Pi^h \nabla \cdot I^h(bW)) \right|$$

$$= \epsilon \left| \sum_{i,j=1}^{J-1} \gamma^2 h \langle \nabla \cdot I^h(bW) \rangle_{i-\frac{1}{2},j-\frac{1}{2}} [\delta_x \bar{W}_x + \delta_y \bar{W}_y]_{i-\frac{1}{2},j-\frac{1}{2}} \right|$$

$$\leq \frac{\epsilon \gamma^2}{\sqrt{2}h} \left\{ 2 \left| \nabla \cdot I^h(bW) \right|_{l_2(\Omega^h)}^2 \right.$$

$$+ \sum_{j=1}^{J-1} h^2 \left[(\bar{W}_x)_{0,j-\frac{1}{2}}^2 + (\bar{W}_x)_{J-1,j-\frac{1}{2}}^2 + 2 \sum_{i=1}^{J-2} (\bar{W}_x)_{i,j-\frac{1}{2}}^2 \right]$$

$$\left. + \sum_{i=1}^{J-1} h^2 \left[(\bar{W}_y)_{i-\frac{1}{2},0}^2 + (\bar{W}_y)_{i-\frac{1}{2},J-1}^2 + 2 \sum_{j=1}^{J-2} (\bar{W}_y)_{i-\frac{1}{2},j} \right] \right\}. \quad (6.125)$$

Note that $\bar{W}_x$ on the left boundary and $\bar{W}_y$ at the bottom are given by the gradient boundary conditions (6.122c).

Rather more manipulation is needed to obtain discrete relations of a form similar to (6.118a). We begin by defining

$$G(x,y) := e^{-(\kappa_1 x_{i-1} + \kappa_2 y_{j-1})} \equiv G_{i-\frac{1}{2},j-\frac{1}{2}} \text{ for } (x,y) \in \Omega_{i-\frac{1}{2},j-\frac{1}{2}}, \quad (6.126)$$

where κ_1, κ_2 are positive constants, and consider the diffusion form first, i.e. $B_d(W, G\Pi^h W)$. Denoting $\mu_x \mu_y W$ by $\bar{W}$ so that the right-hand side of (6.121c) is $-\epsilon(\delta_x^2 + \delta_y^2)\bar{U}$, we consider just the δ_x^2 terms and sum by parts in the x-direction to get, with the j subscript suppressed,

$$- \sum_{i=1}^{J-1} (G\bar{W})_{i-\frac{1}{2}} \delta_x^2 \bar{W}_{i-\frac{1}{2}} = (G\bar{W})_{\frac{1}{2}} \delta_x \bar{W}_0 - (G\bar{W})_{J-\frac{3}{2}} \delta_x \bar{W}_{J-1}$$

$$+ \sum_{i=1}^{J-2} \delta_x (G\bar{W})_i \delta_x \bar{W}_i, \quad (6.127)$$

where from (6.122c) we have $\delta_x \bar{W}_0 = 2\mu_y \delta_x W_{\frac{1}{2}} - \delta_x \bar{W}_1 = 4\bar{W}_{\frac{1}{2}} - \delta_x \bar{W}_1$. Now we have $\delta_x (G\bar{W})_i = G_{i+\frac{1}{2}} \delta_x \bar{W}_i + \bar{W}_{i-\frac{1}{2}} \delta_x G_i$, giving a positive definite sum in $(\delta_x \bar{W})_i^2 = h^2 (\bar{W}_x)_i^2$, from which we obtain the following result.

Lemma 6.5.2 For G given by (6.126) with $e^{\kappa_i} \geq 8$ and $e^{\kappa_i h} \leq 2$ for $i = 1, 2$, *we have*

$$
B_d(W, G\Pi^h W) \geq \tfrac{1}{2}\epsilon e^{-\kappa_1-\kappa_2} \left\{ \sum_{i=0}^{J-1}\sum_{j=1}^{J-1} h^2(\bar{W}_x)^2_{i,j-\frac{1}{2}} \right.
$$

$$
\left. + \sum_{i=1}^{J-1}\sum_{j=0}^{J-1} h^2(\bar{W}_y)^2_{i-\frac{1}{2},j} \right\} - \tfrac{1}{2}\epsilon(\kappa_1^2+\kappa_2^2)|W|^2_{l_2(\Omega^h)}
$$

$$
- \tfrac{1}{8}\epsilon \left\{ \sum_{j=1}^{J-1} G_{J-\frac{3}{2},j-\frac{1}{2}}\, (\mu_y W)^2_{J-1,j-\frac{1}{2}} \right.
$$

$$
\left. + \sum_{i=1}^{J-1} G_{i-\frac{1}{2},J-\frac{3}{2}}(\mu_x W)^2_{i-\frac{1}{2},J-1} \right\}. \tag{6.128}
$$

Proof. As we have seen, for a typical term in (6.127) we have in the x-direction

$$
\delta_x(G\bar{W})_i \delta_x \bar{W}_i = \delta_x \bar{W}_i(G_{i+\frac{1}{2}}\delta_x \bar{W}_i + \bar{W}_{i-\frac{1}{2}}\delta_x G_i)
$$

$$
\geq \tfrac{1}{2}G_{i+\frac{1}{2}}(\delta_x \bar{W}_i)^2 - \tfrac{1}{2}[(\delta_x G_i)^2/G_{i+\frac{1}{2}}](\bar{W}_{i-\frac{1}{2}})^2, \tag{6.129a}
$$

in which $G_{i+\frac{1}{2}} \geq e^{-\kappa_1-\kappa_2}$ and $(\delta_x G_i)^2/G_{i+\frac{1}{2}} \leq (\kappa_1 h)^2$. Similarly

$$
-(G\bar{W})_{J-\frac{3}{2}}\delta_x \bar{W}_{J-1} = G_{J-\frac{3}{2}}\delta_x \bar{W}_{J-1}(\delta_x \bar{W}_{J-1} - \bar{W}_{J-\frac{1}{2}})
$$

$$
\geq \tfrac{1}{2}G_{J-\frac{3}{2}}[(\delta_x \bar{W}_{J-1})^2 - (\bar{W}_{J-\frac{1}{2}})^2] \tag{6.129b}
$$

and

$$
(G\bar{W})_{\frac{1}{2}}\delta_x \bar{W}_0 = \tfrac{1}{4}G_{\frac{1}{2}}\delta_x \bar{W}_0(\delta_x \bar{W}_0 + \delta_x \bar{W}_1)
$$

$$
\geq \tfrac{1}{8}G_{\frac{1}{2}}[(\delta_x \bar{W}_0)^2 - (\delta_x \bar{W}_1)^2]. \tag{6.129c}
$$

Now from the assumptions on κ_i and h, $\tfrac{1}{8}G_{\frac{1}{2},j-\frac{1}{2}} \geq e^{-\kappa_1-\kappa_2}$; and combining the terms in $\delta_x(\delta \bar{W}_1)^2$ we also have $\tfrac{1}{2}G_{\frac{3}{2},j-\frac{1}{2}} - \tfrac{1}{8}G_{\frac{1}{2},j-\frac{1}{2}} \geq \tfrac{1}{4}G_{\frac{3}{2},j-\frac{1}{2}} \geq \tfrac{1}{2}e^{-\kappa_1-\kappa_2}$. This gives all the terms in the first sum of (6.128), observing that $\delta_x \bar{W} = h\bar{W}_x$, and the second is obtained similarly; also the $|W|^2_{l_2(\Omega^h)}$ term follows directly from (6.129a). The final awkward outflow boundary terms follow from the last term in (6.129b) with $W_{J-1,j-\frac{1}{2}} = \tfrac{1}{2}\mu_y W_{J-1,j-\frac{1}{2}}$ and similar terms on the top boundary. □

There remains the consideration of the convection terms. Unfortunately, chequerboard terms of the form $\delta_x\delta_y W_{i-\frac{1}{2},j-\frac{1}{2}}$ are generated which are difficult to control without adding artificial dissipation or making a restriction on the flow field; we take the latter option and in the lemma below impose the condition (6.131). To break up the terms in (6.121b) we use the identities

$$\delta_x(ab) = (\mu_x a)(\delta_x b) + (\delta_x a)(\mu_x b)$$
$$\mu_x(ab) = (\mu_x a)(\mu_x b) + \tfrac{1}{4}(\delta_x a)(\delta_x b),$$

and thence obtain

$$\begin{aligned}
B_c(W, G\Pi^h W) = \sum_{i,j=1}^{J-1} h(G\bar{W})_{i-\frac{1}{2},j-\frac{1}{2}} & \Big\{ [\bar{W}(\mu_y \delta_x b^{(x)} + \mu_x \delta_y b^{(y)})]_{i-\frac{1}{2},j-\frac{1}{2}} \\
& + \tfrac{1}{4}[\delta_x \delta_y W(\mu_x \delta_y b^{(x)} + \mu_y \delta_x b^{(y)})]_{i-\frac{1}{2},j-\frac{1}{2}} \\
& + [\mu_y \delta_x W(\mu_x \mu_y b^{(x)} + \tfrac{1}{4}\delta_x \delta_y b^{(y)})]_{i-\frac{1}{2},j-\frac{1}{2}} \\
& + [\mu_x \delta_y W(\tfrac{1}{4}\delta_x \delta_y b^{(x)} + \mu_x \mu_y b^{(y)})]_{i-\frac{1}{2},j-\frac{1}{2}} \Big\} \\
\equiv S_1 + S_2 + S_3 + S_4. & \qquad\qquad\qquad\qquad\qquad\qquad\qquad (6.130)
\end{aligned}$$

The assumption on the flow field is made so as to set $S_2 = 0$, and so we obtain the following result.

Lemma 6.5.3 *Assume that the flow field satisfies for $i, j = 1, \ldots, J - 1$*

$$[\mu_x \delta_y b^{(x)} + \mu_y \delta_x b^{(y)}]_{i-\frac{1}{2},j-\frac{1}{2}} = 0. \qquad (6.131)$$

Then for G given by (6.126) with $\kappa_i, i = 1, 2$, chosen to satisfy

$$\kappa_i \alpha_i \geq 8(1 + 2\|\nabla b\|_{L_\infty(\Omega)}), \qquad (6.132)$$

and for sufficiently small h, we have

$$\begin{aligned}
B_c(W, G\Pi^h W) \geq {} & 2e^{-\kappa_1 - \kappa_2} |W|^2_{l_2(\Omega^h)} \\
& + \tfrac{1}{8}\alpha_1 \sum_{j=1}^{J-1} hG_{J-\frac{3}{2},j-\frac{1}{2}}(\mu_y W)^2_{J-1,j-\frac{1}{2}} \\
& + \tfrac{1}{8}\alpha_2 \sum_{i=1}^{J-1} hG_{i-\frac{1}{2},J-\frac{3}{2}}(\mu_x W)^2_{i-\frac{1}{2},J-1}. \qquad (6.133)
\end{aligned}$$

Proof. Considering the coefficients in S_3 and S_4, we take h sufficiently small that

$$\begin{aligned}
B^{(x)} &:= \mu_x \mu_y b^{(x)} + \tfrac{1}{4}\delta_x \delta_y b^{(y)} \geq \tfrac{1}{2}\alpha_1, \\
B^{(y)} &:= \mu_x \mu_y b^{(y)} + \tfrac{1}{4}\delta_x \delta_y b^{(x)} \geq \tfrac{1}{2}\alpha_2 \qquad (6.134a)
\end{aligned}$$

and also that

$$\left| \delta_x B^{(x)} \right| \leq 2h \|\nabla b\|_{L_\infty(\Omega)}, \quad \left| \delta_y B^{(y)} \right| \leq 2h \|\nabla b\|_{L_\infty(\Omega)}. \qquad (6.134b)$$

Then, since $\bar{W}\mu_y \delta_x W = \tfrac{1}{2}\delta_x(\mu_y W)^2$ the i-sum in S_3 becomes

$$\sum_{i=1}^{J-1} hG_{i-\frac{1}{2}}\delta_x(\mu_y W)^2_{i-\frac{1}{2}} B^{(x)}_{i-\frac{1}{2}}$$

$$= \tfrac{1}{2}hG_{J-\frac{3}{2}}B^{(x)}_{J-\frac{3}{2}}(\mu_y W)^2_{J-1} - \sum_{i=1}^{J-2} \tfrac{1}{2}h(\mu_y W)^2_i \delta_x(GB^{(x)})_i$$

after summing by parts and using $W_0 = 0$. Now $-\delta_x G_i \geq \tfrac{1}{2}h\kappa_1 G_{i-\frac{1}{2}}$ so that

$$-\delta_x(GB^{(x)})_i = -[B^{(x)}_{i+\frac{1}{2}}\delta_x G_i + G_{i-\frac{1}{2}}\delta_x B^{(x)}_i]$$

$$\geq hG_{i-\frac{1}{2}}[\tfrac{1}{4}\kappa_1\alpha_1 - 2\,\|\nabla\mathbf{b}\|_{L_\infty(\Omega)}],$$

and hence we obtain

$$S_3 \geq [\tfrac{1}{8}\kappa_1\alpha_1 - \|\nabla\mathbf{b}\|_{L_\infty(\Omega)}] \sum_{i=1}^{J-2}\sum_{j=1}^{J-1} h^2 G_{i-\frac{1}{2},j-\frac{1}{2}}(\mu_y W)^2_{i,j-\frac{1}{2}}$$

$$+ \tfrac{1}{4}\alpha_1 \sum_{j=1}^{J-1} hG_{J-\frac{3}{2},j-\frac{1}{2}}(\mu_y W)^2_{J-1,j-\frac{1}{2}}, \tag{6.135}$$

with a similar bound holding for S_4. Now consider combining this with S_1, where the coefficients satisfy

$$\left|\mu_y\delta_x b^{(x)} + \mu_x\delta_y b^{(y)}\right| \leq 2h\,\|\nabla\mathbf{b}\|_{L_\infty(\Omega)},$$

and where $|\bar{W}|^2_{i-\frac{1}{2},j-\frac{1}{2}} \leq \tfrac{1}{2}[(\mu_y W)^2_{i-1,j-\frac{1}{2}} + (\mu_y W)^2_{i,j-\frac{1}{2}}]$. For the i-sum in (6.135), since $G_{i-\frac{1}{2}} > G_{i+\frac{1}{2}}$ we have

$$\sum_{i=1}^{J-2} G_{i-\frac{1}{2}}(\mu_y W)^2_i \geq \sum_{i=1}^{J-1} G_{i-\frac{1}{2}}(W)^2_{i-\frac{1}{2}} - \tfrac{1}{2}G_{J-\frac{3}{2}}(\mu_y W)^2_{J-1}.$$

Hence we obtain

$$\tfrac{1}{2}S_1 + S_3 \geq [\tfrac{1}{8}\kappa_1\alpha_1 - 2\,\|\nabla\mathbf{b}\|_{L_\infty(\Omega)}]\sum_{i=1}^{J-1}\sum_{j=1}^{J-1} h^2 G_{i-\frac{1}{2},j-\frac{1}{2}}(W)^2_{i-\frac{1}{2},j-\frac{1}{2}}$$

$$+ [\tfrac{1}{4}\alpha_1 - h(\tfrac{1}{8}\kappa_1\alpha_1 - \|\nabla\mathbf{b}\|_{L_\infty(\Omega)})]\sum_{j=1}^{J-1} hG_{J-\frac{3}{2},j-\frac{1}{2}}(\mu_y W)^2_{J-1,j-\frac{1}{2}}.$$

With κ_1 chosen to satisfy (6.132) and h sufficiently small to ensure that $h(1 + \|\mathbf{b}\|_{L_\infty(\Omega)}) \leq \tfrac{1}{8}\alpha_1$ we obtain

$$\tfrac{1}{2}S_1 + S_3 \geq \sum_{i=1}^{J-1}\sum_{j=1}^{J-1} h^2 G_{i-\frac{1}{2},j-\frac{1}{2}}(W)^2_{i-\frac{1}{2},j-\frac{1}{2}}$$

$$+ \tfrac{1}{8}\alpha_1 \sum_{j=1}^{J-1} hG_{J-\frac{3}{2},j-\frac{1}{2}}(\mu_y W)^2_{J-1,j-\frac{1}{2}}. \tag{6.136}$$

Combining this with a similar bound for $\frac{1}{2}S_1 + S_4$ gives the desired result.
□

We are now in a position to combine (6.123), (6.125) and Lemmas 6.5.2 and 6.5.3 into the following coercivity result. With b normalised such that $\|b\|_{L_\infty(\Omega)} = 1$, the conditions on κ_i needed in Lemma 6.5.2 are satisfied when (6.132) is satisfied; and we define h_0, the maximum allowed h, to be the minimum of the $\kappa_i^{-1}\ln 2$ required in Lemma 6.5.2, the values given by (6.134a,b) and the value $\frac{1}{8}\alpha_1(1 + \|b\|_{L_\infty(\Omega)})^{-1}$ required at the end of the proof of Lemma 6.5.3.

Theorem 6.5.4 *Assume the flow field satisfies (6.131); also choose κ_i for $i = 1, 2$ to satisfy (6.132) and set $\gamma^2 = e^{-\kappa_1-\kappa_2}h/(2\sqrt{2})$ if $h \geq 2\sqrt{2}\epsilon$, and zero otherwise. Then for all $h \leq h_0$, defined above, and all ϵ such that the mesh Péclet number conditions*

$$\frac{\alpha_i h}{\epsilon} \geq 1 \quad i = 1, 2 \tag{6.137}$$

are satisfied and $\epsilon \leq 2e^{-\kappa_1-\kappa_2}/(\kappa_1 + \kappa_2)$, we have for $\forall W \in S_0^h$

$$B^h(W, G\Pi^h W + \gamma^2 \Pi^h \nabla \cdot I^h(bW) \geq e^{-\kappa_1-\kappa_2}|W|^2_{l_2(\Omega^h)}$$
$$+ \tfrac{1}{2}\gamma^2 |\nabla \cdot I^h(bW)|^2_{l_2(\Omega^h)}. \tag{6.138}$$

Proof. We merely add (6.123), (6.128) and (6.133), and subtract the right-hand side (6.125). The coefficient of $|\nabla \cdot I^h(bW)|^2_{l_2(\Omega)}$ from (6.123) and (6.125) is $\gamma^2[1 - 2\epsilon/(\sqrt{2}h)]$, which we have ensured is at least $\frac{1}{2}\gamma^2$ by setting $\gamma^2 = 0$ if ϵ/h is too large. Also the sums of $(W_x)^2$ and $(W_y)^2$ in (6.125) are dominated by similar sums in (6.128) with our choice of γ^2. The conditions (6.137) are needed to ensure that the averages of W along $i = J - 1$ and $j = J - 1$ that occur in (6.128) are dominated by the corresponding terms in (6.133). The final restriction on ϵ is then required for the $|W|^2_{l_2(\Omega h)}$ term in (6.128) to be dominated by that in (6.133). □

It is worth noting that this key coercivity result has been achieved with a mapping from the trial space to the test space that combines the up-wind association of a node with a cell, that was used in the analysis of the one-dimensional cell vertex method in section 6.4.2, with the exponential weighting that approximates the solution of the dual problem, which was used in establishing the coercivity of the two-dimensional continuous problem in the last section. Now we can derive error bounds from this result.

Theorem 6.5.5 *Suppose the hypotheses of Theorem 6.5.4 hold. Then there is a constant C, independent of ϵ and h, such that*

$$|u - U|_{l_2(\Omega^h)} + \left|\delta^{\frac{1}{2}}\nabla \cdot I^h(b(u - U))\right|_{l_2(\Omega^h)}$$

$$\leq |u - u^I|_{l_2(\Omega^h)} + C\left\{|\nabla\cdot[\mathbf{b}u - I^h(\mathbf{b}u^I)]|_{l_2(\Omega^h)}\right.$$

$$\left. + \epsilon\left[\sum_{i,j=1}^{J-1}[\delta_x(\bar{u}_x - \bar{u}_x^I) + \delta_y(\bar{u}_y - \bar{u}_y^I)]_{i-\frac{1}{2},j-\frac{1}{2}}^2\right]^{\frac{1}{2}}\right\}, \quad (6.139)$$

where $u^I \equiv I^h u$, $(\bar{u}_x)_{i,j-\frac{1}{2}} := h^{-1}\int_{y_{j-1}}^{y_j} u_x(x_i, y)dy$ with a similar definition of $\bar{u}_y$, and $\delta = h$ if $h \geq 2\sqrt{2}\epsilon$ and is zero otherwise.

Thence, when $u \in H^3(\Omega) \cap H_0^1(\Omega)$ and $\mathbf{b} \in (C^3(\bar{\Omega}))^2$, we have

$$|u - U|_{l_2(\Omega^h)} + \left|\delta^{\frac{1}{2}}\nabla\cdot I^h(\mathbf{b}(u - U))\right|_{l_2(\Omega^h)}$$

$$\leq C_1 h^2 |u|_{H^2(\Omega^h)} + C_2 h^2 |u|_{H^3(\Omega^h)}$$

$$+ \epsilon[C_3 h^2 |u|_{H^4(\Omega^h)} + C_4 h |u|_{H^3(B_h)}], \quad (6.140)$$

where $B_h = [(0, 2h) \times (0, 1)] \cup [(0, 1) \times (0, 2h)]$.

Proof. We write $\Theta = G\Pi^h(u^I - U) + \gamma^2 \Pi^h \nabla \cdot I^h(\mathbf{b}(u^I - U))$, the test function to be used in (6.138) to give, since $\delta = 2\sqrt{2}e^{\kappa_1 + \kappa_2}\gamma^2$,

$$e^{-\kappa_1 - \kappa_2}[|u^I - U|_{l_2(\Omega^h)}^2 + \frac{\delta}{4\sqrt{2}}|\nabla\cdot I^h(\mathbf{b}(u^I - U))|_{l_2(\Omega^h)}^2$$

$$\leq B^h(u^I - U, \Theta) = B^h(u^I, \Theta) - (S, \Theta) = B^h(u^I, \Theta) - B(u, \Theta)$$

$$= (\nabla\cdot[I^h(\mathbf{b}u^I) - \mathbf{b}u], \Theta) + B_d(u^I, \Theta) + \epsilon(\nabla^2 u, \Theta)$$

$$\leq |\Theta|_{l_2(\Omega^h)}\left\{|\nabla\cdot[\mathbf{b}u - I^h(\mathbf{b}u^I)]|_{l_2(\Omega^h)} + \epsilon\left[\sum_{i,j=1}^{J-1}\Delta_{i-\frac{1}{2},j-\frac{1}{2}}^2\right]^{\frac{1}{2}}\right\},$$

$$(6.141)$$

where, from the definition of $B_d(\cdot, \cdot)$ in (6.121c), it is clear that in the sum here we have

$$\Delta_{i-\frac{1}{2},j-\frac{1}{2}} = (\nabla^2 u, \chi_{i-\frac{1}{2},j-\frac{1}{2}}) - (\delta_x^2 + \delta_y^2)\bar{u}_{i-\frac{1}{2},j-\frac{1}{2}}^I$$

$$= \delta_x(\bar{u}_x - \bar{u}_x^I) + \delta_y(\bar{u}_y - \bar{u}_y^I);$$

thus the term in braces in (6.141) equals the corresponding term in (6.139). Next, from the observation that $G \leq 1$ and $\gamma^2 \leq \delta^{\frac{1}{2}}$, we have

$$|\Theta|_{l_2(\Omega^h)} \leq |u^I - U|_{l_2(\Omega^h)} + \delta^{\frac{1}{2}}|\nabla\cdot I^h(\mathbf{b}(u^I - U))|_{l_2(\Omega^h)}; \quad (6.142)$$

while the left-hand side of (6.141) is greater than or equal to $e^{-\kappa_1 - \kappa_2}/8\sqrt{2}$ times the square of the expression on the right of (6.142). Finally, then, cancelling one of these factors and observing that $I^h(\mathbf{b}u) = I^h(\mathbf{b}u^I)$ and $|u - U| \leq |u - u^I| + |u^I - U|$ gives (6.139) with $C = 8\sqrt{2}e^{\kappa_1 + \kappa_2}$.

The final bound (6.140) is obtained by estimating the interpolation errors contained in the three expressions on the right of (6.139). The first is bounded using a standard interpolation error estimate, giving the term in (6.140) with coefficient C_1. The second follows from Lemma 6.3.1 with $\mathbf{F} = bu$ and $r = 3$. For the third, let us denote by $\psi_{i-\frac{1}{2},j-\frac{1}{2}}$ the quantity in square brackets in the sum, and by $\Omega^+_{i-\frac{1}{2},j-\frac{1}{2}}$, the cross of five cells on which it has its support. Then for $i > 1$ and $j > 1$ this is zero for cubic functions of x and y, so that a straightforward application of the Bramble–Hilbert lemma yields

$$\left| h^{-1} \chi_{i-\frac{1}{2},j-\frac{1}{2}} \right|^2 \leq \tfrac{1}{5} C_3^2 h^2 \, |u|^2_{H^4(\Omega^+_{i-\frac{1}{2},j-\frac{1}{2}})}, \quad i,j > 1.$$

However, if $i = 1$ or $j = 1$ the use of the extrapolation formulae (6.122c) means that $\psi_{\frac{1}{2},j-\frac{1}{2}}$ and $\chi_{i-\frac{1}{2},\frac{1}{2}}$ are zero only for quadratic functions of x and y; so we have, with the overlapping truncated crosses of cells $\Omega^+_{i-\frac{1}{2},j-\frac{1}{2}}$,

$$\left| h^{-1} \chi_{i-\frac{1}{2},j-\frac{1}{2}} \right|^2 \leq \tfrac{1}{3} C_4^2 \, |u|^2_{H^3(\Omega^+_{i-\frac{1}{2},j-\frac{1}{2}})}, \quad i \text{ or } j = 1.$$

Hence we obtain, after counting the number of overlaps,

$$\sum_{i,j=1}^{J-1} h^2 \left| h^{-1} \chi_{i-\frac{1}{2},j-\frac{1}{2}} \right|^2 \leq C_3^2 h^4 \, |u|^2_{H^4(\Omega^h)} + C_4^2 h^2 \, |u|^2_{H^3(B_h)}, \qquad (6.143)$$

from which the last terms of (6.140) follow immediately. $\qquad \square$

Because $\epsilon = O(h)$ through the assumption (6.137), the error bound in (6.140) is $O(h^2)$, despite the form of the last term. We have left this term as a norm on the inflow boundary strip B_h because the derivatives of u are normally small there; moreover, because meas $B_h = O(h)$, a standard result based on function space interpolation shows that $|u|_{H^3(B_h)} \leq Ch^{\frac{1}{2}} \|u\|_{H^4(\Omega^h)}$. Hence the last term can be regarded as $O(\epsilon h^{\frac{3}{2}})$.

6.6 Two simple nonlinear problems

An extensive study of nonlinear problems would be beyond the scope of the present book; but two examples will serve to demonstrate the applicability of the methods described in this chapter to quite general steady problems. We choose Burgers' equation as a typical scalar convection-diffusion example, and the one-dimensional Euler equations as an example of the generalisations needed to deal with systems of equations. Of the various discretisations, we shall concentrate on the cell vertex scheme and the choice of distribution matrices for it; and some of the main points of interest concern the choice of iteration schemes that are necessary to solve the nonlinear systems of equations, and how these relate to and are motivated by time-stepping techniques.

6.6.1 Burgers' equation

This takes the form

$$-\epsilon u'' + (\tfrac{1}{2}u^2)' = S(x,u) \quad \text{on } (0,1), \tag{6.144a}$$

with the possible Dirichlet boundary conditions

$$u(0) = u_L, \quad u(1) = u_R. \tag{6.144b}$$

We say 'possible' boundary conditions, because if $u_L > 0$, the left-hand boundary condition must be Dirichlet, corresponding to an inflow; however, if a Dirichlet condition is imposed with $u_L < 0$, instead of a homogeneous Neumann condition, there will be a boundary layer on the left and the condition cannot be imposed in the limit $\epsilon = 0$. Similar considerations hold at the right-hand boundary.

For the cell vertex scheme on a general mesh, we denote the nodal flux by

$$F_i(U) \equiv F_i := \tfrac{1}{2}U_i^2 - \epsilon U_i', \tag{6.145a}$$

where U_i' is defined as in (6.71), including the extrapolation to inflow Dirichlet boundaries. A typical cell residual is then given by

$$R_{i-\frac{1}{2}}(U) \equiv R_{i-\frac{1}{2}} := D_- F_i(U) - S_{i-\frac{1}{2}}, \tag{6.145b}$$

where $S_{i-\frac{1}{2}}$ is the cell average of the source function; and if distribution matrices (or residual weights) of the form (6.60) are used, and no artificial dissipation is introduced, we obtain nodal residuals corresponding to all unknowns U_i as

$$N_i(U) := \frac{1}{x_{i+1} - x_{i-1}} \big[(1 + \operatorname{sign} U_{i-\frac{1}{2}})(x_i - x_{i-1})R_{i-\frac{1}{2}} $$
$$+ (1 - \operatorname{sign} U_{i+\frac{1}{2}})(x_{i+1} - x_i)R_{i+\frac{1}{2}} \big], \tag{6.145c}$$

where $U_{i-\frac{1}{2}} := \tfrac{1}{2}(U_i + U_{i-1})$. Normally only one cell residual will be involved in this expression, so that finding a solution U such that $N_i(U) = 0$ implies that either $R_{i-\frac{1}{2}} = 0$ or $R_{i+\frac{1}{2}} = 0$: but if $U_{i-\frac{1}{2}} > 0$ and $U_{i+\frac{1}{2}} < 0$, as occurs when a shock is present between x_{i-1} and x_{i+1}, the two cell residuals are combined with weights which ensure that only the flux difference $F_{i+1} - F_{i-1}$ is used; and when $U_{i-\frac{1}{2}} < 0$ and $U_{i+\frac{1}{2}} > 0$, corresponding to a rarefaction turning point or expansion fan, no cell residual is picked up and this is a situation where either artificial dissipation has to be added or some other procedure introduced. We will consider three cases in order to illustrate the various points involved.

Case (i) First we take $S = 0$ and choose $u_R = -u_L < 0$ so that inflow boundary conditions are applied on both the right and the left. Then it is easy to calculate the exact solution which consists of a stationary shock profile of the form

$$u(x) = a\frac{1 - e^{a(x-\frac{1}{2})/\epsilon}}{1 + e^{a(x-\frac{1}{2})/\epsilon}}, \quad \text{where } u_L = a\frac{1 - e^{-\frac{1}{2}a/\epsilon}}{1 + e^{-\frac{1}{2}a/\epsilon}}, \quad (6.146)$$

see, e.g., Whitham (1974). Note that as $\epsilon \to 0$ the central position of the shock becomes very sensitive to the boundary values, until in the inviscid limit it is not uniquely determined; this results in the position of the shock converging notoriously slowly when ϵ is small, for any iteration scheme. The solution to the unsteady limiting problem is, in fact, uniquely determined by the initial data. Thus it is that when a time-stepping iteration is used to find an approximation to the steady problem, its performance depends particularly strongly on the initial approximation.

Case (ii)   The other two cases are based on examples used by Morton *et al.* (1994), with case (ii) being an 'awkward' example proposed by Roe and van Leer (1988). This uses a source function $S = u(u-1)(2x-1)$ with $u_L = 9.0$ and $u_R = -9.26$, chosen to give a stable stationary shock, and we have added the diffusion term; however, in this case the solution even with $\epsilon = 0$ is uniquely determined by the source function and boundary data.

Case (iii)   With source function $S = \pi \sin \pi(2x-1)$ and initial data $u^0(x) \equiv 0$, the exact unsteady solution either with $\epsilon = 0$ and no boundary conditions, or with $\epsilon > 0$ and homogeneous Neumann boundary conditions, converges to a monotone increasing steady solution with a transcritical expansion at $x = 0$.

For the cell vertex approximation of these three problems we take a uniform mesh of width $h = 1/J$, and denote by $\{U_i^n$ for $i = 0, 1, \ldots, J\}$ the nodal values at the nth iteration and by U^n their linear interpolant. Artificial dissipation will be needed only for case (iii), so we leave that aside for the moment; and we use the trapezoidal rule $\frac{1}{2}[S(x_{i-1}, U_{i-1}) + S(x_i, U_i)]$ to approximate $S_{i-\frac{1}{2}}$, except possibly in the shock cell in case (ii). Then the key ingredients for the approximation after convergence are the definitions of the cell residual and the nodal residual in (6.145).

We will consider just two iteration schemes based on these residuals. The first is essentially Richardson relaxation based on the nodal residual and a local time step parameter, thus becoming Jacobi iteration for the maximal time step. However, some modification to the residual weights is made when $U_{i-\frac{1}{2}}$ is near zero, and we use a simpler notation than that in (6.60). So we write

$$U_i^{n+1} = U_i^n - \frac{1}{2}\Delta t_i^n (D_{i-\frac{1}{2}}^+ R_{i-\frac{1}{2}} + D_{i+\frac{1}{2}}^- R_{i+\frac{1}{2}})^n, \quad (6.147a)$$

where from (6.145) we have, for $i = 2, 3, \ldots, J-2$,

$$R_{i-\frac{1}{2}} = \frac{1}{h}\left[-\frac{\epsilon}{2h}U_{i-2} - (U_{i-\frac{1}{2}} - \frac{\epsilon}{2h})U_{i-1} + (U_{i-\frac{1}{2}} + \frac{\epsilon}{2h})U_i \right.$$

$$-\frac{\epsilon}{2h} U_{i+1}] - S_{i-\frac{1}{2}}; \qquad (6.147b)$$

in addition, at an inflow on the left, we have

$$R_{\frac{1}{2}} = \frac{1}{h}\left[-(U_{\frac{1}{2}} + \frac{\epsilon}{h})U_0 + (U_{\frac{1}{2}} + \frac{2\epsilon}{h})U_1 - \frac{\epsilon}{h}U_2\right] - S_{\frac{1}{2}}, \qquad (6.147c)$$

with a similar expression for inflow at the right; and in case (iii), where there is smooth outflow at the left, we have $U_0' = 0$ and hence set

$$R_{\frac{1}{2}} = \frac{1}{h}\left[-(U_{\frac{1}{2}} - \frac{\epsilon}{2h})U_0 + U_{\frac{1}{2}}U_1 - \frac{\epsilon}{2h}U_2\right] - S_{\frac{1}{2}}, \qquad (6.147d)$$

with a similar formula for $R_{J-\frac{1}{2}}$. Motivated by the coefficients in (6.147b), for $2 \le i \le J - 2$ we define the residual weights and time steps, which together give the relaxation parameters, as follows:

$$\text{if } U_{i-\frac{1}{2}} \le -\frac{\epsilon}{2h} \quad \text{set} \quad D_{i-\frac{1}{2}}^- = 2, \ D_{i-\frac{1}{2}}^+ = 0; \qquad (6.148a)$$

$$\text{if } U_{i-\frac{1}{2}} \ge \frac{\epsilon}{2h} \quad \text{set} \quad D_{i-\frac{1}{2}}^- = 0, \ D_{i-\frac{1}{2}}^+ = 2; \qquad (6.148b)$$

$$\text{if } |U_{i-\frac{1}{2}}| < \frac{\epsilon}{2h} \quad \text{set} \quad D_{i-\frac{1}{2}}^- = D_{i-\frac{1}{2}}^+ = 1; \qquad (6.148c)$$

and, for some global CFL parameter $\nu \in (0,1)$ we set

$$\Delta t_i = \nu h \min\left(\left[|U_{i-\frac{1}{2}}| + \frac{\epsilon}{2h}\right]^{-1}, \left[|U_{i+\frac{1}{2}}| + \frac{\epsilon}{2h}\right]^{-1}\right). \qquad (6.148d)$$

Note that all these parameters change at each iteration and are chosen to make all the coefficients of $\{U_j^n\}$ in (6.147a) non-negative; minor changes are therefore needed near the two boundaries.

The second iteration scheme is a *marching scheme* which can be related to a symmetric Gauss–Seidel iteration. By a marching scheme we mean one in which an attempt is made to set each cell residual to zero successively, starting from the boundary data. In cases (i) and (ii), this means trying to set $R_{i-\frac{1}{2}} = 0$ for $i = 1, 2, \ldots$ until the shock is met, where a combination of cell residuals has to be set to zero; and then sweeping from the right, trying to set $R_{i-\frac{1}{2}} = 0$ for $i = J, J - 1 \ldots$ until the shock is met again. In case (iii), the sequences have, instead, to start from the transcritical expansion. However, such a direct procedure is not possible when $\epsilon \ne 0$, and some modification is needed. As we see from (6.147), $R_{i-\frac{1}{2}}$ involves U_{i+1} through our approximation of U_i'; thus in the left sweep, away from the critical point, $U_i^{n+\frac{1}{2}}$ is calculated so as to set $R_{i-\frac{1}{2}}^{n+\frac{1}{2}} = 0$ where $(U_i')^n$ is used in its definition; while on the right sweep $(U_{i-1}')^{n+\frac{1}{2}}$ is used in the definition of the equation $R_{i-\frac{1}{2}}^{n+1} = 0$ by which U_{i-1}^{n+1} is determined. The modifications that are needed near the critical point will be discussed below for the three separate cases. Since in each case $R_{i-\frac{1}{2}}$ is quadratic in

U_i and U_{i-1}, the equations are solved without iteration with the choice of root being that closest to the current value of $U_{i-\frac{1}{2}}$.

In cases (i) and (ii), both iterations can be designed to ensure that there is a unique index s, corresponding to the *shock point*, such that

$$U^n_{s-\frac{1}{2}} \geq 0 > U^n_{s+\frac{1}{2}}; \tag{6.149a}$$

and in case (iii) there will similarly be a unique index such that

$$U^n_{s-\frac{1}{2}} \leq 0 < U_{s+\frac{1}{2}}, \tag{6.149b}$$

although in this case it may be desirable to find the cell in which the critical point lies. Maintenance of monotonicity in the approximation will clearly ensure these properties and that is the natural approach in cases (i) and (iii); but the solution in case (ii) is not monotone and (6.149a) has to be imposed more directly.

Let us now consider the application of the two iteration schemes to these three cases in turn. In case (i) the exact solution is not only monotone, but also the flux $f(u) \equiv \frac{1}{2}u^2 - \epsilon u'$ is constant, which could form the basis of a special iteration scheme. However, our two schemes perform perfectly satisfactorily for good initial data, with the marching scheme more efficient when $\epsilon < 10^{-2}$. A similar situation holds in case (ii); and for case (iii), where there is no shock to contend with, for the typical value $\epsilon = 10^{-3}$ the marching scheme is ten times more effective. Solutions for cases (ii) and (iii) are shown in Fig. 6.17.

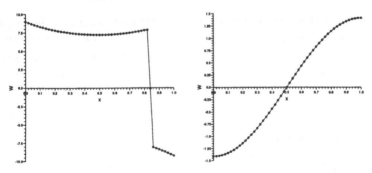

Figure 6.17. *Cell vertex approximation to Burgers' equation (6.144) with* $\epsilon = 10^{-3}$; *case (ii) on the left, case (iii) on the right*

6.6.2 One-dimensional Euler equations

Another model problem used by Morton *et al.* (1994) is for inviscid compressible flow in a nozzle with cross-sectional area

$$\sigma(x) = \pi[1 - \tfrac{1}{10}(1 + \cos \pi x)]^2, \quad -1 \leq x \leq 1. \tag{6.150}$$

For an ideal gas, and using the fact that the enthalpy is constant in steady flow, one can integrate the conservation laws across the nozzle to obtain the pair of equations for $\mathbf{w} \equiv (\rho, \rho u)^T$, in the form $d\mathbf{f}/dx = \mathbf{S}$, namely

$$\frac{d}{dx}\begin{pmatrix} \rho u \\ \rho u^2 + p \end{pmatrix} = \begin{pmatrix} 0 \\ (p/\sigma)d\sigma/dx \end{pmatrix}, \tag{6.151a}$$

where u is the velocity and ρ is the density multiplied by the cross-sectional area; then p is given by

$$p = (\rho/\gamma)(1 - \tfrac{1}{2}(\gamma - 1)u^2) \equiv \rho c^2/\gamma, \tag{6.151b}$$

where γ is the adiabatic gas constant. The Jacobian matrix $\partial \mathbf{f}/\partial \mathbf{w}$ has two real eigenvalues, one of which is always positive and the other is negative for subsonic flow, $|u| < c$, and positive for supersonic flow, $|u| > c$. The most interesting cases are for transonic flow where the flow is subsonic at both the inlet on the left and the outlet on the right, so that one boundary condition has to be given at each end; a typical case is obtained with $(u/c)_L = 0.5$ and $p_R = 0.843$. Then the flow accelerates through a sonic point, $u = c$, which one can show must be at the throat of the nozzle, i.e. at $x = 0$ for the nozzle of (6.150), and has to then achieve the subsonic outlet conditions through a shock.

To approximate this problem by the cell vertex method, the cell residual vectors are first defined as

$$\mathbf{R}_{i-\frac{1}{2}}(\mathbf{W}) \equiv \mathbf{R}_{i-\frac{1}{2}} := D_-\mathbf{f}(\mathbf{W}_i) - \mathbf{S}_{i-\frac{1}{2}}, \tag{6.152a}$$

where the source function is approximated by the trapezoidal rule. Then in each cell the average Jacobian matrix $A_{i-\frac{1}{2}}$ is diagonalised by the eigenvector matrix to give $A = T\Lambda T^{-1}$; with $\mathrm{sign}\Lambda$ defined as the diagonal matrix whose elements are the signs of the elements of Λ, we can define upwind distribution matrices to generalise the scalar case, as in (6.145c), to

$$D^{\pm} := T(I \pm \mathrm{sign}\Lambda)T^{-1}. \tag{6.152b}$$

Hence we can define the nodal residuals as

$$\mathbf{N}_i(\mathbf{W}) := \frac{1}{x_{i+1} - x_{i-1}}\big[D^+_{i-\frac{1}{2}}(x_i - x_{i-1})\mathbf{R}_{i-\frac{1}{2}}$$
$$+ D^-_{i+\frac{1}{2}}(x_{i+1} - x_i)\mathbf{R}_{i+\frac{1}{2}}\big], \tag{6.152c}$$

in the absence of any artificial dissipation. Indeed, if the eigenvalues of A are denoted by $\lambda^{(1)}, \lambda^{(2)}$, corresponding to right eigenvectors $\mathbf{r}^{(1)}, \mathbf{r}^{(2)}$ and left eigenvectors $\mathbf{l}^{(1)}, \mathbf{l}^{(2)}$ normalised to $\mathbf{l}^{(i)T}\mathbf{r}^{(j)} = \delta_{i,j}$, each of these terms can be decomposed to give

$$D^{\pm}\mathbf{R} = (1 \pm \mathrm{sign}\,\lambda^{(1)})(\mathbf{l}^{(1)T}\mathbf{R})\mathbf{r}^{(1)} + (1 \pm \mathrm{sign}\,\lambda^{(2)})(\mathbf{l}^{(2)T}\mathbf{R})\mathbf{r}^{(2)}. \tag{6.152d}$$

Thus in the supersonic region $\mathbf{N}_i$ is just a multiple of $\mathbf{R}_{i-\frac{1}{2}}$, while in the subsonic region it is composed of $(\mathbf{l}^{(1)T}\mathbf{R}\mathbf{r}^{(1)})_{i+\frac{1}{2}}$ and $(\mathbf{l}^{(2)T}\mathbf{R}\mathbf{r}^{(2)})_{i-\frac{1}{2}}$. Boundary conditions are combined with components of boundary cell residuals to give nodal residuals at boundary nodes. At the shock point, corresponding to the index s for which $\lambda^{(1)}_{s-\frac{1}{2}} \geq 0$ and $\lambda^{(1)}_{s+\frac{1}{2}} < 0$, the $\mathbf{r}^{(1)}$ contributions from the cells either side are added together; but at the sonic point there is no such contribution and either residual splitting must be used or the alternative of adding a fourth order dissipation adopted. In *residual splitting*, the position of the sonic point in a cell is calculated from the current approximation, and the two fractions of $\mathbf{l}^{(1)T}\mathbf{R}\mathbf{r}^{(1)}$ used to update the node on either side; details of the technique are given in Morton *et al.* (1994), where its close relationship to second order artificial dissipation is pointed out. Because of its more general utility we will use fourth order artificial dissipation for the results to be given below.

Table 6.1. *Number of iterations required to solve the transonic flow problem*

N	DUW	MTS	SGS(1)	SGS(∞)
17	793	88	85	11
33	1533	191	182	*
65	2905	361	258	15
129	5594	720	660	17
257	10475	1623	911	13

By means of the decomposition given in (6.152d) both of the iteration techniques described in the scalar cases dealt with in section 6.6.1 can be applied to the present problem, as well as other techniques given in Morton *et al.* (1994); this extension to systems of equations corresponds to the use of matrix-valued artificial dissipation by Swanson and Turkel (1992) to extend the idea of upwind difference schemes. The only significant difference from the scalar case is that, in the relaxation method $\delta \mathbf{W}_i = -\omega_i \mathbf{N}_i(\mathbf{W})$, we can either set

$$\omega_i = \nu \min[(x_i - x_{i-1})/\lambda^{(2)}_{i-\frac{1}{2}}, \ (x_{i+1} - x_i)/\lambda^{(2)}_{i+\frac{1}{2}}], \qquad (6.153a)$$

where $\nu \in (0,1)$, because $\lambda^{(2)}$ is positive and the larger of the two eigenvalues, or we can replace ω_i by a matrix so that each wave mode travels at its maximum speed in the update process; that is, we set, on a uniform mesh,

$$\delta \mathbf{W}_i = -\tfrac{1}{2}\nu(x_{i+1} - x_{i-1})|A_i|^{-1}\mathbf{N}_i(\mathbf{W}), \qquad (6.153b)$$

where $|A| := T\,|A|_\delta\,T^{-1}$ and $|A|_\delta$ denotes the fact that $|\lambda^{(1)}|$ is replaced by $|\lambda^{(1)}| + \delta$, to bound it from zero. Table 6.1 shows the number of iterations needed to solve the transonic problem defined above using the distribution matrices given by (6.152d): the first column (diagonalised upwind) corresponds to using (6.153a) with $\nu = 1$ and the second (matrix time-stepping) to using (6.153b); the third used the symmetric Gauss–Seidel iteration described in section 6.6.1 with one update at each stage and using the matrix parameter in (6.153b); and the final column corresponds to using Newton iteration at each point to solve the nodal residual equation before moving to the next point. The vast improvements that are possible with these techniques are obvious. Results for the generalised Lax–Wendroff method have not been included in the table because they depend very much on the amount of artificial dissipation added, and therefore the quality of the resulting solution; they are roughly comparable with those in the DUW column. The last column shows some erratic behaviour and * denotes lack of convergence for one mesh.

The results given in Fig. 6.18 demonstrate the importance of treating both the sonic point and the shock point by the techniques just described.

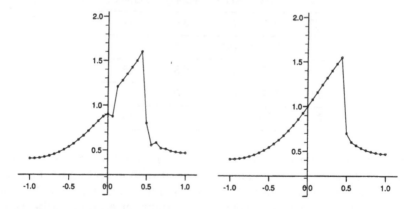

Figure 6.18. *Cell vertex approximation to the transonic nozzle problem; on the left, when the sonic point is ignored and simple Lax–Wendroff distribution matrices used; on the right, with fourth order dissipation at the sonic point and the distribution matrices of (6.152)*

CHAPTER 7

Unsteady problems

7.1 Introduction

There have been very important developments during the last fifteen years in algorithms for approximating unsteady convection-dominated problems, particularly hyperbolic conservation laws. Not all of this is relevant to our present purposes, but many of the main ideas certainly are. Fortunately, there are at least two recent texts which cover the finite difference methodology as well as the basic theory of hyperbolic conservation laws, namely Godlewski and Raviart (1991) and LeVeque (1992); and results obtained with a wide variety of methods for a set of simple convection-diffusion problems are presented in Vreugdenhil and Koren (1993). We shall therefore largely confine the present coverage to finite element and finite volume methods; these are also more naturally suited to multi-dimensional problems in complicated geometries and, in some sense too, to problems in which both diffusion and convection play a significant rôle.

Problems that we shall consider are of the general form

$$\frac{\partial u}{\partial t} + \nabla \cdot (-\epsilon a(\mathbf{x})\nabla u + \mathbf{f}(\mathbf{x}, u)) = S(\mathbf{x}, u) \text{ on } \Omega, \qquad (7.1a)$$

i.e. $\quad \dfrac{\partial u}{\partial t} + Lu = S$ on Ω, with $u(t = 0) = u^0$, $\qquad (7.1b)$

$$\text{and } u = u_L \text{ on } \partial\Omega_D, \quad \partial u/\partial n = 0 \text{ on } \partial\Omega_N, \qquad (7.1c)$$

as in (5.3) with the same constraints on the boundary conditions as in the steady case but with u_L possibly depending on t; the only difference here is that we have generalised the convective flux from $b(\mathbf{x})u$ to the possibly nonlinear $\mathbf{f}(\mathbf{x}, u)$ because many of the methods will be used on such nonlinear equations and have been designed to deal with the phenomena such as shocks that they give rise to. We should note here, too, that the existence, uniqueness and regularity theory for (7.1) with $\epsilon > 0$ is succinctly covered in Godlewski and Raviart (1991) in the process of using the vanishing viscosity method to establish results for the hyperbolic limit. Very commonly, as in earlier chapters, we shall simplify this formulation by setting $a(\mathbf{x}) \equiv 1$

and $S(\mathbf{x}, u) \equiv S(\mathbf{x})$, which is very often zero, as well as $\mathbf{f}(\mathbf{x}, u) \equiv \mathbf{b}(\mathbf{x})u$. We shall also, of course, require that all the methods generalise naturally to dealing with systems of equations for a vector of unknowns $\mathbf{w}$ with a matrix of fluxes $\mathbf{F}(\mathbf{x}, \mathbf{w})$; and we shall often comment on what is entailed in this generalisation.

All of the methods we shall consider will eventually produce approximations $U^n(\mathbf{x})$, at discrete time levels t_n, of the form

$$U^n(\mathbf{x}) = \sum_j U_j^n \phi_j(\mathbf{x}), \qquad (7.2a)$$

for some set of basis functions ϕ_j spanning a trial space S^h. However, especially in the case of piecewise constant basis functions, the resultant schemes can often by expressed in finite difference form; indeed, much of the recent development of difference schemes referred to above has been associated with the interpretation of U_j^n as a cell average. As in the steady case, we shall also use the notation S_E^h to denote the satisfaction of the essential (Dirichlet) boundary conditions, and S_0^h for satisfaction of the corresponding homogeneous conditions. Sometimes the approximation will go through an intermediate formulation, commonly called the semi-discrete problem, giving

$$U(\mathbf{x}, t) = \sum_j U_j(t)\phi_j(\mathbf{x}). \qquad (7.2b)$$

We shall also use various test spaces T^h spanned by basis functions $\psi_j(\mathbf{x})$; and some methods use test functions $\psi_j(\mathbf{x}, t)$ in space-time.

The most common trial functions, in one dimension, are piecewise constants, continuous piecewise linears, discontinuous piecewise linears and discontinuous piecewise quadratics, together with their natural generalisations into multidimensions as in Chapter 4. Higher degree polynomials are also used with some schemes; but since, because of various super-convergence phenomena, fourth order accuracy can be obtained for the pure convection problem with only linear basis functions, this is not very common. At the other extreme it might be wondered how piecewise constants can be used for convection-diffusion problems. However, consider the one-dimensional problem and, with a piecewise constant test function, integrate (7.1a) over an interval $(x_{i-\frac{1}{2}}, x_{i+\frac{1}{2}})$. This will give a difference of fluxes $\nabla_+(-\epsilon a u'_{i-\frac{1}{2}} + f_{i-\frac{1}{2}})$. Now, even an approximation to $f_{i-\frac{1}{2}}$ at the flux boundary has to be recovered from the piecewise constant $U(x)$, i.e. the interval values $\{U_j\}$; and by an extension of the technique, gradient approximations $U'_{i-\frac{1}{2}}$ can be recovered, in a manner similar to that employed in the cell vertex method of the last chapter.

The main distinguishing feature of the various methods we shall describe below is how the time-stepping is effected. We can loosely classify them into three broad classes:

(i) *semi-discrete based methods*, in which a system of ordinary differential equations in the time variable is formed and an O.D.E. solver such as a Runge–Kutta technique is applied to the system. Thus the core of the method is to find formulae for the $\dot{U}_j(t)$ in an expansion of the form (7.2b) for $\partial U/\partial t$;

(ii) *evolution-Galerkin and semi-Lagrangian methods*, based on some approximation E_Δ to the evolution operator $E(\Delta t)$ corresponding to the equation (7.1), particularly in the homogeneous case, that is $E(\Delta t)$: $u(t) \longmapsto u(t + \Delta t)$. Then U^{n+1} is given by the system of equations

$$(U^{n+1} - E_\Delta U^n, \phi_i) = 0, \quad \forall \phi_i \in S_0^h. \tag{7.3a}$$

A Petrov–Galerkin technique may also be applied, giving

$$(U^{n+1} - E_\Delta U^n, \psi_i) = 0, \quad \forall \psi_i \in T^h. \tag{7.3b}$$

Taylor–Galerkin, characteristic-Galerkin, Lagrange–Galerkin and semi-Lagrangian methods are typical of the schemes that fall into this category;

(iii) *generalised Godunov schemes*, in which the operator on the left of (7.1a) is considered as a divergence in space-time and integrated over a space-time cell as in the finite volume methods of the last chapter; the original scheme of this type is that of Godunov (1959).

In many cases, of course, a method uses some combination of techniques from these classes, sometimes by means of the fractional step technique (Yanenko 1971). The Lagrange–Galerkin method, for example, strictly applies to only the operator $\partial/\partial t + \mathbf{b}\cdot\nabla$, which it replaces by a material derivative, with all other terms in (7.11) dealt with in some other way.

Many methods make use of *recovery techniques* to improve the accuracy of basic schemes which are often only first order accurate. The advantage of using these techniques, which are sometimes now called *reconstruction techniques*, rather than going directly to higher order methods through using higher degree polynomials in (7.2), is partly practical and partly because it naturally allows the schemes to be solution-adaptive, as in section 6.2.3. However, there is considerable overlap in the methods based on recovery techniques and those that use some mix of higher and lower order methods, as was pioneered in the *flux-corrected transport* schemes of Boris and Book (1973).

In the following account we leave the semi-discrete based methods until last because there is considerable overlap with the idea of using time-stepping techniques to obtain solutions to steady problems, and this was discussed at some length in the last chapter.

7.2 Evolution-Galerkin and semi-Lagrangian methods

In order to cover both (7.3a) and (7.3b) and to use other than the L_2 projection that is indicated in each case, as well as to include the use of recovery techniques, we will use the following notation for these methods

$$U^{n+1} = (P^h E_\Delta) R^h U^n. \qquad (7.4)$$

Here P^h is some projection onto S_E^h and R^h represents the recovery process; we have bracketed $P^h E_\Delta$ together because in practice this is a single step and $E_\Delta R^h U^n$ is never fully calculated.

7.2.1 Taylor–Galerkin methods

If one merely replaces $\partial u/\partial t$ in (7.1) by some finite difference approximation, such as $(U^{n+1} - U^n)/\Delta t$ or $(U^{n+1} - U^{n-1})/2\Delta t$, and then forms the Galerkin projection to obtain schemes of the form

$$\left(\frac{U^{n+1} - U^n}{\Delta t} + LU^n, \phi_i \right) = (S, \phi_i)$$

$$\text{or} \quad \left(\frac{U^{n+1} - U^{n-1}}{2\Delta t} + LU^n, \phi_i \right) = (S, \phi_i), \qquad (7.5)$$

the result is not very satisfactory for convection-dominated problems. As pointed out by Cullen and Morton (1980) and Morton and Parrott (1980), the use of the full mass matrix that is implied here gives improved accuracy for small values of Δt, but the stability limit is reduced as compared with corresponding finite difference schemes, typically by a factor $\sqrt{3}$. The latter authors introduced the idea that a scheme should satisfy a *unit CFL condition*; that is, for pure one-dimensional convection with constant b the scheme should be exact on a uniform mesh for which $b\Delta t = \Delta x$. The Galerkin schemes do not satisfy this condition, but these authors devised test functions with the same support as the basis functions $\phi_j(x)$ that yielded schemes which do satisfy the condition. Some of the schemes coincide with those given by the Taylor–Galerkin approach, and since the latter is so much more systematic we pass to it immediately.

We consider the Taylor–Galerkin scheme as devised by Donea (1984) — see also Donea *et al.* (1988) — in the framework of evolution-Galerkin schemes (7.4) where E_Δ is given by a Taylor expansion and R^h is the identity. In one dimension, with no source term or diffusion, a Taylor expansion identical to that given in (6.61) to derive the Lax–Wendroff method gives a second order approximation $E_\Delta^{(2)}$; and Donea also uses a third order approximation $E_\Delta^{(3)}$. Thus we have, with $A \equiv \partial f/\partial u$,

$$E_\Delta u \approx u + \Delta t\, u_t + \tfrac{1}{2}(\Delta t)^2 u_{tt} + \tfrac{1}{6}(\Delta t)^3 u_{ttt} + \dots;$$

$$E_\Delta^{(2)} u := u - \Delta t\, f_x + \tfrac{1}{2}(\Delta t)^2 (A f_x)_x; \qquad (7.6a)$$

$$E_\Delta^{(3)} u := u - \Delta t\, f_x + \tfrac{1}{2}(\Delta t)^2 (Af_x)_x + \tfrac{1}{6}(\Delta t)^2 [(Af_x)_x^+ - (Af_x)_x], \quad (7.6b)$$

where we have used the notation $(Af_x)^+$ to denote (Af_x) evaluated for the updated value $E_\Delta^{(3)} u$ rather than for u, so that $E_\Delta^{(3)} u$ is defined implicitly. Using a parameter θ to combine these two schemes, with $\theta = 0$ corresponding to (7.6a) and $\theta = 1$ to (7.6b), after Galerkin projection and integration of the second order terms by parts, the complete scheme for $u_t + f_x = 0$ takes the following form: find $U^{n+1} \in S_E^h$ such that

$$(U^{n+1}, \phi_i) + \tfrac{1}{6}\theta(\Delta t)^2 ((Af_x)^{n+1}, \phi_{i,x})$$
$$= (U^n, \phi_i) - \tfrac{1}{2}(\Delta t)^2 (1 - \tfrac{1}{3}\theta)((Af_x)^n, \phi_{i,x}) - \Delta t(f_x^n, \phi_i) \quad \forall \phi_i \in S_0^h, \quad (7.7)$$

where $\phi_{i,x}$ is used to denote the x-derivative of ϕ_i. The first inner product on each side can be integrated exactly and will give the mass matrix for the basis functions $\{\phi_i\}$; the remaining inner products will generally have to be approximated by some choice of quadrature formulae.

To clarify the rôle and magnitude of each term it is useful to consider the linear advection problem with $f \equiv bu$, $A \equiv b = $ constant, and a continuous linear approximation on a uniform mesh. Then with $\nu := b\Delta t/\Delta x$, and recalling that the mass matrix can be represented by the difference operator $1 + \tfrac{1}{6}\delta^2$, we obtain at interior points the difference scheme

$$[1 + \tfrac{1}{6}(1 - \theta\nu^2)\delta^2]U^{n+1} = [1 + \tfrac{1}{6}(1 - \theta\nu^2 + 3\nu^2)\delta^2 - \nu\Delta_0]U^n. \quad (7.8a)$$

The special choice $\theta\nu^2 = 1$ gives the identity matrix on the left and corresponds to *mass lumping*; the scheme then reduces to the Lax–Wendroff difference method

$$(\theta\nu^2 = 1) \quad U^{n+1} = [1 - \nu\Delta_0 + \tfrac{1}{2}\nu^2\delta^2]U^n. \quad (7.8b)$$

If we set $\theta = 1$ as in (7.6b), the mass matrix is in effect made more and more diagonal as ν increases; like the Lax–Wendroff methods the scheme then satisfies the unit CFL condition, but it is third order accurate for all values of ν. Because of the theoretical justification that these arguments give for the practically convenient recourse to mass lumping, especially when $|\nu|$ is near unity, this is very common practice.

With mass-lumping, schemes based on (7.7) have all the advantages of the Lax–Wendroff method, being explicit, easily generalisable to systems of nonlinear equations in multi-dimensions and having no free parameters to adjust; detailed differences from Lax–Wendroff on a rectangular mesh depend on the quadrature rules used, while (7.7) is just as easily implemented on an unstructured mesh — see Morgan *et al.* (1991) for details. On the other hand, it suffers from the disadvantage of the Lax–Wendroff method in generating spurious oscillations behind sharp variations in the solution; and as in the original presentation of that method (Lax and Wendroff 1960) some artificial dissipation has to be added, preferably in a solution-adaptive manner.

Extending the schemes to add a diffusive term is straightforward if first order accuracy in Δt is adequate for this term. In the expansion for E_Δ, it is then necessary only to replace u_t by $-f_x + \epsilon(au_x)_x$ rather than by $-f_x$ alone. With mass lumping and $f(u) \equiv bu$, (7.7) becomes

$$(U^{n+1}, \phi_i)_L = (U^n, \phi_i)_L + \Delta t(bU^n - \epsilon aU_x^n - \tfrac{1}{2}\Delta t\, b(bU^n)_x, \phi_{i,x}), \quad (7.9a)$$

where $(\cdot, \cdot)_L$ denotes mass lumping. Note how the last term here gives the difference of a flux which corresponds precisely to that in (6.5d) when b is constant. However, when mass lumping is not used it is natural to use the θ-method for modelling the diffusion, giving

$$(U^{n+1}, \phi_i) + \Delta t(\tfrac{1}{6}\theta\Delta t\, b(bU^{n+1})_x + \theta'\epsilon aU_x^{n+1}, \phi_{i,x})$$
$$= (U^n, \phi_i) + \Delta t(bU^n - (1-\theta')\epsilon aU_x^n - \tfrac{1}{2}\Delta t(1-\tfrac{1}{3}\theta)b(bU^n)_x, \phi_{i,x}), \quad (7.9b)$$

where θ' denotes the parameter determining the implicitness of the diffusion terms. There is clearly a lot of overlap in these terms; in particular, when a and b are constants and the mesh is uniform, we can write $\mu := \epsilon a\Delta t/(\Delta x)^2$ and the scheme becomes of the same form as (7.8a), namely

$$(1 + \Phi_1\delta^2)U^{n+1} = (1 + \Phi_0\delta^2 - \nu\Delta_0)U^n,$$

where $\Phi_1 = \tfrac{1}{6}(1 - \theta\nu^2) - \theta'\mu$ and $\Phi_0 = \tfrac{1}{6}(1 - \theta\nu^2) + \tfrac{1}{2}\nu^2 + (1-\theta')\mu$. It is explicit when $\Phi_1 = 0$; and a short calculation shows that it is practically stable (i.e. all Fourier modes are nonincreasing) when $\Phi_0 + \Phi_1 \leq \tfrac{1}{2}$, that is when

$$\nu^2 + 2(1 - 2\theta')\mu + \tfrac{2}{3}(1 - \theta\nu^2) \leq 1. \quad (7.10)$$

Important special cases include: $\theta' = \tfrac{1}{2}$ (Crank–Nicolson) and $\theta\nu^2 = 1$ (mass lumping) which gives the familiar CFL condition $\nu^2 \leq 1$; $\theta' = 0$ and $\theta\nu^2 = 1$ giving the Lax–Wendroff condition $\nu^2 + 2\mu \leq 1$; $\theta' = \tfrac{1}{2}$ and $\theta = 0$ (full mass matrix) giving $\nu^2 \leq \tfrac{1}{3}$, as already referred to; and, the worst case, $\theta' = \theta = 0$ giving $\nu^2 + 2\mu \leq \tfrac{1}{3}$.

In multi-dimensions with $\partial f/\partial u = b$ so that $(Af_x)_x$ is generalised to $\nabla\cdot(b\nabla\cdot f)$, implicit schemes of the form (7.9b) will require the solution at each time step of an algebraic system of the form

$$(U^{n+1}, \phi_i) + \Delta t(\tfrac{1}{6}\theta\Delta t\, \mathbf{b}^{n+1}\nabla\cdot\mathbf{f}^{n+1} + \theta'\epsilon a\nabla U^{n+1}, \nabla\phi_i) = \text{R.H.S.} \quad (7.11)$$

In any of the cases in which $\nabla\cdot\mathbf{f} = \mathbf{b}\cdot\nabla u$ this is a self-adjoint system and is readily solved (or approximately so) by a variety of means — A.D.I., S.O.R., multigrid, conjugate gradient etc., using linearisation where necessary; even if $\mathbf{b}\nabla\cdot\mathbf{f}$ is not of this form, this term is only a small perturbation to a self-adjoint system. Thus, here and subsequently, we shall not comment on the solution of such systems; only when an implicit convection term seriously destroys the self-adjointness shall we have cause to comment.

The best way to include a source term in these schemes is not very obvious; usually a term $\Delta t(S(\cdot, U^n), \phi_i)$ is added to the right-hand side of (7.7), but a similar implicit term could be subtracted from the left. Also, as with the Lax–Wendroff scheme, there is a two-step version which is primarily of interest for systems of equations as it avoids the need to evaluate the Jacobian matrix $A = \partial f / \partial u$.

7.2.2 Lagrange–Galerkin and characteristic-Galerkin methods

This group of schemes includes quite a wide variety introduced independently by a number of different authors, using differing formulations and applying them in various applications areas; a brief survey has been given by Morton (1991b). The common feature is that all or part of the differential operator in (7.1) is approximated by the constancy of the solution along a particle path (*Lagrange–Galerkin method*) or a characteristic (*characteristic-Galerkin method*). For the present convection-diffusion problems these two forms are virtually identical; but, for example, for the one-dimensional Euler equations of inviscid gas dynamics the characteristic speeds are $u - c$, u and $u + c$ where c is the sound speed, and all of these would be used in a characteristic-Galerkin method, while the Lagrange–Galerkin methods would use only the particle trajectories $dX/dt = u$.

In the scalar problem (7.1) suppose for the moment that $\nabla \cdot f \equiv b \cdot \nabla u$ and b is independent of u, so we have

$$\left(\frac{\partial}{\partial t} + b \cdot \nabla\right) u = \nabla \cdot (\epsilon a \nabla u) + S. \tag{7.12}$$

The total derivative on the left is along a trajectory which we denote by $X(\mathbf{x}, s; t)$ and is given by

$$\frac{d\mathbf{X}}{dt} = b \quad \text{with} \quad \mathbf{X}(\mathbf{x}, s; s) = \mathbf{x}. \tag{7.13a}$$

We shall then, in effect, use an approximate evolution operator $\widehat{E}(\Delta t)$ for the equation $u_t + b \cdot \nabla u = 0$, with $u(x, t_n) = u^n(x)$, given by

$$(\widehat{E}(\Delta t)u^n)(\mathbf{x}) := u^n(\mathbf{X}(\mathbf{x}, t_{n+1}; t_n)). \tag{7.13b}$$

We refer to this as an approximate evolution operator because we shall use the same notation when b depends on u and so shocks may form in the true evolution.

The most basic and best known Lagrange–Galerkin algorithm is for pure convection in one dimension with constant b: using a representation of U by piecewise constants U_j on intervals of length h_j, the evolution operator $\widehat{E}(\Delta t)$ just shifts the constants along by a fixed amount $b\Delta t$; hence after the projection as in (7.4), with no recovery and assuming all the CFL numbers

$\nu_j := b\Delta t / h_j$ are less than unity, we obtain

$$h_i U_i^{n+1} = h_i U_i^n - b\Delta t(U_i^n - U_{i-1}^n),$$

$$\text{i.e.} \quad U_i^{n+1} = (1 - \nu_i)U_i^n + \nu_i U_{i-1}^n, \tag{7.14}$$

the familiar upwind difference scheme.

(i)   *Direct formulation.* The direct formulation of the Lagrange–Galerkin method, which is commonly used to approximate the incompressible Navier–Stokes equations (see, for instance, Pironneau (1982) and Süli (1988)), is obtained from (7.13) as follows: we suppose that (7.12) is integrated along trajectories drawn backwards from $(\mathbf{x}, t_{n+1})$ to $\mathbf{X}(\mathbf{x}, t_{n+1}; t_n)$, using some approximation for b; then the integral of the right-hand side is approximated by quadrature. If one-point quadrature, using values at t_n, is used in both cases, we obtain the explicit scheme

$$(U^{n+1}, \phi_i) = (U^n(\mathbf{X}(\cdot, t_{n+1}; t_n)), \phi_i)$$
$$+ \Delta t[-\epsilon(a\nabla U^n, \nabla \phi_i) + (S^n, \phi_i)]; \tag{7.15a}$$

it is more common, however, to use the unconditionally stable implicit scheme based on the same trajectories,

$$(U^{n+1}, \phi_i) + \Delta t[\epsilon(a\nabla U^{n+1}, \nabla \phi_i) - (S^{n+1}, \phi_i)]$$
$$= (U^n(\mathbf{X}(\cdot, t_{n+1}; t_n)), \phi_i). \tag{7.15b}$$

These are the equations away from the boundary; we will discuss boundary conditions later when we have considered other formulations. The critical inner product here is that on the right of (7.15b) and it will normally have to be approximated by some form of quadrature.

However, in our model one-dimensional problem in which $S = 0$, a and b are constants, and a continuous linear approximation is used on a uniform mesh, all the integrals can be performed exactly: $X(x, t_{n+1}; t_n) = x - b\Delta t$ and it is easy to calculate $(U^n(\cdot - b\Delta t), \phi_i)$ by setting

$$\nu := b\Delta t / \Delta x = m + \hat{\nu} \quad \text{with} \quad m \in \mathbb{Z}, \ 0 < \hat{\nu} \le 1, \tag{7.16}$$

and using basic convolution properties of spline functions, as follows. On a uniform mesh, the linear basis function $\phi_j(x)$ is equal to the second order B-spline $\chi^{(2)}(s)$, where $x = (j + s)\Delta x$ and B-splines of any order are generated by the recurrence relation

$$\chi^{(p)}(s) := \int \chi^{(p-1)}(\sigma - s)\chi(\sigma)\mathrm{d}\sigma, \tag{7.17}$$

with $\chi(\sigma) \equiv \chi^{(1)}(\sigma)$ defined as the characteristic function of the interval $[-\frac{1}{2}, \frac{1}{2}]$ — see Fig. 7.1. It follows that

$$(\phi_j(\cdot - \hat{\nu}\Delta x), \phi_i) = \Delta x \int \chi^{(2)}(\sigma - j - \hat{\nu})\chi^{(2)}(\sigma - i)\mathrm{d}\sigma$$

$$= \Delta x \chi^{(4)}(j - i + \hat{\nu}); \tag{7.18}$$

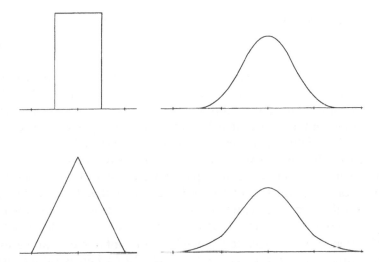

Figure 7.1. *B-spline basis functions on a uniform mesh,* $\phi_j(x) \equiv \chi^{(p)}(s)$, *for* $p = 1, 2$ *on the left and* $p = 3, 4$ *on the right*

and from this cubic spline we can obtain the coefficients in the following expansion,

$$(\Delta x)^{-1}(U^n(\cdot - b\Delta t), \phi_i) = \tfrac{1}{6}\hat{D}^3 U^n_{i-m-2}$$
$$+ (\tfrac{1}{6} + \tfrac{1}{2}\hat{D} + \tfrac{1}{2}\hat{D}^2 - \tfrac{1}{2}\hat{D}^3)U^n_{i-m-1}$$
$$+ (\tfrac{2}{3} - \hat{D}^2 + \tfrac{1}{2}\hat{D}^3)U^n_{i-m}$$
$$+ (\tfrac{1}{6} - \tfrac{1}{2}\hat{D} + \tfrac{1}{2}\hat{D}^2 - \tfrac{1}{6}\hat{D}^3)U^n_{i-m+1}$$
$$= [(1 + \tfrac{1}{6}\delta^2) - \hat{D}\Delta_0 + \tfrac{1}{2}\hat{D}^2\delta^2 - \tfrac{1}{6}\hat{D}^3\delta^2\Delta_-]U^n_{i-m}.$$
$$(7.19)$$

Again with $\mu := \epsilon a \Delta t/(\Delta x)^2$, (7.15b) therefore reduces to the difference scheme

$$[1 + (\tfrac{1}{6} - \mu)\delta^2]U^{n+1}_i = [(1 + \tfrac{1}{6}\delta^2) - \nu\Delta_0$$
$$+ \tfrac{1}{2}\hat{D}^2\delta^2 - \tfrac{1}{6}\hat{D}^3\delta^2\Delta_-]U^n_{i-m}, \qquad (7.20)$$

with (7.15a) obtained by merely replacing $\mu\delta^2 U^{n+1}_i$ by $\mu\delta^2 U^n_i$. Because of the shifting stencil as ν and m change, this scheme is unconditionally stable; and it is third order accurate in the convection terms. The scheme, at least with $m = 0$ and $\mu = 0$, has been derived by many authors from various formulations — see Morton (1991b); the earliest derivation appears to be that of Lesaint (1977). In fact, a one parameter family of schemes such as (7.20) can be obtained by using a mixed inner product in which to carry out the projection, namely

$$(u, v)_\gamma := (u, v) + (\gamma \Delta x)^2 (u', v'). \tag{7.21}$$

As shown in Childs and Morton (1990), where their comparative properties are discussed, these include the Taylor–Galerkin scheme, in which the third difference term is eliminated, and the difference scheme of Warming *et al.* (1973), in which the mass matrix becomes the identity. We shall return to the discussion of these schemes in sections 7.2.7 and 7.2.8.

By the exactly same means, difference schemes for the pure convection problem can be generated from a piecewise constant basis generalising (7.14), or for the complete convection-diffusion problem from higher order splines. In the former case, however, where we take the intervals to be $[x_{j-\frac{1}{2}}, x_{j+\frac{1}{2}}]$ so that $\phi_j(x) \equiv \chi^{(1)}(s)$ with $x = (j + s)\Delta x$, in order to include the diffusion terms we need to recover the gradient of a typical piecewise constant function U. In anticipation of section 7.2.4 where we also consider a nonuniform mesh, we will represent U_x by $D_- U_j$ on $(x_{j-1}, x_j]$ so that we can carry out the following calculation,

$$(U_x, \phi_{i,x}) = -D_+ U_i + D_- U_i \equiv -\delta^2 U_i / \Delta x; \tag{7.22a}$$

then for the piecewise constant trial space it is readily seen that (7.20) is replaced by

$$(1 - \mu \delta^2) U_i^{n+1} = (1 - \hat{\nu} \Delta_-) U_{i-m}^n \equiv \hat{\nu} U_{i-m-1}^n + (1 - \hat{\nu}) U_{i-m}^n. \tag{7.22b}$$

Thus the convection problem is approximated to first order accuracy. In general, for splines of order p this accuracy (on a uniform mesh) is order $2p - 1$ — see Childs and Morton (1990) for more details of these and other schemes, as well as an analysis of their stability properties.

The integrals in (7.15) are also readily evaluated exactly on a nonuniform mesh in one dimension even with variable b (by tracing back the characteristics through each element boundary point), but in higher dimensions this is generally not possible. Even on a rectangular mesh but with a variable $\mathbf{b}$, the transformation $\mathbf{x} \longmapsto \mathbf{X}(\mathbf{x}, t_{n+1}; t_n)$ will in general distort a rectangular cell and rotate it relative to the mesh so that exact integration is impractical. Some form of quadrature is then necessary, but great care has to be taken in its choice if the stability of the scheme is not to be lost. Morton *et al.* (1988) have shown that in one dimension all standard quadrature schemes introduce some intervals of instability; since these often include an interval $0 < \hat{\nu} < \gamma$ they cannot be avoided by decreasing the time step, and if the flow field has a stagnation point they are particularly troublesome. What these authors advocate as an alternative on a rectangular grid is to transform the centre of each cell according to (7.13), but leave the orientation unchanged and dilate the cell only if $\nabla \cdot \mathbf{b} \neq 0$, using the easily established relation for $|J| = \det(\partial X/\partial x)$ that

$$\frac{d|J|}{dt} = (\nabla \cdot b)|J|. \tag{7.23}$$

Then the integrals can be evaluated exactly. It is interesting to note the close link between this technique and the *Second Moment Method* of Egan and Mahoney (1972) which we will describe in more detail in section 7.2.6. In the next section we will describe this *area-weighting* technique in more detail and its stability will be considered in section 7.2.8.

(ii)   *Weak form.* We now consider two alternative formulations of the Lagrange–Galerkin method, of which the first is the *weak form* introduced by Benqué *et al.* (1982) for modelling the Navier–Stokes equations. This uses a test function $\psi(x, t)$ in space-time; multiplying (7.1a) by such a function, integrating by parts over $\Omega \times [t_n, t_{n+1}]$ and assuming the support of ψ does not extend to the boundary, we obtain

$$(u(\cdot, t_{n+1}), \psi(\cdot, t_{n+1})) - (u(\cdot, t_n), \psi(\cdot, t_n))$$
$$= \int_{\Omega \times [t_n, t_{n+1}]} \left[u \frac{\partial \psi}{\partial t} + (-\epsilon a \nabla u + f) \cdot \nabla \psi + S\psi \right]. \tag{7.24}$$

Now suppose ψ_i satisfies the following initial and trajectory conditions,

$$\psi_i(x, t_{n+1}) \equiv \phi_i(x), \quad \frac{\partial \psi_i}{\partial t} + (f/u) \cdot \nabla \psi_i = 0 \quad \text{on } (t_n, t_{n+1}). \tag{7.25a}$$

That is, if we use the same notation as (7.13) even though the trajectory has a different velocity, we have ψ_i constant along the trajectory so that

$$\psi_i(x, t_n) = \phi_i(X(x, t_n; t_{n+1})). \tag{7.25b}$$

Then, if the trapezoidal rule is used for the remaining time integrals, U is determined by

$$(U^{n+1}, \phi_i) + \tfrac{1}{2}\epsilon(a\nabla U^{n+1}, \nabla \phi_i)$$
$$= (U^n, \psi_i(\cdot, t_n)) - \tfrac{1}{2}\epsilon(a\nabla U^n, \nabla \psi_i(\cdot, t_n))$$
$$+ \tfrac{1}{2}[(S^{n+1}, \phi_i) + (S^n, \psi_i(\cdot, t_n))], \tag{7.26}$$

with $\psi_i(\cdot, t_n)$ given by (7.25b).

Notice that there are two related differences from the direct formulation. Firstly, if for example $f \equiv f(u)$, the direct formulation uses the trajectory velocity $b = \partial f/\partial u$ so that $\nabla \cdot f$ can be replaced by $b \cdot \nabla u$; while the weak form works directly with the conservation form and the trajectory velocity is given by the ratio f/u. In the linear case, $f(x, n) \equiv b(x)u$, the two velocities are the same if $\nabla \cdot b = 0$; but if this is not the case and the equation is in conservation form, then the weak form is preferred because it exactly preserves the conservation law. Secondly, the integrals over U^n are of different form: if we use x and y to denote the two ends of a trajectory, so that $X(x, t_n; t_{n+1}) = y$ and $X(y, t_{n+1}; t_n) = x$, the integral in the direct form on the right-hand side of (7.15b) is

$$\int U^n(\mathbf{x})\phi_i(\mathbf{y})d\mathbf{y}; \qquad (7.27a)$$

while that in (7.26) is

$$\int U^n(\mathbf{x})\phi_i(\mathbf{y})d\mathbf{x}. \qquad (7.27b)$$

These are only the same if the trajectories are the same and $dx = dy$. This occurs in the linear case when $\nabla \cdot \mathbf{b} = 0$ so that $\nabla \cdot \mathbf{f} = \nabla \cdot (\mathbf{b}u) = \mathbf{b} \cdot \nabla u$, and in special nonlinear cases, such as the incompressible Navier–Stokes equations, where for the velocity components (u, v) we have $uu_x + vu_y = (u^2)_x + (uv)_y$ and $uv_x + vv_y = (uv)_x + (v^2)_y$ because $u_x + v_y = 0$.

The use of quadrature rules makes for further differences between the two formulations. It has already been noted that the weak form is to be preferred when the equation is in conservative form, and this is emphasised when the area-weighting technique is used. As will be seen in the next section and is indicated by the sketch on the right of Fig. 7.2, approximation of (7.27b) is carried out element by element over the x variable; so the element integrals of U^n are allocated systematically to form the updated U^{n+1}, and the conservation properties of the pure convection problem are preserved exactly. Even when $\nabla \cdot \mathbf{b} = 0$ so that (7.27a) and (7.27b) are exactly the same, differences are introduced by the use of quadrature over y in the former case — see Morton *et al.* (1988) where some numerical examples can be found.

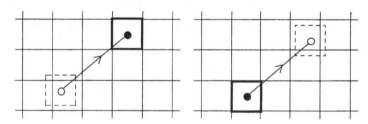

Figure 7.2. *The convective mapping used in the area-weighting technique: the direct formulation is shown on the left and the weak form on the right*

Very often the trajectories in both the direct and the weak formulation will be approximated by straight lines, and we will suppose that this is the case in order to consider the imposition of boundary conditions, starting with the one-dimensional case. To determine $U^{n+1} \in S_E^h$, (7.15) or (7.26) have to be applied for $\forall \phi_i \in S_0^h$, or equivalently (7.27a) or (7.27b) have to be evaluated. Near the left boundary at $x = 0$ with $b > 0$, the situation will be as sketched in Fig. 7.3. For some ϕ_i and some values of y, the trajectory will cross the boundary before it reaches $t = t_n$; but because $b > 0$ a Dirichlet boundary condition $u(0, t) = u_L(t)$ is available and in (7.27a) it is a simple matter to replace $U^n(x)$ by $u_L(\tau)$ where $X(y, t_{n+1}; \tau) = 0$ and continue the integration with respect to y. For (7.27b) it is a little more complicated because the integral over x for $x < 0$ has to be replaced by one over τ for some part of the range (t_n, t_{n+1}); but $dx/d\tau$ is given by the trajectory and, indeed, $u\,dx$ is replaced by $f\,d\tau$ so the boundary condition is applied very naturally. The generalisation to higher dimensions in both cases is now clear and we shall not go into any more detail.

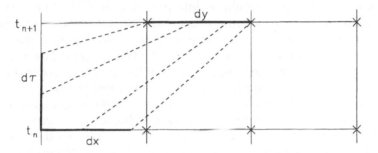

Figure 7.3. *Typical situation for evaluating integrals in (7.27a) or (7.27b) at a left-hand boundary*

(iii) *Euler characteristic Galerkin or ECG scheme.* Finally, we consider this third formulation of the characteristic Galerkin method, introduced by Morton (1982,1983,1985) as a natural solution to the problems met with in designing Petrov–Galerkin methods for unsteady hyperbolic problems. As they were designed for hyperbolic conservation laws of the form $u_t + f_x = 0$ in one dimension, with $f = f(u), \partial f/\partial u = b$, we start with this case. Using the notation of (7.27a) the direct method in this case is very simple, but the ECG scheme considers it in incremental form so that we have

$$(U^{n+1}, \phi_i) - (U^n, \phi_i) = \int U^n(x)[\phi_i(y)dy - \phi_i(x)dx].$$

The ECG scheme also uses the simple trajectory

$$y = x + b(U^n(x))\Delta t, \qquad (7.28a)$$

and introduces the upwind-averaged basis function

$$\Phi_i^n(x) := \frac{1}{y - x} \int_x^y \phi_i(x + s(y - x))ds. \qquad (7.28b)$$

Then we have, away from the boundaries and with $b^n \equiv b(U^n)$,

$$(U^{n+1} - U^n, \phi_i) = \int U^n(x)d(b^n \Delta t\, \Phi_i^n)$$

$$= -\Delta t \int \Phi_i^n b^n dU^n$$

$$= -\Delta t \int \Phi_i^n df(U^n). \tag{7.29a}$$

That is, we have

$$(U^{n+1} - U^n, \phi_i) + \Delta t (f_x(U^n), \Phi_i^n) = 0, \tag{7.29b}$$

where the relationship to a Petrov–Galerkin approximation to $u_t + f_x = 0$ is clearly seen. We have written the inner product here using the standard notation (f_x, Φ), but it should be interpreted as in (7.29a) because in a rigorous analysis of the scheme in Lin *et al.* (1995) it is shown that all the above manipulations are valid for piecewise constant approximations if the integrals are interpreted as Riemann–Stieltjes integrals. The method is designed to model shocked solutions so that the use of piecewise constant or discontinuous linear approximations is very appropriate.

We have not yet considered what happens to the Lagrange–Galerkin or characteristic-Galerkin methods when the characteristics cross, as they generally do to form shocks in the nonlinear case. In its earliest formulations, the ECG scheme was regarded as being based on the evolution operator (7.13b), though with the trajectories given by (7.28a), so that these trajectories were allowed to cross and $\widehat{E}(\Delta t)U^n$ became a multivalued function; then at the projection stage, the Riemann–Stieltjes integrals as in (7.29) were carried out along the graph $[\widehat{E}(\Delta t)U^n, y]$ to give U^{n+1} in S^h — see Fig. 7.4 for an example using the inviscid Burgers' equation. Following the presentation in Childs and Morton (1990), however, the scheme has been based on the *transport-collapse operator* of Brenier (1984), and in Lin *et al.* (1993) the two forms were shown to be strictly equivalent; this has the advantage of giving a direct generalisation into multi-dimensions.

Let us therefore introduce the following definition in respect of the scalar equation $u_t + \nabla \cdot \mathbf{f}(u) = 0$ on $\mathbb{R}^d \times (0, \infty)$, with $u(\mathbf{x}, 0) = u^0(\mathbf{x})$ and $\mathbf{b}(u) \equiv \partial \mathbf{f}/\partial u$. The *transport-collapse* operator $T(t)$ for $u \in L_1(\mathbb{R}^d)$ is given by

$$(T(t)u)(\mathbf{y}) := \int ju(\mathbf{y} - \mathbf{b}(w)t, w) dw \tag{7.30a}$$

where

$$ju(\mathbf{x}, w) = \begin{cases} 1 & \text{if } 0 < w < u(\mathbf{x}) \\ -1 & \text{if } u(\mathbf{x}) < w < 0 \\ 0 & \text{otherwise.} \end{cases} \tag{7.30b}$$

Some of the key properties are contained in the following theorem due to Brenier (1984), and account for the corresponding properties of the ECG scheme.

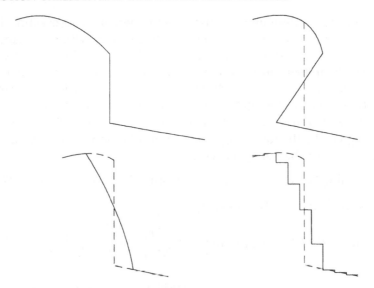

Figure 7.4. *Graphs of initial data u^0 (top left) together with $\widehat{E}(\Delta t)u^0$ (top right), $T(\Delta t)u^0$ (bottom left) and $P^h \widehat{E}(\Delta t)u^0 \equiv P^h T(\Delta t)u^0$ (bottom right), each compared with the exact solution at $\Delta t = \frac{1}{2}$*

Theorem 7.2.1 *Suppose* $f \in C^1(\mathbb{R})$ *and* $u, v \in L_1(\mathbb{R}^d)$. *Then*

(i) $\|T(t)u - T(t)v\|_{L_1} \leq \|u - v\|_{L_1}$ *(L_1 non−expansion)*

(ii) $\int T(t)u(\mathbf{x})\mathrm{d}\mathbf{x} = \int u(\mathbf{x})\mathrm{d}\mathbf{x}$ *(Conservation)*

(iii) $\|T(t)u\|_{L_1} \leq \|u\|_{L_1}$ *(L_1 stability)*

(iv) $TV(T(t)u) \leq TV(u)$ *(total variation non−*

$\qquad\qquad\qquad\qquad\qquad\qquad\qquad\qquad$ *increasing)*

(v) $u \leq v$ *a.e. in* $\mathbb{R}^d \Rightarrow T(t)u \leq T(t)v$.   *(monotonicity)*.

Furthermore, for $u^0 \in L_1(\mathbb{R}^d)$ *the unique entropy solution* $E(t)u^0$ *of the conservation law is given by*

$$E(t)u^0 = \lim_{m \to \infty} [T(t/m)]^m u^0. \qquad (7.31)$$

Proof. See Brenier (1984). □

Now we can define the ECG scheme by

$$(U^{n+1}, \phi_i) = (T(\Delta t)U^n, \phi_i), \qquad (7.32)$$

which generalises the direct formulation of the Lagrange–Galerkin method given in (7.15), where we assumed that b was independent of u so that overturning did not occur. It is clearly inappropriate to pursue this formulation, designed for nonlinear hyperbolic conservation laws, much further here; but we will conclude this section with two examples, both based on piecewise constant basis functions. Consider first the one-dimensional problem on a nonuniform mesh, where the flux function $f(u)$ has a single *sonic point* $\bar{u}$ at which $b(\bar{u}) = 0$. Then from (7.29b) it is clear that the contributions to the update come entirely from the jumps $\Delta_+ f(U_j^n)$ which we split into $f(U_{j+1}^n) - f(\bar{u})$ and $f(\bar{u}) - f(U_j^n)$; so the calculation of U^{n+1} is given by the following simple algorithm in the case that all CFL numbers are less than unity, i.e. when

$$- \min(\Delta x_j, \Delta x_{j-1}) \le b(U_j^n)\Delta t \le \min(\Delta x_j, \Delta x_{j+1}) \quad \forall j. \qquad (7.33a)$$

Initialise U^{n+1} by U^n; then for each j

$$\text{add} - \Delta t[f(U_{j+1}^n) - f(\bar{u})] \text{ to} \begin{cases} U_{j+1}^{n+1}\Delta x_{j+1} \\ U_j^{n+1}\Delta x_j \end{cases} \text{if} \begin{cases} b(U_{j+1}^n) > 0 \\ b(U_{j+1}^n) < 0, \end{cases} \qquad (7.33b)$$

and add $- \Delta t[f(\bar{u}) - f(U_j^n)]$ to $\begin{cases} U_{j+1}^{n+1}\Delta x_{j+1} \\ U_j^n\Delta x_j \end{cases}$ if $\begin{cases} b(U_j^n) > 0 \\ b(U_j^n) < 0. \end{cases} \qquad (7.33c)$

This is equivalent to the well known algorithm of Engquist and Osher (1981) when the mesh is uniform; in Childs and Morton (1990) a simpler, equivalent form is given as well as the generalisation to an arbitrary time step. It is worth noting that even in this general case the algorithm can always be written as

$$(U_i^{n+1} - U_i^n)\Delta x_i + \Delta t(F_{i+\frac{1}{2}}^n - F_{i-\frac{1}{2}}^n) = 0 \qquad (7.34)$$

for appropriately defined cell interface fluxes $F_{i-\frac{1}{2}}^n$.

The other example is on a uniform square mesh in two dimensions where both components of $\mathbf{b}(U_{ij}^n)$ are everywhere non-negative and the CFL numbers in each direction are less than unity. Then, as given in Morton and Childs (1989) and derived and generalised in Lin *et al.* (1995), if we write $\mathbf{f} = (f, g)$ so that $f \equiv \int b^{(x)}du$ and $g = \int b^{(y)}du$, and introduce

$$q(u) \equiv \int b^{(x)}b^{(y)}du, \qquad (7.35a)$$

we obtain the scheme

$$U_{ij}^{n+1} = U_{ij}^n - (\Delta t/h)[\Delta_{-x}f(U_{ij}^n) + \Delta_{-y}g(U_{ij}^n)]$$
$$+ (\Delta t/h)^2\Delta_{-x}\Delta_{-y}q(U_{ij}^n). \qquad (7.35b)$$

Thus this is the nonlinear generalisation of the tensor product scheme which will be given for the linear problem and arbitrary time steps in (7.37). A similar difference scheme has recently also been given by LeVeque (1995).

As has been noted already, the Lagrange–Galerkin method readily extends to systems of equations as it uses the same particle trajectories for all the equations. The ECG scheme has been applied to the one-dimensional Euler equations by flux-difference splitting in Morton and Sweby (1987) and Morton and Childs (1989), and by local diagonalisation in Lin and Morton (1993); but the best approach in multi-dimensions is probably through the methods to be described in section 7.3.

7.2.3 Area-weighting techniques

To describe these techniques in more detail we use them to generalise (7.20) and (7.22b) to a two-dimensional square mesh of size h, using a variable velocity field b with $\nabla \cdot b = 0$. The integrals on the right of the direct formulation (7.15b) are to be evaluated element by element at the new time level; and, using the notational abbreviation $X(y, t_{n+1}; t_n) \equiv x$ as in (7.27), they can be written as

$$\sum_{(\alpha)} \int_{\Omega_\alpha} U^n(x)\phi_i(y)dy, \tag{7.36a}$$

where Ω_α is a typical element. Denoting the centroid of the element by y_α and its image by x_α, this integral in the area-weighting method is replaced by

$$\sum_{(\alpha)} \int_{-\frac{1}{2}h}^{\frac{1}{2}h} \int_{-\frac{1}{2}h}^{\frac{1}{2}h} U^n(x_\alpha + \xi)\phi_i(y_\alpha + \xi)d\xi \tag{7.36b}$$

which is readily evaluated exactly.

In the case of piecewise constants, and using the more usual double suffix notation for the mesh points which are the centroids of the elements, suppose the centroid y_{ij} is mapped back to x_{ij}, where

$$y_{ij} - x_{ij} = b_{ij}\Delta t = h(r + \hat{\nu}_x, s + \hat{\nu}_y), \quad r, s \in \mathbb{Z}$$

with $\hat{\nu}_x, \hat{\nu}_y \in (0, 1]$, as shown on the left of Fig. 7.2. Then the contributions to the updated value U_{ij}^{n+1} provided by the integral corresponding to (7.36b) come from $U_{i-r,j-s}^n, U_{i-r-1,j-s}^n, U_{i-r,j-s-1}^n$ and $U_{i-r-1,j-s-1}^n$ with the respective coefficients $(1 - \hat{\nu}_x)(1 - \hat{\nu}_y), \hat{\nu}_x(1 - \hat{\nu}_y), (1 - \hat{\nu}_x)\hat{\nu}_y$ and $\hat{\nu}_x\hat{\nu}_y$. In particular, when b is a constant vector so that $\hat{\nu}_x$ and $\hat{\nu}_y$ are constants and the area-weighting technique is exact, the two-dimensional generalisation of (7.22b) is seen to be

$$\begin{aligned}(1 - \mu\delta_x^2 - \mu\delta_y^2)U_{ij}^{n+1} &= (1 - \hat{\nu}_x\Delta_{-x} - \hat{\nu}_y\Delta_{-y} + \hat{\nu}_x\hat{\nu}_y\Delta_{-x}\Delta_{-y})U_{i-r,j-s}^n \\ &= (1 - \hat{\nu}_x\Delta_{-x})(1 - \hat{\nu}_y\Delta_{-y})U_{i-r,j-s}^n.\end{aligned} \tag{7.37}$$

The important feature of this scheme to notice is the *corner term* equal to $\hat{\nu}_x\hat{\nu}_y\Delta_{-x}\Delta_{-y}U_{i-r,j-s}^n$ which distinguishes it from the simple upwind scheme even when $r = s = 0$. It is this term which is responsible for maintaining the unconditional stability of the scheme: thus Fourier analysis shows that the upwind scheme without the corner term is stable only if $\nu_x + \nu_y \leq 1$; while the product form (7.37) is clearly stable for $0 \leq \hat{\nu}_x, \hat{\nu}_y \leq 1$, and hence for all ν_x and ν_y. Note too that were the original equation to be the conservation law $u_t + \nabla \cdot (\mathbf{b}u) = 0$ with $\nabla \cdot \mathbf{b} \neq 0$, in order to write it in the form (7.12) we would need to introduce a source term $S \equiv -(\nabla \cdot \mathbf{b})u$. In the discrete form (7.15b) this would lead to U^{n+1} being multiplied by $(1 + \Delta t \nabla \cdot \mathbf{b})$ and this could easily be incorporated in the area-weighting update. It is easily checked that this would largely compensate for the difference between $\sum h^2 U_{ij}^{n+1}$ and $\sum h^2 U_{ij}^n$ introduced in the procedure described above; but we have seen that the alternative weak formulation is to be preferred in these circumstances.

For the linear, or rather bilinear, approximation the centroid of a typical element is at $\mathbf{y}_{i-\frac{1}{2},j-\frac{1}{2}}$ and is mapped back by the vector $\mathbf{b}_{i-\frac{1}{2},j-\frac{1}{2}} \Delta t$ to pick up the values of U^n. Using the same notation $h(r + \hat{\nu}_x, s + \hat{\nu}_y)$ for this as in the piecewise constant case, i.e. suppressing the element dependence, we now have to integrate products of bilinear functions over the four rectangles into which the element is subdivided, as shown on the left of Fig. 7.2: for example, and in the notation used for local basis functions in section 4.4.1, a contribution is obtained to U_{ij}^{n+1} from its local basis function $\frac{1}{4}(1 + \xi)(1 + \eta)$ in the element, multiplied by U^n evaluated in each of the rectangles; that involving $U_{i-r,j-s}^n$ in the top right rectangle is given by

$$\frac{h^2}{64} \int_{-1}^{-1+2\hat{\nu}_x} (1 + \xi)(1 + \xi - 2\hat{\nu}_x)d\xi \int_{-1}^{-1+2\hat{\nu}_y} (1 + \eta)(1 + \eta - 2\hat{\nu}_y)d\eta$$

$$= h^2(\tfrac{1}{3} - \tfrac{1}{2}\hat{\nu}_x + \tfrac{1}{6}\hat{\nu}_x^3)(\tfrac{1}{3} - \tfrac{1}{2}\hat{\nu}_y + \tfrac{1}{6}\hat{\nu}_y^3), \quad (7.38)$$

and this is the only rectangle giving a contribution involving $U_{i-r,j-s}^n$. However, the support of the basis function for U_{ij}^{n+1} comprises three further squares and generally there will be nine contributions of this type, corresponding to the three integrals making up the coefficient of U_{i-m}^n in (7.19); but notice that generally $\hat{\nu}_x$ and $\hat{\nu}_y$ will have different values for the four squares. For a constant vector $\mathbf{b}$, however, all these terms can be assembled to give an exact update formula which generalises (7.20), namely

$$[(1 + \tfrac{1}{6}\delta_x^2)(1 + \tfrac{1}{6}\delta_y^2) - \mu(1 + \tfrac{1}{6}\delta_y^2)\delta_x^2 - \mu(1 + \tfrac{1}{6}\delta_x^2)\delta_y^2]U_{ij}^{n+1}$$

$$= L_x^C L_y^C U_{i-r,j-s}^n, \quad (7.39a)$$

$$\text{where} \quad L_x^C = (1 + \tfrac{1}{6}\delta_x^2) - \hat{\nu}_x \Delta_{0x} + \tfrac{1}{2}\hat{\nu}_x^2\delta_x^2 - \tfrac{1}{6}\hat{\nu}_x^3\delta_x^2\Delta_{-x}, \quad (7.39b)$$

with a corresponding expression for L_y^C — c.f. (4.45). The precise form of this formula is relatively unimportant: the key point is that area weighting allows similar approximate schemes to be generated for variable $\mathbf{b}$ and on nonuniform rectangular meshes. Furthermore, the technique can be applied to the weak formulation by approximating integrals of the form (7.27b) using the mapping shown on the right of Fig. 7.2.

7.2.4 Adaptive recovery techniques

At each time step of the evolution-Galerkin schemes just described, the approximation U^{n+1} is given as the L_2 projection onto S^h of the evolved solution $E_\Delta U^n$. If S^h is the space of continuous piecewise linear functions and $E_\Delta U^n$ has a steep front, U^{n+1} is likely to have an oscillatory over-shoot which might violate some physical constraint, such as positivity (of concentration, density, temperature, for example) or monotonicity. This would not happen with a piecewise constant S^h, but then the order of accuracy could drop from three to one. The phenomenon is similar to that found with higher order as compared with first order difference schemes, and embodied in *Godunov's Theorem* that a nonadaptive monotone difference scheme in conservation law form is at most first order accurate (see Godlewski and Raviart (1991), Theorem 3.1, p125).

It might be argued that the fault here lies with the L_2 projection, with its known propensity for giving oscillatory approximations and, at least in one dimension, an escape from this dilemma might lie in using a projection in the Dirichlet norm used in many of the Petrov–Galerkin methods of Chapter 5, or more generally in the mixed norm $\|u\|_\gamma^2$ of (7.27). Schemes using such norms are considered in Childs and Morton (1990), where some are shown to be equivalent to well known difference schemes, and this approach is also another way of validating mass lumping. However, it is shown there that L_2 projection generally gives the best accuracy; and, in any case, this device does not always eliminate unwanted oscillations, especially in higher dimensions. Thus we shall here use only L_2 projection and overcome this problem by using adaptive recovery techniques.

The term recovery was first used in the present context by Morton (1982) in order to indicate the link to the field of *optimal recovery* in approximation theory; in a typical situation there, one is given some discrete data about an unknown function, such as its projection onto a finite dimensional space, together with other more qualitative information, such as positivity, monotonicity or smoothness, and one seeks estimates of other quantities, such as point values of the unknown function, which make use of both sets of information. This is precisely our present situation: for the solution of most systems of partial differential equations one usually has much *a priori* information; for example, in inviscid fluid dynamics in two dimensions one knows that the solution is smooth between a finite number of curves across which it has a jump discontinuity. For many elliptic problems one has a maximum principle, a variational principle and interior smoothness. The recovery process should make use of as much of this information as is practically convenient. We have already referred to the widespread use in finite element methods of gradient recovery; and in Barrett *et al.* (1988a)(1988b) other examples of recovery techniques as applied to steady problems are given.

By far the most common adaptive recovery techniques for unsteady problems consist of constructing a discontinuous piecewise linear approximation in one dimension from a piecewise constant approximation, as first used by van Leer (1977). This is now done in a variety of ways using different criteria, and we shall concentrate on this case. However, first we should note a general result using B-splines which shows that there need be no loss of accuracy in using recovery techniques, rather than using a higher order method directly.

Lemma 7.2.2 *Suppose the characteristic-Galerkin method is applied to $u_t + bu_x = 0$ on the real line, with constant b. If, on a uniform mesh, the trial space uses splines of order m and nonadaptive recovery is applied with splines of order $m + r$, the resulting accuracy is $O(h^{2m+r-1})$. If recovery is by splines of order $m + 2r$, the result is exactly equivalent to a method using a trial space of splines of order $m + r$ with no recovery.*

Proof. See Childs and Morton (1990). ☐

More limited results of this kind that can be proved on a nonuniform mesh and with a variable velocity b can be found in Morton and Süli (1992).

Suppose $U \in S^h$ approximates u, and the recovery procedure R^h yields a function $\tilde{u}$. If the recovered function were exact we would insist that U be its projection. Thus in general we shall suppose that

$$P^h R^h V = V \quad \forall V \in S^h, \tag{7.40a}$$

so that

$$(U - \tilde{u}, \phi_j) = 0 \quad \forall \phi_j \in S^h, \tag{7.40b}$$

is the defining relation for $\tilde{u} \equiv R^h U$ when U and $\tilde{u}$ have the same number of free parameters, as in the spline case. However, in general $\tilde{u}$ will have more free parameters and these will be used to recover adaptively in order to have $\tilde{u}$ satisfy various desirable criteria. Note that (7.40) also ensures that conservation properties of U are preserved in $\tilde{u}$.

Now let us consider recovery by discontinuous linear functions from piecewise constants, on a nonuniform mesh in one dimension. The constraint (7.40) reduces to requiring that on each interval we set

$$\tilde{u}(x) \equiv U_j + s_j(x - x_j) \text{ for } x \in (x_{j-\frac{1}{2}}, x_{j+\frac{1}{2}}), \tag{7.41}$$

where $x_j = \frac{1}{2}(x_{j-\frac{1}{2}} + x_{j+\frac{1}{2}})$ and the $\{s_j\}$ are free parameters to be chosen so that $\tilde{u}$ has some desirable properties; note that if $\tilde{u}$ were required to be continuous, we would revert to the spline case, so generally we would then have $\tilde{u}(x_j) \neq U_j$, the recovery process would be nonlocal and there would be no free parameters. For the discontinuous linear recovery, on the other hand, it is natural, and common practice, to determine s_j locally, as some function of the neighbouring divided differences D_+U_j and D_-U_j. Thus, suppose U were the projection onto piecewise constants of a smooth function which in the neighbourhood of x_j has the quadratic form

$$u(x) \approx a + b(x - x_j) + c(x - x_j)^2. \qquad (7.42a)$$

Then, by this projection assumption, it is clear that a, b and c can be determined from the values U_{j-1}, U_j and U_{j+1}; moreover, if $\tilde{u}$ is to be the corresponding piecewise linear projection of u, it is easy to see that we must have $s_j = b$. Indeed, if we denote by h_j the length of the interval $(x_{j-\frac{1}{2}}, x_{j+\frac{1}{2}})$ so that $D_-U_j = 2(U_j - U_{j-1})/(h_j + h_{j-1})$, a little further calculation shows that

$$s_j = b = \frac{(h_{j-1} + \frac{1}{2}h_j)D_+U_j + (\frac{1}{2}h_j + h_{j+1})D_-U_j}{h_{j-1} + h_j + h_{j+1}} \qquad (7.42b)$$

$$U_j = a + \tfrac{1}{12}ch_j^2, \quad \text{and} \quad ch_j = D_+U_j - D_-U_j. \qquad (7.42c)$$

On a uniform mesh (7.42b) reduces to $s_j = D_0U_j$ and is exactly the same as either of the gradient recovery procedures used in the cell vertex method and given by (6.71c). However, we note that on a nonuniform mesh the two are different, because of the assumption here that U_j is the cell average of the underlying function (7.42a), while for the gradient recovery of method B in (6.71c) it was assumed that the underlying quadratic was interpolated; note too that it is this assumption that leads to (7.42c) instead of $U_j = a$.

The freedom in the choice of the slopes is most commonly used to ensure that the recovery is *monotonicity-preserving*, which is not always the case with the formula (7.42b). Indeed, it is clear that the following requirement holds.

Lemma 7.2.3 *Recovery by discontinuous linear functions is monotonicity-preserving if and only if $D_-U_j \geq 0 \, \forall j$ implies that*

$$s_j \geq 0 \quad \text{and} \quad U_j - \tfrac{1}{2}s_jh_j \geq U_{j-1} + \tfrac{1}{2}s_{j-1}h_{j-1} \quad \forall j. \qquad (7.43)$$

Then the test data consisting of $\{U_j = 0, j < 0; U_j = 1, j \geq 0\}$, for which $\Delta_-U_j = 0$ except for $\Delta_-U_0 = 1$, shows that the formula (7.42b) violates these criteria by having negative jumps at $x_{-\frac{3}{2}}$ and $x_{\frac{1}{2}}$. To satisfy the first condition of (7.43) and to include (7.42b), we can describe the adaptive recovery task as to find $\theta_{j-}, \theta_{j+} \in [0, 1]$ so we can write

$$s_j = \theta_{j-}D_-U_j + \theta_{j+}D_+U_j, \qquad (7.44)$$

to give a result as close as possible to (7.42b) while satisfying the second condition of (7.43). Unfortunately, the latter is an implicit constraint on our choice. However, it can also be written as $D_-U_j \geq (s_{j-1}h_{j-1} +$

$s_j h_j)/(h_{j-1} + h_j)$, from which we can easily deduce the following useful conditions when all slopes are non-negative:

$(sufficient\ condition)\quad s_j \leq \min(D_-U_j, D_+U_j);$ \hfill (7.45a)

$(necessary\ condition)\quad s_j \leq \min\left(\dfrac{h_{j-1} + h_j}{h_j}D_-U_j, \dfrac{h_j + h_{j+1}}{h_j}D_+U_j\right),$

i.e. \hfill $s_j h_j \leq 2\min(\Delta_-U_j, \Delta_+U_j).$ \hfill (7.45b)

To deal with extrema when U is not monotone, it is common practice, though rather restrictive, to insist that the recovery process is TVD: this implies the further condition

$(from\ TVD\ condition)\quad s_j = 0 \text{ if } (D_-U_j)(D_+U_j) \leq 0.$ \hfill (7.45c)

Before looking at particular explicit formulae for s_j, however, we consider the effect of the choice on a simple evolution Galerkin algorithm; more will be said about the resulting algorithms in the next section, where we shall also consider evolution with the full quadratic of (7.42).

Linear constant coefficient advection in one dimension will be sufficient for our present purposes; and we can suppose that the CFL numbers ν_i lie in $(0, 1)$. Then we have seen, as in (7.14) or (7.33), that the Lagrange–Galerkin or characteristic-Galerkin method, with piecewise constants and no recovery, gives an update that reduces to the first order upwind scheme $U_i^{n+1} = (1 - \nu_i)U_i^n + \nu_i U_{i-1}^n$. Employing the same construction with the discontinuous linear function, we find that a total flux equal to

$$b\Delta t[U_i + \tfrac{1}{2}s_i(h_i - b\Delta t)]$$ \hfill (7.46a)

passes out of the ith cell through the boundary at $x_{i+\frac{1}{2}}$. Thus the update corresponding to (7.14) is given by

$$h_i(U_i^{n+1} - U_i^n) + \Delta t\Delta_- F_{i+\frac{1}{2}} = 0$$ \hfill (7.46b)

$$\text{with } F_{i+\frac{1}{2}} := b[U_i + \tfrac{1}{2}s_i h_i(1 - \nu_i)].$$

We can compare this with a second-order Lax–Wendroff scheme, which we can deduce from a finite volume approach or by suitably interpreting the gradients in the Taylor–Galerkin scheme of (7.9a). Thus with U_x at $x_{j+\frac{1}{2}}$ interpreted as D_+U_j, we have

$$h_i(U_i^{n+1} - U_i^n) = -\Delta t\Delta_-[bU - \tfrac{1}{2}\Delta t\, b(bU)_x]_{i+\frac{1}{2}}$$
$$= -b\Delta t\Delta_-[U_{i+\frac{1}{2}} - \tfrac{1}{2}b\Delta t D_+U_i]$$
$$= -b\Delta t\Delta_-[U_i + \tfrac{1}{2}s_i h_i - \tfrac{1}{2}b\Delta t D_+U_i], \hfill (7.47)$$

if we set $U_{i+\frac{1}{2}} = U_i + \tfrac{1}{2}s_i h_i$. This scheme is clearly the same as (7.46) if we set $s_i = D_+U_i$; such a relation, for difference schemes on a uniform mesh, is used by LeVeque (1992) to relate *flux-limiter methods* with *slope-limiter methods*.

To obtain sufficient conditions for the scheme (7.46) to be TVD, we use the ratio $r_j := \Delta_- U_j / \Delta_+ U_j$ introduced in (6.19) and define a coefficient ϕ_j such that we can write

$$s_j h_j =: \phi_j \Delta_+ U_j \equiv (\phi_j / r_j) \Delta_- U_j. \qquad (7.48)$$

Then by substituting for $s_i h_i$ and $s_{i-1} h_{i-1}$ we can write (7.46) as

$$U_i^{n+1} = U_i^n - \nu_i [1 + \tfrac{1}{2}(1 - \nu_i)(\phi_i / r_i) - \tfrac{1}{2}(1 - \nu_{i-1})\phi_{i-1}] \Delta_- U_i^n. \qquad (7.49a)$$

This is TVD if the expression in square brackets always lies in the interval $[0, 1]$; and, with the assumptions on the CFL numbers, this is assured if

$$0 \le \frac{\phi_j}{r_j} \le 2 \text{ and } 0 \le \phi_j \le 2 \quad \forall j. \qquad (7.49b)$$

Substituting (7.48) into (7.49b) and comparing with (7.45b), we see that, *for monotone data, the necessary conditions for the recovery to be monotonicity-preserving are precisely these sufficient conditions for the resultant Lagrange–Galerkin algorithm to be TVD.*

For simplicity we will now specialise to a uniform mesh in order to bring together the adaptive recovery techniques used to define $s_j h_j$, the finite difference flux-limiter and slope-limiter methods and the solution-adaptive finite volume schemes described in section 6.2.3. With no mesh ratios involved, it is clear that $s_j h_j$ should just depend on $\Delta_+ U_j$ and $\Delta_- U_j$: if it depends only on their ratio, the recovery is determined by specifying $\phi_j = \phi(r_j)$ for some function ϕ, as is the case with the slope-limiter and flux-limiter difference schemes; and in the first place, in order to satisfy the TVD recovery condition (7.45c), we will assume that $\phi(r) = 0$ for $r \le 0$. In Fig. 7.5 we shade the region given by the TVD condition (or necessary monotonicity-preserving condition)(7.49b), together with darker shading for the corresponding sufficient monotonicity-preserving condition which by (7.45a) becomes $0 \le \phi/r \le 1$ and $0 \le \phi \le 1$. A further important condition is that to ensure the second order accuracy of the resultant scheme (7.46), which clearly only affects the choice near $r = 1$. We have seen that the three-point Lax–Wendroff or Taylor–Galerkin scheme results from the choice $s_j h_j = \Delta_+ U_j$ or $\phi(r) \equiv 1$; the corresponding Beam–Warming scheme (using the three values U_j, U_{j-1}, U_{j-2}) results from $s_j h_j = \Delta_- U_j$ or $\phi(r) \equiv r$; and the Lagrange–Galerkin scheme (7.46) with (7.42b) results from $\phi(r) \equiv \tfrac{1}{2}(1+r)$, giving a four-point scheme — note that this arithmetic mean of the Lax–Wendroff and Beam–Warming schemes, was advocated by Fromm (1968) nearly thirty years ago. These are all shown in Fig. 7.5, from which we see that they all eventually escape from the TVD region, although Beam–Warming lies in the region for $r \le 2$, Lax–Wendroff for $r \ge \tfrac{1}{2}$, and Fromm for $\tfrac{1}{3} \le r \le 3$. The main idea in solution-adaptive methods based on discontinuous linear recovery is therefore to switch between these second order methods according to the value of r.

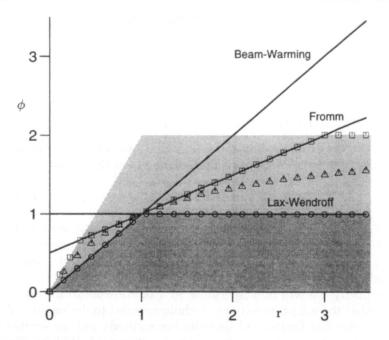

Figure 7.5. *Various discontinuous linear recovery schemes, defined by $s_j h = \phi \Delta_+ U_j$, and their corresponding linear advection schemes:* $\phi \equiv 1$ *(Lax–Wendroff),* $\phi \equiv r$ *(Beam-Warming) and* $\phi \equiv \frac{1}{2}(1 + r)$ *(Fromm). A sufficient condition for monotonicity-preserving recovery is satisfied in the region which is heavily shaded; a necessary condition, equivalent to a condition for the resulting scheme to be TVD, is satisfied in the whole shaded region. Recovery schemes shown are: minmod* ∘ ∘ ∘*, harmonic* △△△*, and Davis* ☐☐☐

Three particular recovery schemes are also shown in Fig. 7.5. The first is given by the upper limit of the sufficient condition for monotonicity preservation, corresponding to the equality sign in (7.45a); it was proposed by van Leer (1974), is often referred to as the *minmod* slope-limiter and, for $r \geq 0$, is given by

$$(minmod\ limiter) \quad \phi(r) = \min(r, 1). \tag{7.50a}$$

Thus it leads to the use of the Lax–Wendroff flux for $r \geq 1$ but 'limits' this by a factor r for $r < 1$. The second choice, due to Davis (1984), follows the arithmetic mean slope, $sh = \Delta_0 U$ or $\phi = \frac{1}{2}(1 + r)$, that gives the Lagrange–Galerkin or Fromm scheme while it lies in the TVD region, and otherwise uses the upper limit of the region; for $r \geq 0$ it is therefore

$$(Davis\ limiter) \quad \phi(r) = \min(2r, \tfrac{1}{2}(1 + r), 2). \tag{7.50b}$$

The slope limiting embodied in (7.50b) was also used by van Leer (1979) in his MUSCL scheme — see section 7.3.1. The third choice uses the harmonic mean slope, which lies between the first two and always lies in the TVD region; this is usually called the *van Leer limiter*, but because this author could be associated with all three limiters we use the alternative name; it is given by

$$(harmonic \; limiter) \quad \phi(r) = \frac{2r}{1+r}. \tag{7.50c}$$

We reiterate that each of these formulae are for use only when $r \geq 0$; by (7.45c), to maintain TVD properties we need to set $\phi = 0$ for $r < 0$.

As has been pointed out by Jameson (1995), however, the first and last of these recovery schemes can be combined in the form

$$(Jameson \; limiters) \quad \phi(r) = \tfrac{1}{2}(1+r)\left[1 - \left|\frac{1-r}{1+|r|}\right|^q\right], \quad q > 0, \tag{7.51}$$

which is valid for all r, since $r < 0$ implies $\phi = 0$. Taking $q = 1$ gives (7.50a) and $q = 2$ gives (7.50c). This is the largest value that always satisfies the TVD condition; but note that the larger the value of q that is used, the closer the scheme comes to the Fromm or Lagrange–Galerkin choice $\phi(r) = \tfrac{1}{2}(1+r)$.

On the other hand, Osher and Chakravarthy (1984) pointed out that difference schemes revert to first order accuracy at extremal points if the TVD property is maintained; and Shu (1987) showed that the *total variation bounded (or TVB)* property, which is all that is needed to establish convergence, can be maintained by allowing an $O(\Delta t)$ growth at extrema in one time step. We see from the Lagrange–Galerkin update (7.46) that this merely requires that the slopes s_i be uniformly bounded: more generally, we know that evolution and projection do not increase the variation (c.f. Theorem 7.2.1), so that it is only the increase during the recovery that needs to be controlled and hence the slopes bounded. Thus at any extremum, even the recovered slope of (7.42b) can be used so long as both D_+U_j and D_-U_j are bounded: this merely excludes such recovery in the neighbourhood of a shock or other form of discontinuity, since at smooth extrema $D_\pm U_j$ should be $O(h)$. Thus the key practical difficulty is to distinguish smooth extrema from a corner at the end of a constant section; the switches near $r = 0$ and $r = \infty$ shown in Fig. 7.5 are designed to prevent unacceptable overshoots at just such points.

The limitations of using only r and $\phi(r)$ as a basis for the recovery are shown up by these situations; and at this point it is helpful to compare with the procedures used in section 6.2.3 to recover inter-cell convected fluxes, and with the use of artificial dissipation. Strict comparison with the former is not possible because the algorithms are different and the underlying approximations also differ; but in (7.47) we identified $U_{i+\frac{1}{2}}$ with $U_i + \tfrac{1}{2}s_i h_i = U_i + \tfrac{1}{2}\phi_i \Delta_+ U_i$ and Fig. 6.3 was presented in terms of a normalised $\widehat{U}_{i+\frac{1}{2}}$. Thus we can make the link

$$\widehat{U}_{\cdot+\frac{1}{2}} = \frac{1}{1+r}[\tfrac{1}{2}\phi(r) + r], \tag{7.52}$$

so that, for example, upwinding corresponds to both $\phi = 0$ and $\widehat{U} = r/(1+r)$. It follows that the two approaches treat the case $r < 0$ in the same way, and therefore generally cover the same range of choice.

To the extent that an adaptive scheme based on artificial dissipation monitors only $\delta^2 U_j$, it is equivalent to the schemes just described; but if $\delta^4 U_j$ is also monitored it should be able to recognise smooth extrema. Thus application of second and fourth order dissipation to a central difference approximation of pure convection, as in (6.41a), is equivalent to writing

$$U_{i+\frac{1}{2}} = \tfrac{1}{2}(U_i + U_{i+1}) - (\epsilon^{(2)} - \epsilon^{(4)}\delta^2)\Delta_+ U_i. \tag{7.53a}$$

Thus $\phi \equiv 1$ gives the same result as setting $\epsilon^{(2)} = \epsilon^{(4)} = 0$ in this formula, and any scheme based on a choice of ϕ can be obtained by setting $\epsilon^{(4)} = 0$ and $\epsilon^{(2)} = \tfrac{1}{2}(1 - \phi)$, though any value $\phi > 1$ will entail using negative values of $\epsilon^{(2)}$. But as we saw in section 6.3.2, a much more typical choice of second order dissipation is given by

$$\epsilon^{(2)}_{i+\frac{1}{2}} = \alpha^{(2)} \max\left(\frac{|1 - r_i|}{1 + |r_i|}, \frac{|1 - r_{i+1}|}{1 + |r_{i+1}|}\right), \tag{7.53b}$$

and the extra adaptivity here means that the coefficient of second order dissipation is $O(h^3)$ in smooth parts of the solution and $O(h)$ where needed. Furthermore, as in (7.51) higher powers of these ratios may be used; and (7.53b) could be replaced by $\alpha^{(2)}|1 - r_i r_{i+1}|/(1 + |r_i r_{i+1}|)$, as in the SLIP scheme of Jameson (1995). Thus it is clear that the wide stencils used here, together with the fourth order terms, give considerable scope for recognising smooth extrema, which we shall demonstrate in the next section.

We have treated this particular case of one-dimensional recovery by discontinuous linear functions at great length, and almost to the exclusion of other possibilities, for several reasons. Firstly, the alternative of replacing the discontinuities at cell boundaries by linear ramp functions, which was chosen instead by Morton (1982) and Childs and Morton (1990) as an adaptive recovery procedure, has proved to be of less general utility: as shown in Morton and Sweby (1987) it can give more accurate results, but as the recovery is nonlocal it is more complicated and more difficult to generalise; and in Lin *et al.* (1995) it is demonstrated that the convergence theory for ECG schemes using discontinuous linear recovery is much simpler than that for this continuous linear recovery. Secondly, discontinuous linear recovery is readily and conveniently generalised to two and three dimensions in various ways, although much more research is needed on slope-limiting criteria: on triangular or tetrahedral meshes linear recovery is natural — see Sonar (1993) for an example; but on quadrilateral or hexahedral meshes multilinear recovery could also be used. And thirdly, generalisations using discontinuous higher order polynomials can be developed in a similar way: the best known is the *piecewise parabolic method* (PPM) of Colella and Woodward (1984), while the more recent *discontinuous Galerkin methods* of Cockburn and Shu (1989) and the *ENO schemes* of Harten (1987), Harten et al. (1987) and Shu and Osher (1988) are in the same spirit.

We end this section by outlining a technique, often used in deriving these schemes, to recover key quantities from cell averages in one dimension. Thus suppose we wish to recover directly the intercell value $U_{i+\frac{1}{2}}$ from U_{i-1}, U_i and U_{i+1}, where we assume the mesh is uniform. Then we introduce the underlying primitive function $w(x) = \int^x u(s)\mathrm{d}s$, for which we can assume $w(x_{i-\frac{3}{2}}) = 0$ and obtain $w(x_{i-\frac{1}{2}}) = hU_{i-1}, w(x_{i+\frac{1}{2}}) = h(U_{i-1} + U_i)$ and $w(x_{i+\frac{3}{2}}) = h(U_{i-1} + U_i + U_{i+1})$. From these four values we can construct a cubic and estimate $U_{i+\frac{1}{2}}$ by $w'(x_{i+\frac{1}{2}})$; a short calculation gives

$$U_{i+\frac{1}{2}} = \tfrac{1}{2}(U_i + U_{i+1}) - \tfrac{1}{6}\delta^2 U_i = \tfrac{1}{6}(2U_{i+1} + 5U_i - U_{i-1}). \qquad (7.54)$$

This is directly comparable with the formula (6.5e) giving the QUICK scheme, which is based on the assumption that U interpolates u.

It is useful to note here that there is a very large literature on Hermite cubic interpolants and how they may be constructed so as to maintain monotonicity — see Huynh (1993).

7.2.5 Examples of PERU schemes

In sections 7.2.2 and 7.2.3 various evolution-Galerkin schemes have been given that made no use of an adaptive recovery technique. Using piecewise constant basis functions in one dimension gave the first order algorithms of (7.14), (7.22b) and (7.33); and in two dimensions these became (7.35) and (7.37). On the other hand, using continuous piecewise linear basis functions gave the scheme (7.20) in one dimension and (7.39) in two dimensions; on a uniform mesh these are third order accurate algorithms for the convection process. Now, by means of a few examples, we want to show how the adaptive recovery techniques described in the preceding section can be used in the general PERU procedure given by (7.4) to generate second and third order accurate schemes that also avoid spurious oscillatory phenomena. With the recovered approximation $\tilde{u}$ satisfying (7.40b), we recall that the general PERU scheme for the scalar problem in one dimension generalises (7.29b) to give the formula

$$(U^{n+1} - U^n, \phi_i) + \Delta t(f_x(\tilde{u}^n), \tilde{\Phi}_i^n) = 0, \qquad (7.55)$$

where $\bar{\Phi}_i^n$ is given by (7.28b) with $y = x + b(\tilde{u}^n)\Delta t$.

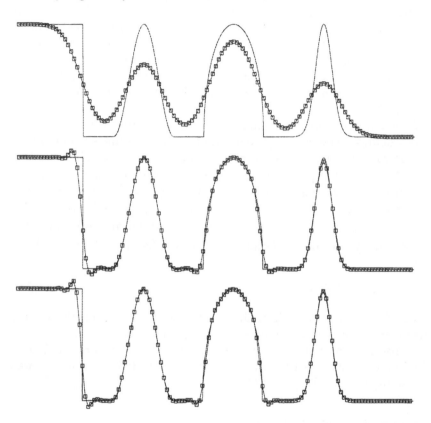

Figure 7.6. *Pure convection approximated by, from the top, the first order upwind scheme, Taylor–Galerkin and Lagrange–Galerkin with piecewise linears*

Since pure convection poses the most demanding test of the procedures, we will consider only this case. This also has the advantage that, although we compare schemes in this PERU framework, the methods covered include many important difference schemes as well as several of the Godunov schemes considered in section 7.3; moreover, in Leonard (1991) and Leonard and Niknafs (1991) we have a linear advection test problem and many comparative results that provide an admirable basis for comparing algorithms. In the latter paper, four profiles of varying smoothness, with width $20\Delta x$ and spacing $30\Delta x$, are advected through 100 time steps with the CFL number $\nu = 0.45$. We begin in Fig. 7.6 by showing the exact result compared with approximations given by the first order upwind scheme (7.14), the Taylor–Galerkin scheme (7.8a) with the full mass matrix $\theta = 1$, which renders it third order accurate, and the Lagrange–Galerkin scheme (7.20) using a continuous piecewise linear basis. These first order and third order results provide the extremes between which we wish to plot a middle course.

Any PERU scheme for linear advection, that is based on recovery by discontinuous linear functions from piecewise constants, can be written as in (7.46) in the form

$$U_i^{n+1} = U_i^n - \nu_i \Delta_-[U_i + \tfrac{1}{2} s_i h_i (1 - \nu_i)], \qquad (7.56)$$

when $\nu_i = b \Delta t / h_i \leq 1$ and s_i is the slope in the ith interval. In the top picture in Fig. 7.7 we show results obtained with the Davis limiter, $s_i h_i = \phi(r_i) \Delta_+ U_i$ with ϕ given by (7.50b); because this is strictly TVD, there are no overshoots or undershoots but the errors at the peaks are some eight times those of the best third order schemes of Fig. 7.6. This is mainly due to poor discrimination by the limiter between smooth extrema and sharp corners in the solution; but in addition the higher order scheme is only second order accurate and needs to be third order.

We know from the general result given in Lemma 7.2.2 that recovery by quadratic B-splines from piecewise constants will give the third order accurate Lagrange–Galerkin scheme shown at the bottom of Fig. 7.6. However, this has no free parameters available to provide for adaptivity. We could, instead, use the full quadratic (7.42) that has been the basis of our discontinuous linear recovery. Then, by calculating the total flux through a cell boundary as in (7.46) and (7.56), and using the fact that $b\,dt = dx = h\,d\xi$, say, together with the values of the coefficients given in (7.42c), we obtain on a uniform mesh from this discontinuous parabolic recovery

$$U_i^{n+1} - U_i^n = -\Delta_- \int_{\frac{1}{2} - \nu}^{\frac{1}{2}} [(1 - \tfrac{1}{12}\delta^2) + \xi \Delta_0 + \xi^2 \delta^2] U_i^n \, d\xi$$

$$= -\nu \Delta_- \{ U_i + \tfrac{1}{6}(1 - \nu)[(2 - \nu)\Delta_+ + (1 + \nu)\Delta_-] \} U_i. \quad (7.57)$$

It is easily checked that this is the same as the Warming–Kutler–Lomax difference scheme that has already been referred to, and whose accuracy will be discussed in section 7.2.8. The result of its application to the present model problem is shown as the second plot of Fig. 7.7.

Alternatively, we could recover with parabolas in each cell that are also continuous from cell to cell, and still have parameters free for adaptation. This is the starting point of the PPM method which, however, also includes a sophisticated set of adaptive controls. Introducing intercell values $\tilde{u}_{i+\frac{1}{2}}$, we write in each cell

$$\tilde{u}_i(\xi) = (\tfrac{1}{2} - \xi)\tilde{u}_{i-\frac{1}{2}} + (\tfrac{1}{2} + \xi)\tilde{u}_{i+\frac{1}{2}} + c_i(\tfrac{1}{4} - \xi^2), \qquad (7.58a)$$

and to maintain conservation, or the projection condition (7.40), we have

$$U_i = \tfrac{1}{2}(\tilde{u}_{i-\frac{1}{2}} + \tilde{u}_{i+\frac{1}{2}}) + \tfrac{1}{6}c_i. \qquad (7.58b)$$

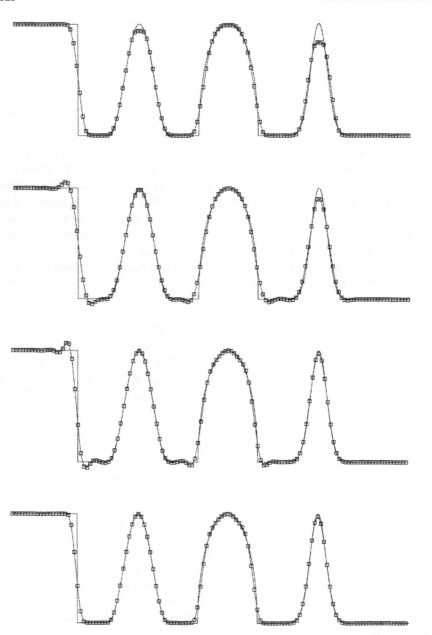

Figure 7.7. *As for Fig. 7.6, with from the top, discontinuous linear recovery (Davis limiter), discontinuous parabolic and continuous parabolic recovery, and with an adaptive switch between the first and third.*

Thus only the set of $\bar{u}_{i+\frac{1}{2}}$ need to be recovered; Colella and Woodward (1984) use the technique described at the end of the last section for this purpose, interpolating the indefinite integral with a quartic rather than a cubic, before applying their limiting procedures. The result of this continuous piecewise parabolic recovery without any limiting is shown as the third plot from the top of Fig. 7.7.

This at last gives an approximation of the peaks which is sufficiently close to that of the Lagrange–Galerkin method that it provides a basis for obtaining a TVD solution without too much peak clipping. The final plot of Fig. 7.7 realises this objective with a scheme which switches between the recovered flux obtained with the continuous parabolic method and that of the Davis-limited discontinuous linear method. The discrimant is related to that described in section 6.3.3 and elsewhere, and used as a coefficient for second order artificial dissipation; however, instead of applying the discrimator $|\delta^2 p|/(4 + \delta^2 p)$ to the solution U, it is applied to the differences $\Delta_- U$. The result speaks for itself, and is comparable to results obtained by Leonard and Niknafs (1991) using similar ideas.

To conclude this section, it should be emphasised that the approach to high accuracy that has been followed here is readily generalised to nonlinear equations, nonuniform meshes, larger CFL numbers and the addition of diffusion; moreover, it leads to the most practical approach to the accurate solution of systems of equations in more dimensions, through the generalised Godunov schemes of section 7.3.

7.2.6 Semi-Lagrangian methods

We shall not attempt to describe in this volume the wide variety of *particle methods, Lagrangian* or *free-Lagrange methods, moving point methods, moving finite element methods* and the like which are sometimes used to approximate unsteady convection-diffusion problems. These all make use of a free or moving (Lagrangian) mesh instead of or in addition to the fixed (Eulerian) mesh that we have used — see Hockney and Eastwood (1981), Rees and Morton (1991) and Baines (1994) for some descriptions, comparisons and further references. However, *semi-Lagrangian methods*, like the Lagrange-Galerkin methods we have already described, make use of some of the same ideas in intermediate calculations that take place in carrying out one evolutionary step on the fixed mesh. Corresponding to the direct and weak formulation of the Lagrange–Galerkin methods, there are *forward* and *backward* semi-Lagrangian methods.

A good example of the former class is the *Second Moment Method* of Egan and Mahoney (1972), which we have already referred to as using similar ideas to the area-weighting techniques of section 7.2.3. In two dimensions, each grid cell has associated with it a distribution of u which is parametrised by its zeroth, first and second moments; and it is this which is convected and diffused in one time step before it is redistributed to the grid cells which it then intersects. We will describe a one-dimensional version. For the ith cell of width h_i, define U_i, X_i and H_i by the following integrals over the cell,

$$\int U \mathrm{d}x = h_i U_i, \quad \int U x \mathrm{d}x = h_i U_i X_i,$$

$$\int U x^2 \mathrm{d}x = h_i U_i (X_i^2 + \tfrac{1}{12} H_i^2). \tag{7.59}$$

That is, with the distribution regarded as a square pulse, $h_i U_i$ is the total 'mass' in the pulse, X_i the position of its centre and H_i its width; by projecting the initial data onto the usual piecewise constant approximation we would start with $X_i = x_i$ and $H_i = h_i$. Convection through one time step moves the distribution centre X_i along the characteristic path (7.13a), typically by a second order Runge-Kutta scheme. Diffusion is effected by widening the pulse, $H_i^2 \rightarrow H_i^2 + 24\epsilon\Delta t$; it is easily checked that this is exactly consistent with using an explicit finite difference scheme to transform a set of values $\ldots, 0, 1, 0, \ldots$ to $\ldots, \nu, 1 - 2\nu, \nu, \ldots$, where $\nu = \epsilon\Delta t/h^2$, and reinterpreting the result as a single pulse. The time step ends with a redistribution of all the pulses to the grid cells; thus in a short time step the total mass $h_i U_i$ may be redistributed to cells $i - 1, i$ and $i + 1$ because of the changes to X_i and H_i. Note that the effect of having $\nabla \cdot \mathbf{b} \neq 0$ can also be taken account of by changing H_i, and several other extensions to this basic algorithm are possible — see de Kok (1993) for more details and references.

The backward semi-Lagrangian methods are of greater general utility and they are now being widely used in meteorological and environmental computations; one of the earliest applications is that by Krishnamurti (1962) and a more recent one that by Staniforth and Côté (1991). In simple situations there is almost a complete overlap with the Lagrange–Galerkin methods based on continuous piecewise linear basis functions, typified by the schemes (7.20) and (7.39), after the mass matrix has been taken account of. The differences in more demanding problems arise firstly from the interpolation at the foot of the drawn back characteristics that replaces evaluation of the projection integral (7.27b) and, secondly, the greater attention given to how the characteristic path is approximated.

For linear constant coefficient advection, where the characteristic path is exact, the Lagrange–Galerkin scheme of (7.15) reduces to

$$(U^{n+1}, \phi_i) = \int U^n(\mathbf{x})\phi_i(\mathbf{x} + \mathbf{b}\Delta t)\mathrm{d}x; \tag{7.60a}$$

while the semi-Lagrangian scheme uses interpolatory polynomials that we can denote by $P_i(\mathbf{x}; U^n)$ to obtain

$$U_i^{n+1} = P_i(\mathbf{x}_i - \mathbf{b}\Delta t; U^n). \tag{7.60b}$$

In one dimension, with a continuous piecewise linear basis on a uniform mesh and with $b\Delta t/\Delta x = m + \hat{\nu}$, we can deduce from (7.17) and (7.18) that the Lagrange–Galerkin scheme (7.60a) can be written as

$$(1 + \tfrac{1}{6}\delta^2)U_i^{n+1} = \sum_{p=-2}^{p=1} U_{i-m+p}^n \chi^{(4)}(\hat{\nu} + p), \qquad (7.61a)$$

where $\chi^{(4)}$ is the cubic B-spline. In particular, we can deduce from this by setting $m = \hat{\nu} = 0$ that $(1 + \tfrac{1}{6}\delta^2)U_i^n$ is the nodal value at x_i of a cubic B-spline with expansion coefficients $\{U_j^n\}$; and this gives the clue to identifying (7.61a) with a semi-Lagrange method. To avoid confusion, we denote the nodal values of the piecewise linear Lagrange–Galerkin method (7.60a) and (7.61a) by $^{LG}U_i^n$ and the semi-Lagrangian nodal values in (7.60b) by $^{SL}U_i^n$; then we relate them, and introduce cubic spline recovery coefficients $^{SL}\tilde{U}_i^n$, by setting

$$^{SL}U_i^n = (1 + \tfrac{1}{6}\delta^2)^{SL}\tilde{U}_i^n, \qquad ^{SL}\tilde{U}_i^n = {}^{LG}U_i^n. \qquad (7.61b)$$

Hence we can rewrite (7.61a) as

$$^{SL}U_i^{n+1} = \sum_{p=-2}^{p=1} {}^{SL}\tilde{U}_{i-m+p}^n \chi^{(4)}(\hat{\nu} + p) \qquad (7.61c)$$

which is now precisely in the semi-Lagrangian form (7.60b); that is, the Lagrange–Galerkin nodal values are the expansion coefficients in a cubic spline recovery procedure for the semi-Lagrange method. The two schemes then carry different parameters which have the same time evolution. We restate the result as follows.

Lemma 7.2.4 *On a uniform mesh in one dimension, and for linear constant coefficient advection, the Lagrange–Galerkin method based on continuous piecewise linear basis functions is equivalent to a semi-Lagrangian method using cubic spline interpolation.*

With this identification we have in effect extended the recurrence relation (7.17) to include delta functions, and also extended Lemma 7.2.2 to say that a linear spline characteristic-Galerkin method without recovery is equivalent to using a trial space of delta functions at the nodes and then recovering with cubic splines to give a PERU scheme; it is the latter that is identified with the semi-Lagrangian method, with the cubic spline expansion used to interpolate for the feet of the characteristics, i.e.

$$^{SL}U^n(x) = \sum_j {}^{SL}\tilde{U}_j^n \chi^{(4)}((x - x_j)/\Delta x) \qquad (7.62)$$

is the appropriate interpolation to use in (7.60b).

In contrast to the Lagrange–Galerkin methods, many semi-Lagrangian methods suffer from a lack of conservation, and considerable ingenuity has been devoted to overcoming this deficiency — see for instance, Rasch and Williamson (1990), Garcia-Navarro and Priestley (1994). Part of the problem, of course, stems from the fact that the correct definition of $\int U^n(x)dx$ is less clear when only the nodal values U_i^n are given. However, the above relationship can be exploited to show what might be done to define a conservative semi-Lagrangian method, on a nonuniform mesh in more than one dimension, and with a nonconstant velocity field. Suppose the equation is in conservation form, $u_t + \nabla \cdot (bu) = 0$, and boundary effects can be neglected; then, as in (7.27b), conservation for the weak form of the Lagrange–Galerkin method, for any set of basis functions that span the unit constant, follows from merely summing the relation

$$\int {}^{LG}U^{n+1}(\mathbf{y})\phi_i(\mathbf{y})d\mathbf{y} = \int {}^{LG}U^n(\mathbf{x})\phi_i(\mathbf{y})d\mathbf{x}, \qquad (7.63)$$

over the index i. On a tensor product mesh we use multi-linear basis functions; and let M be the mass matrix, with entries $M_{i,j} = (\phi_i, \phi_j)$, and D the lumped mass matrix with (diagonal) entries $D_{i,i} = (\phi_i, 1)$. Then suppose we generalise (7.61b) by setting

$$D_{i,i}{}^{SL}U_i^n = (M {}^{SL}\tilde{\mathbf{U}}^n)_i = (M {}^{LG}\mathbf{U}^n)_i = \int {}^{LG}U^n(\mathbf{y})\phi_i(\mathbf{y})d\mathbf{y}; \qquad (7.64a)$$

and define the integral of the semi-Lagrange approximation by

$$\sum_i D_{i,i}{}^{SL}U_i^n = \sum_{i,j} \int {}^{SL}\tilde{U}_j^n \phi_j(\mathbf{x})\phi_i(\mathbf{x})d\mathbf{x} =: \int {}^{SL}U^n(\mathbf{x})d\mathbf{x}, \qquad (7.64b)$$

where $\mathbf{U}$ denotes the column vector with components U_i. Thus the $D_{i,i}$ are the correct weights to use when summing the nodal values ${}^{SL}U_i^n$ to give the integral. Now to obtain exact conservation, the mass matrix equation in (7.64a) has to be solved to give the parameters ${}^{SL}\tilde{U}_i^n$, so that these can then be used to 'interpolate' at the feet of the characteristics. Continuing to exploit the relationship with the Lagrange–Galerkin method, suppose we define the coefficients

$$B_{i,j} := \int \phi_i(\mathbf{y})\phi_j(\mathbf{x})d\mathbf{x}, \qquad (7.65a)$$

so that, by substituting from (7.64a), the conservation relation (7.63) can be written as

$$D_{i,i}{}^{SL}U_i^{n+1} = \sum_j B_{i,j}{}^{SL}\tilde{U}_j^n. \qquad (7.65b)$$

This is a conservative update which we would like to identify as a semi-Lagrangian method. If approximations are made in calculating $B_{i,j}$, it is clear that the key relation to ensure exact conservation is to have

$$\sum_i B_{i,j} = \sum_i M_{i,j} = D_{j,j}. \tag{7.65c}$$

Now suppose that we write $x_i' \equiv X(x_i, t_{n+1}; t_n)$ for the foot of the characteristic drawn back from x_i. Then this construction in (7.65b) can be recognised as a semi-Lagrangian method if the right-hand side is an expansion for $U^n(x_i')$, and each $B_{i,j}$ is the value of a basis function. If b is constant, we can indeed write

$$\psi_{i,j}(\mathbf{z}) := \int \phi_i(\mathbf{x} + \mathbf{x}_i - \mathbf{z})\phi_j(\mathbf{x})d\mathbf{x}, \quad B_{i,j} = \psi_{i,j}(x_i');$$

$\psi_{i,j}(\mathbf{z})$ is piecewise cubic in each coordinate direction but it is not a spline unless the mesh is uniform; and, similarly, in order for the expansion $\sum_{(j)} {}^{SL}\tilde{U}_j^n \psi_{i,j}(\mathbf{z})$ to be interpolatory we need

$$\int \phi_i(\mathbf{x} + \mathbf{x}_i - \mathbf{x}_k)\phi_j(\mathbf{x})d\mathbf{x} = \int \phi_k(\mathbf{x})\phi_j(\mathbf{x})d\mathbf{x},$$

which is only true on a uniform mesh. If b is not constant, we cannot even define the function $\psi_{i,j}(\mathbf{z})$. Nevertheless, a practical semi-Lagrangian method might be based on calculating the ${}^{SL}\tilde{U}_j^n$ from the ${}^{SL}U_j^n$, using these as coefficients in a nonuniform mesh cubic B-spline expansion, and then for each i for which the interpolated value $B_{i,j}$ at x_i' is nonzero, scaling the $B_{i,j}$ so that (7.65c) is satisfied.

We conclude this section by describing the computation of the characteristic paths that has been found to be very effective by Staniforth and Côté (1991); since they consider pollutant transport in a system of equations which includes the shallow water equations, the technique includes the case where b depends on the solution u. Each trajectory is approximated by a straight line drawn back from a mesh point x_i; with the above notation x_i' for the foot and δx_i for $x_i - x_i'$ we write

$$\delta x_i = \Delta t\mathbf{b}(x_i - \tfrac{1}{2}\delta x_i, t + \tfrac{1}{2}\Delta t); \tag{7.66a}$$

and, if necessary, b is extrapolated by

$$\mathbf{b}(\mathbf{x}, t + \tfrac{1}{2}\Delta t) = \tfrac{3}{2}\mathbf{b}(\mathbf{x}, t) - \tfrac{1}{2}\mathbf{b}(\mathbf{x}, t - \Delta t). \tag{7.66b}$$

A few iterations of (7.66) serves to give δx_i sufficiently accurately. Note that, in the application to the shallow water equations, only the advective terms are approximated in this way, all other terms being treated in a semi-implicit manner, as advocated by Kwizak and Robert (1971).

7.2.7 Use of a fundamental solution

In the locally exact schemes of section 3.3.2 and those using local Green's functions in section 5.8, we presented methods based on exact solutions of $-\epsilon u_{xx} + bu_x = 0$; and in the preceding sections here we have used exact solutions of $u_t + bu_x = 0$. Now we make use of exact solutions to $u_t + bu_x = \epsilon u_{xx}$, as presented by Morton and Sobey (1993). On the whole real line and with constant b, it is easy to see that the exact solution for initial data $u^0(x)$ is

$$u(x,t) = \frac{1}{2\sqrt{(\pi \epsilon t)}} \int_{-\infty}^{\infty} e^{-s^2/4\epsilon t} u^0(x - bt + s)ds. \tag{7.67}$$

This was used by Morton and Sobey to derive finite difference and finite element approximations and also to provide rigorous error bounds. Substitution of a polynomial approximation for the initial data in (7.67) leads to a set of standard integrals that we shall need,

$$a_r := \frac{1}{\sqrt{\pi}} \int_{-\infty}^{\infty} \zeta^r e^{-\zeta^2} d\zeta; \tag{7.68}$$

clearly $a_{2m+1} = 0$ and $a_{2m+2} = \frac{1}{2}(2m + 1)a_{2m}$ so that the leading terms in the sequence are $1, 0, \frac{1}{2}, 0, \frac{3}{4}, \ldots$.

Consider first a finite difference approximation on a uniform mesh, generated in a similar way to the semi-Lagrangian scheme (7.60b). That is, we substitute a polynomial expansion about x_i into the formula (7.67) applied over a single time step, to obtain

$$
\begin{aligned}
U_i^{n+1} &= \frac{1}{2\sqrt{(\pi \epsilon \Delta t)}} \int_{-\infty}^{\infty} e^{-s^2/4\epsilon \Delta t} \sum_p b_{i,p}(s - b\Delta t)^p ds \\
&= \sum_p \frac{b_{i,p}}{\sqrt{\pi}} \int_{-\infty}^{\infty} e^{-\zeta^2} [2\sqrt{(\epsilon \Delta t)}\zeta - b\Delta t]^p d\zeta \\
&= b_{i,0} - b_{i,1}b\Delta t + b_{i,2}[(b\Delta t)^2 + 2\epsilon\Delta t] - b_{i,3}[(b\Delta t)^3 + 6b\epsilon(\Delta t)^2] \\
&\quad + b_{i,4}[(b\Delta t)^4 + 12b^2\epsilon(\Delta t)^3 + 12\epsilon^2(\Delta t)^2] + \cdots.
\end{aligned} \tag{7.69}
$$

If we use a centrally positioned quadratic interpolant, we obtain

$$b_{i,0} = U_i^n, \quad b_{i,1} = D_0 U_i^n, \quad b_{i,2} = \tfrac{1}{2}D_+ D_- U_i^n,$$

and hence the Lax–Wendroff scheme, which with $\nu = b\Delta t/\Delta x$ and $\mu = \epsilon \Delta t/(\Delta t)^2$ becomes

$$U_i^{n+1} = [1 - \nu \Delta_0 + (\tfrac{1}{2}\nu^2 + \mu)\delta^2]U_i^n.$$

This is stable in the practical sense for $\nu^2 + 2\mu \leq 1$ and, as we have already seen, the same scheme can be derived in many other ways. More interestingly, if we use a cubic interpolant centred on the upwind interval

(x_{i-1}, x_i), we obtain $b_{i,3} = \frac{1}{6} D_+ D_-^2 U_i^n$ and derive the QUICKEST scheme, which was proposed by Leonard (1979a) and has been widely used in many application areas, namely

$$U_i^{n+1} = [1 - \nu \Delta_0 + (\tfrac{1}{2}\nu^2 + \mu)\delta^2 + \tfrac{1}{6}\nu(1 - \nu^2 - 6\mu)\delta^2 \Delta_-]U_i^n. \qquad (7.70)$$

When $\mu = 0$ this is a scheme we have generated in several ways — as a PERU scheme with piecewise quadratic recovery from piecewise constants (7.57), as a characteristic-Galerkin method with a continuous piecewise linear basis and a mixed norm projection (7.21) or mass lumping, or originally as the finite difference scheme due to Warming $et\ al.$ (1973). The addition of the diffusion term in (7.70), written as

$$\mu \delta^2 [(1 - \nu)U_i^n + \nu U_{i-1}^n]$$

to show how it is based on the linearly interpolated foot of the characteristic, is distinctive and very natural. Indeed, this QUICKEST scheme has a good claim to be the key explicit finite difference scheme for unsteady convection-diffusion: as shown by Morton and Sobey (1993), it satisfies the practical stability condition for $\mu \leq \frac{1}{2}$ and for all $\nu \in [0, 1]$, with the stability limit extending to $\mu = \frac{9}{8}$ at $\nu = \frac{1}{2}$; applying the Peano kernel theorem to the integrals in (7.67) and (7.69), these authors also give error bounds in the form $C(\mu, \nu)(\Delta x)^3 |u|_{4,\infty}$, and plot the amplitude and phase errors of typical Fourier modes which clearly show the advantage of QUICKEST over the Lax–Wendroff scheme; the various ways of generating the scheme show how it can be readily generalised to arbitrary time steps, to nonuniform meshes in several dimensions, and with nonlinear convection terms; and with the leading error being a fourth order dissipative term, it is an ideal starting point for deriving solution-adaptive schemes with positivity-preserving, TVB or other properties.

Alternatively, substituting a finite element approximation into (7.67), and carrying out all the integrals exactly, would define the convection-diffusion evolution operator $E(\Delta t)$ to be used in a PERU scheme. However, integrating piecewise polynomials multiplied by the exponential kernel is impractical. Morton and Sobey (1993) therefore defined an approximate evolution operator E_Δ by supposing that the polynomial forms holding in the interval containing the foot of the characteristic extended to the whole real line. With piecewise linear basis functions, using the mixed norm (7.21) and with $\nu = m + \hat{\nu}$, they then obtained the following scheme which contains both (7.20) and (7.70) as special cases,

$$[1 + (\tfrac{1}{6} - \gamma^2)\delta^2](U_i^{n+1} - U_{i-m}^n)$$
$$= [-\hat{\nu}\Delta_0 + (\tfrac{1}{2}\hat{\nu}^2 + \mu)\delta^2 - \hat{\nu}(\tfrac{1}{6}\hat{\nu}^2 + \mu - \gamma^2)\delta^2 \Delta_-]U_{i-m}^n. \qquad (7.71)$$

In analysing and displaying the errors for such schemes, they also show that when the mesh Péclet number, which is now $\beta = \hat{\nu}/\mu$, is small it is the explicit QUICKEST scheme with $\gamma^2 = \frac{1}{6}$ which is to be preferred; while, for large β, as we have seen in section 7.2.2 it is the implicit scheme with $\gamma = 0$ which has the edge in accuracy. It is clear that both (7.71) and (7.70) can be followed by higher order accurate schemes by increasing the degree of the polynomials used in the two definitions.

In the ECG scheme for the pure convection problem that was described in section 7.2.2, an upwind-averaged basis function (7.28b) was introduced as a test function; such a function can also be associated with the scheme (7.71). On the other hand, optimal test functions were introduced for the steady convection-diffusion problem in section 4.3.1. The consideration that we have given in this section to approximating the unsteady convection-diffusion problem by means of the fundamental solution (7.67) provides the means to relate these two types of test function. In Morton (1992) it is shown that if the mixed norm used in (7.71) corresponds to taking $\gamma = \epsilon/b\Delta x \equiv 1/\beta$, the two test functions have exactly the same defining relation. Note that this choice of γ is in accord with the choice indicated in the previous paragraph on accuracy grounds.

7.2.8 Some stability and convergence results

The most distinctive property of the Lagrange–Galerkin, characteristic-Galerkin and related schemes described in sections 7.2.2–7.2.7 is their unconditional linear stability, and their high accuracy over the large time steps that are therefore allowed. Moreover, for nonlinear scalar conservation laws, basing the schemes on the transport collapse operator brings other properties which lead quickly to convergence results. The following two theorems exemplify their theoretical properties for such problems.

Theorem 7.2.5 *Suppose the (x, t)-mesh is quasiregular, in the sense that $h \leq \Delta x_j \leq Ch, h \leq \Delta t_n \leq Ch$ for some constant C as $h \to 0$; and suppose the scalar problem $u_t + f_x(u) = 0$, with $f(\cdot) \in C^2$, is approximated on $\mathbb{R} \times (0, T]$ by the ECG scheme of (7.55), namely*

$$(U^{n+1} - U^n, \phi_i) + \Delta t_n(f_x(\bar{u}^n), \bar{\Phi}_i^n) = 0, \qquad (7.72)$$

using piecewise constant basis functions $\{\phi_i\}$ and recovery by discontinuous linear functions with the minmod limiter (7.50a). Then we have

$$\|U^n\|_{L_\infty} = \|\bar{u}^n\|_{L_\infty} \leq \|U^0\|_{L_\infty} \quad \forall n, \qquad (7.73a)$$

$$TV(U^n) \leq TV(U^0) \quad \forall n, \qquad (7.73b)$$

$$\sum_j |U_j^n - U_j^m| \Delta x_j \leq K(t_n - t_m) \quad 0 \leq t_m \leq t_n \leq T, \qquad (7.73c)$$

where K is a constant independent of m and n.

Proof. This follows directly from the properties of the transport collapse operator set out in Theorem 7.2.1 and the fact that the recovery using (7.50a) is also TVD. □

Theorem 7.2.6 *The approximations obtained with the hypotheses of Theorem 7.2.5 are compact in $L_\infty([0,t]; L_1(\mathbb{R}))$ and converge to a solution of the conservation law satisfying the entropy inequality*

$$\int \int [\eta(u)\psi_t + q(u)\psi_x]dxdt \geq 0, \quad 0 \leq \psi \in C_0^\infty(\mathbb{R} \times (0, \infty)), \quad (7.74)$$

for any entropy pair (η, q), i.e. η is any Lipschitz continuous convex function and $q(u) = \int_0^u f'(\lambda)\eta'(\lambda)d\lambda$.

Proof. See Lin *et al.* (1993), Theorem 6.1. □

Because the transport collapse operator and its key properties are equally valid for scalar problems in multi-dimensions, to generalise these two theorems to such cases requires only an appropriate generalisation of the ECG scheme (7.72) and extension of the TVD linear recovery process. The former generalisation is given in Lin *et al.* (1995), based on the transport collapse operator, and how to carry out the recovery is discussed in section 7.2.4.

The analysis in these scalar conservation law problems is simplified by the fact that all the characteristics are straight lines, but precise error estimates are made difficult by the nonlinear nature of the problems. So let us consider next a linear convection-diffusion problem with an incompressible velocity field $b(x)$ in d dimensions; and we approximate this by a direct Lagrange–Galerkin method, as in (7.13) and (7.15b) except that we assume exact integration of a source term $S(x, t)$ and we may introduce a recovery stage. Unconditional stability follows directly from (7.15b) for the implicit scheme: suppose U^n and V^n are two solutions with the same source term, but differing initial data, and we multiply (7.15b) by $U_i^{n+1} - V_i^{n+1}$ and sum over i; using the simplified notation of (7.27), we have

$$\|U^{n+1} - V^{n+1}\|^2 + \epsilon\Delta t \left\|a^{\frac{1}{2}}\nabla(U^{n+1} - V^{n+1})\right\|^2$$

$$= \int [U^n(x) - V^n(x)][U^{n+1}(y) - V^{n+1}(y)]dy$$

$$\leq \|U^n - V^n\| \|U^{n+1} - V^{n+1}\|, \quad (7.75)$$

because of the assumed incompressibility; this implies immediately that $\|U^{n+1} - V^{n+1}\| \leq \|U^n - V^n\|$.

The dramatic improvement in accuracy over a conventional Eulerian difference scheme is highlighted by the error estimates given first by Douglas and Russell (1982). For a scheme without recovery and using polynomials of degree $k \geq 1$, they obtain in the one-dimensional case the estimate

$$\|u - U\|_{l_\infty(0,T;L_2(\mathbb{R}))} \leq C_1\Delta t \left\|\frac{D^2u}{Dt^2}\right\|_{L_2(0,T;L_2(\mathbb{R}))}$$

$$+ C_2 h^{k+1} \left[\|u\|_{L_\infty(0,T;H^{k+1}(\mathbb{R}))} + \left\| \frac{\partial u}{\partial t} \right\|_{L_2(0,T;H^{k+\theta}(\mathbb{R}))} \right], \qquad (7.76)$$

where D/Dt is the material or Lagrangian derivative $\partial/\partial t + b\partial/\partial x$ and $\theta = 1$ when $k = 1$ but is zero for $k \geq 2$. Thus when the solution changes slowly along the characteristic trajectories the full order of accuracy of the finite element approximation can be maintained.

The constants C_1 and C_2 in (7.76) depend on ϵ; and if this dependency is removed the second term becomes $O(h^k)$, i.e. it is sub-optimal. To consider this further we simplify to the pure convection problem but introduce the recovery operator R^h. Thus we can write, with $U^0 = P^h u^0$,

$$U^{n+1} = P^h \widehat{E}(\Delta t) R^h U^n + \int_{t_n}^{t_{n+1}} P^h \widehat{E}(t_{n+1} - \tau) S(\cdot, \tau) d\tau. \qquad (7.77)$$

With a finite element trial space S^h consisting of piecewise polynomials of degree $k, k \geq 0$, we assume that the recovery process satisfies the usual condition (7.40a) and, by using piecewise polynomials of degree $l \geq \max(1, k)$, has the approximation property that there exists an integer $r \geq 1$, such that for $\forall v \in H^{k+r}(\mathbb{R}^d)$

$$\|v - R^h P^h v\|_{H^s(\mathbb{R}^d)} \leq C_s h^{k+r-s} |v|_{H^{k+r}(\mathbb{R}^d)}, \quad s = 0, 1, \qquad (7.78)$$

for some constants C_s. Then we have the following error bound when a constant time step Δt is used.

Theorem 7.2.7 *Suppose* $u \in C([0, T]; H^{k+r}(\mathbb{R}^d))$ *and the continuous polynomial approximation* $v \longmapsto R^h P^h v$ *satisfies* (7.78), *including the case of no recovery in which* $R^h \equiv I$ *and* $r = 1$. *Then the Lagrange–Galerkin approximation* U *of* (7.77) *satisfies*

$$\|u - R_h U\|_{l_\infty(0,T;L_2(\mathbb{R}^d))}$$
$$\leq C h^{k+r} [1 + (T/\Delta t) \min(1, \nu)] |u|_{L_\infty(0,T;H^{k+r}(\mathbb{R}))} \qquad (7.79)$$

for some constant C, *where* $\nu := (\Delta t/h) \max(\|\mathbf{b}\|_\infty, \|\nabla \mathbf{b}\|_\infty)$.

Proof. We decompose the global error into the projection error η and an evolutionary error ξ by

$$e^n := u^n - R^h U^n = (u^n - R^h P^h u^n) + (R^h P^h u^n - R^h U^n)$$
$$=: \eta^n + \xi^n; \qquad (7.80a)$$

and comparing (7.77) with the corresponding expression for u^{n+1} we have

$$\xi^n = R^h P^h \widehat{E}(\Delta t)(u^{n-1} - R^h U^{n-1})$$
$$= R^h P^h \widehat{E}(\Delta t)(\eta^{n-1} + \xi^{n-1}) \qquad (7.80b)$$
$$= R^h P^h \widehat{E}(\Delta t) \xi^{n-1} + R^h P^h (\widehat{E}(\Delta t) - I) \eta^{n-1}. \qquad (7.80c)$$

This is the starting point for most error estimates for evolution-Galerkin methods. Because both $R^h P^h$ and $\widehat{E}(\Delta t)$ are nonexpansive, we have from (7.80b) that $\|\xi^n\| \leq \|\xi^{n-1}\| + \|\eta^{n-1}\|$, which gives the large time step estimate in (7.79) corresponding to $\nu > 1$. Such an estimate was first given by Pironneau (1982).

For small time steps, we define the mapping $\mathbf{x} \longmapsto \mathbf{H}^\theta(\mathbf{x}, s; t)$ by

$$\mathbf{H}^\theta(\mathbf{x}, s; t) := \theta \mathbf{X}(\mathbf{x}, s; t) + (1 - \theta)\mathbf{x},$$

for $0 \leq \theta \leq 1$. Then for sufficiently small h and $\nu < 1$, we can deduce that the Jacobian of this mapping is bounded from below by some quantity J_H, and obtain

$$\left\| (\widehat{E}(\Delta t) - I)\eta^m \right\|^2 = \int_{\mathbb{R}^d} |\eta^m(\mathbf{X}(\mathbf{x}, \Delta t; 0)) - \eta^m(\mathbf{x})|^2 \, d\mathbf{x}$$

$$= \int_{\mathbb{R}^d} \left| \int_0^1 \nabla \eta^m (\mathbf{H}^\theta(\mathbf{x}, \Delta t; 0)) \cdot (\mathbf{X}(\mathbf{x}, \Delta t; 0) - \mathbf{x}) d\theta \right|^2 \, d\mathbf{x}$$

$$\leq J_H^{-1}(\Delta t \|\mathbf{b}\|_\infty)^2 \|\nabla \eta^m\|^2; \tag{7.81}$$

for details of this analysis the reader is referred to Morton and Süli (1992) where it is shown that $J_H \geq (1 - \nu h e^{\nu h})^d$. Applying (7.78) with $s = 1$ and introducing ν then gives the alternative case of (7.79).   □

This theorem does not cover the case of piecewise constant S^h and no recovery, or recovery by discontinuous linear functions, in which we assume only that the recovery space consists of piecewise polynomials of degree $l \geq k \geq 0$. To cover such cases we work with functions of bounded variation, extending the definition of total variation to

$$\mathrm{Var}(w) := \sup_{\mathbf{v} \in [C_0^1(\mathbb{R}^d)]^d} \frac{(w, \nabla \cdot \mathbf{v})}{\|\mathbf{v}\|_\infty}, \tag{7.82a}$$

and introducing the norms, for the spaces $V^s(\mathbb{R}^d)$,

$$\|w\|_{V^s(\mathbb{R}^d)} := \|v\|_{L_1} + \sum_{|\alpha| \leq s-1} \mathrm{Var}(D^\alpha w), \quad s \geq 1, \tag{7.82b}$$

together with the corresponding semi-norms; we also work in $L_1(\mathbb{R}^d)$ rather than $L_2(\mathbb{R}^d)$ and define P^h as the L_1 projector onto S^h, to obtain the following result.

Theorem 7.2.8 *Suppose $U \in C([0, T]; V^{k+r}(\mathbb{R}^d))$ and the approximation given by $R^h P^h$ satisfies (7.78) with $H^s(\mathbb{R}^d)$ and $H^{k+r}(\mathbb{R}^d)$ replaced by $V^s(\mathbb{R}^d)$ and $V^{k+r}(\mathbb{R}^d)$ respectively. Then the Lagrange–Galerkin approximation U of (7.77) satisfies*

$$\|u - R^h U\|_{l_\infty(0, T; L_1(\mathbb{R}^d))}$$
$$\leq Ch^{k+r}[1 + (T/\Delta t)\min(1, \nu)] |u|_{L_\infty(0, T; V^{k+r}(\mathbb{R}^d))} \tag{7.83}$$

for some constant C and with ν as in Theorem 7.2.7.

Proof. The proof proceeds as for Theorem 7.2.7, except that (7.81) follows directly only by assuming $\eta^m \in W^{1,1}(\mathbb{R}^d)$ and we have only $\eta^m \in V^1(\mathbb{R}^d)$. However, if $v \in W^{1,1}(\mathbb{R}^d)$ then $\|\nabla v\| = \mathrm{Var}(v) = |v|_{V^1(\mathbb{R}^d)}$ and in Morton and Süli (1992) a density argument is used to extend the bound to $V^1(\mathbb{R}^d)$.
$\square$

Various generalisations of these two results are possible. If $\nabla \cdot \mathbf{b} \neq 0$ the term $T/\Delta t$ in the estimates is merely replaced by an exponential factor arising from the relation (7.23) for the Jacobian J of the mapping $\mathbf{x} \to \mathbf{X}(\mathbf{x}, \Delta t; 0)$. Similar estimates have also been obtained by Süli (1988) for the Navier–Stokes equations, so it is clear that they are unaffected by the addition of diffusion terms and the use of an approximate velocity field.

Unfortunately, these results are still not sharp. For example, in the case of piecewise constant S^h and no recovery, we have $k = 0, r = 1$ in the approximation estimate corresponding to (7.78) with norms $V^0(\mathbb{R}^d)$ and $V^1(\mathbb{R}^d)$, but (7.83) does not imply convergence to any order. As shown in Childs and Morton (1990), the basic analysis can be modified to obtain an extra power of h in this case, for convection in one dimension, when b is constant and the CFL numbers are less than unity. The device that is used consists of introducing elements Δ^n of S^h given by

$$\Delta_j^n := \frac{1}{b\Delta t} \int_{x_{j+1/2}-b\Delta t}^{x_{j+1/2}} (u^n - \Pi^h u^n)\mathrm{d}x, \qquad (7.84\mathrm{a})$$

where Π^h is the projection onto piecewise constants. Then it is clear that, using the notation η^n for $u^n - \Pi^h u^n$ and ϕ_j for the piecewise constant basis function,

$$\begin{aligned}
([E(\Delta t) - I]\eta^n, \phi_j) &= \int_{x_{j-1/2}-b\Delta t}^{x_{j+1/2}-b\Delta t} \eta^n \mathrm{d}x - \int_{x_{j-1/2}}^{x_{j+1/2}} \eta^n \mathrm{d}x \\
&= \int_{x_{j-1/2}-b\Delta t}^{x_{j-1/2}} \eta^n \mathrm{d}x - \int_{x_{j+1/2}-b\Delta t}^{x_{j+1/2}} \eta^n \mathrm{d}x \\
&= b\Delta t(\Delta_{j-1}^n - \Delta_j^n) \\
&= ([E(\Delta t) - I]\Delta^n, \phi_j). \qquad (7.84\mathrm{b})
\end{aligned}$$

Hence, instead of (7.80c) we can write

$$\xi^n - \Delta^n = \Pi^h \widehat{E}(\Delta t)(\xi^{n-1} - \Delta^{n-1}) - (\Delta^n - \Delta^{n-1})$$

so that

$$\|\xi^n - \Delta^n\| \leq \|\xi^{n-1} - \Delta^{n-1}\| + \|\Delta^n - \Delta^{n-1}\|. \qquad (7.85)$$

If $\partial^2 u/\partial x^2$ is bounded, then so is $\partial^2 u/\partial x \partial t$ and $\Delta^n - \Delta^{n-1} = O(h\Delta t)$ while $\Delta^n = O(h)$, so we obtain the result

$$\|u - U\|_{l_\infty(0,T;L_2(\mathbb{R}))} \leq Ch. \qquad (7.86)$$

This is a typical *supraconvergence* result, and it is worth noting that the identification of the quantity Δ^n is equivalent to the identification of a special component in the truncation error, which was the approach used by Manteuffel and White (1986) in their original treatment of this phenomenon. In Morton and Süli (1992) the definition of Δ^n is extended to the case of a variable b, and a similar construction is devised for the continuous piecewise linear trial space to show that this gives $O(h^2)$ accuracy, a result which we merely state in the following theorem.

Theorem 7.2.9 *Suppose* $\partial^3 u/\partial x^3 \in L_\infty(0, T; L_\infty(\mathbb{R}))$,

$$0 < c_0 \le b\Delta t/\Delta x_j \le 1 \text{ and } \Delta x_j/\Delta x_{j-1} \le 5, \ \forall j.$$

Then the Lagrange–Galerkin approximation to $u_t + bu_x = S$ *based on a piecewise linear trial space satisfies*

$$\|u - U\|_{l_\infty(0,T;L_2(\mathbb{R}))} \le Ch^2. \tag{7.87}$$

Moreover, Morton and Süli show that the key requirement to establishing the full order of accuracy of a Lagrange–Galerkin scheme, with recovery, in this way is the following: suppose that Q^h like P^h is a continuous projector from $L_p(\mathbb{R}^d) \cap C(\mathbb{R}^d)$ onto S^h, so that for $\xi^n := Q^h u^n - R^h U^n$ we can write as in (7.80c)

$$\xi^n = R^h P^h \widehat{E}(\Delta t) \xi^{n-1} + P^h [Q^h \widehat{E}(\Delta t) - \widehat{E}(\Delta t) Q^h] u^{n-1}; \tag{7.88a}$$

then we want to be able to represent the last term as

$$R^h P^h [Q^h \widehat{E}(\Delta t) - \widehat{E}(\Delta t) Q^h] u^{n-1}$$
$$= R^h P^h [\widehat{E}(\Delta t) - I] V^{n-1} + \Delta t \psi^{n-1}, \tag{7.88b}$$

where $V^n \in S^h, \psi^n \in S^h$ and the L_p norms of $(\Delta t)^{-1}(V^n - V^{n-1}), V^n$ and ψ^n are all $O(h^{k+1})$. It is an open question whether such representations are possible in two dimensions and for CFL numbers greater than unity.

In special cases, specifically by assuming uniform Cartesian meshes, even higher order accuracy can be achieved, as we have observed in earlier sections. Consider, for example, one-dimensional convection with constant b, using a continuous piecewise linear trial space. As shown by Childs and Morton (1990), the difference scheme (7.20) (with $\mu = 0$) that is obtained using L_2 projection, is one of a family generated using the mixed inner product of (7.21) and corresponding to (7.71) with $\mu = 0$, namely

$$[1 + (\tfrac{1}{6} - \gamma^2)\delta^2](U_i^{n+1} - U_{i-m}^n)$$
$$= \hat{\nu}[-\Delta_0 + \tfrac{1}{2}\hat{\nu}\delta^2 - (\tfrac{1}{6}\hat{\nu}^2 - \gamma^2)\delta^2 \Delta_-]U_{i-m}^n, \tag{7.89}$$

where $b > 0$ and $\nu = m + \hat{\nu}, m \in \mathbb{Z}, \hat{\nu} \in (0, 1]$. These schemes are all unconditionally stable, as the stability range for (7.89) contains $0 \le \hat{\nu} \le 1$ for all choices of γ^2; and their Fourier analysis also shows that the leading term in the truncation error is given by

$$T.E. \sim \tfrac{1}{24}\hat{\nu}(1-\hat{\nu})[\hat{\nu}(1-\hat{\nu})+12\gamma^2]\frac{(k\Delta x)^4}{\Delta t}e^{ik(x-m\Delta x)}. \qquad (7.90)$$

Thus, on this basis, the L_2 projection Lagrange–Galerkin scheme is four times more accurate than the Taylor–Galerkin scheme ($\gamma^2 = \tfrac{1}{6}\hat{\nu}^2$) and 36 times more accurate than the explicit Warming–Kutler–Lomax difference scheme ($\gamma^2 = \tfrac{1}{6}$) — as is indicated by comparing the right-hand peaks in the plotted results of Figs. 7.6 and 7.7.

These accuracy estimates extend in various directions. Fourth order rather than third order accuracy can be achieved by more central time differencing; for example, combining (7.89) with the corresponding 'backwards' scheme that gives U^{n-1} in terms of U^n yields for $\gamma^2 = \tfrac{1}{6}\nu^2$ and $\nu^2 \leq 1$ the scheme

$$[1 + \tfrac{1}{6}(1-\nu^2)\delta^2](U_i^{n+1} - U_i^{n-1}) = 2\nu\Delta_0 U_i^n,$$

which was called the leapfrog-Petrov–Galerkin scheme by Morton and Parrott (1980), and this can clearly be extended to arbitrary values of ν. Corresponding schemes are also obtained from B-splines of any order, with splines of order m giving accuracy of order $(\Delta x)^{2m-1}$ in the one step form (7.89), consistent with the original estimates of Strang (1971) which were later extended to variable coefficient problems by Thomée and Wendroff (1974). Also tensor product approximations on uniform Cartesian meshes extend the schemes to multi-dimensions, as in (7.39).

A comment is in order here on the interpretation of these high order *superconvergence results*; for example, how can piecewise linear approximations give third and fourth order accuracy? The term superconvergence is appropriate because the higher order is achieved only at the mesh points and in the evolutionary error ξ, not of course in the projection error η. It comes about on a uniform mesh because the evolution operator commutes with the convolution defining the B-spline recurrence, as we saw in (7.17) and (7.18); when interpreting a finite element method as a finite difference scheme there is an implied use of a restriction operator, and the importance of commutator relationships between this and the evolution operator is seen in the error analysis by Cullen and Morton (1980) and in (7.88). Numerical experiments confirm that this superconvergence survives the introduction of variable velocity fields but not the use of nonuniform meshes, as was also seen in the analysis of section 7.2.6.

The sharpest estimates for the convection-diffusion problem are based on the use of a fundamental solution as in the last section. There the family of schemes (7.71) was given which generalises that given by (7.89) to include diffusion, with $\mu = \epsilon\Delta t/(\Delta x)^2$. It includes the QUICKEST scheme ($\gamma^2 = \tfrac{1}{6}$), which was generated by Morton and Sobey (1993) by using a cubic interpolant through $U_{i-m-2}^n, U_{i-m-1}^n, U_{i-m}^n$ and U_{i-m+1}^n, and applying the exact evolution operator in the form given by (7.67). If Q^h is used to denote this interpolatory projector, they used the Peano kernel theorem to express the interpolatory error as

$$(I - Q^h)u = \int_{-\infty}^{\infty} K(\cdot, \eta)u^{iv}(y)\mathrm{d}y;$$

and hence the evolutionary error committed in one time step, namely $E(\Delta t)(I - Q^h)u^n$, was given at the point x_i by $\Delta t T_i$, where

$$\Delta t T_i = \int_{-\infty}^{\infty} K_E(x_i, y)u^{iv}(y)\mathrm{d}y \qquad (7.91a)$$

and

$$K_E(x_i, y) := \frac{1}{2\sqrt{(\pi\epsilon\Delta t)}} \int_{-\infty}^{\infty} K(x_i - b\Delta t + s, y)e^{-s^2/4\epsilon\Delta t}\mathrm{d}s. \qquad (7.91b)$$

Morton and Sobey gave expressions for the Peano kernel function and hence were able to make explicit estimates of T_i which is effectively the truncation error. We summarise the result for this important and typical scheme.

Theorem 7.2.10 *On a uniform mesh the (shifted) QUICKEST scheme for approximating the one-dimensional, constant coefficient, convection-diffusion problem is stable for any $\nu := b\Delta t/\Delta x$ and for $\mu := \epsilon\Delta t/(\Delta x)^2 \leq \frac{1}{2}$; and the error satisfies, for compact data,*

$$\|u - U\|_{l_\infty(0,T;l_\infty(\mathbb{R}))} \leq C(\mu, \hat{\nu})(T/\Delta t)(\Delta x)^4 \left|\frac{\partial^4 u}{\partial x^4}\right|_{l_\infty(0,T;L_\infty(\mathbb{R}))} \qquad (7.92)$$

Proof. See Morton and Sobey (1993), where an explicit expression is given for $C(\mu, \hat{\nu})$ as well as an asymptotic form consistent with (7.90); the full stability region in the $(\mu, \hat{\nu})$-plane is also given, extending from $0 \leq \mu \leq \frac{1}{2}$ for $\hat{\nu} = 0, 1$ to $0 \leq \mu \leq \frac{9}{8}$ at $\hat{\nu} = \frac{1}{2}$. $\square$

Finally, it must be emphasised that all of the results in this section are based on the presumption that all integrals in inner products and over source terms are carried out exactly. This caveat is serious only in regard to the inner products and the effect of quadrature on the unconditional stability of the exact schemes. In section 7.2.2 reference has already been made to Morton *et al.* (1988), where it is shown that for pure constant coefficient convection on a uniform mesh it is necessary to bound the CFL number away from zero to obtain stability when many standard quadrature rules are used; Gauss–Lobatto rules are the important exceptions to this situation. We summarise the necessary stability conditions in the following theorem where, except when noted to the contrary, it is assumed the same rule is used to evaluate the mass matrix, M.M., and the integrals in (7.27b), R.H.S.

Theorem 7.2.11 *For linear, constant coefficient, convection on a uniform mesh, the necessary stability conditions contained in the following table apply when the inner products in a Lagrange–Galerkin method based on linear elements are evaluated by quadrature. In the last two lines the quadrature rules are assumed*

M.M.	R.H.S.	nec. stab. condtn.
centroid	centroid	uncond. unstable
exact	centroid	$0 \leq \nu \leq 1/\sqrt{6}$
vertex	vertex	uncond. stable
exact	vertex	uncond. unstable
Gauss–Legendre		$1 - x_m < \nu$
Gauss–Lobatto		$0 \leq \nu \leq 2/(m+1)(m+2)$

to integrate quadratics exactly: so, for Gauss-Legendre, $m \geq 2$ and x_m is the last quadrature point in a scheme on the unit interval; and, for Gauss-Lobatto, $m \geq 1$.

Proof. See Morton *et al.* (1988). It is also shown there that lumping the mass matrix generally stabilises the schemes.   □

The unconditional stability of the vertex or trapezoidal rule is attractive; but even when applied to subdivided elements it leads to a significant loss of accuracy. Hence high order Gauss rules are often used in practice, so that $1 - x_m$ is small and stagnation points are either avoided or treated very carefully. The alternative is to use the area-weighting technique of section 7.2.3, or other similar schemes that are based on exact integration of the inner products arising from special approximations to the evolution operator.

7.3 Generalised Godunov schemes

Let us write the equation for a vector of unknowns **w** in the divergence form

$$\frac{\partial \mathbf{w}}{\partial t} + \nabla \cdot \mathcal{F} = \mathbf{S}, \quad \mathcal{F} \equiv \mathcal{F}(\mathbf{w}, \nabla \mathbf{w}), \tag{7.93}$$

where the fluxes $\mathcal{F}, \equiv (\mathbf{f}, \mathbf{g})$ in two dimensions, contain both convective and diffusive terms. Writing E_τ for an approximate evolution operator over a time $\tau \in [0, \Delta t]$, and letting Ω_α denote an element of the mesh (commonly a quadrilateral or triangle in two dimensions), a generalised Godunov scheme has the form

$$\int_{\Omega_\alpha} (\mathbf{W}^{n+1} - \mathbf{W}^n) d\Omega + \int_0^{\Delta t} d\tau \int_{\partial \Omega_\alpha} \mathcal{F}(E_\tau \mathbf{w}, \nabla E_\tau \mathbf{w}) \cdot \mathbf{n} d\Gamma$$

$$= \int_0^{\Delta t} d\tau \int_{\Omega_\alpha} S d\Omega. \tag{7.94}$$

That is, the evolution operator is used to predict the fluxes through the sides of the conservation cell, rather than to predict directly the new values $\mathbf{W}^{n+1}$ as in the preceding sections.

7.3.1 Explicit schemes

The original scheme of Godunov (1959) was devised for the inviscid Euler equations in one dimension: with the vector $\mathbf{W}_i^n$ representing the average density, momentum and energy in cell i, each supposed to be constant over the cell, what is called a *Riemann problem* is posed by the discontinuous data at each cell boundary; Godunov proposed that these be solved exactly, by a combination of shocks, contact discontinuities and rarefaction waves, which gives E_τ exactly up to a time when solutions from neighbouring boundaries interact. In many ways this is the ideal approximation; the disadvantages are the computational cost of solving the Riemann problems, and the limitation to first order accuracy arising from the implied piecewise constant trial space. Thus a great many schemes have been derived from this starting point, using *approximate Riemann solvers*, recovery techniques such as those described in section 7.2.4, and higher order trial spaces.

Our concern here is more with the scalar problem, but with diffusion and in more than one dimension, and we seek new algorithms from the formulation (7.94). Without diffusion and in one dimension, the piecewise constant trial space without recovery will lead to very little that is new: indeed, it is clear that for any of the evolution-Galerkin schemes we have already discussed, if the exact evolution operator can be used the generalised Godunov approach (7.94) will give the same algorithm; and Lin (1995) has shown for the Brenier transport collapse operator, too, the two formulations when used with piecewise constants give the same result. So our interest in this short section is with new algorithms that can be obtained by approximating the flux integrals in (7.94).

We therefore begin with schemes based on the discontinuous linear recovery in one dimension described in section 7.2.4. In the scalar case without diffusion, $u_t + f_x = 0$, one can in fact derive an exact expression for the flux integral by solving a generalised Riemann problem. Consider the interface at $x_{i+\frac{1}{2}}$ and suppose the flux across it comes from the ith cell; and we also suppose that a shock does not form in the cell during the time step. Thus in the cell we have initially

$$\tilde{u}(x) = U_i + s_i(x - x_i); \qquad (7.95a)$$

and for $\forall \tau \in [0, \Delta t]$ we assume there is a unique solution $x_\tau \in [x_{i-\frac{1}{2}}, x_{i+\frac{1}{2}}]$ to the trajectory equation

$$x_{i+\frac{1}{2}} - x_\tau = \tau f'(\tilde{u}(x_\tau)), \qquad (7.95b)$$

which implies a CFL condition. Hence for the flux integral we obtain

$$\int_0^{\Delta t} f(\tilde{u}(x_\tau))d\tau = \Delta t\, f(\tilde{u}(x_{\Delta t})) - \int_0^{\Delta t} \tau df$$

$$= \Delta t\, f(\tilde{u}(x_{\Delta t})) - \int_0^{\Delta t}(x_{i+\frac{1}{2}} - x_\tau)d\tilde{u}$$

$$= \Delta t\, f(\tilde{u}(x_{\Delta t})) + \tfrac{1}{2}s_i(x_{i+\frac{1}{2}} - x_{\Delta t})^2. \qquad (7.96)$$

It is easy to check that this is consistent with the expression in (7.46a) obtained in the linear case, $f \equiv bu$. Alternative situations can be treated similarly; and for the shock case we can either reduce the time step to exclude it or treat it as in the transport collapse approximation.

Exact integrals like (7.96) do not exist for systems of equations, so it is usual to make some approximation at this point. The first two terms in a Taylor expansion were used for the Euler equations by Ben-Artzy and Falcovitz (1984), and the stability and accuracy of various choices have been discussed by Vila (1989). Suppose we denote by $\mathbf{f}^{RP}(\mathbf{W}_L, \mathbf{W}_R)$ the interface flux given by the Riemann problem with constant states $\mathbf{W}_L$ and $\mathbf{W}_R$ on the left and right respectively; and similarly let $\mathbf{f}^{GRP}(\mathbf{W}_i, \mathbf{W}_{i+1})$ denote the sought after generalisation of (7.96), after division by Δt. Then two·useful approximations are motivated by properties of (7.96): for $f \in C^2(\mathbb{R})$ we have from (7.95)

$$\frac{1}{\Delta t}\int_0^{\Delta t} f(\tilde{u}(x_\tau))d\tau = f(\tilde{u}(x_{\Delta t})) + \tfrac{1}{2}f'(\tilde{u}(x_{\Delta t}))[\tilde{u}(x_{i+\frac{1}{2}} - 0) - \tilde{u}(x_{\Delta t})]$$

$$= f(\tilde{u}(x_{\frac{1}{2}\Delta t})) + O(h^2) \qquad (7.97a)$$

$$= f(\tilde{u}(x_{i+\frac{1}{2}} - 0)) + O(h), \qquad (7.97b)$$

where $\tilde{u}(x_{i+\frac{1}{2}} - 0)$ is the value on the left of the cell interface at the beginning of the time step and $\tilde{u}(x_{\frac{1}{2}\Delta t})$ the value midway through the time step. Thus a first order approximation to the general interface flux (or one that might be used with Runge–Kutta time-stepping, as in the next section) is

$$\bar{\mathbf{f}}^{GRP}(\mathbf{W}_i^n, \mathbf{W}_{i+1}^n) \approx \mathbf{f}^{RP}(\bar{\mathbf{w}}^n(x_{i+\frac{1}{2}} - 0), \bar{\mathbf{w}}^n(x_{i+\frac{1}{2}} + 0)); \qquad (7.98a)$$

and a second order approximation is

$$\bar{\mathbf{f}}^{GRP}(\mathbf{W}_i^n, \mathbf{W}_{i+1}^n) \approx \mathbf{f}^{RP}(\bar{\mathbf{w}}^{n+\frac{1}{2}}(x_{i+\frac{1}{2}} - 0), \bar{\mathbf{w}}^{n+\frac{1}{2}}(x_{i+\frac{1}{2}} + 0)), \qquad (7.98b)$$

where various time-stepping algorithms may be used to approximate $\bar{\mathbf{w}}^{n+\frac{1}{2}}$ at the interface. Denoting the resulting flux approximation by $\mathbf{F}_{i+\frac{1}{2}}^{n+\frac{1}{2}}$, we obtain the underlying form of approximation to $\mathbf{w}_t + \mathbf{f}_x = 0$,

$$h_i(\mathbf{W}_i^{n+1} - \mathbf{W}_i^n) + \Delta t(\mathbf{F}_{i+\frac{1}{2}}^{n+\frac{1}{2}} - \mathbf{F}_{i-\frac{1}{2}}^{n+\frac{1}{2}}) = 0. \qquad (7.99)$$

Of course, stated baldly like this, (7.99) covers all explicit, conservative schemes and its utility depends on the calculation of the fluxes.

Having shown how the generalisation to a system of equations may be achieved, let us return to the scalar case and add in diffusion to give a total flux $f(u) - \epsilon u_x$. For substitution into the convective flux, from (7.95a) we have that $\tilde{u}^n(x_{i+\frac{1}{2}} - 0)$ is given by $U_i^n + \frac{1}{2}s_i h_i$ and so, from a sequence of natural approximations, we get

$$\tilde{u}^{n+\frac{1}{2}}(x_{i+\frac{1}{2}} - 0) \approx U_i^n + \tfrac{1}{2}s_i h_i + \tfrac{1}{2}\Delta t[\tilde{u}_t = -f'\tilde{u}_x]$$
$$= U_i^n + \tfrac{1}{2}s_i(h_i - \Delta t f'); \qquad (7.100)$$

here f' might be evaluated with an interface value of $\tilde{u}$, but a more natural choice is the upwind value $f'(U_i)$; similarly, we will approximate the diffusive flux by $-\epsilon s_i$. Extrapolated variables of this kind can be substituted into Riemann solvers or approximate Riemann solvers through the formula (7.98b) so as to give the interface fluxes; for systems of hyperbolic equations this leads to the *MUSCL* class of schemes, after the *Monotone Upwind-centred Scheme for Conservation Laws* pioneered by van Leer (1979). For our standard linear convection-diffusion problem with $b(x) > 0$ it gives, in the form (7.94),

$$h_i(U_i^{n+1} - U_i^n) + \Delta t\, \Delta_- \left\{ b_{i+\frac{1}{2}}[U_i^n + \tfrac{1}{2}s_i(h_i - \Delta t\, b_i)] - \epsilon s_i \right\}$$
$$= \int_0^{\Delta t} dt \int_{x_{i-1/2}}^{x_{i+1/2}} S dx. \quad (7.101)$$

Notice the similarity with the PERU schemes of section 7.2.5.

Extension to two dimensions follows naturally, although differences in detail are common. The value U_i^n is associated with the barycentre of the cell whether this is a quadrilateral, a triangle or a polygon in some dual to a primary triangulation; the backward difference in (7.101) is replaced by a sum over the edges of the cell; and the flux is again extrapolated to the midpoint of the edge at the time level $n + \frac{1}{2}$. Suppose V_i denotes the measure of a cell with barycentre at x_i; suppose a typical edge has midpoint at m_e, where the velocity field is b_e, and l_{ie} is in the direction of the normal to the edge from cell i and has a length equal to the length of the edge; and let the subscript ue denote the cell which is upwind of the midpoint m_e. Then the update procedure consists of three steps: first, a recovery process is carried out to give u-gradients s_i in each cell, using some form of slope limiter; then values of U are extrapolated to the midpoint of each edge at time level $n + \frac{1}{2}$, using U_{ue}^n, s_{ue} and b_{ue}; and finally it is readily checked that (7.101) becomes

$$V_i(U_i^{n+1} - U_i^n) + \Delta t \sum_{(e)} \{b_e \cdot l_{ie}[U_{ue}^n + s_{ue} \cdot (m_e - x_{ue} - \tfrac{1}{2}\Delta t b_{ue}]$$

$$- \epsilon l_{ie} \cdot s_{ue} \} = \int_0^{\Delta t} d\tau \int_{\Omega_i} S d\Omega. \qquad (7.102)$$

The generalisation of this formula to a system of nonlinear conservation laws is achieved through (7.98b) by means of an approximate Riemann solver.

These schemes, (7.101) and (7.102), are designed for time steps in which characteristics do not cross more than one cell; and it is worth noting what they reduce to on a square uniform mesh with a constant velocity field having positive components. Suppose also that the gradients s_i are obtained by central differencing. Then it is clear that the one-dimensional (7.101) reduces to

$$U_i^{n+1} = U_i^n - \Delta_- \{\nu[1 + \tfrac{1}{2}(1 - \nu)\Delta_0] - \mu\Delta_0\}U_i^n, \qquad (7.103a)$$

where $\nu = b\Delta t/\Delta x$ and $\mu = \epsilon\Delta t/(\Delta x)^2$. This is Fromm's difference scheme with diffusion added, as already referred to in connection with the scheme in (7.50b). Stability is limited to $0 \le \nu \le 1$ and $0 \le \mu \le \mu_{\max}(\nu)$, where $\mu_{\max}(0) = \tfrac{1}{2} = \mu\max(1)$ and $\mu\max(\tfrac{1}{2}) = \tfrac{9}{8}$; and note how the diffusion difference is centred at $U_{i-\frac{1}{2}}$, as in the QUICKEST scheme (7.70) when $\nu = \tfrac{1}{2}$. Under similar circumstances (7.102) reduces to

$$U_i^{n+1} = U_i^n - \Delta_{-x} \left\{ \nu_x \left[1 + \tfrac{1}{2}\Delta_{0x} - \tfrac{1}{2}(\nu_x\Delta_{0x} + \nu_y\Delta_{0y})\right] - \mu\Delta_{0x}\right\}U_i^n$$
$$- \Delta_{-y}\{\nu_y \left[1 + \tfrac{1}{2}\Delta_{0y} - \tfrac{1}{2}(\nu_x\Delta_{0x} + \nu_y\Delta_{0y})\right] - \mu\Delta_{0y}\}U_i^n. \qquad (7.103b)$$

This has some mixed differences, corresponding to the important corner terms in the Lagrange–Galerkin methods (7.37) and (7.39); but it is clear from the derivation that more would result if the edge integrations were approximated by a more accurate formula than the midpoint rule.

Piecewise quadratic recovery on each cell will give very similar schemes, as we saw in sections 7.2.4 and 7.2.5; the PPM scheme of Colella and Woodward (1984) is the best known scheme of this class. Higher order approximations can also be used as the basic trial space for U, such as linear on triangles and bilinear on quadrilaterals; but with the implicitness resulting from the mass matrices it is more convenient to consider these along with the truly implicit schemes of the next section. A useful survey of both types of scheme is to be found in Hirsch (1990).

7.3.2 Generalised box schemes

The whole of the operator on the left of (7.93) can be regarded as in divergence form, and both of the integrals on the left of (7.94) as being over faces in a space-time discretisation. So the same variety of finite volume discretisations are available as in the steady case, which are illustrated in Fig. 6.1. The schemes of the preceding section are best regarded as being consistently of the cell edge type; while the box scheme is of cell vertex type. In the one-dimensional box scheme $\mathbf{W}$ is parametrised by its nodal values, and is bilinear in (x, t) in each rectangle $[x_{i-1}, x_i] \times [t_n, t_{n+1}]$, so that without diffusion (7.94) becomes

$$\tfrac{1}{2}h_i[(\mathbf{W}_i^{n+1} - \mathbf{W}_i^n) + (\mathbf{W}_{i-1}^{n+1} - \mathbf{W}_{i-1}^n)] + \tfrac{1}{2}\Delta t \Delta_-[\mathbf{F}_i^{n+1} + \mathbf{F}_i^n]$$

$$= \int_0^{\Delta t} d\tau \int_{x_{i-1}}^{x_i} S dx, \tag{7.104}$$

where $h_i := x_i - x_{i-1}$ and $\mathbf{F}_i^n := \mathbf{f}(\mathbf{W}_i^n)$. Note that the trapezoidal rule used here is exact for $\mathbf{W}$ but an approximation for $\mathbf{f}(\mathbf{W})$.

It seems that this scheme was first advocated by Thomas (1937) for modelling flood waves in rivers; its subsequent rediscovery by many authors, including Wendroff (1960), Preissmann (1961) and Thomée (1962), and its continued widespread use in this field, is eloquent testimony to its effectiveness. In such an application, and for general quasilinear hyperbolic systems, a coupled system of nonlinear equations has to be solved at each time step: since in most river flow situations the flow is subcritical, this is readily achieved by a global Newton iteration using a tridiagonal solver on each linearised system. When critical points occur the situation will be similar to that for the steady cell vertex equations applied to a transonic nozzle flow in section 6.6.2, and similar methods can be used. The justification for this effort is that the CFL number can be taken to be 10–20, with no loss of accuracy because the physically important flood wave travels much more slowly and the mesh can be adjusted to its speed. It should also be pointed out here that it is common to increase the implicitness in (7.104) when solving river flow problems, in order to introduce some damping of the chequer-board mode, that is, one uses $\theta \mathbf{F}_i^{n+1} + (1 - \theta)\mathbf{F}_i^n$ with $\theta > \tfrac{1}{2}$.

Adding diffusion to (7.104) by means of the cell vertex method is straightforward: one merely has to approximate the gradients $(\mathbf{W}')_i^n$ as in (6.71); however, it was seen in section 6.6.1 that this does make the iteration more difficult to carry out, especially when there are critical points to deal with. An alternative is to use the formulation of Keller (1971), already referred to in section 6.1, in which $\mathbf{W}'$ is introduced as a further variable and approximated in the same manner; this has the effect of averaging the scheme (7.104) over two neighbouring cells so that a centred three-point scheme results for the diffusion.

Similarly, extending (7.104) to two and three dimensions is very simple to do in principle, using the techniques presented in the sections of Chapter 6 on the cell vertex method, but the solution of the implicit equations at each time level is very much more difficult in the general case. Runge–Kutta and Lax–Wendroff time-stepping, coupled with the use of multigrid techniques, were given as the main methods for solving the steady cell vertex equations. But this involves a heavy investment of effort, and the relative difficulty of solving the cell vertex equations because of the counting problem as compared with the cell centre equations, was referred to as a disincentive to their wider adoption. Fortunately, for the scalar and linear convection-diffusion problem the symmetric Gauss–Seidel iteration described in section 6.3.6 is both straightforward and highly effective.

Unlike virtually all the other methods discussed in this chapter, the box method makes no use of the characteristics or flow direction, being based solely on the conservation properties implied by the form of the equations (7.93). A reasonably natural generalisation and interpretation of the *angled derivative scheme*, introduced by Roberts and Weiss (1966), does however form a link between the box scheme and the explicit Godunov methods of the last section. Consider the approximation of $w_t + f_x = 0$ in the form (7.99), and interpolate $\mathbf{W}_{i+\frac{1}{2}}^{n+\frac{1}{2}}$ between $\mathbf{W}_i^{n+1}$ and $\mathbf{W}_{i+1}^n$ to obtain

$$h_i(\mathbf{W}_i^{n+1} - \mathbf{W}_i^n) + \Delta t \Delta_- f(\tfrac{1}{2}\mathbf{W}_i^{n+1} + \tfrac{1}{2}\mathbf{W}_{i+1}^n) = 0. \qquad (7.105)$$

One could instead average $f(\mathbf{W}_i^{n+1})$ and $f(\mathbf{W}_{i+1}^n)$ which would then exactly match the more usual form for the angled derivative scheme, as well as the corrector step in a MacCormack scheme.

These schemes are implicit and pose similar problems to the box scheme as regards the solution of the equations for $\mathbf{W}^{n+1}$; but they are not unconditionally stable. A linearisation of the equations to be solved for $\mathbf{W}_i^{n+1}$ readily shows that in the scalar case they are linearly stable only for $\nu \geq -1$, consistent with the domain of dependence CFL condition. So they are quite directionally dependent. This may seem surprising since the difference stencil forms a parallelogram, or just a box which is pushed over in the upwind direction at the new level, but in the downwind direction at the old time level; and in fact the seeming contrast with the box scheme is misleading. Consider linear, constant coefficient convection for which one obtains the equations

$$U_i^{n+1} - U_i^n + \tfrac{1}{2}\nu(U_i^{n+1} - U_{i-1}^{n+1} + U_{i+1}^n - U_i^n) = 0,$$

i.e. $\quad (1 + \tfrac{1}{2}\nu)U_i^{n+1} = \tfrac{1}{2}\nu U_{i-1}^{n+1} + (1 + \tfrac{1}{2}\nu)U_i^n - \tfrac{1}{2}\nu U_{i+1}^n. \qquad (7.106a)$

Then clearly unless $\left|\tfrac{1}{2}\nu\right| \leq \left|1 + \tfrac{1}{2}\nu\right|$, or $\nu \geq -1$, there is unbounded growth of the solution with increasing i. On the other hand, a Fourier analysis gives for the amplification factor

$$\lambda = (1 - \gamma e^{i\xi})/(1 - \gamma e^{-i\xi}), \quad \gamma = \tfrac{1}{2}\nu/(1 + \tfrac{1}{2}\nu),$$

so that $|\lambda| = 1$ for all ν, just as with the box scheme. The fact is, of course, that the box scheme is directional in the same sense, in that in this scalar problem the boundary condition has to be applied at the inflow boundary; and the system that has to be solved is

$$(1 + \nu)U_i^{n+1} = (-1 + \nu)U_{i-1}^{n+1} + (1 - \nu)U_i^n + (1 + \nu)U_{i-1}^n, \qquad (7.106b)$$

which is only stable if $|1 - \nu| \leq |1 + \nu|$, or $\nu \geq 0$. The advantage of the box scheme is that the switch in required marching direction is related more naturally to the switch of boundary condition, and this shows up in the comparative behaviour of the two schemes in regard to numerical boundary conditions.

A much greater variety of generalised box schemes can be derived by using characteristic trajectories to derive fluxes in the formula (7.104), or in formulae obtained by combining two such cells as in the generalisation by Keller (1961) of the explicit scheme due to Courant *et al.* (1952). Although unconditionally stable apart from the boundary effects just referred to, the box scheme has a notoriously large phase lag error for $\nu > 1$. So the question arises whether one can correct this deficiency by using the characteristics to modify the flux integrals. Again for $w_t + f_x = 0$, let us use the correction to the trapezoidal rule provided by the Euler–Maclaurin sum formula to get

$$\int_{t_n}^{t_{n+1}} \mathbf{f} \, dt = \tfrac{1}{2}\Delta t(\mathbf{f}^{n+1} + \mathbf{f}^n) - \tfrac{1}{12}(\Delta t)^2(\mathbf{f}_t^{n+1} - \mathbf{f}_t^n) + O((\Delta t)^2) \quad (7.107a)$$

and then use the differential equation to write

$$\mathbf{f}_t = \mathbf{f}' \mathbf{w}_t = -\mathbf{f}' \mathbf{f}_x. \quad (7.107b)$$

For simplicity let us suppose all the eigenvalues of $\mathbf{f}'$ are non-negative; then in (7.104) we replace the flux definitions as follows,

$$\mathbf{F}_i^{n+1} \rightarrow \mathbf{f}(\mathbf{W}_i^{n+1}) - \tfrac{1}{6}\Delta t \, \mathbf{f}'(\mathbf{W}_i^{n+1})D_+\mathbf{f}(\mathbf{W}_i^{n+1}) \quad (7.108a)$$

$$\mathbf{F}_i^n \rightarrow \mathbf{f}(\mathbf{W}_i^n) + \tfrac{1}{6}\Delta t \, \mathbf{f}'(\mathbf{W}_i^n)D_-\mathbf{f}(\mathbf{W}_i^n). \quad (7.108b)$$

This replaces the usual four-point scheme by a six-point scheme. Much remains to be done to explore the properties of such schemes; the survey already referred to (Hirsch 1990) forms a valuable starting point.

7.4 Semi-discrete based schemes

In the preceding section, at least two of the schemes needed to make use of flux functions evaluated at other than the initial time level: the higher order Godunov scheme (7.99) required approximations to $\mathbf{F}_{i+\frac{1}{2}}^{n+\frac{1}{2}}$, which in some cases were obtained from Taylor series expansions; and in the box scheme (7.104) a Newton iteration was presumed in order to obtain approximations to $\mathbf{F}_i^{n+1}$. Both cases therefore come close to being low order Runge–Kutta schemes. The logical next step is to use the full range of time-stepping procedures as outlined in sections 6.2.2 and 6.3.3 of the last chapter, but now have them designed to be time accurate. The advantage is that the whole problem is then decomposed into two distinct parts: the design of a spatial discretisation that as in (6.32) yields a nodal residual

$$N_i(U) \equiv L_h U_i - S_i; \tag{7.109a}$$

and the choice of a time-stepping procedure that approximates the semi-discrete equations

$$\frac{\partial U_i}{\partial t} = N_i(U), \quad \forall i. \tag{7.109b}$$

The disadvantage of this approach is that one does not take account of the interaction between the two discretisations, as in the Taylor–Galerkin and Lagrange–Galerkin schemes.

It is clear from previous sections of this chapter, particularly section 7.2.7, that the penalty for this semi-discrete approach is quite small at low mesh Péclet numbers. Thus in the case of predominantly parabolic problems, it is now common to reverse the usual procedure of choosing the time-stepping after the spatial discretisation, and to design special spatial discretisations to be used with a method of lines approach based on standard ODE solvers — see for example Skeel and Berzins (1990). Even for hyperbolic problems, the Taylor–Galerkin scheme is commonly used in a two-step form which is close to a semi-discrete approach; the first step is usually to a staggered mesh so that two spatial discretisations are needed, but this fits comfortably into the format for predictor-corrector schemes used for ODE's. Thus it is only when compared with the use of Lagrange–Galerkin or semi-Lagrangian schemes at large CFL numbers that the semi-discrete approach would seem to have any real disadvantage.

Whether using multi-level or Runge–Kutta time-stepping, it is clearly important not only to maintain time accuracy but also to maintain key properties of the spatial discretisation. One such property that has received much attention is the TVD property, both because of the quality of solution obtained and the resulting nonlinear stability; and Shu (1988) has shown how simple multi-level schemes can be made TVD, or TVB. For the two scalar problems in one dimension, $u_t \pm f_x = 0$, he supposes that one has two r^{th} order accurate nodal residuals which are TVD: that is,

$$N_i^{(+)}(u) = \frac{\partial f(u)}{\partial x} + O(h^r), \quad N_i^{(-)}(u) = -\frac{\partial f(u)}{\partial x} + O(h^r), \tag{7.110a}$$

and both give simple TVD updates if $(\Delta t/h) \leq \lambda_0$. Then the time discretisation that he proposes is given by

$$U_i^{n+1} = \sum_{k=0}^{m} [\alpha_k U_i^{n-k} - \Delta t |\beta_k| N_i^{(\mathrm{sign}\beta_k)}(U^{n-k})], \tag{7.110b}$$

where $\alpha_k \geq 0, \sum \alpha_k = 1$ and $\alpha_k = 0 \Rightarrow \beta_k = 0$. He shows that this scheme is TVD if the CFL condition

$$(\Delta t/h) \leq \lambda_0 \min_k (\alpha_k/|\beta_k|) \tag{7.111a}$$

is satisfied; and it is s^{th} order accurate in space and time, with $1 \leq s \leq r$, if the coefficients satisfy the conditions

$$\sum_{k=0}^{m}(-k)^{l-1}(l\beta_k - k\alpha_k) = 1, \quad l = 1, 2, \ldots, s. \qquad (7.111b)$$

These last conditions are just those obtained from a Taylor series expansion for a linear multi-step ODE scheme. Shu shows that schemes of any order, with a positive CFL limit, can be constructed in this way and gives many examples; the best, though, has a CFL coefficient of $\frac{3}{4}$ multiplying λ_0 in (7.111a).

The same schemes also maintain the TVB property, a relaxation of TVD which as we have already remarked allows for a uniformly high order of accuracy in the spatial discretisation to be carried through solution extrema. However, in general Runge–Kutta schemes are preferred over multi-level schemes, because of their smaller storage requirements. Shu and Osher (1988) have proposed several such schemes, which use the same device as in (7.110b) to ensure that a convex combination of TVB (or TVD) approximations maintains the TVB property. Thus they have the form, for an m-step method,

$$U_i^{(k)} = \sum_{l=0}^{k-1}[\alpha_{lk}U_i^{(l)} - \Delta t|\beta^{lk}|N_i^{(\text{sign}\beta_{lk})}(U^{(l)})] \qquad (7.112)$$

for $k = 1, 2, \ldots, m$ with $U_i^{(0)} = U_i^n$ and $U_i^{(m)} = U_i^{n+1}$. The coefficients satisfy $\alpha_{lk} \geq 0, \sum \alpha_{lk} = 1 \ \forall k$, plus the algebraic conditions needed for a given order of accuracy. Among the particular schemes given, is one of third order accuracy with $m = 2$, a CFL coefficient of unity and all the $\beta_{lk} \geq 0$; but $N_i^{(-)}$ as well as $N_i^{(+)}$ is needed for a higher order of accuracy.

Thus, compared with the similar schemes given in section 6.3.3 for the steady problem, it is clear that the time accuracy constraint makes these Runge–Kutta schemes much less efficient. When the two discretisations $N_i^{(+)}$ and $N_i^{(-)}$ are needed, and possibly also some implicitness introduced because of the wide range of time scales or the diffusion terms, then the coupling of the time and space approximations that is the basis of the methods in section 7.3 and 7.4 appears to be preferable.

References

Abrahamsson, L.R. (1977), *A priori* estimates for solutions of singular perturbations with a turning point, *Stud. Appl. Math* **56**, 51–69.

Abrahamsson, L.R., Keller, H.B. and Kreiss, H.O. (1974), Difference approximations for singular perturbations of systems of ordinary differential equations, *Numer. Math.* **22**, 367–391.

Allen, D. and Southwell, R. (1955), Relaxation methods applied to determining the motion, in two dimensions, of a viscous fluid past a fixed cylinder, *Quart. J. Mech. Appl. Math.* **8**, 129–145.

Aubin, J.P. (1967), Behaviour of the error of the approximate solutions of boundary value problems for linear elliptic operators by Galerkin's and finite difference methods, *Ann. Scuola Norm. Sup. Pisa* **21**, 599–637.

Babuška, I. and Rheinboldt, W. (1978a), A posteriori error estimates for the finite element method, *Intern. J. Numer. Methods Engrg.* **12**, 1597–1615.

Babuška, I. and Rheinboldt, W. (1978b), Error estimates for adaptive finite element computations, *SIAM J. Numer. Anal.* **15**(4), 736–754.

Baines, M.J. (1994), *Moving finite elements*, Monographs on Numerical Analysis, Clarendon Press, Oxford.

Balland, P. and Süli, E. (1994), Analysis of the cell vertex finite volume method for hyperbolic problems with variable coefficients, Technical Report NA94/01, Oxford University Computing Laboratory. Submitted for publication.

Barrett, J. W. and Morton, K. W. (1980), Optimal finite element solutions to diffusion-convection problems in one dimension, *Intern. J. Numer. Methods Engrg.* **15**, 1457–1474.

Barrett, J.W. and Morton, K.W. (1981), Optimal Petrov-Galerkin methods through approximate symmetrization, *IMA J. Numer. Anal.* **1**, 439–468.

Barrett, J.W. and Morton, K.W. (1982), Optimal finite element approximation for diffusion-convection problems, *in* J.R. Whiteman, ed., The Mathematics of Finite Elements and Applications IV MAFELAP 1981, Academic Press, London, pp. 403–411.

Barrett, J.W. and Morton, K.W. (1984), Approximate symmetrization and Petrov-Galerkin methods for diffusion-convection problems, *Comput. Methods Appl. Mech. Engrg.* **45**, 97–122.

Barrett, J.W., Moore, G. and Morton, K.W. (1988a), Optimal recovery in the finite-element method, part 1: Recovery from weighted L^2 fits, *IMA J. Numer. Anal.* **8**, 149–184.

Barrett, J.W., Moore, G. and Morton, K.W. (1988b), Optimal recovery in the finite element method, part 2: Defect correction for ordinary differential equations, *IMA J. Numer. Anal.* **8**, 527–540.

Bear, J. and Verruijt, A. (1987), *Modeling groundwater flow and pollution*, D. Reidel, Dordrecht.

Ben-Artzy, M. and Falcovitz, J. (1984), A second order Godunov type scheme for compressible fluid dynamics, *J. Comput. Phys.* **55**, 1–32.

Bender, C.M. and Orszag, S.A. (1978), *Advanced mathematical methods for scientists and engineers*, McGraw-Hill Inc, New York.

Benqué, J.P., Labadie, G. and Ronat, J. (1982), A new finite element method for the Navier-Stokes equations coupled with a temperature equation, *in* T. Kawai, ed., Proceedings of the Fourth International Symposium on Finite Element Methods in Flow Problems, North-Holland, Amsterdam, pp. 295–301.

Benson, D.J. (1991), Finite volume solution of the Stokes and Navier-Stokes equations, PhD thesis, University of Oxford.

Berger, A.E., Solomon, J.M. and Ciment, M. (1981), An analysis of a uniformly accurate difference method for a singular perturbation problem, *Math. Comp.* **37**, 79–94.

Berger, A.E., Solomon, J.M., Ciment, M., Leventhal, S.H. and Weinberg, B.C. (1980), Generalized OCI schemes for boundary layer problems, *Math. Comp.* **35**(151), 695–731.

Black, F. and Scholes, M. (1973), The pricing of options and corporate liabilities, *J. Pol. Econ.* **81**, 637–659.

Boris, J.P. and Book, D.L. (1973), Flux-corrected transport, I. SHASTA, a fluid transport algorithm that works, *J. Comput. Phys.* **11**, 38–69.

Boyd, T.J.M. and Sanderson, J.J. (1969), *Plasma dynamics*, Nelson, London.

Bramble, J.H. and Hilbert, S.R. (1970), Estimation of linear functionals on Sobolev spaces with application to Fourier transforms and spline interpolation, *SIAM J. Numer. Anal.* **7**, 113–124.

Brenier, Y. (1984), Average multivalued solutions for scalar conservation laws, *SIAM J. Numer. Anal.* **21**, 1013–1037.

Childs, P.N. and Morton, K.W. (1990), Characteristic Galerkin methods for scalar conservation laws in one dimension, *SIAM J. Numer. Anal.* **27**, 553–594.

Christie, I., Griffiths, D.F., Mitchell, A.R. and Zienkiewicz, O.C. (1976), Finite element methods for second order differential equations with significant first derivatives, *Intern. J. Numer. Methods Engrg.* **10**, 1389–1396.

Ciarlet, P.G. (1978), *The finite element method for elliptic problems*, North Holland, Amsterdam.

Ciment, M. and Leventhal, H. (1975), Higher order compact implicit schemes for the wave equation, *Math. Comp.* **29**(132), 985–994.

Cockburn, B. and Shu, C.W. (1989), TVB Runge-Kutta local projection discontinuous Galerkin finite element method for scalar conservation laws II: General framework, *Math. Comp.* **52**, 411–435.

Colella, P. and Woodward, P. (1984), The piecewise parabolic method (PPM) for gas-dynamical simulations, *J. Comput. Phys.* **54**, 174–201.

Collatz, L. (1960), *The numerical treatment of differential equations*, 3rd edn, Springer-Verlag, Berlin.

Collatz, L. (1973), Hermitean methods for initial value problems in partial differential equations, *in* J.J.H. Miller, ed., Topics in Numerical Analysis, Academic Press, London–New York, pp. 41–61.

Courant, R., Isaacson, E. and Rees, M. (1952), On the solution of nonlinear hyperbolic differential equations by finite differences, *Comm. Pure Appl. Math.* **5**, 243–249.

Craig, A. and Perella, A.J. (1995), Highly accurate solution of the stationary convection-diffusion equation, Preprint.

Crandall, M.G. and Lions, P.L. (1983), Viscosity solutions of Hamiliton–Jacobi equations, *Trans. Amer. Math. Soc.* **277**, 1–42.

Crank, J. and Gupta, R.S. (1972), A moving boundary problem arising from the diffusion of oxygen in absorbing tissue, *J. Inst. Math. Appl.* **10**, 19–33.

Crumpton, P.I. and Shaw, G.J. (1994), A vertex-centred finite volume method with shock detection, *Internat. J. Numer. Methods Fluids* **18**, 605–625.

Crumpton, P.I., Mackenzie, J.A. and Morton, K.W. (1993), Cell vertex algorithms for the compressible Navier-Stokes equations, *J. Comput. Phys.* **109**(1), 1–15.

Cullen, M.J.P. and Morton, K.W. (1980), Analysis of evolutionary error in finite element and other methods, *J. Comput. Phys.* **34**, 245–67.

Davis, S.F. (1984), A rotationally biased upwind difference scheme for the Euler equations, *J. Comput. Phys.* **56**, 65–92.

de Kok, J.M. (1993), Forward semi-Lagrangian methods: the second moment method, *in* C.B. Vreugdenhil and B. Koren, eds, Numerical Methods for Advection-Diffusion Problems, Vol. 45 of *Notes on Numerical Fluid Mechanics*, Vieweg, Braunschweig, pp. 243–260.

Dick, E. (1988), A flux-difference splitting method for steady Euler equations, *J. Comput. Phys.* **76**, 19–32.

Donea, J. (1984), A Taylor–Galerkin method for convective transport problems, *Intern. J. Numer. Methods Engrg.* **20**, 101–119.

Donea, J., Selmin, V. and Quartapelle, L. (1988), Recent developments of the Taylor–Galerkin method for the numerical solution of hyperbolic problems, *in* K.W. Morton and M.J. Baines, eds, Numerical Methods for Fluid Dynamics III, Clarendon Press, Oxford, pp. 171–185.

Doolan, E.P., Miller, J.J.H. and Schilders, W.H.A. (1980), *Uniform numerical methods for problems with initial and boundary layers*, Boole Press, Dublin.

Dorr, F.W. (1970), The numerical solution of singular perturbations of boundary value problems, *SIAM J. Numer. Anal.* **7**, 281–313.

Douglas, J. and Russell, T.F. (1982), Numerical methods for convection–dominated diffusion problems based on combining the method of characteristics with finite element or finite difference procedures, *SIAM J. Numer. Anal.* **19**, 871–885.

Duff, I.S., Erisman, A.M. and Reid, J.K. (1986), *Direct methods for sparse matrices*, Clarendon Press, Oxford.

Eberle, A., Rizzi, A. and Hirschel, E.H. (1992), *Numerical solutions of the euler equations for steady flow problems*, Vol. 34 of *Notes on Numerical Fluid Mechanics*, Vieweg, Braunschweig.

Eckhaus, W. (1972), Boundary layers in linear elliptic singular perturbation problems, *SIAM Rev.* **14**(2), 225–270.

Eckhaus, W. (1979), *Asymptotic analysis of singular perturbations*, North-Holland, Amsterdam.

Egan, B.E. and Mahoney, J.R. (1972), Numerical modelling of advection and diffusion of urban area source pollutants, *J. Appl. Meteorology* **11**, 312–322.

El-Mistikawy, T.M. and Werle, M.J. (1978), Numerical method for boundary layers with blowing — the exponential box scheme, *AIAA J.* **16**, 749–751.

Elman, H.C. and Chernesky, M.P. (1993), Ordering effects on relaxation methods applied to the discrete one-dimensional convection-diffusion equation, *SIAM J. Numer. Anal.* **30**(5), 1268–1290.

Engquist, B. and Osher, S. (1981), One-sided difference approximations for nonlinear conservation laws, *Math. Comp.* **36**, 321–352.

Eriksson, K., Estep, D., Hansbo, P. and Johnson, C. (1995), *Introduction to adaptive methods for differential equations*, in Acta Numerica 1995, Cambridge University Press, pp. 105–158.

Field, M.R. (1994), The setting up and solution of the cell vertex equations, PhD thesis, University of Oxford.

Fornberg, B. (1993), Computing steady incompressible flows past blunt bodies — a historical overview, *in* M.J. Baines and K.W. Morton, eds, Numerical Methods for Fluid Dynamics IV, Clarendon Press, Oxford, pp. 115–133.

Fromm, J.E. (1968), A method for reducing dispersion in convective difference schemes, *J. Comput. Phys.* **3**(2), 176–189.

Garabedian, P.R. (1964), *Partial differential equations*, J. Wiley & Sons, New York.

García-Archilla, B. and Mackenzie, J.A. (1991), Analysis of a supraconvergent cell vertex finite volume method for one-dimensional convection-diffusion problems, Technical Report NA91/13, Oxford University Computing Laboratory. Published in IMA J. Num. Analysis Vol 15 pages 101-115 (1995).

Garcia-Navarro, P. and Priestley, A. (1994), A conservative and shape-preserving semi-Lagrangian method for the solution of the shallow water equations, *Internat. J. Numer. Methods Fluids* **18**, 273–294.

Gartland, E.C. (1991), On the stability of compact discretizations on singularly perturbed differential equations, *in* H.-G. Roos, A. Felgenhauer and L. Angermann, eds, Proc. of ISAM 91, Technical University of Dresden, pp. 63–70.

Gaskell, P.H. and Lau, A.K.C. (1988), Curvature compensated convective transport: SMART, a new boundedness preserving transport algorithm, *Internat. J. Numer. Methods Fluids* **8**(6), 617–641.

Gilbarg, D. and Trudinger, N.S. (1983), *Elliptic partial differential equations of second order*, 2nd edn, Springer, Berlin.

Girault, V. and Raviart, P.-A. (1979), *Finite element approximation of the Navier-Stokes equations*, Vol. 749 of *Lecture Notes in Mathematics*, Springer-Verlag, Berlin.

Godlewski, E. and Raviart, P.-A. (1991), *Hyperbolic systems of conservation laws*, Vol. 3/4, Ellipses, Paris.

Godunov, S.K. (1959), A finite difference method for the numerical computation of discontinuous solutions of the equations of fluid dynamics, *Mat. Sb.* **47**, 271–306.

Gresho, P. and Lee, R.L. (1981), Don't suppress the wiggles — they're telling you something, *Comput. & Fluids* **9**, 223–253.

Griffiths, D.F. and Mitchell, A.R. (1979), On generating upwind finite element methods, *in* T.J.R. Hughes, ed., Finite element methods for convection dominated flows, Vol. AMD 34, ASME, pp. 91–104.

Grisvard, P. (1985), *Elliptic problems in nonsmooth domains*, Pitman, London.

Gushchin, V.A. and Shchennikov, V.V. (1974), A monotonic difference scheme of second-order accuracy, *U.S.S.R. Computational Math. and Math. Phys.* **14**, 252–256.

Hall, M.G. (1985), Cell-vertex multigrid schemes for solution of the Euler equations, *in* K.W. Morton and M.J. Baines, eds, Proceedings of the Conference on Numerical Methods for Fluid Dynamics, University of Reading, Clarendon Press, Oxford, pp. 303–345.

Hall, M.G. (1991), A vertex centroid scheme for improved finite volume solution of the Navier-Stokes equations, AIAA Paper 91-1540.

Harten, A. (1983), High resolution schemes for conservation laws, *J. Comput. Phys.* **49**, 357–393.

Harten, A. (1987), ENO schemes with subcell resolution, *J. Comput. Phys.* **83**, 148–184.

Harten, A., Engquist, B., Osher, S. and Chakravarthy, S.R. (1987), Uniformly high order accurate essentially non-oscillatory schemes, III, *J. Comput. Phys.* **71**, 231–303.

Hegarty, A.F., E.O'Riordan and Stynes, M. (1993), A comparison of uniformly convergent difference schemes for two-dimensional convection-diffusion problems, *J. Comput. Phys.* **105**, 24–42.

Hegarty, A.F., Miller, J.J.H. and O'Riordan, E. (1980), Uniform second order difference schemes for singular perturbation problems, *in* J.J.H. Miller, ed., Boundary and interior layers — computational and asymptotic methods, Boole Press, Dublin, pp. 301–305.

Heinrich, J. C., Huyakorn, P. S., Mitchell, A. R. and Zienkiewicz, O. C. (1977), An upwind finite element scheme for two-dimensional convective transport equations., *Intern. J. Numer. Methods Engrg.* **11**, 131–143.

Heinrich, J.C. and Zienkiewicz, O.C. (1979), The finite element method and 'upwinding' techniques in the numerical solution of convection dominated flow problems, *in* T.J.R. Hughes, ed., Finite element methods for convection dominated flows, Vol. AMD 34, ASME, pp. 105–136.

Hemker, P.W. (1977), A numerical study of stiff two-point boundary problems, PhD thesis, Mathematisch Centrum, Amsterdam.

Hemker, P.W. and Miller, J.J.H., eds (1979), *Numerical analysis of singular perturbation problems*, Academic Press, London-New York.

Hindmarsh, A.C., Gresho, P.M. and Griffiths, D.F. (1984), The stability of explicit Euler time–integration for certain finite difference approximations of the multidimensional advection–diffusion equation, *Internat. J. Numer. Methods Fluids* **4**, 853–897.

Hirsch, C. (1988), *Numerical computation of internal and external flows*, Vol. 1, John Wiley & Sons, Chichester.

Hirsch, C. (1990), *Numerical computation of internal and external flows*, Vol. 2, John Wiley & Sons, Chichester.

Hockney, R.W. and Eastwood, J.W. (1981), *Computer simulation using particles*, McGraw-Hill, New York.

Holmes, D.G. and Connell, S.D. (1989), Solution of the 2D Navier-Stokes equations on unstructured adaptive grids, AIAA Paper 89-1932-CP.

Hughes, T.J.R. and Brooks, A. (1979), A multi-dimensional upwind scheme with no crosswind diffusion, *in* T.J.R. Hughes, ed., Finite element methods for convection dominated flows, Vol. AMD 34, ASME, pp. 19–35.

Hughes, T.J.R. and Brooks, A. (1982), A theoretical framework for Petrov-Galerkin methods with discontinuous weighting functions: application to the streamline-upwind procedure, *in* R.H. Gallagher, D.H. Norrie, J.T. Oden and O.C. Zienkiewicz, eds, Finite Elements in Fluids, Vol. 4, Wiley, New York, pp. 47–65.

Huynh, H.T. (1993), Accurate monotone cubic interpolation, *SIAM J. Numer. Anal.* **30**, 57–100.

Il'in, A.M. (1969), Differencing scheme for a differential equations with a small parameter affecting the highest derivative, *Mat. Zametki* **6**, 237–248.

Jameson, A. (1979), Acceleration of transonic potential flow calculations on arbitrary meshes by the multiple grid method, AIAA Paper 79-1458.

Jameson, A. (1995), Analysis and design of numerical schemes for gas dynamics: Part 1 artificial diffusion, upwind biasing, limiters and their effect on accuracy and multigrid convergence, Accepted for publication in Inter. J. of CFD.

Jameson, A. and Caughey, D.E. (1977), A finite volume method for transonic potential flow calculations, *in* Proc. AIAA Third Computational Fluid Dynamics Conferenence, Albuqueque, pp. 35–54.

Jameson, A., Baker, T.J. and Weatherill, N.P. (1986), Calculation of inviscid transonic flow over a complete aircraft, AIAA Paper 86–0103.

Jameson, A., Schmidt, W. and Turkel, E. (1981), Numerical solutions of the Euler equations by finite volume methods with Runge–Kutta time stepping schemes, AIAA Paper 81-1259.

Johnson, C. and Nävert, U. (1981), An analysis of some finite element methods for advection-diffusion problems, *in* O. Axelsson, L.S. Frank and A. van der Sluis, eds, Analytical and Numerical Approaches to Asymptotic Problems in Analysis, North-Holland, Amsterdam, pp. 99–116.

Johnson, C., Schatz, A.H. and Wahlbin, L.B. (1987), Crosswind smear and pointwise errors in streamline diffusion finite element methods, *Math. Comp.* **49**(179), 25–38.

Keller, H.B. (1961), On the solution of semi-linear hyperbolic systems by unconditionally stable difference methods, *Comm. Pure Appl. Math.* **XIV**(3), 447–456.

Keller, H.B. (1971), A new difference scheme for parabolic problems, *in* B. Hubbard, ed., Numerical solution of partial differential equations II, Academic Press, New York, pp. 327–350.

Keller, H.B. and Cebeci, T. (1971), Accurate numerical methods for boundary layer flow I: two-dimensional laminar flows, *in* Lecture Notes in Physics, Proceedings of Second International Conference on Numerical Methods in Fluid Dynamics, Springer-Verlag, Berlin, pp. 92–100.

Keller, H.B. and Cebeci, T. (1972), Accurate numerical methods for boundary layer flows II: two-dimensional turbulent flows, *AIAA J.* **10**(9), 1193–1199.

Kellogg, R.B. and Tsan, A. (1978), Analysis of some difference approximations for a singular perturbation problem without turning points, *Math. Comp.* **32**, 1025–1039.

Kreiss, H. O., Manteuffel, T. A., Swartz, B., Wendroff, B. and White, A. B. (1986), Supra-convergent schemes on irregular grids, *Math. Comp.* **47**, 537–554.

Krishnamurti, T.N. (1962), Numerical integration of primitive equations by a quasi-Lagrangian advective scheme, *J. Appl. Meteorology* **1**, 508–521.

Křížek, M. (1991), On semiregular families of triangulations and linear interpolation,

Appl. Math. **36**(3), 223–232.

Kwizak, M. and Robert, A.J. (1971), A semi-implicit scheme for grid-point atmospheric models of the primitive equations, *Monthly Weather Rev.* **99**, 32–36.

Launder, B.E. and Spalding, B. (1972), *Mathematical models of turbulence*, Academic Press, New York.

Lax, P.D. and Wendroff, B. (1960), Systems of conservation laws, *Comm. Pure Appl. Math.* **13**, 217–237.

Leonard, B.P. (1979*a*), A stable and accurate convective modelling procedure based on quadratic upstream interpolation, *Comput. Methods Appl. Mech. Engrg.* **19**, 59–98.

Leonard, B.P. (1979*b*), A survey of differences of opinion on numerical muddling of the incomprehensible defective confusion equation, *in* T.J.R. Hughes, ed., Finite Element Methods for Convection Dominated Flows, Vol. 34 of *AMD*, ASME, New York, pp. 1–17.

Leonard, B.P. (1991), The ULTIMATE conservative difference scheme applied to unsteady one-dimensional advection, *Comput. Methods Appl. Mech. Engrg.* **88**, 17–74.

Leonard, B.P. and Niknafs, H.S. (1991), Sharp monotonic resolution of discontinuities without clipping of narrow extrema, *Comput. & Fluids* **19**(1), 141–154.

Lesaint, P. (1977), Numerical solution of the equation of continuity, *in* J.J.H. Miller, ed., Topics in Numerical Analysis III, Academic Press, London, pp. 199–222.

Lesaint, P. and Zlámal, M. (1979), Superconvergence of the gradient of finite element solutions, *Rev. Fr. Auto. Inf. Rech. opérat. Sér. Rouge Analyses Num. (RAIRO)* **13**, 139–166.

LeVeque, R.J. (1992), *Numerical methods for conservation laws*, Lectures in Mathematics ETH Zürich, 2nd edn, Birkhäuser Verlag, Basel.

LeVeque, R.J. (1995), High-resolution conservative algorithms for advection in incompressible flow, To appear in the SIAM J. Numer. Anal.

Levine, N. (1985), Superconvergent recovery of the gradient from piecewise linear finite-element approximations, *IMA J. Numer. Anal.* **5**, 407–427.

Lin, P. (1995), Godunov type of the Euler characteristic Galerkin scheme for one-dimensional scalar conservation laws, Technical Report NA95/02, Oxford University Computing Laboratory, OX1 3QD.

Lin, P. and Morton, K.W. (1993), An upwind finite element method for nonlinear hyperbolic systems of conservation laws, Technical Report NA93/10, Oxford University Computing Laboratory.

Lin, P., Morton, K.W. and Süli, E. (1993), Euler characteristic Galerkin scheme with recovery, M^2AN **27**(7), 863–894.

Lin, P., Morton, K.W. and Süli, E. (1995), Characteristic Galerkin schemes for conservation laws in two and three space dimensions, To appear in the SIAM J. Numer. Anal.

Liu, F. and Jameson, A. (1990), Multigrid Euler calculations for three-dimensional cascades, AIAA Paper 90–0688.

Mackenzie, J.A. (1991), Cell vertex finite volume methods for the solution of the compressible Navier-Stokes equations, PhD thesis, University of Oxford.

Mackenzie, J.A. and Morton, K.W. (1992), Finite volume solutions of convection-diffusion test problems, *Math. Comp.* **60**(201), 189–220.

Mackenzie, J.A., Sonar, T. and Süli, E. (1994), Adaptive finite volume methods for hyperbolic problems, *in* J.R. Whiteman, ed., The Proceedings of the Conference on the Mathematics of Finite Elements and Applications MAFELAP VIII, Wiley, pp. 288–297.

MacNeal, R.H. (1953), An asymmetrical finite difference network, *Quart. Appl. Math.* **11**, 295–310.

Manteuffel, T.A. and White, Jr., A.B. (1986), The numerical solution of second-order boundary value problems on nonuniform meshes, *Math. Comp.* **47**(176), 511–535.

Marcum, D.L. and Weatherill, N.P. (1994), Unstructured grid generation using iterative point insertion and local reconnection, AIAA Paper 94-1926.

Mavriplis, D.J. (1995), Multigrid techniques for unstructured meshes, *in* H. Deconinck, ed., Computational Fluid Dynamics, Lecture Series 1995-02, von Karman Institute, Belgium. von Karman Institute Lecture Notes, March.

McDonald, P.W. (1971), The computation of transonic flow through two-dimensional gas turbine cascades, Paper 71-GT-89, ASME, New York.

Miller, J.J.H., O'Riordan, E. and Shishkin, G.I. (1995), *Fitted numerical methods for singular perturbation problems — error estimates in the maximum norm for linear problems in one and two dimensions*, World Scientific, Singapore.

Mock, M.S. (1983), *Analysis of mathematical models of semiconductor devices*, Boole Press, Dublin.

Moore, J. and Moore, J. (1983), Calculation of horseshoe vortex flow without numerical mixing, Technical Report JM/83–11, Virginia Polytechnic Inst. and State University, Blacksburg, Virginia 24061.

Morgan, K., Peraire, J., Peiró, J. and Hassan, O. (1991), The computation of three dimensional flows using unstructured grids, *Comput. Methods Appl. Mech. Engrg.* **87**, 335–352.

Morton, K.W. (1982), Shock capturing, fitting and recovery, *in* E. Krause, ed., Proceedings of the Eighth International Conference on Numerical Methods in Fluid Dynamics, Aachen, Vol. 170 of *Lecture Notes in Physics*, Springer-Verlag, Berlin, pp. 77–93.

Morton, K.W. (1983), Characteristic Galerkin methods for hyperbolic problems, *in* M. Pandolfi and R. Piva, eds, Proceedings of Fifth GAMM Conference on Numerical Methods in Fluid Dynamics, Rome, Vieweg, Braunschweig, pp. 243–250.

Morton, K.W. (1985), Generalised Galerkin methods for hyperbolic problems, *Comput. Methods Appl. Mech. Engrg.* **52**, 847–871.

Morton, K.W. (1988), Finite volume and finite element methods for the steady Euler equations of gas dynamics, *in* J.R. Whiteman, ed., The Mathematics of Finite Elements and Applications VI MAFELAP 1987, Academic Press, London, pp. 353–378.

Morton, K.W. (1991*a*), Finite volume methods and their analysis, *in* J.R. Whiteman, ed., The Mathematics of Finite Elements and Applications VII MAFELAP 1990, Academic Press, Englewood Cliffs, pp. 189–214.

Morton, K.W. (1991*b*), Lagrange–Galerkin and Characteristic–Galerkin methods and their applications, *in* B. Engquist and B. Gustafsson, eds, Third International Conference on Hyperbolic Problems. Theory, Numerical Methods and Applications, Vol. II, Studentlitteratur, pp. 742–755. Proceedings, Uppsala, Sweden, June 11–15, 1990.

Morton, K.W. (1992), Upwinded test functions for finite element and finite volume methods, *in* D.F. Griffiths and G.A. Watson, eds, Numerical analysis 1991 Proceedings of the 14th Dundee Conference, June 1991, number 260 *in* 'Pitman Research Notes in Mathematics', Longman, Harlow, pp. 128–141.

Morton, K.W. (1994), Finite volume methods for convection-diffusion problems, *in* M. Křížek, P. Neittaanmäki and R. Stenberg, eds, Finite Element Methods: Fifty Years of the Courant Element, Vol. 164 of *Lecture Notes in Pure and Applied Mathematics*, Marcel Dekker, Inc., New York, pp. 301–313.

Morton, K.W. and Childs, P.N. (1989), Characteristic Galerkin methods for hyperbolic systems, *in* J. Ballmann and R. Jeltsch, eds, Nonlinear Hyperbolic Equations — Theory, Computation Methods, and Applications., Vol. 24 of *Notes on Numerical Fluid Mechanics*, Vieweg, pp. 435–455.

Morton, K.W. and Mayers, D.F. (1994), *Numerical solution of partial differential equations – an introduction*, Cambridge University Press, Cambridge.

Morton, K.W. and Paisley, M.F. (1989), A finite volume scheme with shock fitting for the steady Euler equations, *J. Comput. Phys.* **80**, 168–203.

Morton, K.W. and Parrott, A.K. (1980), Generalised Galerkin methods for first-order hyperbolic equations, *J. Comput. Phys.* **36**, 249–270.

Morton, K.W. and Scotney, B.W. (1985), Petrov-Galerkin methods and diffusion-convection problems in 2D, *in* J.R. Whiteman, ed., The Mathematics of Finite Elements and Applications V MAFELAP 1984, Academic Press, London, pp. 343–366.

Morton, K.W. and Sobey, I.J. (1993), Discretization of a convection-diffusion equation, *IMA J. Numer. Anal.* **13**(1), 141–160.

Morton, K.W. and Stynes, M. (1994), An analysis of the cell vertex method, M^2AN **28**(6), 699–724.

Morton, K.W. and Süli, E. (1992), Evolution-Galerkin methods and their supraconvergence, Technical Report NA92/15, Oxford University Computing Laboratory. To appear in Numer. Math.

Morton, K.W. and Sweby, P.K. (1987), A comparison of flux limited difference methods and characteristic Galerkin methods for shock modelling, *J. Comput. Phys.* **73**, 203–230.

Morton, K.W., Murdoch, T. and Süli, E. (1992), Optimal error estimation for Petrov-Galerkin methods in two dimensions, *Numer. Math.* **61**, 359–372.

Morton, K.W., Priestley, A. and Süli, E. (1988), Stability analysis of the Lagrange-Galerkin method with non-exact integration, M^2AN **22**(4), 625–653.

Morton, K.W., Rudgyard, M.A. and Shaw, G.J. (1994), Upwind iteration methods for the cell vertex scheme in one dimension, *J. Comput. Phys.* **114**(2), 209–226.

Murray, W.D. and Landis, F. (1959), Numerical and machine solutions of transient heat-conduction problems involving melting or freezing, *Trans. ASME* **81**, 106–112.

Ni, R.-H. (1981), A multiple grid scheme for solving the Euler equations, *AIAA Journal* **20**(11), 1565–1571.

Niijima, K. (1990), Pointwise error estimates for a streamline diffusion finite element scheme, *Numer. Math.* **56**, 707–719.

Nitsche, J.A. (1968), Ein kriterium für die quasi-optimalitat des Ritzchen Verfahrens, *Numer. Math.* **11**, 346–348.

O'Riordan, E. and Stynes, M. (1986), An analysis of a superconvergence result for a singularly perturbed boundary value problem., *Math. Comp.* **46**, 81–92.

Osher, S. and Chakravarthy, S. (1984), High resolution schemes and the entropy condition, *SIAM J. Numer. Anal.* **21**, 955–984.

Pironneau, O. (1982), On the transport-diffusion algorithm and its application to the Navier-Stokes equations, *Numer. Math.* **38**, 309–332.

Prandtl, L. (1905), Über Flüssigkeitsbewegung bei sehr kleiner Reibung, *in* Verhandlungen des III. Internationalen Mathematiker-Kongresses, Heidelberg 1904, pp. 484–491.

Preissmann, A. (1961), Propagation des intumescences dans les canaux et rivières, *in* 1st Congrès de l'Assoc. Française de Calc., AFCAL, Grenoble, pp. 433–442.

Protter, M.H. and Weinberger, H. (1967), *Maximum principles in differential equations*, Prentice-Hall, Englewood Cliffs.

Pudykiewicz, J. (1989), Simulation of the Chernobyl dispersion with a 3-D hemispheric tracer model, *Tellus* **41B**, 391–412.

Quarteroni, A. and Valli, A. (1994), *Numerical approximation of partial differential equations*, Vol. 23 of *Springer Series in Computational Mathematics*, Springer-Verlag, Berlin.

Raithby, G.D. (1976), Skew upstream differencing schemes for problems involving fluid flow, *Comput. Methods Appl. Mech. Engrg.* **9**, 153–164.

Rasch, P.J. and Williamson, D.L. (1990), On shape-preserving interpolation and semi-Lagrangian transport, *SIAM J. Sci. Statist. Comput.* **11**, 656–687.

Reddy, S.C. and Trefethen, L.N. (1994), Pseudospectra of the convection-diffusion operator, *SIAM J. Appl. Math.* **54**(6), 1634–1649.

Rees, M.D. and Morton, K.W. (1991), Moving point, particle and free Lagrange methods for convection-diffusion equations, *SIAM J. Sci. Statist. Comput.* **12**(3), 547–572.

Reinhardt, H.J. (1982), *A posteriori* error analysis and adaptive finite element methods for singularly perturbed convection-diffusion equations, *Math. Methods Appl. Sci.* **4**, 529–548.

Richtmyer, P.D. (1957), *Difference methods for initial-value problems*, 1st edn, Interscience, New York.

Richtmyer, R.D. and Morton, K.W. (1967), *Difference methods for initial-value problems*, 2nd edn, Wiley-Interscience, New York. Reprint edn (1994) Krieger Publishing Company, Malabar.

Riesz, F. and Nagy, B.Sz. (1955), *Functional analysis*, F. Ungar Publishing Co., New York. (Translated from 2nd French edition by L.F. Boron).

Roberts, K.V. and Weiss, N.O. (1966), Convective difference schemes, *Math. Comp.* **20**(94), 272–299.

Roe, P.L. and van Leer, B. (1988), Non-existence, non-uniqueness and slow convergence in discrete conservation laws, *in* K.W. Morton and M.J. Baines, eds, Numerical Methods for Fluid Dynamics III, Clarendon Press, Oxford, pp. 520–529.

Samarskii, A.A. (1977), *Theory of finite difference schemes*, Nauka, Moscow.

Scotney, B.W. (1985), An analysis of the Petrov-Galerkin finite element method, PhD thesis, University of Reading.

Shishkin, G.I. (1992a), A difference approximation of a singularly perturbed boundary-value problem for quasilinear elliptic equations degenerating into first-order equations, *Comput. Math. Math. Phys.* **32**(4), 467–480.

Shishkin, G.I. (1992*b*), A difference scheme for a singularly perturbed parabolic equation degenerating on the boundary, *Comput. Math. Math. Phys.* **32**(5), 621–636.

Shokin, Yu.I. (1983), *The method of differential approximations*, Springer-Verlag, New York.

Shu, C.-W. (1987), TVB uniformly high-order schemes for conservation laws, *Math. Comp.* **49**, 105–121.

Shu, C.-W. (1988), Total-variation-diminishing time discretisations, *SIAM J. Sci. Statist. Comput.* **9**, 1073–1084.

Shu, C.-W. and Osher, S. (1988), Efficient implementation of essentially non-oscillatory shock-capturing schemes, *J. Comput. Phys.* **77**, 439–471.

Skeel, R.D. and Berzins, M. (1990), A method for the spatial discretization of parabolic equations in one space variable, *SIAM J. Sci. Statist. Comput.* **11**(1), 1–32.

Smith, R.M. and Hutton, A.G. (1982), The numerical treatment of advection — a performance comparison of current methods, *J. Numer. Heat Transfer* **5**, 439–461.

Sonar, T. (1993), On the design of an upwind scheme for compressible flow on general triangulations, *Numer. Algorithms* **4**, 135–149.

Sonar, T. and Süli, E. (1994), A dual graph-norm refinement indicator for finite volume approximations of the Euler equations, Technical Report NA94/09, Oxford University Computing Laboratory. Submitted for publication.

Spalding, D.B. (1972), A novel finite difference formulation for differential expressions involving both first and second derivatives, *Intern. J. Numer. Methods Engrg.* **4**, 551–559.

Spijker, M.N. (1991), On a conjecture by LeVeque and Trefethen related to the Kreiss matrix theorem, *BIT* **31**, 551–555.

Staniforth, A. and Côté, J. (1991), Semi-Lagrangian integration schemes and their application to environmental flows, *Monthly Weather Rev.* **119**(9), 2206–2223.

Stoyan, G. (1979), Monotone difference schemes for diffusion-convection problems, *ZAMM* **59**, 361–372.

Stoyan, G. (1984), Explicit error estimates for difference schemes solving the stationary constant coefficient diffusion-convection-reaction equation, *Z. Angew. Math. Mech.* **64**, 173–191.

Strang, G. (1971), The finite element method and approximation theory, *in* B. Hubbard, ed., Numerical Solution of Partial Differential Equations II (Synspade 1970), Academic Press, New York, pp. 547–583.

Strang, G. and Fix, G. (1973), *An analysis of the finite element method*, Prentice-Hall, Englewood Cliffs.

Stynes, M. and O'Riordan, E. (1986), A finite element method for a singularly perturbed boundary value problem, *Numer. Math.* **50**(1), 1–15.

Süli, E. (1988), Convergence and nonlinear stability of the Lagrange-Galerkin method for the Navier-Stokes equations, *Numer. Math.* **53**, 459–483.

Süli, E. (1992), The accuracy of cell vertex finite volume methods on quadrilateral meshes, *Math. Comp.* **59**(200), 359–382.

Swanson, R.C. and Turkel, E. (1992), On central-difference and upwind schemes, *J. Comput. Phys.* **101**, 297–306.

Swartz, B.K. (1974), The construction of finite difference analogs of some finite element schemes, *in* C. de Boor, ed., Mathematical Aspects of Finite Elements in Partial Differential Equations, Academic Press, New York, pp. 279–312.

Tadmor, E. (1984), Numerical viscosity and the entropy condition for conservative difference schemes, *Math. Comp.* **32**, 369–382.

Tadmor, E. (1988), Convenient total variation diminishing conditions for nonlinear difference schemes, *SIAM J. Numer. Anal.* **25**, 1002–1014.

Thomas, H.A. (1937), Hydraulics of flood movements in rivers, Technical report, Carnegie Institute of Technology, Pittsburgh, Pennsylvania.

Thomée, V. (1962), A stable difference scheme for the mixed boundary problem for a hyperbolic first order system in two dimensions, *J. Soc. Indust. Appl. Math.* **10**, 229–245.

Thomée, V. and Wendroff, B. (1974), Convergence estimates for Galerkin methods for variable coefficient initial value problems, *SIAM J. Numer. Anal.* **11**, 1059–1068.

Trefethen, L.N. (1992), Pseudospectra of matrices, *in* D.F. Griffiths and G.A. Watson, eds, Numerical Analysis 1991. Proceedings of the 14th Dundee Conference, number 260 *in* 'Pitman Research Notes in Mathematics', Longman, Harlow, pp. 234–266.

van der Houwen, P.J. (1977), *Construction of integration formulas for initial value problems*, North Holland, Amsterdam.

van Dyke, M. (1994), Nineteenth-century roots of the boundary-layer idea, *SIAM Rev.* **36**(3), 415–424.

van Leer, B. (1974), Towards the ultimate conservative difference scheme. II Monotonicity and conservation combined in a second order scheme, *J. Comput. Phys.* **14**, 361–370.

van Leer, B. (1977), Towards the ultimate finite difference scheme. III Upstream-centred finite-difference schemes for ideal compressible flow, *J. Comput. Phys.* **23**, 263–275.

van Leer, B. (1979), Towards the ultimate conservative difference scheme. V A second-order sequel to Godunov's method, *J. Comput. Phys.* **32**, 101–136.

van Roosbroeck, W. (1950), Theory of flow of electrons and holes in germanium and other semiconductors, *Bell Syst. Tech. J.* **29**, 560–607.

Varga, R.S. (1962), *Matrix iterative analysis*, Prentice–Hall, Englewood Cliffs.

Vila, J.P. (1989), An analysis of a class of second-order accurate Godunov-type schemes, *SIAM J. Numer. Anal.* **26**(4), 830–853.

Vreugdenhil, C.B. and Koren, B., eds (1993), *Numerical methods for advection-diffusion problems*, Vol. 45 of *Notes on Numerical Fluid Mechanics*, Vieweg, Braunschweig.

Warming, R.F. and Hyett, B.J. (1974), The modified equation approach to the stability and accuracy analysis of finite-difference methods, *J. Comput. Phys.* **14**, 159–179.

Warming, R.F., Kutler, P. and Lomax, H. (1973), Second- and third-order non-centred difference schemes for nonlinear hyperbolic equations, *AIAA J.* **11**, 189–196.

Wathen, A. and Silvester, D. (1993), Fast iterative solution of stabilised Stokes systems. Part I: Using simple diagonal preconditioners, *SIAM J. Numer. Anal.* **30**(3), 630–649.

Weatherill, N.P. and Hassan, O. (1994), Efficient three-dimensional Delaunay triangulation with automatic point creation and imposed boundary constraints, *Intern. J. Numer. Methods Engrg.* **37**, 2005–2039.

Wendroff, B. (1960), On centered difference equations for hyperbolic systems, *J. Soc. Indust. Appl. Math.* **8**, 549–555.

Whitham, G.B. (1974), *Linear and nonlinear waves*, John Wiley & Sons, New York.

Yanenko, N.N. (1971), *The method of fractional steps*, Springer-Verlag, Berlin.

Zienkiewicz, O.C. and Zhu, J.Z. (1992*a*), The superconvergent patch recovery and *a posteriori* error estimates. Part 1: The recovery technique, *Intern. J. Numer. Methods Engrg.* **33**, 1331–1364.

Zienkiewicz, O.C. and Zhu, J.Z. (1992*b*), The superconvergent patch recovery and *a posteriori* error estimates. Part 2: Error estimates and adaptivity, *Intern. J. Numer. Methods Engrg.* **33**, 1365–1382.

Zienkiewicz, O.C., Gallagher, R.H. and Hood, P. (1975), Newtonian and non-Newtonian viscous incompressible flow, temperature induced flows : finite element solutions, *in* J.R. Whiteman, ed., The Mathematics of Finite Elements and Applications II MAFELAP, Academic Press, London, pp. 235–267.

Zlámal, M. (1978), Superconvergence and reduced integration in the finite element method, *Math. Comp.* **32**(143), 663–685.

Zlámal, M. (1994), A box finite element method giving solution gradients with a higher order accuracy, *in* M. Křížek, P. Neittaanmäki and R. Stenberg, eds, Finite Element Methods: Fifty Years of the Courant Element, Vol. 164 of *Lecture Notes in Pure and Applied Mathematics*, Marcel Dekker, Inc., New York, pp. 501–504.

Index